Geophysical Monograph Series

Including

IUGG Volumes

Maurice Ewing Volumes
Mineral Physics Volumes

GEOPHYSICAL MONOGRAPH SERIES

Geophysical Monograph Volumes

1. Antarctica in the International Geophysical Year *A. P. Crary, L. M. Gould, E. O. Hulburt, Hugh Odishaw, and Waldo E. Smith (Eds.)*
2. Geophysics and the IGY *Hugh Odishaw and Stanley Ruttenberg (Eds.)*
3. Atmospheric Chemistry of Chlorine and Sulfur Compounds *James P. Lodge, Jr. (Ed.)*
4. Contemporary Geodesy *Charles A. Whitten and Kenneth H. Drummond (Eds.)*
5. Physics of Precipitation *Helmut Weickmann (Ed.)*
6. The Crust of the Pacific Basin *Gordon A. Macdonald and Hisashi Kuno (Eds.)*
7. Antarctic Research: The Matthew Fontaine Maury Memorial Symposium *H. Wexler, M. J. Rubin, and J. E. Caskey, Jr. (Eds.)*
8. Terrestrial Heat Flow *William H. K. Lee (Ed.)*
9. Gravity Anomalies: Unsurveyed Areas *Hyman Orlin (Ed.)*
10. The Earth Beneath the Continents: A Volume of Geophysical Studies in Honor of Merle A. Tuve *John S. Steinhart and T. Jefferson Smith (Eds.)*
11. Isotope Techniques in the Hydrologic Cycle *Glenn E. Stout (Ed.)*
12. The Crust and Upper Mantle of the Pacific Area *Leon Knopoff, Charles L. Drake, and Pembroke J. Hart (Eds.)*
13. The Earth's Crust and Upper Mantle *Pembroke J. Hart (Ed.)*
14. The Structure and Physical Properties of the Earth's Crust *John G. Heacock (Ed.)*
15. The Use of Artificial Satellites for Geodesy *Soren W. Henricksen, Armando Mancini, and Bernard H. Chovitz (Eds.)*
16. Flow and Fracture of Rocks *H. C. Heard, I. Y. Borg, N. L. Carter, and C. B. Raleigh (Eds.)*
17. Man-Made Lakes: Their Problems and Environmental Effects *William C. Ackermann, Gilbert F. White, and E. B. Worthington (Eds.)*
18. The Upper Atmosphere in Motion: A Selection of Papers With Annotation *C. O. Hines and Colleagues*
19. The Geophysics of the Pacific Ocean Basin and Its Margin: A Volume in Honor of George P. Woollard *George H. Sutton, Murli H. Manghnani, and Ralph Moberly (Eds.)*
20. The Earth's Crust: Its Nature and Physical Properties *John C. Heacock (Ed.)*
21. Quantitative Modeling of Magnetospheric Processes *W. P. Olson (Ed.)*
22. Derivation, Meaning, and Use of Geomagnetic Indices *P. N. Mayaud*
23. The Tectonic and Geologic Evolution of Southeast Asian Seas and Islands *Dennis E. Hayes (Ed.)*
24. Mechanical Behavior of Crustal Rocks: The Handin Volume *N. L. Carter, M. Friedman, J. M. Logan, and D. W. Stearns (Eds.)*
25. Physics of Auroral Arc Formation *S.-I. Akasofu and J. R. Kan (Eds.)*
26. Heterogeneous Atmospheric Chemistry *David R. Schryer (Ed.)*
27. The Tectonic and Geologic Evolution of Southeast Asian Seas and Islands: Part 2 *Dennis E. Hayes (Ed.)*
28. Magnetospheric Currents *Thomas A. Potemra (Ed.)*
29. Climate Processes and Climate Sensitivity (Maurice Ewing Volume 5) *James E. Hansen and Taro Takahashi (Eds.)*
30. Magnetic Reconnection in Space and Laboratory Plasmas *Edward W. Hones, Jr. (Ed.)*
31. Point Defects in Minerals (Mineral Physics Volume 1) *Robert N. Schock (Ed.)*
32. The Carbon Cycle and Atmospheric CO_2: Natural Variations Archean to Present *E. T. Sundquist and W. S. Broecker (Eds.)*
33. Greenland Ice Core: Geophysics, Geochemistry, and the Environment *C. C. Langway, Jr., H. Oeschger, and W. Dansgaard (Eds.)*
34. Collisionless Shocks in the Heliosphere: A Tutorial Review *Robert G. Stone and Bruce T. Tsurutani (Eds.)*
35. Collisionless Shocks in the Heliosphere: Reviews of Current Research *Bruce T. Tsurutani and Robert G. Stone (Eds.)*
36. Mineral and Rock Deformation: Laboratory Studies —The Paterson Volume *B. E. Hobbs and H. C. Heard (Eds.)*
37. Earthquake Source Mechanics (Maurice Ewing Volume 6) *Shamita Das, John Boatwright, and Christopher H. Scholz (Eds.)*
38. Ion Acceleration in the Magnetosphere and Ionosphere *Tom Chang (Ed.)*
39. High Pressure Research in Mineral Physics (Mineral Physics Volume 2) *Murli H. Manghnani and Yasuhiko Syono (Eds.)*
40. Gondwana Six: Structure, Tectonics, and Geophysics *Gary D. McKenzie (Ed.)*
41. Gondwana Six: Stratigraphy, Sedimentology, and Paleontology *Garry D. McKenzie (Ed.)*
42. Flow and Transport Through Unsaturated Fractured Rock *Daniel D. Evans and Thomas J. Nicholson (Eds.)*
43. Seamounts, Islands, and Atolls *Barbara H. Keating, Patricia Fryer, Rodey Batiza, and George W. Boehlert (Eds.)*

44 **Modeling Magnetospheric Plasma** *T. E. Moore and J. H. Waite, Jr. (Eds.)*

45 **Perovskite: A Structure of Great Interest to Geophysics and Materials Science** *Alexandra Navrotsky and Donald J. Weidner (Eds.)*

46 **Structure and Dynamics of Earth's Deep Interior (IUGG Volume 1)** *D. E. Smylie and Raymond Hide (Eds.)*

47 **Hydrological Regimes and Their Subsurface Thermal Effects (IUGG Volume 2)** *Alan E. Beck, Grant Garven, and Lajos Stegena (Eds.)*

48 **Origin and Evolution of Sedimentary Basins and Their Energy and Mineral Resources (IUGG Volume 3)** *Raymond A. Price (Ed.)*

49 **Slow Deformation and Transmission of Stress in the Earth (IUGG Volume 4)** *Steven C. Cohen and Petr Vaníček (Eds.)*

50 **Deep Structure and Past Kinematics of Accreted Terranes (IUGG Volume 5)** *John W. Hillhouse (Ed.)*

51 **Properties and Processes of Earth's Lower Crust (IUGG Volume 6)** *Robert F. Mereu, Stephan Mueller, and David M. Fountain (Eds.)*

52 **Understanding Climate Change (IUGG Volume 7)** *Andre L. Berger, Robert E. Dickinson, and J. Kidson (Eds.)*

53 **Plasma Waves and Instabilities at Comets and in Magnetospheres** *Bruce T. Tsurutani and Hiroshi Oya (Eds.)*

54 **Solar System Plasma Physics** *J. H. Waite, Jr., J. L. Burch, and R. L. Moore (Eds.)*

55 **Aspects of Climate Variability in the Pacific and Western Americas** *David H. Peterson (Ed.)*

56 **The Brittle-Ductile Transition in Rocks** *A. G. Duba, W. B. Durham, J. W. Handin, and H. F. Wang (Eds.)*

57 **Evolution of Mid Ocean Ridges (IUGG Volume 8)** *John M. Sinton (Ed.)*

58 **Physics of Magnetic Flux Ropes** *C. T. Russell, E. R. Priest, and L. C. Lee (Eds.)*

59 **Variations in Earth Rotation (IUGG Volume 9)** *Dennis D. McCarthy and Williams E. Carter (Eds.)*

60 **Quo Vadimus Geophysics for the Next Generation (IUGG Volume 10)** *George D. Garland and John R. Apel (Eds.)*

61 **Cometary Plasma Processes** *Alan D. Johnstone (Ed.)*

62 **Modeling Magnetospheric Plasma Processes** *Gordon R. Wilson (Ed.)*

63 **Marine Particles: Analysis and Characterization** *David C. Hurd and Derek W. Spencer (Eds.)*

64 **Magnetospheric Substorms** *Joseph R. Kan, Thomas A. Potemra, Susumu Kokubun, and Takesi Iijima (Eds.)*

65 **Explosion Source Phenomenology** *Steven R. Taylor, Howard J. Patton, and Paul G. Richards (Eds.)*

66 **Venus and Mars: Atmospheres, Ionospheres, and Solar Wind Interactions** *Janet G. Luhmann, Mariella Tatrallyay, and Robert O. Pepin (Eds.)*

67 **High-Pressure Research: Application to Earth and Planetary Sciences (Mineral Physics Volume 3)** *Yasuhiko Syono and Murli H. Manghnani (Eds.)*

68 **Microwave Remote Sensing of Sea Ice** *Frank Carsey, Roger Barry, Josefino Comiso, D. Andrew Rothrock, Robert Shuchman, W. Terry Tucker, Wilford Weeks, and Dale Winebrenner*

69 **Sea Level Changes: Determination and Effects (IUGG Volume 11)** *P. L. Woodworth, D. T. Pugh, J. G. DeRonde, R. G. Warrick, and J. Hannah*

70 **Synthesis of Results from Scientific Drilling in the Indian Ocean** *Robert A. Duncan, David K. Rea, Robert B. Kidd, Ulrich von Rad, and Jeffrey K. Weissel (Eds.)*

71 **Mantle Flow and Melt Generation at Mid-Ocean Ridges** *Jason Phipps Morgan, Donna K. Blackman, and John M. Sinton (Eds.)*

72 **Dynamics of Earth's Deep Interior and Earth Rotation (IUGG Volume 12)** *Jean-Louis Le Mouël, D.E. Smylie, and Thomas Herring (Eds.)*

73 **Environmental Effects on Spacecraft Positioning and Trajectories (IUGG Volume 13)** *A. Vallance Jones (Ed.)*

74 **Evolution of the Earth and Planets (IUGG Volume 14)** *E. Takahashi, Raymond Jeanloz, and David Rubie (Eds.)*

75 **Interactions Between Global Climate Subsystems: The Legacy of Hann (IUGG Volume 15)** *G. A. McBean and M. Hantel (Eds.)*

76 **Relating Geophysical Structures and Processes: The Jeffreys Volume (IUGG Volume 16)** *K. Aki and R. Dmowska (Eds.)*

77 **The Mesozoic Pacific: Geology, Tectonics, and Volcanism—A Volume in Memory of Sy Schlanger** *Malcolm S. Pringle, William W. Sager, William V. Sliter, and Seth Stein (Eds.)*

78 **Climate Change in Continental Isotopic Records** *P. K. Swart, K. C. Lohmann, J. McKenzie, and S. Savin (Eds.)*

79 **The Tornado: Its Structure, Dynamics, Prediction, and Hazards** *C. Church, D. Burgess, C. Doswell, R. Davies-Jones (Eds.)*

80 **Auroral Plasma Dynamics** *R. L. Lysak (Ed.)*

81 **Solar Wind Sources of Magnetospheric Ultra-Low Frequency Waves** *M. J. Engebretson, K. Takahashi, and M. Scholer (Eds.)*

82 **Gravimetry and Space Techniques Applied to Geodynamics and Ocean Dynamics (IUGG Volume 17)** *Bob E. Schutz, Allen Anderson, Claude Froidevaux, and Michael Parke (Eds.)*

83 **Nonlinear Dynamics and Predictability of Geophysical Phenomena (IUGG Volume 18)** *William I. Newman, Andrei Gabrielov, and Donald L. Turcotte (Eds.)*

84 **Solar System Plasmas in Space and Time** *J. Burch, J. H. Waite, Jr. (Eds.)*

85 The Polar Oceans and Their Role in Shaping the Global Environment   *O. M. Johannessen, R. D. Muench, and J. E. Overland (Eds.)*

86 Space Plasmas: Coupling Between Small and Medium Scale Processes   *Maha Ashour-Abdalla, Tom Chang, and Paul Dusenbery (Eds.)*

87 The Upper Mesosphere and Lower Thermosphere: A Review of Experiment and Theory   *R. M. Johnson and T. L. Killeen (Eds.)*

88 Active Margins and Marginal Basins of the Western Pacific   *Brian Taylor and James Natland (Eds.)*

89 Natural and Anthropogenic Influences in Fluvial Geomorphology   *John E. Costa, Andrew J. Miller, Kenneth W. Potter, and Peter R. Wilcock (Eds.)*

90 Physics of the Magnetopause   *Paul Song, B.U.Ö. Sonnerup, and M.F. Thomsen (Eds.)*

91 Seafloor Hydrothermal Systems: Physical, Chemical, Biological, and Geological Interactions   *Susan E. Humphris, Robert A. Zierenberg, Lauren S. Mullineaux, and Richard E. Thomson (Eds.)*

92 Mauna Loa Revealed: Structure, Composition, History, and Hazards   *J. M. Rhodes and John P. Lockwood (Eds.)*

93 Cross-Scale Coupling in Space Plasmas   *James L. Horwitz, Nagendra Singh, and James L. Burch (Eds.)*

94 Double-Diffusive Convection   *Alan Brandt and H.J.S. Fernando (Eds.)*

95 Earth Processes: Reading the Isotopic Code   *Asish Basu and Stan Hart (Eds.)*

96 Subduction Top to Bottom   *Gray E. Bebout, David Scholl, Stephen Kirby, and John Platt (Eds.)*

Maurice Ewing Volumes

1 Island Arcs, Deep Sea Trenches, and Back-Arc Basins   *Manik Talwani and Walter C. Pitman III (Eds.)*

2 Deep Drilling Results in the Atlantic Ocean: Ocean Crust   *Manik Talwani, Christopher G. Harrison, and Dennis E. Hayes (Eds.)*

3 Deep Drilling Results in the Atlantic Ocean: Continental Margins and Paleoenvironment   *Manik Talwani, William Hay, and William B. F. Ryan (Eds.)*

4 Earthquake Prediction—An International Review   *David W. Simpson and Paul G. Richards (Eds.)*

5 Climate Processes and Climate Sensitivity   *James E. Hansen and Taro Takahashi (Eds.)*

6 Earthquake Source Mechanics   *Shamita Das, John Boatwright, and Christopher H. Scholz (Eds.)*

IUGG Volumes

1 Structure and Dynamics of Earth's Deep Interior   *D. E. Smylie and Raymond Hide (Eds.)*

2 Hydrological Regimes and Their Subsurface Thermal Effects   *Alan E. Beck, Grant Garven, and Lajos Stegena (Eds.)*

3 Origin and Evolution of Sedimentary Basins and Their Energy and Mineral Resources   *Raymond A. Price (Ed.)*

4 Slow Deformation and Transmission of Stress in the Earth   *Steven C. Cohen and Petr Vaníček (Eds.)*

5 Deep Structure and Past Kinematics of Accreted Terranes   *John W. Hillhouse (Ed.)*

6 Properties and Processes of Earth's Lower Crust   *Robert F. Mereu, Stephan Mueller, and David M. Fountain (Eds.)*

7 Understanding Climate Change   *Andre L. Berger, Robert E. Dickinson, and J. Kidson (Eds.)*

8 Evolution of Mid Ocean Ridges   *John M. Sinton (Ed.)*

9 Variations in Earth Rotation   *Dennis D. McCarthy and William E. Carter (Eds.)*

10 Quo Vadimus Geophysics for the Next Generation   *George D. Garland and John R. Apel (Eds.)*

11 Sea Level Changes: Determinations and Effects   *Philip L. Woodworth, David T. Pugh, John G. DeRonde, Richard G. Warrick, and John Hannah (Eds.)*

12 Dynamics of Earth's Deep Interior and Earth Rotation   *Jean-Louis Le Mouël, D.E. Smylie, and Thomas Herring (Eds.)*

13 Environmental Effects on Spacecraft Positioning and Trajectories   *A. Vallance Jones (Ed.)*

14 Evolution of the Earth and Planets   *E. Takahashi, Raymond Jeanloz, and David Rubie (Eds.)*

15 Interactions Between Global Climate Subsystems: The Legacy of Hann   *G. A. McBean and M. Hantel (Eds.)*

16 Relating Geophysical Structures and Processes: The Jeffreys Volume   *K. Aki and R. Dmowska (Eds.)*

17 Gravimetry and Space Techniques Applied to Geodynamics and Ocean Dynamics   *Bob E. Schutz, Allen Anderson, Claude Froidevaux, and Michael Parke (Eds.)*

18 Nonlinear Dynamics and Predictability of Geophysical Phenomena   *William I. Newman, Andrei Gabrielov, and Donald L. Turcotte (Eds.)*

Mineral Physics Volumes

1 Point Defects in Minerals   *Robert N. Schock (Ed.)*

2 High Pressure Research in Mineral Physics   *Murli H. Manghnani and Yasuhiko Syona (Eds.)*

3 High Pressure Research: Application to Earth and Planetary Sciences   *Yasuhiko Syono and Murli H. Manghnani (Eds.)*

Geophysical Monograph 97

Radiation Belts: Models and Standards

J. F. Lemaire
D. Heynderickx
D. N. Baker
Editors

American Geophysical Union

Published under the aegis of the AGU Books Board.

Cover illustration based on a schematic cross section of the trapped radiation belts surrounding the Earth. (Courtesy of Richard A. Mewaldt)

Library of Congress Cataloging-in-Publication Data
Radiation belts: models and standards / J.F. Lemaire, D. Heynderickx,
 D.N. Baker, editors.
  p. cm. -- (Geophysical monograph, ISSN 0065-8448 ; 97)
  Includes bibliographical references.
  ISBN 0-87590-079-8 (alk. paper)
  1. Van Allen radiation belts. 2. Magnetosphere. I. Lemaire, J.
 II. Heynderickx, D. III. Baker, D. N. IV. Series.
 QC809.V3R34 1996
 538'.766--dc21   96-48740
                                                                  CIP

ISBN 0-87590-079-8
ISSN 0065-8448

Copyright 1996 by the American Geophysical Union
2000 Florida Avenue, N.W.
Washington, DC 20009

Figures, tables, and short excerpts may be reprinted in scientific books and journals if the source is properly cited.

  Authorization to photocopy items for internal or personal use, or the internal or personal use of specific clients, is granted by the American Geophysical Union for libraries and other users registered with the Copyright Clearance Center (CCC) Transactional Reporting Service, provided that the base fee of $1.50 per copy plus $0.35 per page is paid directly to CCC, 222 Rosewood Dr., Danvers, MA 01923. 0065-8448/96/$01.50+0.35.
  This consent does not extend to other kinds of copying, such as copying for creating new collective works or for resale. The reproduction of multiple copies and the use of full articles or the use of extracts, including figures and tables, for commercial purposes requires permission from AGU.

Printed in the United States of America.

CONTENTS

Preface
J. F. Lemaire and M.I. Panasyuk xi

Introduction
J. G. Roederer xiii

Theoretical Radiation Belt Models

Source and Loss Processes for Radiation Belt Particles
M. Walt 1

Processes Acting Upon Outer Zone Electrons
C. E. McIlwain 15

Dynamic Physical Modelling of Trapped Particles for Satellite Survey
S. Bourdarie, D. Boscher, and T. Beutier 27

Anomalous Cosmic Rays: The Principal Source of High Energy Heavy Ions in the Radiation Belts
R. A. Mewaldt, R. S. Selesnick, and J. R. Cummings 35

Formation of the Radiation Belts by Anomalous Cosmic Rays and Similar Phenomena
A. V. Dmitriev, V. D. Ilyin, S. N. Kuznetsov, and B.Yu. Yushkov 43

Jovian, Solar, and Other Possible Sources of Radiation Belt Particles
D. N. Baker, S. G. Kanekal, M. D. Looper, J. B. Blake, and R.A. Mewaldt 49

MHD/Particle Simulations of Radiation Belt Formation During a Storm Sudden Commencement
M. K. Hudson, S. R. Elkington, J. G. Lyon, V. A. Marchenko, I. Roth, M. Temerin, and M. S. Gussenhoven 57

Cross Field Entry of High Charge State Energetic Heavy Ions Into the Earth's Magnetosphere
W. N. Spjeldvik 63

CRRES Observations and Radial Diffusion Theory of Radiation Belt Protons
J. M. Albert 69

Dynamic Models of the Energetic Ions in the Earth's Radiation Belts
V. F. Bashkirov 73

Physical Radiation Belt Models: Report of Discussion Group A
T. Beutier 77

Empirical Radiation Belt Models

Recent Development in the NASA Trapped Radiation Models
S. F. Fung 79

Phillips Laboratory Space Physics Division Radiation Models
M. S. Gussenhoven, E. G. Mullen, and D. H. Brautigam 93

A New Empirical Electron Model
D. J. Rodgers 103

Experimental Validation of South Atlantic Anomaly Motion Using a Two-Dimensional Cross-Correlation Technique
M. Lauriente, A. L. Vampola, and K. Gosier 109

CONTENTS

Low Altitude Trapped Radiation Model Using TIROS/NOAA Data
S. L. Huston, G. A. Kuck, and K. A. Pfitzer 119

Modelling He and H Isotopes in the Radiation Belts
R. S. Selesnick and R. A. Mewaldt 123

Electrons With Energy Exceeding 10 MeV in the Earth's Radiation Belt
A. M. Galper, V. V. Dmitrenko, V. M. Gratchev, Yu.V. Efremova, V. G. Kirillov-Ugryumov, S. V. Koldashov, L. V. Maslennikov, V. V. Mikhailov, Yu.V. Ozerov, A. V. Popov, N. I. Shvets, S. E. Ulin, and S. A. Voronov 129

Low Altitude Models of Radiation Belts Based on Data From Russian Satellites
Yu.V. Mineev and E. D. Tolstaya 135

Comparison Between NASA and INP/MSU Radiation Belt Models
A. A. Beliaev and J. F. Lemaire 141

Empirical Radiation Belt Models: Report of Discussion Group B
D. J. Rodgers 147

Coordinates and Indices

Introduction to Trapped Particle Flux Mapping
J. G. Roederer 149

Canonical Coordinates for Radiation-Belt Modeling
M. Schulz 153

Magnetic Field Models in the Inner Magnetosphere
T. I. Pulkkinen 161

A Quantitative Test of Different Magnetic Field Models Using Conjunctions Between DMSP and Geosynchronous Orbit
G. D. Reeves, L. A. Weiss, M. F. Thomsen, and D. J. McComas 167

A New Tool for Calculating Drift Shell Averaged Atmospheric Density
D. Heynderickx, M. Kruglanski, J. F. Lemaire, and E. J. Daly 173

Radiation Conditions Modelling at the Geostationary Orbit
G. V. Popov, V. I. Degtjarev, and S. S. Sheshukov 179

Dynamics of Energetic Electrons in the Radiation Belts
L. V. Tverskaya 183

Field Modeling Methods for the Inner Magnetosphere
D. P. Stern 189

Use of (B, L) Coordinates in Radiation Dose Models
M. Kruglanski 195

Coordinates and Indices: Report of Discussion Group C
D. Heynderickx 201

Missions and Data Acquisition

Availability of Radiation Belt Data and the Need for New Sources
A. D. Johnstone 203

CONTENTS

First Results and Perspectives of Monitoring Radiation Belts
M. I. Panasyuk, E. N. Sosnovets, O. S. Grafodatsky, V. I. Verkhoturov, and Sh. N. Islyaev 211

Current and Future Data Available in Japan
T. Kohno 217

UARS PEM Contribution to Radiation Belt Modelling
J. R. Sharber, J. D. Winningham, R. Link, R. A. Frahm, D. L. Chenette, and E. E. Gaines 223

Radiation Belt Observations From CREAM and CREDO
C. Dyer, A. Sims, and C. Underwood 229

Los Alamos Geosynchronous Space Weather Data for Radiation Belt Modeling
G. D. Reeves, R. D. Belian, T. C. Cayton, M. G. Henderson, R. A. Christensen, P. S. McLachlan, and J. C. Ingraham 237

Outer Zone Relativistic Electron Flux Variations Observed by SAMPEX During Nov. 1–8, 1993
X. Li, D. N. Baker, M. Temerin, J. B. Blake, and S. G. Kanekal 241

ISEE Measurements for Radiation Belt Modeling
R. H. W. Friedel, E. Keppler, G. Loidl, and A. Korth 247

Measurement of Radiation Belt Particles With ETS-6 Onboard Dosimeter
T. Goka, H. Matsumoto, T. Fukuda, and S. Takagi 251

Global Distributions of Trapped He Fluxes From OHZORA Satellite During the Geomagnetically Quiet Period of 1984–1987
N. Hasebe, A. Ryowa, M. Kobayashi, K. Kondoh, J. Hamada, Y. Mishima, K. Nagata, K. Kohno, J. Kikuchi, and T. Doke 255

Energetic Particle Data Archived at IEP SAS
K. Kudela and M. Slivka 259

Monitoring of the Radiation Belts With the Radiation Environment Monitor REM
P. Bühler, L. Desorgher, A. Zehnder, L. Adams, and E. Daly 265

Some Characteristics of Hot Magnetospheric Plasma at Geostationary Orbit
T. A. Ivanova, Yu.V. Kutuzov, B. V. Marjin, N. N. Pavlov, I. A. Rubinshtein, E. N. Sosnovets, M. B. Teltsov, L. V. Tverskaya, and N. A. Vlasova 269

Internal Charging in the Outer Zone and Operational Anomalies
G. L. Wrenn and A.J. Sims 275

Missions and Data Acquisition: Report of Discussion Group D
R. Friedel 279

Computer Models and Tools

Global Imaging by Energetic Neutral Particles
T. Beutier, J.-A. Sauvaud, D. Boscher, and S. Bourdarie 281

Global Imaging and Radio Remote Sensing of the Magnetosphere
S. F. Fung and J. L. Green 285

Artificial Neural Network (ANN) Forecasting of Energetic Electrons at Geosynchronous Orbit
G. A. Stringer, I. Heuten, C. Salazar, and B. Stokes 291

CONTENTS

ESA Update of AE-8 Using CRRES Data and a Neural Network
A. L. Vampola 297

The Trapped Radiation Software Package UNIRAD
D. Heynderickx, M. Kruglanski, J. Lemaire, E. J. Daly, and H. D. R. Evans 305

EnviroNET Space Environment Information via the WWW: A Computer Based Demonstration
P. J. Messore and M. Lauriente 311

Radiation Belt Models for the PC: RADMODLS
A. L. Vampola 315

Computer Animation of the TIROS/NOAA Observations of the Low-Altitude (850 km) Radiation Environment
H. H. Sauer and D.C. Wilkinson 321

PREFACE

The exciting new results of CRRES and SAMPEX show that there are additional physical sources of energetic electrons and ions trapped in the Van Allen belts, some of which were completely unexpected.

The NASA and Russian empirical models of the radiation belts need to be updated and extended. To outline different ways to achieve this task and to identify the less well known aspects of physical and empirical models of the radiation belts were the objectives of a workshop held in Brussels, October 17-20, 1995, entitled "Radiation Belts: Models and Standards." It was attended by over 60 delegates from all major laboratories involved in developing new physical as well as empirical models of the energy and spatial distributions of energetic electrons and ions trapped in the geomagnetic field.

This volume is based on the invited and contributed papers/posters which were presented at this international workshop and is organized as five sections:

Physical Radiation Belt Models
Empirical Radiation Belt Models
Coordinates and Indices
Missions and Data Acquisition
Computer Models and Tools

The first four of these sections correspond to the four subgroups which met in parallel discussion sessions. The subgroups have prepared summaries and recommendations for future development in this area, which is currently experiencing a renaissance after more than a decade of lethargy.

The recommendations of the subgroups as well as the abstracts of papers included in this volume are also displayed on the WWW (http://magnet.oma.be/wrb.html). The major issues raised at these discussion sessions were summarized by the editors of this monograph and by the conveners of the workshop in an article entitled "Researchers chart course for updating radiation belt models," by *Baker et al.* [*Eos Trans. AGU*, 77(23), June 4, 1996].

Although the early pioneers in this field of research could not all attend this workshop, some of them contributed to this monograph either comprehensive review papers (M. Walt; M. Schulz) or unpublished material from the 1970s (C. E. McIlwain) or stimulating remarks (J. G. Roederer). We hope that this volume will remain a reference book for the space physics community interested in studies of the Earth's radiation belts.

These events would not have been so fruitful without the participation of the world specialists in this field. We wish to thank them all for the time they have given to prepare the meeting as well as the manuscripts. Both conveners wish also to point out the outstanding work of the chairpersons, vice-chairpersons and reporters of the four subgroups.

We would like to extend our special thanks to Baron Ackerman, Director of the Belgian Institute for Space Aeronomy (BISA), for his generous logistic and financial support. We also received substantial financial support from the Fonds National de la Recherche Scientifique—Nationaal Fonds voor Wetenschappelijk Onderzoek (FNRS-NFWO), as well as from the Services Fédéraux des Affaires Scientifiques, Techniques et Culturelles—Federale Diensten voor Wetenschappelijke, Technische en Culturele Aangelegenheden (SSTC-DWTC). At the request of E. J. Daly, ESA/ESTEC/WMA, the Space System Environment Analysis section of the European Space Agency (ESA), also provided financial support. We are grateful to all these organizations for their substantial support. We wish also to acknowledge NASA/Space Physics Division, JPL, AIAA and ISO for their formal support of the international workshop.

The directors of the Belgian Royal Observatory, P. Pâquet, and of the Royal Institute of Meteorology, H. Malcorps, are also acknowledged for the facilities they offered to the local organisers.

The members of the International Programme Committee were D. N. Baker (U.S.A.), E. J. Daly (The Netherlands), M. S. Gussenhoven (U.S.A.), A. D. Johnstone (U.K.), T. Kohno (Japan), A. Konradi (U.S.A.), A. Korth (Germany) and W. N. Spjeldvik (U.S.A.). We thank them all for their advice during the preparation of the workshop.

The success of the meeting also relied on the dedication of the participants and on their ability to exchange ideas during the poster sessions, parallel discussion sessions and oral presentations. The questions and answers after the oral presentations have been added at the end of the corresponding papers.

The convenors acknowledge all the many other persons who contributed to the preparation of this meeting, especially Jean Palange, Daniel Heynderickx, Michel Kruglanski and Viviane Pierrard from the Belgian Institute for Space Aeronomy, Brussels, who constituted with one of the conveners the Local Organizing Committee.

J. F. Lemaire (BISA, Brussels)
M. I. Panasyuk (INP/MSU, Moscow)

Introduction

Juan G. Roederer

Geophysical Institute, University of Alaska-Fairbanks

The editors have asked me to write an introduction to this volume, no doubt because of my involvement in theoretical/numerical radiation belt research during those exciting, early "good old days" of space physics.

Indeed, the Van Allen radiation belts are the "oldest" part of the magnetosphere in terms of scientific discovery. An impressive amount of research was carried out during the 1960s: the average fluxes of high-energy trapped protons and electrons were mapped, the principal sources, sinks and diffusion mechanisms were identified, and the theory of radiation belt dynamics was developed. The main motivation was not limited to scientific curiosity: strategic realities of the time made the effects of nuclear explosions in near-Earth space a subject of intense theoretical and experimental inquiry, with the dynamics of natural and artificially injected trapped particles an important part of it.

In the 1970s, interest shifted toward lower-energy particles that had become accessible to measurement with more sophisticated detectors, and magnetospheric plasma physics was developed. In addition, the distant regions of the magnetosphere, where no stable trapping can occur, became a center of attention. By the end of the 1970s, radiation belt research had de facto moved out from planet Earth to planet Jupiter and beyond! But many problems subsisted, including:

- injection mechanisms from the solar wind, ionosphere and cosmic rays;
- acceleration mechanisms and the transitions from tail plasma storage to the ring current and to the radiation belt;
- diffusion processes;
- substorm effects;
- precipitation in the South Atlantic anomaly and other loss processes, including escape into the solar wind.

The topics of energetic particle damage to solar cells and spacecraft electronics, as well as radiation hazards to astronauts/cosmonauts, became subjects of great practical importance.

During the mid-1980s, there was a revival of interest in experimental radiation belt physics, mainly triggered by the availability of new detailed and comprehensive energetic particle measurements on spacecraft. It was a time, too, to also revive the theoretical and numerical studies of geomagnetically trapped particles, given the availability of good data, more accurate and flexible magnetic field models, and fast supercomputers. Of particular importance was—and continues to be the use of the new data and computational tools for practical purposes, in particular, to revamp the procedures for trapped particle flux mapping and to develop new, more comprehensive and general models of geomagnetically trapped radiation in some appropriate, easily usable coordinate system.

The three adiabatic invariants are the "natural" coordinates to represent a distribution of radiation belt particles that is isotropic in all phases of periodicity. However, there are situations, particularly involving static configurations of the magnetic field, in which some appropriate functions of M, J, Φ can be used as more "intuitive" coordinates for the representation of trapped particle fluxes, e.g., McIlwain's I value and B, L coordinate system, Kaufmann's $K = I/\sqrt{B}$ invariant, Roederer's generalized L^* value.

Having a credible flux map for a reference state of the magnetosphere, a measurement made at a different time can be transformed back to the reference state for comparative studies, for instance, to identify irreversible changes that could have occurred, or to test external or internal field models by using the trapped particles as "remote sensing" probes. While general characteristics of the dynamic behavior of trapped particles may be determined using only crude approximations and simplifications, the determination of dynamic effects, such as particle precipitation in the South Atlantic anomaly, trapping boundary changes, the effects of an electric field on the lower-energy particles, the effects of internal magnetic field distortions, and, most importantly, the test of the new theories of source and loss mechanisms, all require a more accurate approach. Last but not least, accurate flux mapping and accurate adiabatic transformation algorithms are of critical practical importance to develop user-friendly, trustworthy radiation flux models for the quantitative prediction of radiation effects on technological systems and humans in space.

The Workshop on Radiation Belts, on which this volume is based, was a crucial first step toward achieving the above goals, with the new data, new magnetic field models and awesome computer power now available in major research centers. In particular, the implementation of some of the specific recommendations formulated by the four panels of the Workshop will eventually lead to a greatly improved understanding of radiation belt physics, and provide a quantitative basis for the development of more realistic and more practical models of geomagnetically trapped particles. The following recommendations struck me as particularly significant:

1. To encourage spacecraft operators to place standard radiation monitoring packages on their spacecraft.

2. To develop instruments with greater pitch angle information for future radiation belt study missions.

3. To compile a global radiation belt data base with current data and to organize a pilot project to assemble flux maps.

4. To follow a modular approach in model design, in particular, for the low-altitude regions where the atmosphere acts as a longitude dependent, moving boundary, and for the geostationary orbit region, where field asymmetries and fast time variations are important.

5. To request that "modelers" routinely provide detailed information on how they constructed their models (the so-called "metadata").

6. To seek a competent international body that could coordinate the activities needed to develop a standard radiation belt model.

I wish I had been able to attend this workshop—but other obligations toward international science (the simultaneously occurring meeting of the ICSU Advisory Council on Global Change) prevented me from doing so. I look forward very much to reading the present volume to learn what my colleagues—especially those from the younger generation—have achieved in recent years!

Source and Loss Processes for Radiation Belt Particles

M. Walt

Stanford University, Stanford, California

The search for source and loss mechanisms of radiation belt particles began immediately after the discovery of the belts and continues to the present time. Identified loss processes include collisions of radiation belt particles with atmospheric constituents and deflections by a variety of plasma waves, some of which are produced by the particles themselves. Sources of radiation belt particles include the decay of cosmic ray albedo neutrons, the solar wind, the ionosphere, and anomalous cosmic rays. The routes by which solar wind and ionospheric plasma become trapped and accelerated are still unclear but probably involve temporary storage in the near-Earth tail, convective injection into the radiation belts during magnetic storms and substorms, and subsequent radial diffusion driven by global fluctuations in the Earth's magnetic and electric fields. This radial diffusion acts as both a source and a loss mechanism as it redistributes trapped particles throughout the magnetosphere. A recently discovered source process is the acceleration/redistribution of solar flare particles by collisionless shocks passing through the magnetosphere. These processes are discussed with emphasis on the uncertainties in our understanding of the mechanisms involved, on the incomplete information of the physical conditions within the magnetosphere, and on the approximations currently used to describe them in radiation belt modeling.

1. INTRODUCTION

Shortly after the discovery of the radiation belts the search for the particle source and loss mechanisms began. During the 37 years of intense research the "principal sources and losses" have been discovered many times, and the search still continues. One reason for the lack of closure on this fundamental question is that in some regions of phase space, several source and loss mechanisms have comparable magnitudes. Furthermore, these magnitudes are often uncertain by at least a factor of 10. Hence, after evaluating a process using the best experimental information available, one often finds the process may be very important or quite negligible within the accuracy of available information. Many processes have mistakenly been identified as important or even dominant based on optimistic estimates of the relevant parameters.

Radiation Belts: Models and Standards
Geophysical Monograph 97
Copyright 1996 by the American Geophysical Union

A second major difficulty in identifying important sources and losses is the difficulty in evaluating source and loss processes separately. If the radiation belts are in equilibrium,

$$\frac{df}{dt} = \sum \text{sources} - \sum \text{losses}, \qquad (1)$$

where f is the distribution function of the species in question, and the sources and losses are functions of f, position, and particle energy, but not of time. Ignoring for the moment the problem of selecting an equilibrium flux where large time variations are observed, one cannot experimentally obtain the strength of either the sources or losses from observations of the equilibrium distributions. Occasionally a time dependent event occurs, for example a momentary injection of particles by a magnetic substorm or by a high altitude nuclear detonation. After such an event the source terms in Eq. (1) are zero, and the loss rate can be evaluated directly from measurements of df/dt.

Finally, the dynamic nature of the radiation belts has been an obstacle to understanding their formation and to constructing radiation belt models. The important processes, with

2 SOURCE AND LOSS PROCESSES

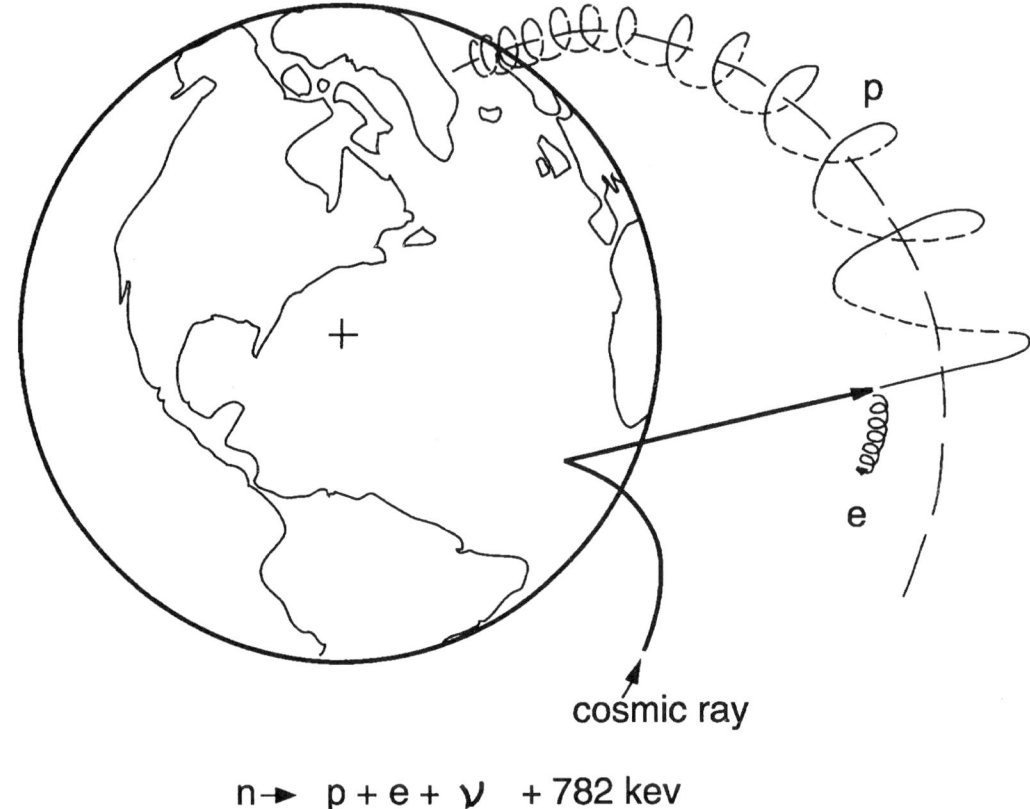

$$n \rightarrow p + e + \nu + 782 \text{ kev}$$

Figure 1. Cosmic ray albedo neutron decay (CRAND) source of trapped protons and electrons

few exceptions, vary with time in irregular and unpredictable ways. Also, the radiation belts are a system of partially interacting components each of which depends on external and internal parameters, in some cases non-linearly. Under these circumstances it is not surprising that understanding the fundamental causes of the radiation belts has been such a difficult task.

2. COSMIC RAY ALBEDO NEUTRON SOURCE

The first source process to be identified was proposed independently by at least three groups [*Singer*, 1958; *Kellogg*, 1958; *Vernov et al.*, 1959] within weeks of van Allen's discovery of the radiation belts. Energetic neutrons produced by collisions of primary cosmic rays with atmospheric nuclei sometimes escape the atmosphere and pass through the geomagnetic field. A small fraction of the energetic neutrons decay in flight while still within the earth's field, and the charged products of the decay can become trapped (see Figure 1).

A neutron will decay into a proton, an electron, and a neutrino with a half life of 10.5 minutes. Most of the momentum of the neutron remains with the newly born proton; and if its initial direction and position are favorable, it will be trapped. The electron will be emitted in a random direction (in the center of mass system) with an average energy of a few hundred keV and can also be trapped. The proton will have approximately the energy of the escaping neutron whose energy can be as high as several hundred MeV. Assuming that the protons are lost only by slowing down through collisions with bound and free electrons in the atmosphere, this weak neutron source (and weak loss rate) can maintain a radiation belt comparable in intensity and energy to the observed high energy proton belt. At lower energies the high intensity and rapid time variations of the observed trapped proton fluxes indicate that other source and loss mechanisms must be present. The energy region where cosmic ray albedo neutrons are the principal proton source was once believed to be all energies above 10 MeV. Better measurements of the neutron flux [*Preszler et al.*, 1972] and a more sophisticated analysis raised this energy to about 30 MeV. Recent data from the CRRES satellite [*Blake et al.*, 1992] show that protons well above 50 MeV are accelerated and injected by shock waves, and this process may be as important as the neutron source between 50 and 100 MeV.

In the case of the electrons, it was quickly found that neutron decay was not an important source. The energy spectrum of electrons from neutron decay did not agree with trapped electron measurements [*Kellogg*, 1960; *Walt*, 1961], and the intensity of the electron belt was much larger than could be sustained by decaying neutrons [*O'Brien*, 1962]. Also, the

3. PARTICLE LOSSES FROM COLLISIONS WITH ATMOSPHERIC CONSTITUENTS

An energetic charged particle will be scattered in pitch angle and will lose energy when it collides with atmospheric atoms. The energy loss rate for a particle of kinetic energy E, charge Z and velocity v passing through a medium containing various species i of number density N_i and atomic number z_i is given by the familiar dE/dx formula [Bethe, 1933]:

$$\Delta E = v \frac{dE}{dx} = -\frac{e^4 Z^2}{4\pi\varepsilon_0^2 m_0 c\beta} \sum_i z_i N_i \ln \frac{E(E/m_0 c^2 + 2)^{1/2}}{I_i}, \quad (2)$$

where I_i is the average electronic excitation energy of atomic constituent i. Because of the factor $v = c\beta$ in the denominator, low energy particles slow down and are lost more rapidly than higher energy ones.

The cross section for scattering a charged particle of mass M and charge Z through angle η by a massive nucleus of charge z is given by the Rutherford formula

$$\sigma(\eta, E) = \frac{z^2 Z^2 e^4}{64\pi^2 \varepsilon_0^2 M^2 c^4} \frac{1-\beta^2}{\beta^4} \frac{1}{\sin^4(\eta/2)}. \quad (3)$$

Because the mass M of the scattered particle is in the denominator, protons and heavier ions do not scatter appreciably in angle while slowing down. The particle energy enters this equation through the factor containing β, and it is clear that low energy electrons are scattered more effectively. The $\sin^{-4}(\eta/2)$ term describes the angular distribution of scattered particles and indicates that the scattered particles are strongly peaked in the forward direction. Although the cross section is limited at small angles by the shielding of the scattering nucleus by its orbital electrons, the predominant scattering is at small angles, and many collisions are required to deflect a particle appreciably. This fact is useful in that it allows pitch angle scattering of electrons by atmospheric atoms to be treated by diffusion theory.

For the highest energy protons ($E > 50$ MeV) collisions with the nucleus of the scattering center are also important; such collisions effectively remove the proton from the radiation belts. Lower energy protons are also removed by charge exchange collisions in which the proton picks up an electron from the scattering atom, becomes a neutral hydrogen atom, and escapes the geomagnetic field. Charge exchange collisions are also important for heavier ions which may have multiple charges. In a single collision the ion may lose or gain electrons, depending largely on its velocity.

Calculating the effects of the atmosphere on trapped particles is a straightforward but tedious exercise. It is necessary to construct averages of the atmospheric density over the particle's trajectory, and for the case of electron scattering, one needs to know the average density encountered at each pitch angle. Since the magnetic field is irregular near the earth where atmospheric encounters are most important, one must use an accurate magnetic field model as well as models of atmospheric densities for each constituent.

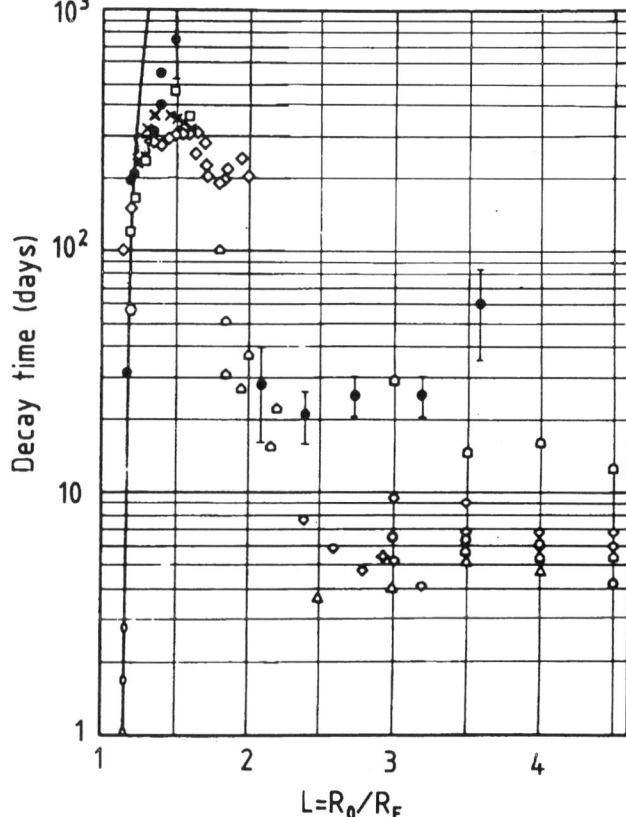

Figure 2. Decay time constants for energetic trapped electrons. The solid line shows expected atmospheric collision decay time, and symbols give observed decay times for various energies and times.

Calculations of electron loss rates from atmospheric collisions were completed in the early 1960's and showed that the calculated and measured loss rates were in agreement only for $L < 1.3$ [Walt, 1964]. At higher L values electrons were lost much more rapidly than could be explained by atmospheric collisions. This situation is illustrated in Figure 2 which shows the observed lifetimes of relativistic electrons compared to the expected decay rate due to atmospheric collisions. The particle lifetime, or the time for the particle flux to decay by a factor 1/e, has become the parameter characterizing the loss rate. Most of the experimental data of Figure 2 were obtained following high altitude nuclear detonations although similar results have been derived from electron flux decays following magnetic storms and substorms.

The lifetime of relativistic electrons colliding with atmospheric atoms increases rapidly with L, is approximately a year at $L = 1.25$ and increases to about 10 years at $L = 1.5$. Experimental lifetimes are a maximum of about 1 year at $L = 1.5$ and fall to a few weeks above $L = 2$. While the scatter in the experimental data is large and the lifetimes de-

4 SOURCE AND LOSS PROCESSES

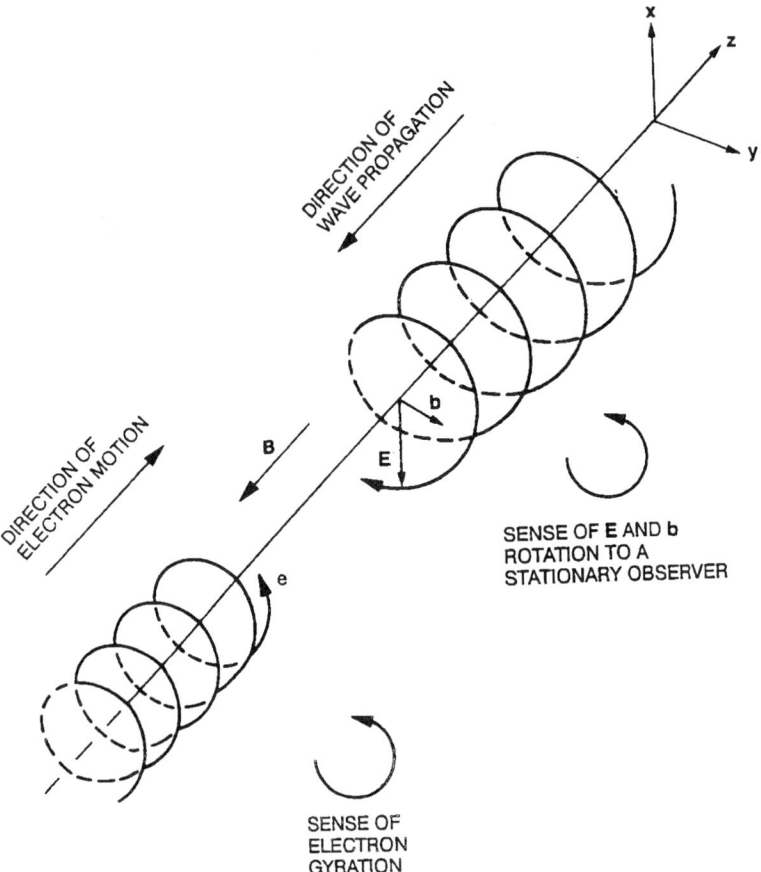

Figure 3. Interaction of an electron with a circularly polarized wave moving parallel to the magnetic field. Resonance occurs if the Doppler shifted wave frequency equals the electron gyrofrequency.

pend on the energy of the electrons being measured, the data clearly show that losses in addition to atmospheric collisions are dominant above $L \approx 1.3$.

4. WAVE PARTICLE INTERACTIONS

To explain the observed loss rate above $L \approx 1.3$ [*Dungey*, 1965] suggested that whistler mode waves could deflect the electrons and might be responsible for their rapid disappearance. He pointed out that if the particle experienced the Doppler shifted wave frequency at its own gyrofrequency, that is, if

$$\frac{\Omega_e}{\gamma} = \omega - \mathbf{k} \cdot \mathbf{v} \qquad (4)$$

the interaction over many gyroperiods would deflect the particle. Cumulative deflections from many interactions with waves would force the electrons into the loss cone, and collisions with the atmosphere would then remove them. In Eq. (4) Ω_e/γ is the energetic electron gyrofrequency, γ is the usual relativistic factor, ω and $\mathbf{k}$ are the whistler wave frequency and wave number, and $\mathbf{v}$ is the particle velocity.

For whistler mode waves $\Omega_e > \omega$; therefore, unless γ is large, $\mathbf{k}$ and $\mathbf{v}$ must be in opposite directions to achieve resonance. The geometry of electron interactions with a circularly polarized electromagnetic wave propagating parallel to the magnetic field is illustrated in Figure 3.

This concept was examined by *Roberts* [1966] using wave frequencies and intensities derived from ground measurements of ducted whistler waves. Ducted whistlers are guided along the field lines by field-aligned enhancements in the thermal electron density. Roberts concluded that this wave particle interaction mechanism was not adequate to explain the observed pitch angle scattering. Basically, electrons trapped in the equatorial plane have no velocity parallel to the field line, and the resonance condition of Eq. (4) cannot be satisfied.

The wave-particle interaction theory was revived and extended by *Kennel and Petschek* [1966] who pointed out that as the particle pitch angle decreased, energy would be transferred to the wave. Thus, if the electron distribution had a maximum at large pitch angles, the diffusion of electrons to smaller pitch angles through interactions with waves would amplify the waves. These arguments led Kennel and Petschek

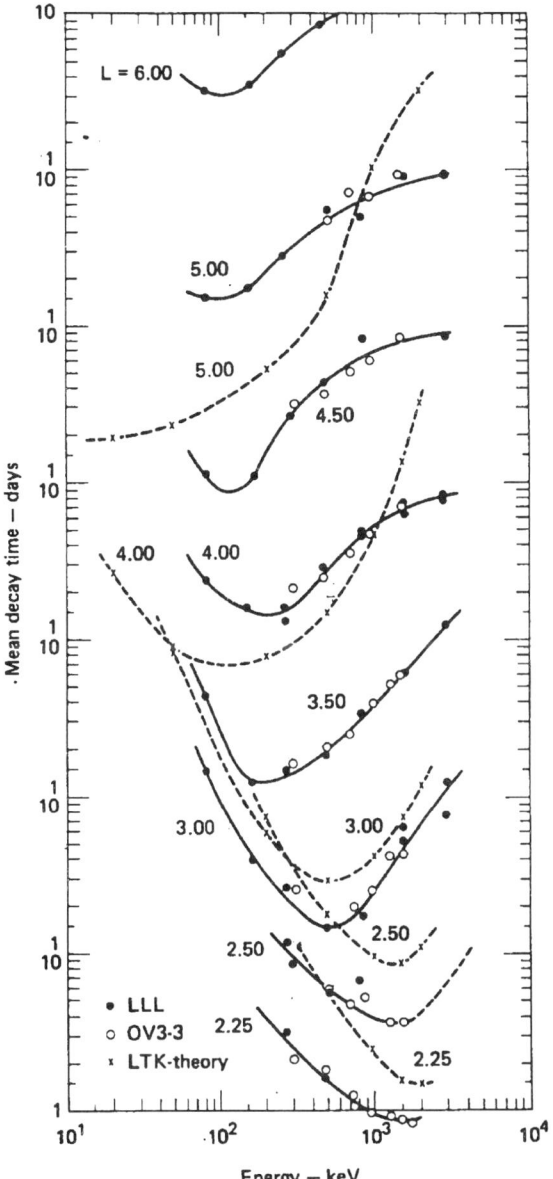

Figure 4. Comparison of observed electron flux decay times with those expected from interactions with plasmaspheric hiss

to the concept of a limited trapped flux; too large a particle population would stimulate more waves which in turn would remove the excess electrons. Similar considerations with ion cyclotron waves would apply to the trapped protons.

The Kennel-Petschek flux limit requires that the waves are propagating parallel to the magnetic field and are partially reflected at the ionosphere. However, subsequent experiments have shown that these conditions are not generally present. Also, experiments have found cases where the measured fluxes exceed the Kennel-Petschek trapping limit, suggesting that the Kennel-Petschek limit does not apply locally. The waves which interact with the electrons at a given location may be produced somewhere else and by other populations of particles. The particle lifetime question then becomes one of finding the intensity and character of waves throughout the magnetosphere and computing the effect of this wave field on the local electron and proton fluxes.

In the plasmasphere a principal form of electromagnetic wave is plasmaspheric hiss. This whistler mode wave field is concentrated at frequencies of a few hundred hz and propagates at a wide distribution of angles with respect to the magnetic field. Since the waves are not field aligned, the interaction with particles can occur at all harmonics of the particle gyration frequency. This extended interaction increases the diffusion rate and in particular allows interactions of the wave with electrons having pitch angles near 90°. The pitch angle diffusion of electrons from plasmaspheric hiss was first extensively studied by *Lyons et al.* [1972] and more recently extended by *Albert* [1994]. With estimated values for the intensities, frequency spectra, and wave normal distributions of the waves, these calculations give electron lifetimes which for $L > 2$ are comparable with experimental values. A comparison of observed electron decay times and the *Lyons et al.* [1972] results is given in Figure 4 [*West et al.*, 1981]. The solid curves are from experiment and the dotted curves were taken from the *Lyons et al.* [1972] theory. The vertical scale is displaced for each L value to prevent confusion. The agreement is quite good, although the absolute values depend on wave characteristics which are not well known. More importantly the computed electron lifetimes exhibit the same energy and L dependence as the observed values, indicating that the nature of the interaction is correct even if the average wave intensity could not be estimated accurately.

Lyons et al. [1972] did not treat the region $L < 2$ directly, but extrapolations of their values to that region give very long lifetimes. Since the observed electron lifetimes at $1.3 < L < 2$ (see Figure 2) are much shorter than expected either from atmospheric collisions or from interactions with plasmaspheric hiss, an additional loss mechanism must be present in the outer part of the inner electron zone. A possible loss mechanism is electron interactions with man-made waves from high powered VLF communication transmitters and power line harmonics which enter the magnetosphere [*Vampola and Kuck*, 1978; *Imhof et al.*, 1981; *Koons et al.*, 1981]. These waves have been observed in the magnetosphere although their entry through the ionosphere is sporadic. Occasionally the trapped electron spectra in the drift loss cone is quite monoenergetic and is consistent with the scattering of trapped electrons by fixed frequency waves from high powered VLF transmitters. However, this mechanism is still somewhat controversial, and a better understanding of the average transmitter wave intensities and spatial distributions in the magnetosphere is needed to clarify its importance.

An experiment specifically designed to study the precipitation of electrons by VLF transmitters was carried out by *Imhof et al.* [1982]. In this experiment Navy VLF transmitters were pulsed in a coded sequence while a low altitude (200 km) satellite passing above the transmitter searched for an identical time sequence in precipitating electrons. During the 6 months of operation five passes were found in which

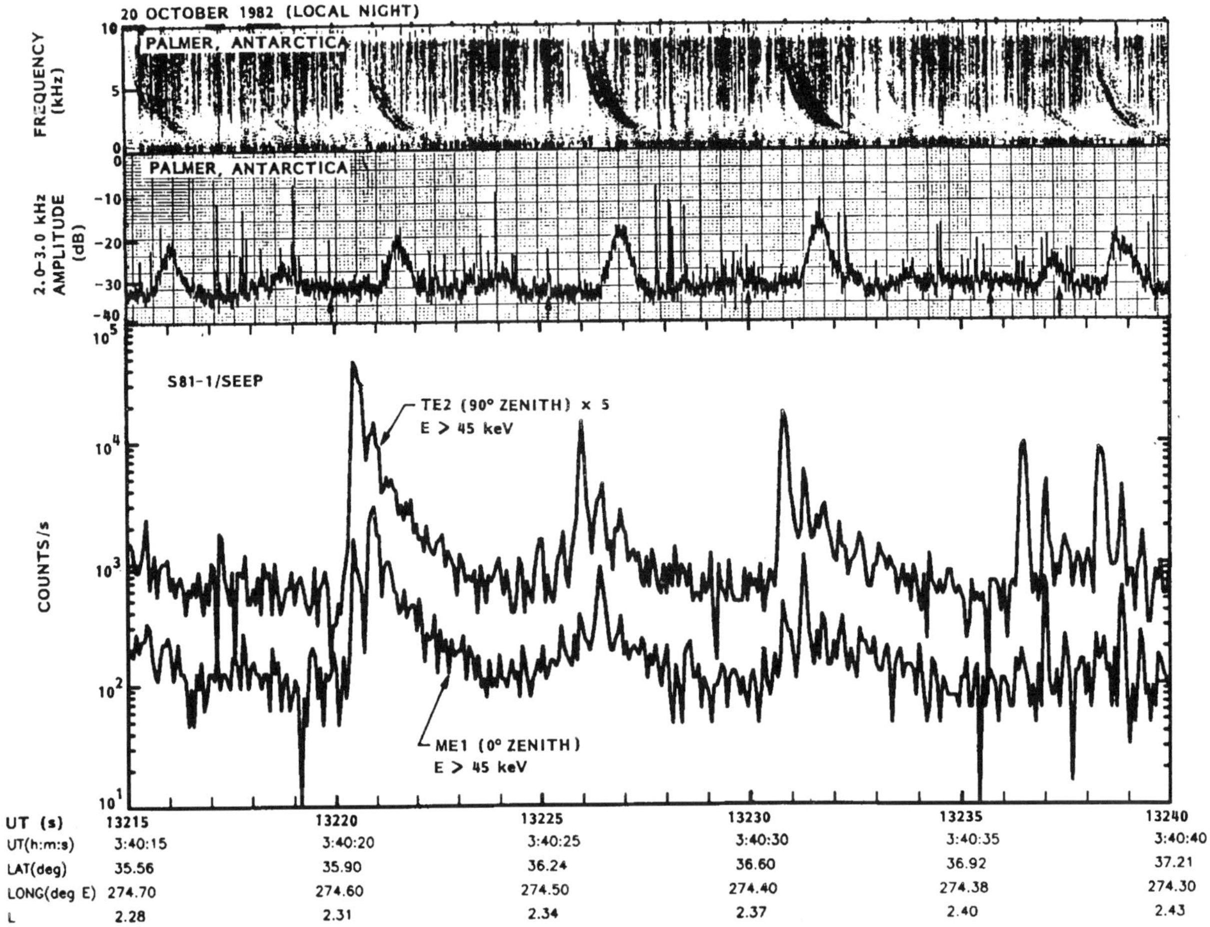

6 SOURCE AND LOSS PROCESSES

Figure 5. Coincidence of electron precipitation observed by a satellite in the northern hemisphere (bottom panel) with ducted whistlers detected at Palmer, Antarctica (top panel)

electrons were precipitated from the radiation belts in coincidence with the transmitter pulses. This experiment demonstrated the viability of the mechanism, but the efficiency of precipitation was low; most passes showed no measurable precipitation. The importance of transmitters in removing trapped electrons does not appear to be large, at least in the L region investigated here ($2 < L < 3$).

In searching for transmitter induced precipitation events many cases were found where electrons were precipitated by the VLF waves from lightning flashes [*Voss et al.*, 1985]. In these events ducted whistlers were detected by ground receiving stations in coincidence with electron bursts being observed by a satellite (see Figure 5). Other evidence for lightning-induced precipitation comes from VLF propagation experiments in which a momentary increase in D-region electron density caused by the precipitation is detected by its effect on VLF transmissions in the earth-ionosphere waveguide. Thousands of such cases have now been recorded, indicating that lightning regularly precipitates bursts of electrons.

Analysis of the satellite measurements of electron precipitation [*Inan et al.*, 1985] has confirmed the mechanism as a wave-particle event initiated by a ducted whistler. Estimates of the importance of this process depend on the wave intensities, the occurrence rates of ducted whistlers, and on the volume fraction of the plasmasphere occupied by ducts. None of these factors is well known, especially the duct sizes and their numbers. Estimates of the precipitation rates of electrons by ducted whistlers give electron lifetimes comparable to experiment in the region $2 < L < 3$, but the reliability of this result is low due to the uncertain factors mentioned above.

In addition to ducted whistlers, lightning also introduces considerable wave energy into the magnetosphere in the unducted mode. While these waves cannot be detected from the ground, they have been observed from satellites. Their wave normal angles are large, and they will interact with electrons at all harmonics of the gyrofrequency. Furthermore, an individual lightning stroke will introduce non-ducted waves over a much larger fraction of the magnetosphere than is possible for the ducted mode. Calculations of the effects of non-ducted waves on trapped electrons are being done [*Lauben et*

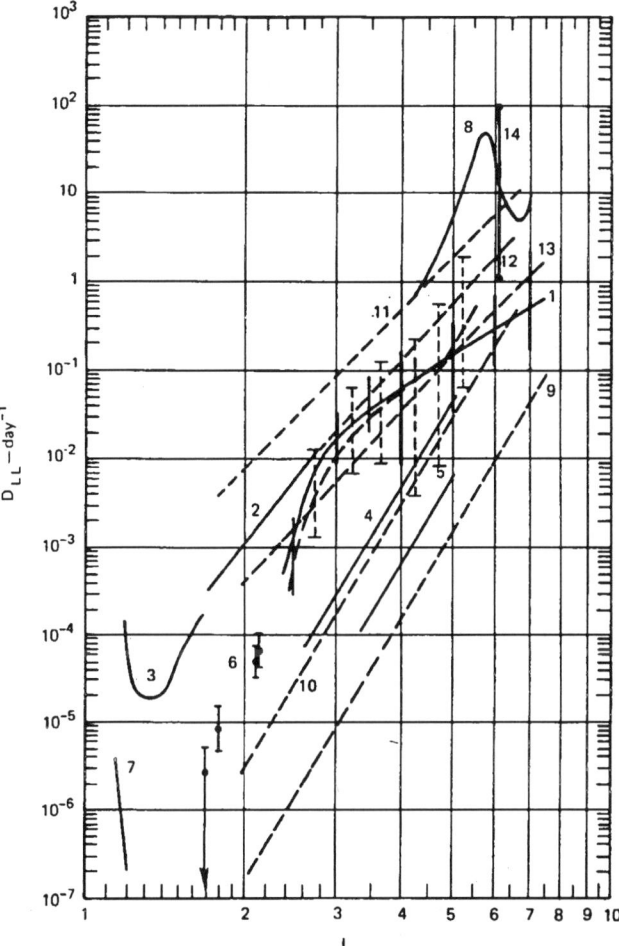

Figure 6. Radial diffusion coefficients obtained by a variety of methods. Dashed lines are based on theory using Eqs. (5) and (6). Solid curves and points are based on fitting the diffusion equation to experimental flux values.

al., 1995], but the overall importance of the process is uncertain because of a lack of information on the average intensity and occurrence rate of non-ducted waves.

5. RADIAL DIFFUSION

Solar flare particles and solar wind plasma were recognized immediately as potential sources of ions and electrons for the radiation belts. However, to populate the belts these particles must somehow migrate across magnetic shells and reach the innermost regions of the radiation belts. This process of radial transport, or radial diffusion as it is normally termed, was suspected to occur in order to explain the motion of transient features in the radial profiles of the radiation belts. Also, the systematic increase in particle energy with decreasing L was compatible with the motion of particles across L shells, changing the third adiabatic invariant but conserving the first two.

Analytic expressions for the expected radial diffusion coefficients have been derived based on idealized models of magnetic and electric field fluctuations [*Falthammar*, 1968]. Assuming that only the lowest two terms of the multipole expansion of the disturbance magnetic field are important and assuming that the perturbing magnetic fields are small compared to the normal dipole field, one obtains a diffusion coefficient for equatorially trapped particles:

$$D_{LL}^M = \frac{\pi^2}{2}\left(\frac{5}{7}\right)^2 \frac{R_E^2 L^{10}}{B_0^2} \nu_{\text{drift}}^2 P_A(\nu_{\text{drift}}), \quad (5)$$

where R_E is the radius of the earth, B_0 is the surface magnetic field on the equator, ν_{drift} is the longitudinal drift frequency, and $P_A(\nu_{\text{drift}})$ is the power spectrum of the asymmetric part of the magnetic perturbation evaluated at the particle drift frequency. Note that D_{LL}^M increases rapidly with L, the L^{10} factor overwhelming the weak L dependence of the factors containing ν_{drift}. If the particles have equatorial pitch angles less than $90°$, the diffusion coefficient is reduced substantially.

For diffusion driven by electric potential fields a similar calculation yields a diffusion coefficient of

$$D_{LL}^E(L, \nu_{\text{drift}}) = \frac{L^6}{8R_E^2 B_0^2} \sum_{n=1}^{N} P_n(L, n\nu)_{\nu=\nu_{\text{drift}}}, \quad (6)$$

where P_n is the nth Fourier component of the global electric field perturbation. For electric perturbations in which the field lines are equipotentials, the diffusion coefficient is independent of equatorial pitch angle.

These equations are convenient to use although it is difficult to obtain accurate values of the power spectra of the electric and magnetic perturbations. Several important assumptions were made in these derivations. For example it was assumed that the radial motion of a particle during an individual perturbation was so small that the change in its drift velocity could be neglected. It was also assumed that the distribution in longitude of the particles was uniform at the start of the perturbation. Hence, individual perturbations must be small, and the time interval between disturbance pulses must be long enough for the dispersion in particle drift velocity to restore the uniform distribution in longitude. Neither of these conditions is valid during substorms; hence the theory as represented by Eqs. (5) and (6) does not apply to these major events. Particle simulations have been done by *Chen et al.* [1994] and by *Riley and Wolf* [1992] to investigate the limitations of diffusion theory in treating the radial displacements caused by field perturbations. They find that on average the diffusion approach yields adequate results, but in individual cases of prescribed electric field perturbations, the simulation results are quite different from the diffusion theory predictions. Agreement improves as the field perturbations become smaller.

Radial diffusion is a redistribution mechanism that is both a source and loss mechanism depending on the overall radial distribution. For example it is a source mechanism for bringing solar wind particles into the inner regions of the magnetosphere, but it is a loss mechanism for the protons

8 SOURCE AND LOSS PROCESSES

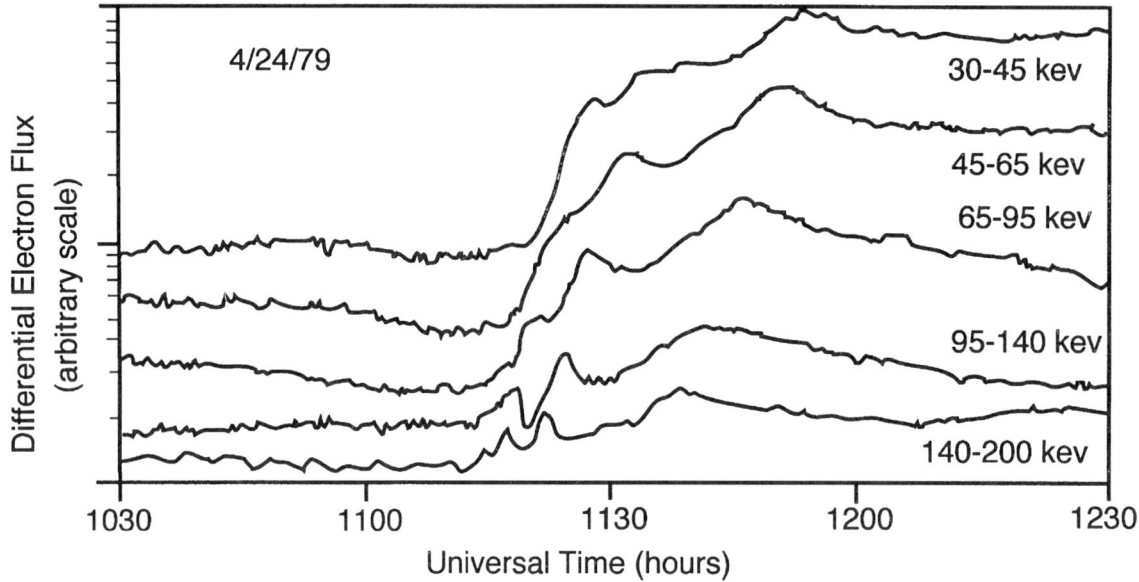

Figure 7. Substorm injection of trapped electron fluxes observed from geosynchronous orbit

injected into the inner belt by decaying neutrons. Efforts to determine values for the radial diffusion coefficients have been made using the above equations and introducing measured or estimated values for the power spectra of the various components of the electric and magnetic field fluctuations. A more empirical approach to finding D_{LL} is to adjust the diffusion coefficient in the time dependent form of the diffusion equation to fit measured time variations in the radial profiles of trapped particles. In cases where one is confident of the source and loss processes, the steady-state radial diffusion equation can be used and D_{LL} selected to fit experimental data.

Examples of diffusion coefficients obtained by all these methods are shown in Figure 6. The two extreme curves labeled 9 and 11 are based on early estimates of electric and magnetic field variations and are quite uncertain in magnitude. While there is a large spread in the absolute values of the remaining curves, all values of D_{LL} show the expected strong variation with L. Some of the scatter in D_{LL} can also be attributed to the time variations in D_{LL} since the diffusion rate depends on geomagnetic activity. The agreement between the diffusion coefficients derived at different times and by a variety of methods and their rough correspondence to values obtained by perturbation theory argues for the reality of the radial diffusion process. However, the assumptions required for Eqs. (5) and (6) are quite restrictive and the results are only approximate for most actual circumstances. Also, the use of a diffusion equation itself to treat radial transport is only valid in the limit of small individual displacements, and the extraction of a D_{LL} from data will be an illusion if the diffusion process is not the dominant one.

Although the radial diffusion process provides a means to bring solar wind particles into the radiation belts, it is not sufficient to accelerate particles to the observed energies. If the magnetic moment is conserved, the ratio of initial to final particle energies is approximately the ratio of the initial and final mirror magnetic fields. In the earth's magnetosphere, the field ratio is not large enough to accelerate solar wind electrons to the observed energies of > 5 MeV. Either the initial energy must be higher or a larger acceleration factor is needed.

Higher initial energies of electrons have been proposed to account for the multi MeV trapped electrons. Sources of high energy electrons in interplanetary space include electrons leaking from Jupiter's radiation belts, solar flare electrons, and electrons which have been accelerated by interplanetary shock waves. Which, if any, of these electron sources is responsible for the high energy tail of trapped electrons is not clear at present.

Larger accelerations can be obtained by radial diffusion if multiple modes of radial diffusion and pitch angle diffusion are present and these modes accelerate particles in different ways. For example consider a two mode situation in which one mode involves radial diffusion at constant first and second adiabatic invariant with pitch angle scattering at constant energy, while the second mode allows diffusion in L at constant energy. A particle could diffuse inward, conserving its first and second invariant and gaining energy as it decreases its radial distance. It might then drift outward via the alternate mode of radial diffusion retaining its energy. A second inward diffusion under the first mode would further increase its energy, and subsequent round trips could raise its energy to any arbitrary value. Most particles would not follow this linear process, but would drift inward and outward at random, losing and gaining energy. Only the very lucky particles would experience repeated accelerations, but this bimodal diffusion would satisfy the need for greater acceleration than can be achieved by third invariant diffusion alone. This process was suggested by *Theodoridis et al.* [1969] and has recently been revived by *Fujimoto and Nishida* [1990]. Although this

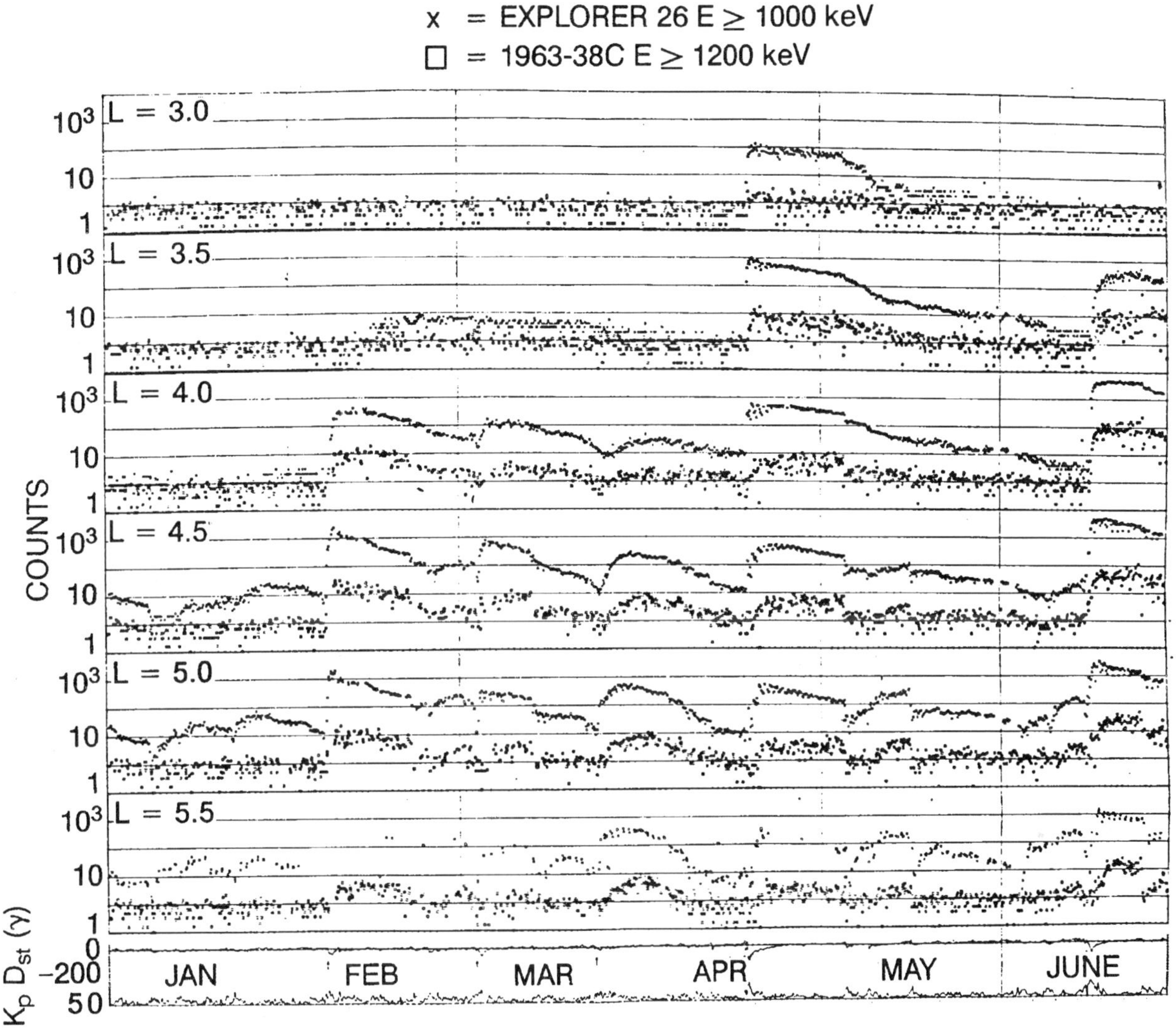

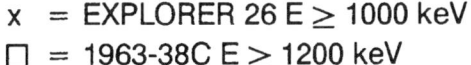

Figure 8. Magnetic storm effects on > 1 MeV electrons trapped at various L values. Flux variations are largest and most frequent at higher L.

process is appealing in principle, one must identify all the mechanisms involved and their magnitudes to evaluate the importance of bimodal diffusion in the earth's radiation belts.

6. STORM AND SUBSTORM INJECTION EVENTS

The source and loss mechanisms discussed thus far have been of a steady state character, or in the case of radial diffusion, the process involves slowly varying averages over many small episodes of change. However, the radiation belts are not benign; very large and rapid time variations in trapped particle intensities occur, particularly in the outer regions where sudden increases in sources and losses dominate the character of the radiation belts.

Particle injections during magnetic activity were noted in the 1960's. The injection process accompanying substorms was modeled by *DeForest and McIlwain* [1971], who showed that the observed substorm fluxes were consistent with the sudden appearance of a trapped population of electrons and ions deep in the magnetophere on the night side. Subsequent work has explored the geometry of the electric and magnetic fields causing the injection and the particle drifts [*Stern,* 1990; *Spiro,* 1988]. Basically, the substorm process brings particles from the reservoir in the plasma sheet and accelerates them to form an outer radiation belt and storm-time ring current. The population of particles in the near earth plasma sheet contains plasma of both solar wind and atmospheric origin, so the injected particles include electrons and protons as well as helium and oxygen ions. Higher Z ions from the sun are also present in reduced concentrations, but once they are injected they remain trapped and are moved by radial diffusion into the heart of the belts.

Figure 7 is an example of a substorm injection as seen from geosynchronous orbit [*Reeves et al.*, 1992]. Fluxes of accelerated electrons are shown in several energy intervals, and it is apparent that the more energetic particles arrive first at the satellite. Drift dispersion is generally seen in substorm injections and becomes larger as the local time of the satellite position advances from the midnight sector, where the injection takes place. Flux increases of > 1 MeV electrons accompanying major storms are shown in Figure 8 [*Williams et al.*, 1968] which depicts six months of data at a selection of L values. The lowest panel shows the K_p and D_{st} magnetic indices. Major injections accompany each D_{st} depression. The electron flux variations are larger and more frequent at higher L values and only major storms, such as the one in April enhance the belts below $L = 3$. This figure also illustrates the difficulty of treating the outer belt as a steady state balance of sources and losses with a well-defined average value.

Clues to the source of the injected particles can be found in their charge state if the particles are observed shortly after their injection. For example, singly charged He and O originate in the atmosphere while He^{++} and multiply charged oxygen ions probably originate in the solar corona. However, this reasoning is not always valid, since it has been shown by *Spjeldvik and Fritz* [1978] that trapped particles eventually reach an equilibrium charge distribution which is independent of their initial state.

The substorm injection process has many uncertain features. Although particle simulations have traced paths for the injected particles and have found their origin, these trajectories depend on assumed magnetic and electric fields. The global electrodynamics of a typical substorm, and the variations among substorms, are not well known. Until the electric and magnetic fields throughout the magnetosphere are known at all times during a storm or substorm, the particle simulation results will be incomplete.

7. ACCELERATION DURING A STORM SUDDEN COMMENCEMENT

In March of 1991 a large magnetic storm caused profound changes in the Earth's radiation belts, greatly increasing the intensities of electrons, protons, and heavier ions in the slot region. By a fortunate coincidence, the CRRES satellite was at $L = 2.3$ and recorded the acceleration process in great detail, including the longitudinal dependence of the newly accelerated particles [*Blake et al.*, 1992]. Surprisingly, electron energies above 10 MeV, and proton energies above 50 MeV were observed, indicating that a powerful acceleration process, thus far unnoticed, was present in the magnetosphere.

The newly injected proton and electron fluxes are shown in Figure 9 (protons on top, electrons below). The initial acceleration process was limited in longitude so that a maximum in the counting rate occurred each time the accelerated bunch drifted in longitude past the satellite. Dispersion in drift period within the bunch eventually smooths out the distribution, leaving a higher residual flux level.

Subsequent detailed analysis and particle simulations by *Li et al.* [1993] showed that the observations were completely consistent with an interaction of trapped particles and untrapped solar flare particles with the electric and magnetic

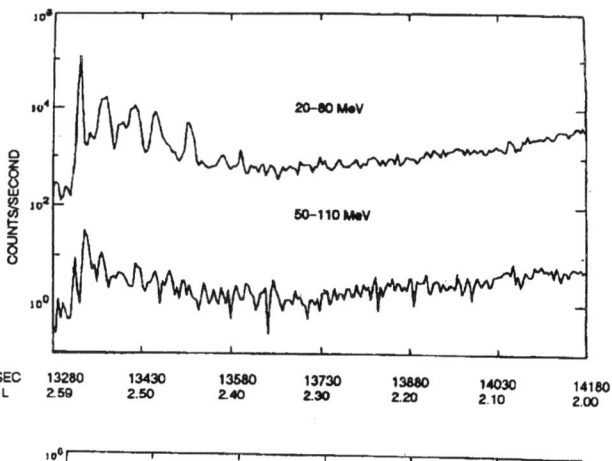

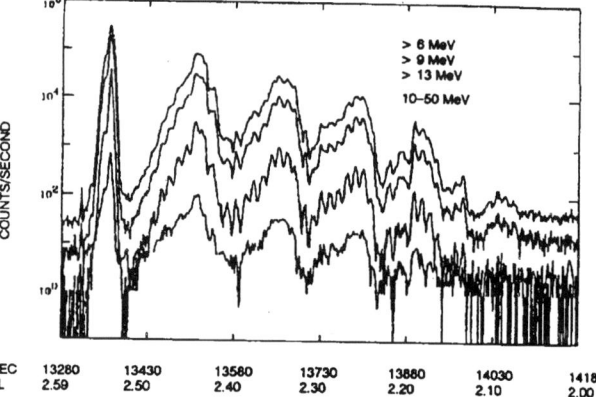

Figure 9. Counting rates of protons (top) and electrons (bottom) following the storm sudden commencement of 24 March, 1991. Damped oscillations are drift echoes of the original injection which occured near 13,326 s.

fields of a collisionless shock passing through the magnetosphere. The shock was produced in interplanetary space by a solar coronal mass ejection, propagated through interplanetary space and impacted the magnetosphere at a time when the solar flare particles were also present. As confirmed by detailed particle simulations those electrons and protons which were on the dayside at the time the shock arrived could experience an accelerating electric field during an appreciable portion of their drift orbit about the earth. Solar flare protons within the magnetosphere could also have their orbits modified and become trapped. A subsequent search of records [*Gussenhoven et al.*, 1994] has shown that this process has occurred at other times, although not as dramatically as in the March 31 event.

Perhaps the most important lesson from the March 1991 event was the realization that even after decades of experimental and theoretical effort, all the important processes producing trapped radiation had not even been described, let alone understood. The acceleration of solar flare protons to 50 MeV and their expected diffusion into the inner radiation belt has also called into question the long held belief that all

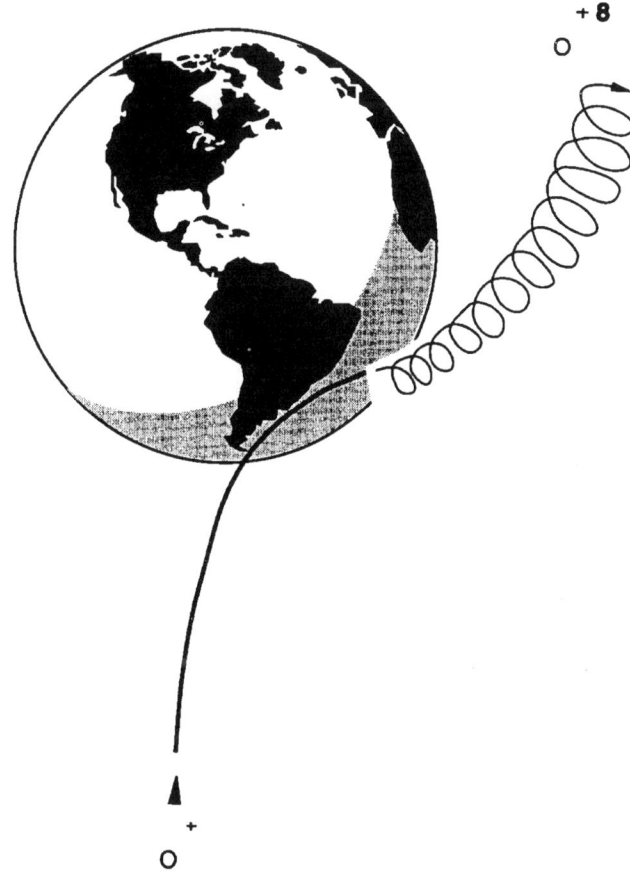

Figure 10. Conversion of an anomalous cosmic ray of singly charged oxygen to a trapped multiply charged oxygen ion by the removal of 7 electrons in an atmospheric collision

the trapped protons above 30 MeV are produced by decaying neutrons.

8. OTHER TRAPPING PROCESSES

Recent measurements by the SAMPEX satellite have established the existence and characteristics of a weak, but persistent belt of multiply charged nitrogen, oxygen and neon nuclei near $L = 2$ [*Selesnick et al.*, 1995]. The satellite was in a low altitude (500 km) polar orbit and the heavy nuclei were detected only when the satellite was not in the drift loss cone. The energy spectrum of the heavy nuclei was a decreasing function of energy extending out beyond 50 MeV/nucleon. These particles appear to be a distinct population with an entirely different origin than other radiation belt particles.

The source of these trapped ions is believed to be the anomalous cosmic rays. Anomalous cosmic rays are formed from interplanetary neutral particles which are ionized by solar ultra violet radiation or by collisions with energetic interplanetary particles. The newly formed ions are then swept away from the sun by the electric field associated with the solar wind. At the heliospheric boundary the ions are further accelerated and diffuse inward, scattering from magnetic irregularities in the interplanetary field. At earth orbit anomalous cosmic rays represent a population of singly charged ions with energies below several hundred MeV, low by normal cosmic ray standards. These ions penetrate the magnetosphere in Störmer trajectories which cannot directly lead to trapping. However, occasionally an anomalous cosmic ray inside the magnetosphere will collide with another particle, usually a neutral atmospheric molecule, and will lose additional electrons. Such a collision will instantly decrease the particle's rigidity and can result in the anomalous cosmic ray being trapped. Analysis of the intensity, energy spectra and composition of this belt confirms the trapping mechanism. The process is illustrated schematically in Figure 10, which depicts a singly charged oxygen ion being stripped of 7 more electrons near the earth and becoming trapped.

Studies of this trapped group of particles is leading to a better understanding of the requirements for permanent trapping. The quantity $\rho|\nabla B|/B$ is a measure of the relative change in the magnetic field over distances of a gyroradius, ρ. This parameter must be small (≈ 0.1) for a particle to be trapped. However, it must be large for a particle which is initially outside the magnetosphere to penetrate to the SAMPEX orbit at mid latitudes. In the case of anomalous cosmic rays, the initial single charge allows the particle to enter the magnetosphere and the subsequent collision reduces the gyroradius by a factor between 2 and the atomic number of the nucleus. A knowledge of the charge distribution of the trapped ions can lead to a better understanding of the limits of the trapping parameter.

Although SAMPEX measurements were only made at low altitude, these multiply charged ions should also exist at high altitude. Those with mirroring points at low altitude will pass through higher altitudes during their bounce motion, but others should be produced which mirror above SAMPEX. Since both the source strength and the loss rate are proportional to the average atmosphere encountered by the particle in its trapped orbit, the intensity at various locations inside the magnetosphere should be relatively uniform after allowing for the intensities of anomalous cosmic rays at various locations. Thus, a study of the distribution of these trapped ions could reveal the presence of loss processes other than atmospheric collisions.

9. OTHER HEAVY IONS

The magnetosphere also contains heavy ions originating in the solar wind and solar cosmic rays which become trapped in the outer magnetosphere, diffuse inward and are accelerated. Phase space distributions of the particles reveal their source is at the outer boundary and their intensities and energy spectra are in general agreement with these assumptions. The charge state distributions of these ions have been calculated [*Spjeldvik and Fritz*, 1978] although the results are somewhat uncertain because all the relevant charge exchange cross sections are not known. One interesting feature derived from these calculations is that the charge state distribution approaches equilibrium and depends only on the velocity of the ion. Thus, the prospect of determining the origin of the ions by their charge state must be questioned for ions which have been trapped for long periods.

However, the singly charged oxygen and helium ions which are present in the ring current during the main phase of a magnetic storm must come from the upper atmosphere. These ions have been observed to flow out of the atmosphere at high latitudes during magnetically active periods.

10. SUMMARY

Electrons and ions trapped in the geomagnetic field arise from several distinct sources and encounter multiple loss mechanisms. Many of these mechanisms are not quantitatively understood at present, and it may well be that additional source and loss processes exist but have not been recognized. Equilibrium calculations of radiation belt populations based on the known source and loss processes give reasonable agreement with measurements. However, errors in the parameters upon which these models are based could easily conceal additional source and loss processes.

This paper has treated the source and loss processes which are well understood at least in principle. However, even for these well recognized processes, there are substantial uncertainties in the quantitative values of the parameters involved. Similarly, there are often mathematical approximations used in the evaluation of a process which limit the conclusions which can be drawn from a comparison with experiment. Thus, substantial additional work must be done before one can with confidence claim that we understand the source and loss processes of the terrestrial radiation belts.

REFERENCES

Albert, J.M., Quasi-linear pitch angle diffusion coefficients: retaining high harmonics, *J. Geophys. Res., 99*, 23,741, 1994.

Bethe, H., Handbuch der Physik, Vol. 29, Chapter 1, Part 1, edited by A. Smekal, Springer, Berlin, 1933.

Blake, J.B., W.A. Kolasinski, R.W. Fillius and E.G. Mullen, Injection of electrons and protons with energies of tens of MeV into $L < 3$ on March, 1991, *Geophys. Res. Lett., 19*, 821, 1992.

Chen, M.W., L.R. Lyons and M. Schulz, Simulations of phase space distributions of storm time proton ring current, *J. Geophys. Res., 99*, 5745, 1994.

Dungey, J.W., Loss of Van Allen electrons due to whistlers, *Planetary Space Sci., 11*, 591, 1963.

DeForest, S.E. and C.E. McIlwain, Plasma clouds in the magnetosphere, *J. Geophys. Res., 76*, 3587, 1971.

Falthammar, C.-G., Radial diffusion by violation of the third adiabatic invariant, in Earth's Particles and Fields, edited by B.M. McCormac, pp. 157, Reinhold, 1968.

Fujimoto, M. and A. Nishida, Energization and anisotropization of energetic electrons in the earth's radiation belt by the recirculation process, *J. Geophys. Res., 95*, 4265, 1990.

Gussenhoven, M.S., E.G. Mullen and M.D. Violet, Solar particle events as seen on CRRES, *Adv. Space Res., 14*, 619, 1994.

Imhof, W.L., R.R. Anderson, J.B. Reagan and E.E. Gains, The significance of VLF transmitters in the precipitation of inner belt electrons, *J. Geophys. Res., 86*, 11,225, 1981.

Inan, U.S., M. Walt, H.D. Voss and W.L. Imhof, Energy spectra and pitch angle distributions of lightning-induced electron precipitation: Analysis of an event observed on the S81-1 (SEEP) satellite, *J. Geophys. Res., 94*, 1,379, 1989.

Kellogg, P., Possible explanation of the radiation observed by Van Allen at high altitude in satellites, *Nuovo cimento, 11*, 48, 1959.

Kellogg, P., Electrons of the Van Allen radiation, *J. Geophys. Res., 65*, 2,705, 1960.

Kennel, C.F. and H.E. Petschek, Limit on stably trapped particle fluxes, *J. Geophys. Res., 71*, 1, 1966.

Koons, H.C., B.C. Edgar and A.L. Vampola, Precipitation of inner zone electrons by whistler mode waves from the VLF transmitters UMS and NWC, *J. Geophys. Res., 86*, 640, 1981.

Lauben, D.S., U.S. Inan and T.F. Bell, Precipitation of radiation belt electrons by nonducted whistler mode VLF wave pulses, *EOS, 76*, F495, 1995.

Li, X., I. Roth, M. Temerin, J.R. Wygant, M.K. Hudson and J.B. Blake, *J. Geophys. Res., 20*, 2,423, 1993.

Lyons, L.R., R.M. Thorne and C.F. Kennel, Pitch-angle diffusion of radiation belt electrons within the plasmasphere, *J. Geophys. Res., 77*, 3455, 1972.

O'Brien, B.J., Lifetimes of outer-zone electrons and their precipitation into the atmosphere, *J. Geophys. Res., 67*, 3687, 1962.

Preszler, A.M., G.M. Simnett and R.S. White, Earth albedo neutrons from 10 to 100 MeV, *Phys. Rev. Lett., 28*, 982, 1972.

Reeves, G.D., G. Kettmann, T.A. Fritz and R.D. Belian, Further investigation of the CDAW 7 substorm using geosynchronous particle data: multiple injections and their implications, *J. Geophys. Res., 97*, 6417, 1992.

Riley, P. and R.A. Wolf, Comparison of diffusion and particle drift descriptions of radial transport in the earth's inner magnetosphere, *J. Geophys. Res., 97*, 16865, 1992.

Roberts, C.S., Electron loss from the Van Allen zones due to pitch angle scattering by electromagnetic disturbances, in Radiation Trapped in the Earth's Magnetic Field, edited by B.M. McCormac, Reidel Pub. Dordrecht, Holland, 1966.

Selesnick, R.S., A.C. Cummings, J.R. Cummings, R.A. Mewaldt, and E.C. Stone, Geomagnetically trapped anomalous cosmic rays, *J. Geophys. Res., 100*, 9503, 1995.

Singer, S.F., Trapped albedo theory of the radiation belt, *Phys. Rev. Lett., 1*, 181, 1958.

Spjeldvik, W.N. and T.A. Fritz, Theory of charge states of energetic oxygen ions in the earth's radiation belts, *J. Geophys. Res., 83*, 1583, 1978.

Spiro, R.W., R.A. Wolf and B.G. Fejer, Penetration of high latitude electric field effects to low latitudes during SUNDIAL 1984, *Ann. Geophys., 6*, 39, 1988.

Stern, D.P., Substorm electrodynamics, *J. Geophys. Res., 95*, 12,057, 1990.

Theodoridis, G.C., R.R. Paolini and S. Frankenthal, Acceleration of trapped electrons and protons through bimodal diffusion in the Earth's radiation belts, *J. Geophys. Res., 74*, 1,238, 1969.

Vampola, A.L. and G.A. Kuck, Induced precipitation of inner zone electrons, 1. Observations, *J. Geophys. Res., 83*, 2543, 1978.

Vernov, S.N., N.L. Grigorov, I.P. Ivanenko, A.I. Lebedinskii, V.W. Murzin and A.E. Chudakov, Possible mechanism of production of terrestrial corpuscular radiation under the action of cosmic rays, *Soviet Physics*, Doklady 4, 154, 1959.

Voss, H.D., W.L. Imhof, M. Walt, J. Mobilia, E.E. Gaines, J.B. Reagan, U.S. Inan, R.A. Helliwell, D.L. Carpenter, J.P. Katsufrakis, and H.C. Chang, Lightning-induced electron precipitation, *Nature, 312*, 740, 1985.

Walt, M. and W.M. MacDonald, Energy spectrum of electrons trapped in the geomagnetic field, *J. Geophys. Res., 66*, 207, 1961.

Walt, M., The effects of atmospheric collisions on geomagnetically trapped electrons, *J. Geophys. Res., 69*, 3947, 1964.

Williams, D.J., J.F. Arens and L.J. Lanzerotti, Observations of trapped electrons at low and high altitudes, *J. Geophys. Res.*, 73, 5673, 1968.

West, H.I., R.M. Buck and G.T. Davidson, The dynamics of energetic electrons in the earth's outer radiation belt during 1968 as observed by the Lawrence Livermore National Laboratory's spectrometer on OGO 5, *J. Geophys. Res.*, 86, 2111, 1981.

M. Walt, Stanford University, Stanford, CA. 94305

DISCUSSION

Q: M. Schulz. What is the present thinking regarding (a) solar energetic particle events and (b) Jupiter, as sources of relativistic trapped radiation for Earth?

A: M. Walt. These issues are likely to be discussed (a) by members of Mary Hudson's group and (b) by Dan Baker later on at this workshop.

Q: J. Albert. Electron-hiss pitch-angle scattering lifetimes (after Lyons et al.) have been recalculated keeping all harmonics (non artificial cut-off).

A: M. Walt. There are lot of assumptions behind the calculations that also should be reconsidered. For example the wave normal distribution assumed by Lyons et al. is only approximate. Better values of electron lifetimes due to hiss will help evaluate the importance of this wave field.

Q: M.I. Panasyuk. In the middle of the 70-s Stanford people published a paper devoted to the problem of the artificial nature of the gap between the electron radiation belt which can be produced by the industrial activity of mankind in the northern hemisphere. Do you have any comments now?

A: M. Walt. The source you refer to is radiation from electric power grids at harmonics of the power line frequency. While this radiation is often present, its average value is not known, and definitive calculations of its effect on trapped electrons are not known.

Q: G. Ginet. What is the primary source of artificial VLF waves? Can the total power be easily estimated?

A: M. Walt. The primary sources are VLF communication transmitter and power line harmonics. The strength of these emissions in the magnetosphere, above the ionosphere, is variable and subject to amplification by trapped particles. The average values are not known well enough for electron lifetime determinations.

Q: M.S. Gussenhoven. We do not seem to understand fully the entry of solar particles and their ability to contribute to radiation belts. Solar protons can penetrate deeply into the magnetosphere and then recede. A question J. Albert stumped me with is: "Is a 30 MeV solar proton at $L = 3$ and with a 90° pitch angle trapped? It certainly didn't get there by being trapped, e.g. it conserved energy, not μ."

A: M. Walt. Yes. Störmer theory in a real, time dependent, magnetosphere is poorly understood.

Processes Acting Upon Outer Zone Electrons[1]

C.E. McIlwain

Department of Physics, University of California, San Diego

Five distinct processes have been found which cause time variations in the energetic electron fluxes in the outer zone. It is shown how the combined action of these processes can produce the observed time dependencies. One process has been definitely identified as being due to a specific physical mechanism, namely, adiabatic betatron acceleration. It is found that changes in both the ring current field and the magnetospheric boundary current field produce predictable changes in the particle fluxes. An example of how trapped particle measurements can be used to compute relatively accurate D_{st} values is presented. Since the adiabatic effects are predictable, they can be removed to exhibit the non-adiabatic effects more clearly. Using this technique, an occurrence of enhanced loss has been found which may be due to the instability predicted by *Andronov and Trakhtengerts* [1964] and by *Kennel and Petschek* [1966].

1. INTRODUCTION

It has been shown in many papers [*Forbush et al.*, 1962; *Hoffman et al.*, 1962; *McIlwain*, 1963; *Frank*, 1965; *Williams*, 1966; *Davis and Williamson*, 1966] that the outer zone electron fluxes exhibit large temporal variations which are correlated with variations in the earth's magnetic field. In the present paper it is shown that the variations in the electron fluxes can be ascribed to the simultaneous action of at least five distinct processes. One of these processes can be definitely associated with a particular and well understood physical mechanism, namely adiabatic betatron acceleration. It was predicted some years ago that this should be an important mechanism acting upon trapped particles by *Dessler and Karplus* [1961] but the first experimental verification was only recently made [*McIlwain*, 1966].

The present paper is primarily concerned with the betatron acceleration process. The other four processes will be treated in more detail in future papers.

2. DETECTOR

Most of the data presented here was obtained by a directional scintillation detector which is shielded by at least $2.5\,\mathrm{g\,cm^{-2}}$ in all directions except for a $\pm 8°$ cone for which the absorber thickness is $0.048\,\mathrm{g\,cm^{-2}}$. For a wide range of electron spectra the efficiency versus energy for the lower electronic threshold (which corresponds to $0.28\,\mathrm{MeV}$ energy loss) is well represented by a step function which rises from zero to 0.62 at $0.50\,\mathrm{MeV}$.

The detector points perpendicular to the satellite spin axis. Since the satellite spin period was short compared with the accumulation time, the counting rates obtained correspond to the directional flux averaged over the plane perpendicular to the spin axis. The angular distribution of the outer zone electrons near the magnetic equator is such that only relatively small changes are required to convert the spin average counting rates into rates which correspond to the average over all directions and therefore the omnidirectional intensities. The function used for this conversion is

$$\frac{1.0}{1.25 - 0.5\varphi/90}, \qquad (1)$$

---

[1]This paper was presented at the September 1966 Inter-Union Symposium on Solar-Terrestrial Physics in Belgrade, Yugoslavia. The paper was accepted for publication by the *Journal of Geophysical Research* on October 10, 1966. Unfortunately, the paper was never re-submitted after the author performed the minor modifications suggested by a reviewer. The correction to Eq. (17) provided by Leo Davis has been included, as have the extensive rewordings suggested by Dr. Sidney Chapman. Other than these changes (which were penciled in during 1966) and the conversion to LaTeX, this is the original version of the paper.

16 PROCESSES ACTING UPON OUTER ZONE ELECTRONS

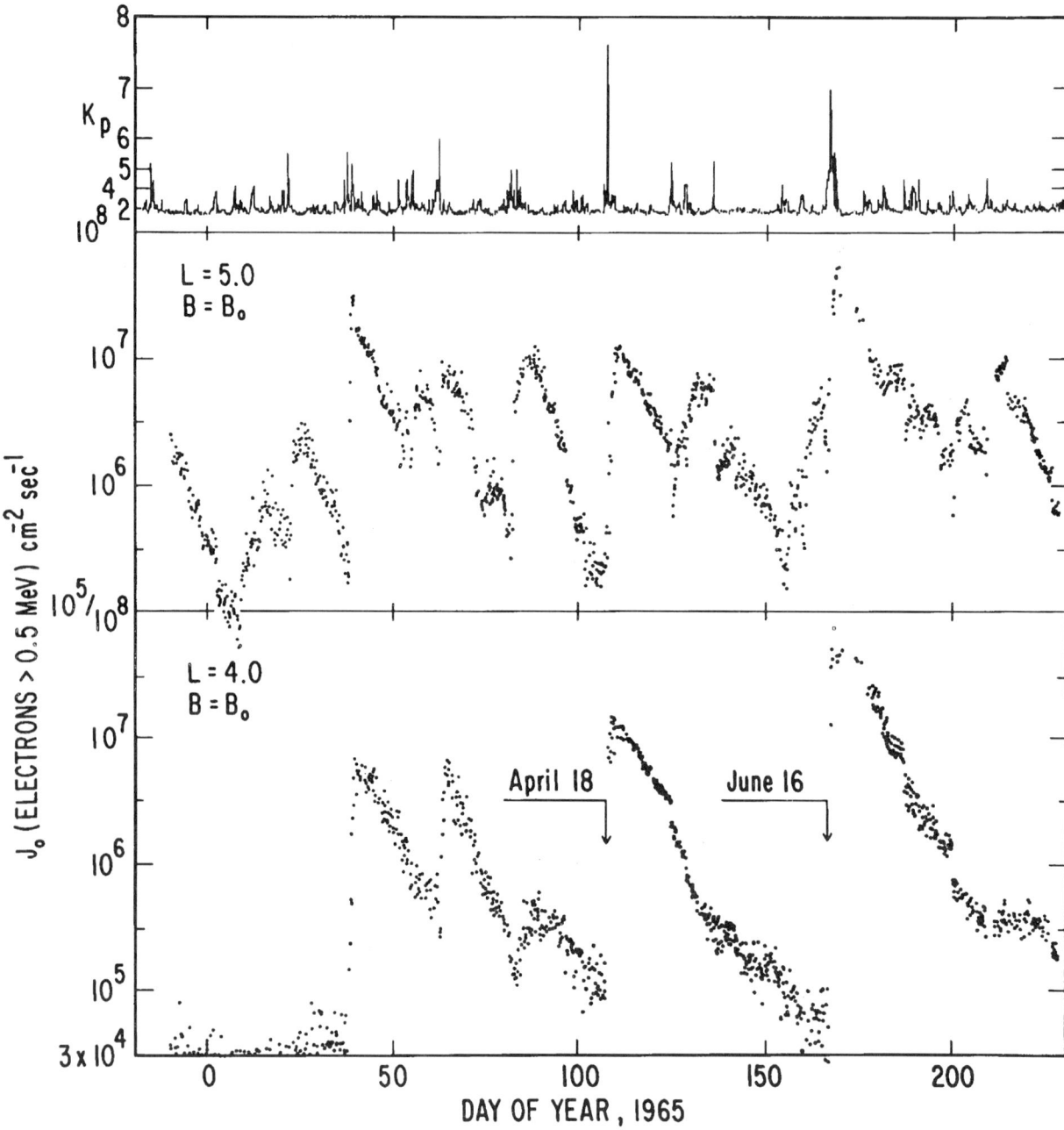

Figure 1. Time variations in the omnidirectional intensities of electrons with energies greater than 0.5 MeV at $L = 4.0$ and 5.0 measured during 1965 by the Explorer 26 satellite. The rapid increases at times of magnetic disturbances and the tendency to decay with about a two-week time constant are easily seen.

where φ is the angle between the spin axis of the satellite and the computed local **B** vector in degrees. Multiplication of the counting rates by this factor and by $(\varepsilon G)^{-1} = 25,000$ yields intensities of electrons with energies greater than 0.5 MeV with absolute errors of less than ±20% and relative errors which are typically less than 7%.

3. NORMALIZATION TO $B = B_0$

Fortunately one of the processes (pitch angle scattering?) in the outer zone is such that the relative variation of the electron intensities along lines of force near the magnetic equator is kept constant in time. The measured variation along lines of

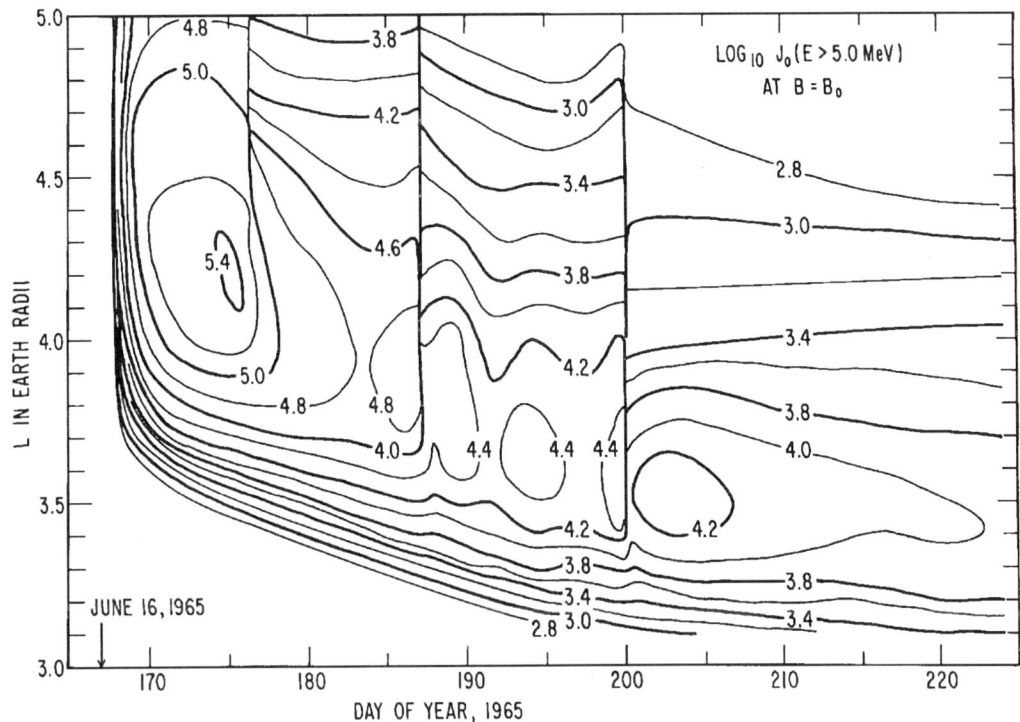

Figure 2. Isointensity contours of high energy electrons following the rapid acceleration on June 16, 1965. The inward motion apparently due to radial diffusion and the rapid loss on days 187 and 200 are of particular interest.

force is well represented by

$$J(B) = kB^{-N}, \qquad (2)$$

with $N = 0.3$ to 0.4.

The measurements reported here are confined to the region $B/B_0 = 1.0$ to 3.0, thus normalization to $B = B_0$ involves multiplication by numbers between 1.0 and $3^{0.4} = 1.55$.

Except for short periods immediately following large non-adiabatic perturbations, the B dependence along lines of force (i.e., the latitude dependence) can be ignored thus reducing the important spatial variables to the radial distance (i.e., L) and local time. The orbits of the Explorer 15 and 26 satellites are such that only a narrow range of local time is covered while they are in the outer zone during any particular observation period, thus the local time dependence usually does not need to be considered explicitly in the study of time variations covering a period of only a few months. Most of the data presented here were taken within ±6 hours of local noon.

4. THE FIVE PROCESSES

Figure 1 shows the time variations in the fluxes of electrons with energies greater than 0.5 MeV at $L = 4.0$ and 5.0 during 1965 as measured with the Explorer 26 satellite. The most obvious features of this data are the rapid increases every month or so and the persistent exponential decay with about a two-week time constant. These features are so easily perceived that little further analysis is required to establish the existence of the first two processes:

Process 1—Rapid Non-Adiabatic Acceleration,
Process 2—Persistent Decay.

Detailed examination of the April 18, 1965 event reveals that Process 1 can cause a large acceleration within only a few hours time period. There is some indication that the cases in which the acceleration appears to continue over a period of several days actually consist of a series of discrete events each of which lasts only a few hours. There can be no doubt that Process 1 involves non-adiabatic acceleration, because there is no available reservoir with an adequate supply of such energetic electrons, and because the electron fluxes remain high long after the magnetic field perturbations have subsided.

There is much evidence to indicate that the physical mechanism responsible for Process 2 is pitch angle scattering of the electrons into the loss cone: namely the persistent precipitation of electrons at low altitudes [*O'Brien*, 1962; *O'Brien*, 1964; *Paulikas and Freden*, 1964; *Paulikas et al.*, 1966], the theoretical prediction of several different mechanisms which cause pitch angle perturbations [*Dungey*, 1963; *Cornwall*, 1964; *Dungey*, 1965; *Chang and Pearlstein*, 1965; *Kennel and Petschek*, 1966; *Cornwall*, 1966; *Eviatar*, 1966; *Pearlstein et al.*, 1966; *Chang*, 1966], the strong tendency to maintain a particular pitch angle distribution, and the apparent increase in the decay time constants with increasing electron energy. Energy loss and scattering due to interaction with the atmosphere is of course important at low altitudes and is

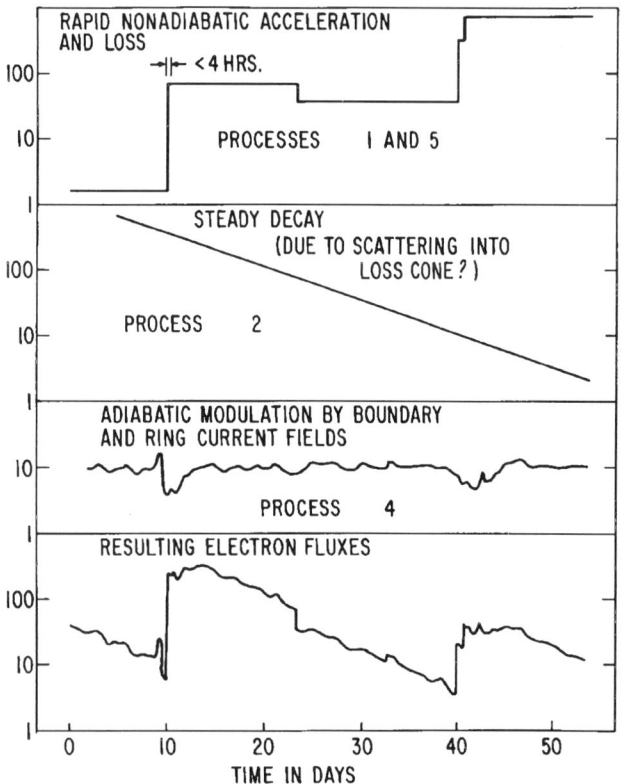

Figure 3. Characteristic effects of Processes 1, 2, 4, and 5 and their combined effect upon the energetic electrons in the outer zone.

almost certain to be an important element in any theory which can properly explain Process 2.

Figure 2 shows the time and L dependence of the electrons with energies greater than 5 MeV following the rapid acceleration which occurred on June 16, 1965. In this figure it can be seen that the lower boundary appears to move inward with time. Considerable theoretical work [*Kellogg*, 1959, *Parker*, 1960; *Herlofson*, 1960; *Davis and Chang*, 1962; *Dungey*, 1965; *Falthammar*, 1966] has been published which predict radial diffusion due to the breakdown of the third adiabatic invariant. Following the suggestion made by *Frank* [1965] when he published the first evidence that this mechanism is important for trapped electrons we label the third process radial diffusion:

Process 3—Radial Diffusion.

Ring current magnetic fields cause an adiabatic acceleration of the inner zone protons [*McIlwain*, 1966]; this process also acts upon the outer zone electrons [*Dessler and Karplus*, 1961]. Supporting observational evidence is given later in this paper, thus we have:

Process 4—Adiabatic Acceleration.

Trapped proton fluxes sometimes exhibit rapid non-adiabatic decreases [*McIlwain*, 1964; *McIlwain*, 1966]. Figures 1 and 2 reveal that the electron fluxes also exhibit rapid non-adiabatic decreases e.g., on days 187 and 200 in these figures. Since the effect seems to occur at times when the magnetic field is distended by ring current particles, it is tempting to ascribe the decreases to a loss of particles into the magnetospheric tail region [*Williams and Ness*, 1966]. There is no evidence, however, that lines which normally cross the equator as low as 3 earth radii are ever drawn into the tail region, thus the fifth process is given a noncommittal label:

Process 5—Rapid Loss,

where the loss may be a loss in energy or in number of particles.

Processes 1, 2, 4 and 5 are illustrated in Figure 3 to demonstrate how the net effect of all four can produce the typical time dependence of outer zone electrons. Process 3, radial diffusion, superimposes a gradual increase on the time variation fluxes when there is a large positive gradient with respect to L.

5. THE RAPID NON-ADIABATIC PROCESSES

Processes 1 and 5 appear to occur only during times of magnetic storms which in turn often appear to occur when the earth's magnetic field is depressed by the presence of ring current particles. It now seems to be safe to assume that the magnetic field fluctuations are due to plasma instabilities which occur when the magnetic field is loaded with an excessive energy density of trapped particles. Since the average magnetic field depression at the earth, i.e., D_{st}, can be used as a measure of the total kinetic energy of the trapped particles, it is of interest to examine whether there is any correlation between the maximum D_{st} values during a magnetic storm, and the maximum magnetic field at which instabilities are manifest. Now Processes 1 and 5 are probably due to the time and longitude dependent electric and magnetic fields created by the instabilities. They can therefore be used to determine how deep into the earth's magnetic field the instabilities penetrated during any given magnetic storm.

A detailed study of the correlation between the maximum D_{st} values and the innermost lines of force on which Processes 1 and 5 take place will shortly be made. A preliminary survey however has yielded the following important result:

> There is a high probability that Process 1 will act upon the 0.5 MeV electrons and that Process 5 will act upon the 40 MeV protons on a line of force when the minimum magnetic field along the line of force is less than 10 ± 3 times the average magnetic field depression.

Low latitude aurorae are probably another manifestation of instabilities. A study to determine the relationship between D_{st} and the minimum latitude of auroral emissions should prove interesting though possibly a little difficult to interpret due to the nondipole shape of the field lines when they are heavily laden with charged particles.

6. OBSERVATIONS OF ADIABATIC ACCELERATIONS

Figure 4 shows the time dependence of the omnidirectional

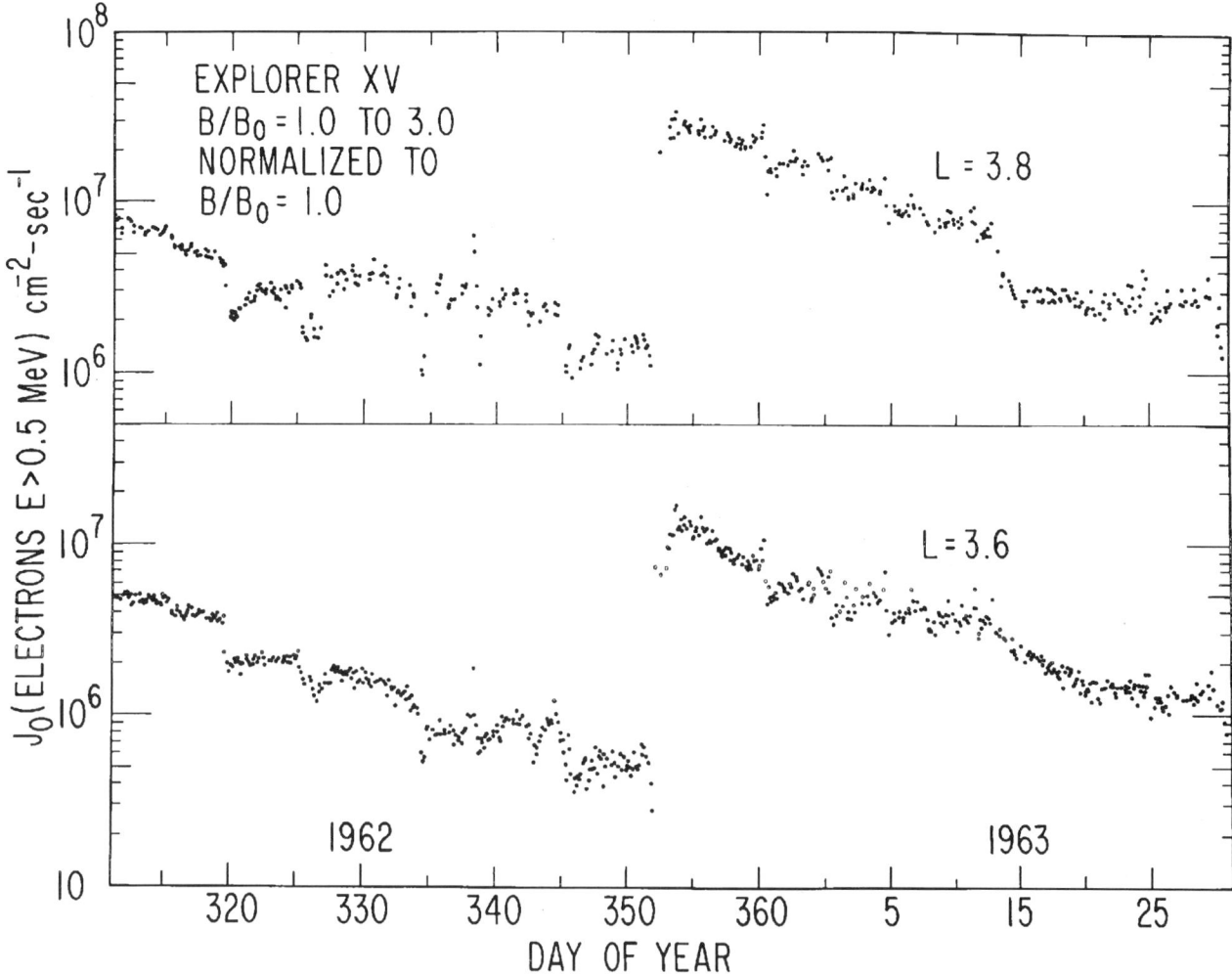

Figure 4. Time dependence of the 0.5 MeV electron fluxes at $L = 3.6$ and $L = 3.8$ during the operating lifetime of Explorer 15. In addition to the large increase on day 352 of 1962 and the general tendency to decay, many nonstatistical fluctuations can be seen to occur.

intensity of electrons with energies greater than 0.5 MeV at the magnetic equator at $L = 3.6$ and 3.8 earth radii during the operating lifetime of Explorer 15. This is the same data which was published earlier [*McIlwain*, 1963] except for the improved normalization of the data discussed in Sect. 3. The dependence of the intensities upon L at three different times is shown in Figure 5.

Figure 4 shows that there was a rapid non-adiabatic acceleration (Process 1) on day 352 of 1962, and that there was a general tendency for the fluxes to decrease in time (Process 2). It shows also many other time variations which are far outside the scatter of the data which is typically less than about ±10%. We proceed to show that most of these variations are due to betatron acceleration (Process 4), that is, the acceleration due to the variation of the magnetic flux inside the shells upon which the electrons would stay during their drift motion about the earth if the magnetic field were constant.

If the decay removes a fixed fraction of the electrons per unit time, then multiplication of the data by $\exp(\text{time}/\tau)$ should completely remove the effects of this process providing the decay time constant τ is properly chosen.

Radial diffusion is not readily discernible in the 0.5 MeV electron data. If it is in fact important for these electrons, then the effects of this process upon the electron fluxes can apparently be included as part of the exponential time dependence ascribed to persistent decay.

It was found that $\tau = 16 \pm 1$ days provided a good fit of the data in Figure 4. The time variations in this data remaining after multiplication by $\exp(t/16)$ are shown in Figures 6 and 7. In Figure 7, the normalization was shifted by factors of 25 and 30 for $L = 3.6$ and 3.8, respectively in order to remove the effects of Process 1 on day 352 of 1962. The continuous lines in these figures are the D_{st} values computed by *Sugiura and Hendricks* [1966] of the Goddard Space Flight Center

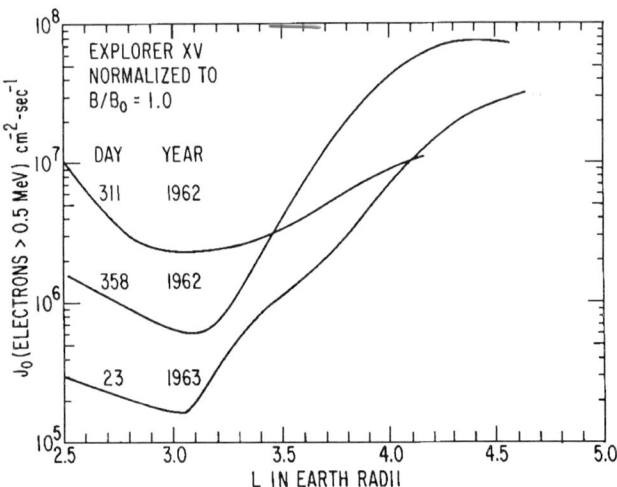

Figure 5. Radial dependence of the 0.5 MeV electron fluxes at three different times as measured by Explorer 15.

using the equation

$$D_{st} = \frac{1}{n} \sum_{i=1}^{n} (\Delta H_i - S_{qi}) \sec \lambda_i, \quad (3)$$

where λ is the magnetic latitude, ΔH is the deviation of the horizontal magnetic field component from the average quiet time field and S_q is the average daily variation at each of the magnetic observatories. For the time period shown here, the hourly mean values from the three stations, Hermanus, San Juan, and Honolulu were used. The D_{st} values are plotted on linear scales while the electron data are plotted on logarithmic scales thus implying the relationship

$$J_0 = k \exp(D_{st}/\beta), \quad (4)$$

where $\beta = +54\,\gamma$ for $L = 3.6$ and $+43\,\gamma$ for $L = 3.8$.

It is easy to find places in Figures 6 and 7 where the correspondence between D_{st} and the particle data is not particularly good, but, there is clearly an intimate relationship between the two quantities. The early points at $L = 3.8$ in Figure 6 are below the D_{st} trace, which is normalized to the later data. A Process 1 event must have occurred on this line of force on day 327 of 1962, though this is not very evident in Figure 4.

The close correspondence between D_{st} and the electron fluxes implies that if the D_{st} effects were removed, the fluxes would exhibit a smooth exponential decay thus indicating that Process 2 is independent of time. One notable deviation from uniform decay can be found in Figure 7 during the period from days 353 to 359 of 1962. Here the D_{st} values suggest increases in the particle fluxes which did not occur, perhaps because of the instability predicted by *Kennel and Petschek* [1966] and by *Cornwall* [1966], which causes a rapid loss of particles when the particle fluxes exceed certain limits. This effect might be labeled as a distinct process, but for now it will be considered as an enhancement of Process 2.

7. THEORETICAL PREDICTIONS

In their paper predicting the betatron effect upon trapped particles, *Dessler and Karplus* [1961] computed the motion and energy change of the trapped particles mirroring at the magnetic equator due to ring current magnetic field changes. The equations for computing the change in particle intensity (i.e., counting rates) are given in a recent paper [*McIlwain*, 1966]:

$$p_2^2 = \frac{p_1^2 B_2}{B_1} \quad (5)$$

$$j_2(B_2, E_2) = \left(\frac{p_2}{p_1}\right)^2 j_1(B_1, E_1) \quad (6)$$

$$B_1 = \left[(B_2 + K)^{\frac{1}{3}} + \frac{K}{2}(B_2 + K)^{-2/3}\right]^3 \quad (7)$$

or for small K

$$B_1 \cong B_2 + 2.5K, \quad (8)$$

where it was assumed that the magnetic field change has the same value of $-K$ everywhere, and where p is the particle momentum, B is the magnetic field at the particle's location, E the particle kinetic energy, and j the directional particle intensity (differential in energy) with the subscripts 1 and 2 corresponding to the values before and after the field change.

Using Eq. (5) and (8) and the relativistically correct relationship between momentum and energy we find

$$E_1 = \left[(E_2^2 + 2E_r E_2)\left(1 + \frac{2.5K}{B_2}\right) + E_r^2\right]^{\frac{1}{2}} - E_r, \quad (9)$$

where E_r ($= m_0 c^2$) is the particle rest energy. A useful approximation to Eq. (9) is

$$E_1 \cong E_2 \left(1 + \frac{2.5KA}{B_2}\right), \quad (10)$$

where

$$A = 1 - \frac{0.5E_2}{E_2 + E_r}. \quad (11)$$

If the original spatial and energy dependence can be represented by

$$j_i(B, E) = g(B) \exp(-E/E_0), \quad (12)$$

then the integral intensity measured before the field change by a detector sensitive to particles with energies greater than E_d is

$$J_i = g(B_i) \int_{E_d}^{\infty} \exp(-E/E_0)\, dE$$
$$= g(B_i) E_0 \exp(-E_d/E_0). \quad (13)$$

If the detector remains at the same location in space, then after the field change the B value at that location will be

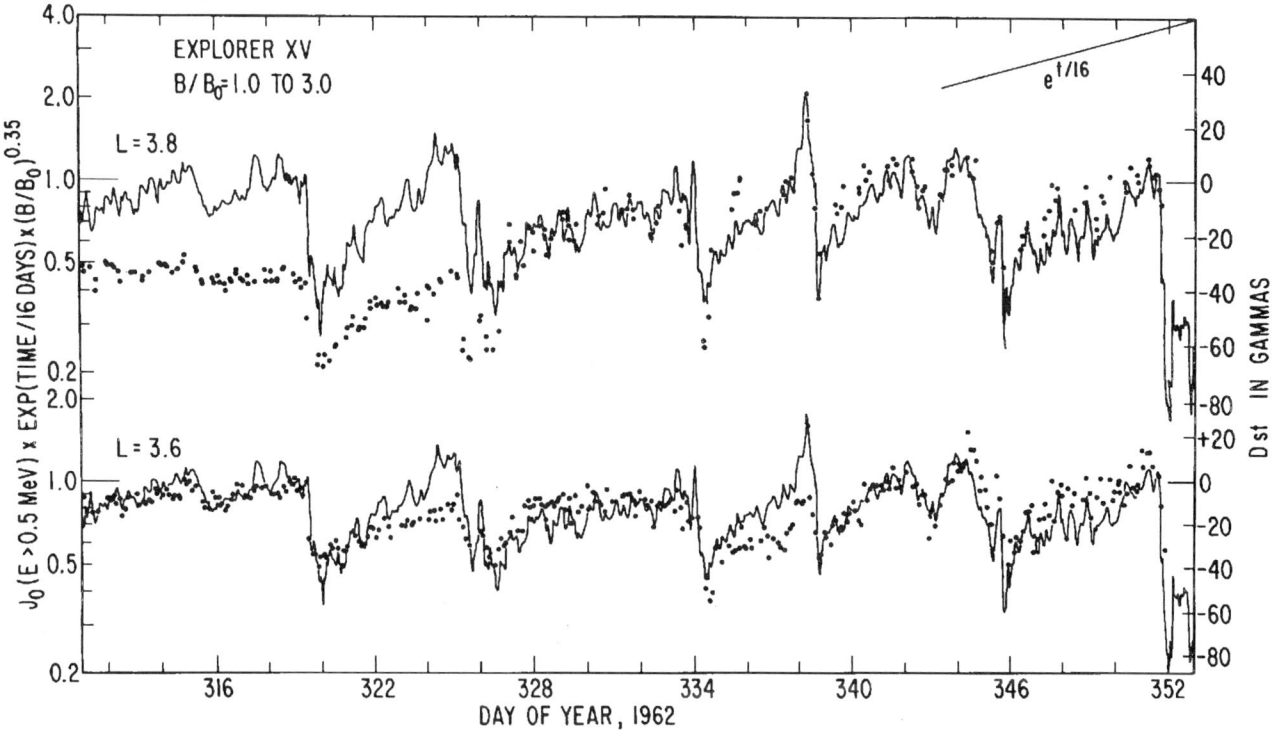

Figure 6. The full line shows D_{st}, the data represent the first half of the data shown in Figure 4 after multiplication by $\exp(t/16)$ to remove the effects of Process 2 (persistent decay). At $L = 3.8$ there was a non-adiabatic acceleration on day 327.

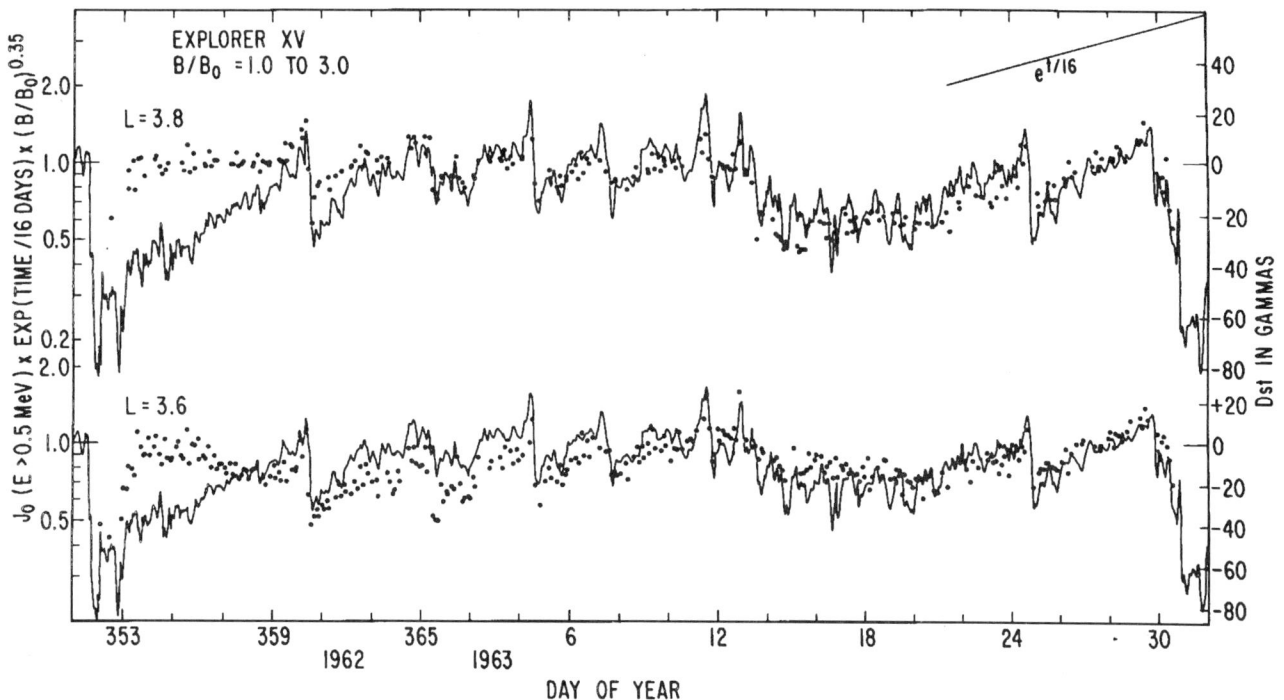

Figure 7. A continuation of Figure 6 with the data renormalized to remove the effects of the Process 1 event on day 352 of 1962. The intensities following this event did not increase as predicted by D_{st} thus implying an enhanced loss rate.

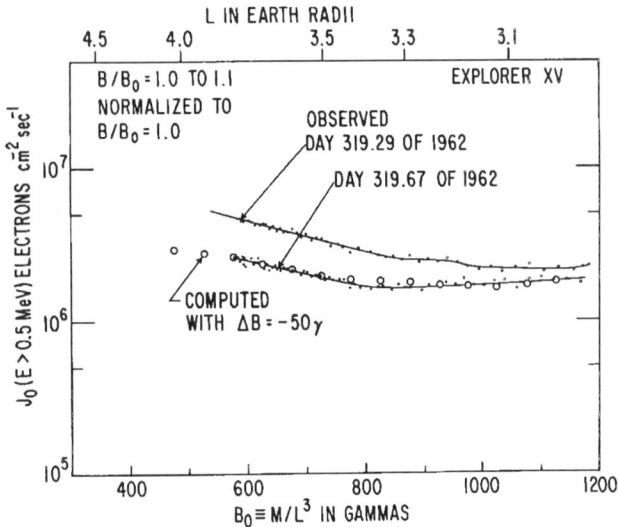

Figure 8. The dependences upon B_0 before and after the decrease in D_{st} on day 319 of 1962 compared with the predicted dependence computed from the upper curve assuming a field change of $-50\,\gamma$.

$B_i - K$ (note: it is still assumed that the field change is the same everywhere in space), thus by Eq. (8), it will measure the particles that were at $B_i + 1.5K$. Taking

$$B_1 = B_i + 1.5K \qquad (14)$$
$$B_2 = B_i - K, \qquad (15)$$

Eq. (6) gives the differential flux after the change to be

$$j_f(E_2) = g(B_1)\frac{B_2}{B_1}\exp\left(-\frac{E_1}{E_0}\right). \qquad (16)$$

If E_0 is not large compared to E_d, then the variation of the factor A [see Eq. (11)] with energy can be neglected and the integral flux after the change will be

$$\begin{aligned}
J_f &= \int_{E_d}^{\infty} j_f(E_2)\,dE_2 \\
&\cong g(B_1)\frac{B_2 E_0}{B_1 + 2.5KA} \\
&\quad \times \exp\left[-\frac{E_d}{E_0}\left(1 + \frac{2.5KA}{B_2}\right)\right],
\end{aligned} \qquad (17)$$

where A might be evaluated at $E_d + E_0/2$.

Dividing Eq. (17) by Eq. (13) gives the counting rate of a detector sensitive to particles with energies greater than E_d at a fixed location in space to be

$$\begin{aligned}
r &= J_f/J_i \\
&= \frac{B_2}{B_1 + 2.5KA}\frac{g(B_1)}{g(B_i)}\times\exp\left[-\frac{2.5KAE_d}{B_2 E_0}\right] \\
&= \frac{B_i - K}{B_i + K(1.5 + 2.5A)}\frac{g(B_i + 1.5K)}{g(B_i)}
\end{aligned}$$

$$\times \exp\left[-\frac{2.5KAE_d}{E_0(B_i - K)}\right] \qquad (18)$$

after a field decrease of K relative to the initial counting rate at the initial field value of B_i.

If we represent the initial B dependence by

$$g(B) = k\exp(aB + bB^2), \qquad (19)$$

then

$$\begin{aligned}
r = \frac{B_i - K}{B_i + K(1.5 + 2.5A)} &\times \exp\Big[1.5aK + 3bB_i K \\
&+ 2.25bK^2 - \frac{2.5AE_d K}{E_0(B_i - K)}\Big].
\end{aligned} \qquad (20)$$

This equation gives the change in the directional flux of particles which mirror at the magnetic equator. But no serious error is likely if it is used to compute the changes in omni-directional fluxes near the equator providing the B values used correspond to the equatorial field on the lines of force.

8. COMPARISON WITH OBSERVATIONS

Equation (20) gives the kind of relationship implied by Figures 6 and 7, i.e. that given by Eq. (4). For simplicity let $b = 0$, then the spatial dependence over the three month time period gives values for $1/a$ ranging from about -90 to $-370\,\gamma$ at $L = 3.6$, while the energy dependence varies little from $E_0 = 0.4$ MeV. Thus with $E_d = 0.5$ MeV giving $A = 0.71$ and with $L = 3.6$ giving $B_i = M/L^3 = 668\,\gamma$ we have

$$r = \frac{668 - K}{668 + 3.3K}\exp\left(-\frac{1.5K}{90} - \frac{2.2K}{668 - K}\right) \qquad (21)$$

to

$$r = \frac{668 - K}{668 + 3.3K}\exp\left(-\frac{1.5K}{370} - \frac{2.2K}{668 - K}\right), \qquad (22)$$

which for small K approximate respectively to

$$r = \exp(-K/38) \qquad (23)$$

and

$$r = \exp(-K/74). \qquad (24)$$

If $K = -D_{st}$ the predicted dependence of the counting rates is of the same form as implied in Figures 6 and 7, and the predicted sensitivity to D_{st} is also similar: β (predicted) $=38$ to $74\,\gamma$ compared with β (implied by the figures) $= 54\,\gamma$.

It is of interest to see whether the change in the spatial dependence is the same as predicted. As can be seen in Figure 6, there was a substantial change in D_{st} during day 319 of 1962. In Figure 8 the fluxes measured before and after this change are plotted versus $B_0 = M/L^3$ which is the predicted equatorial field with no contributions from external current systems. As before it is assumed that the true value

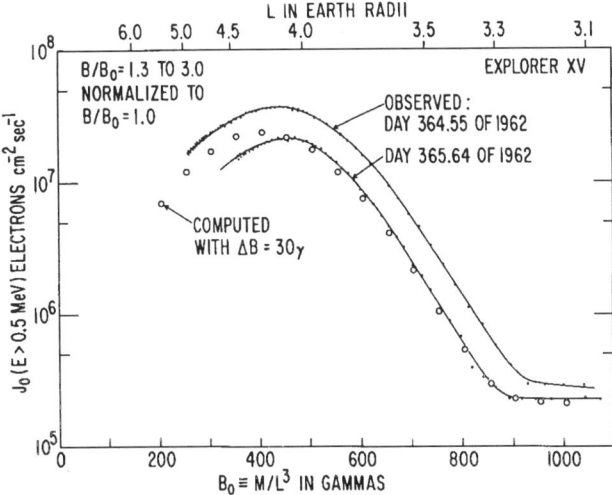

Figure 9. A second example of the change in the B_0 dependence due to Process 4. See the text for a possible explanation of the discrepancies at low B_0 values.

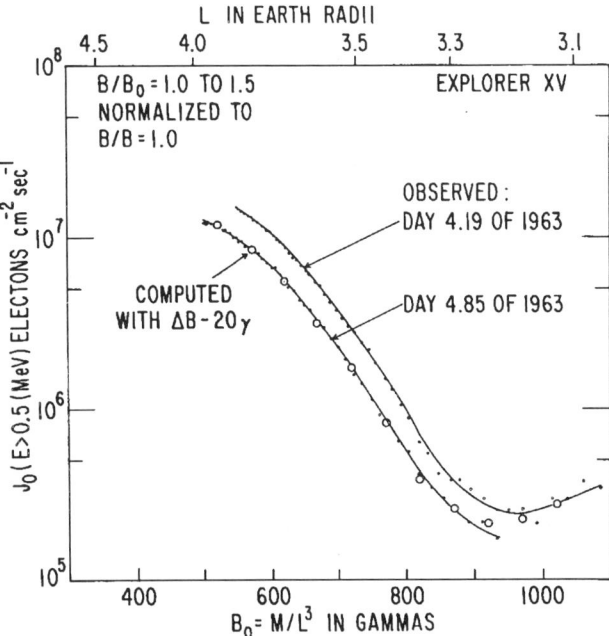

Figure 10. A third example of the change in the B_0 dependence due to process No. 4.

of the magnetic field is $B_0 - K$. This figure also shows the B_0 dependence predicted by Eq. (16) if K is taken to be zero initially and 50 γ after the change in D_{st}. The predictions lie within about ±10% of the measured values which are up to a factor of 2 lower than the initial fluxes. The difference in the D_{st} values between the times of these two sets of data was only −28 γ but the particle measurements probably provide a more accurate determination of the spatial average of the field change than the D_{st} values computed from the field measured at only three ground stations.

Two other examples of changes in the B_0 dependencies are shown in Figures 9 and 10 where again, values for $\Delta B = -K$ can be chosen which will fit the observations. The departure of the computed from the observed values in Figure 9 in the region where $B_0 < 450\,\gamma$ may be because the satellite was at a magnetic latitude of about 30° at this time, and the actual equatorial magnetic field on these lines may be considerably less than the assumed values $(M/L^3 - K)$, since the field lines will be non-dipolar when K is not zero.

The magnetospheric boundary current as well as the ring current can cause predictable particle acceleration; see the large increase during day 338 of 1962 (Figure 6) following a sudden commencement.

9. OTHER OUTER ZONE OBSERVATIONS

Many sets of data have been published, which demonstrate clear correlations with magnetic disturbances. One early attempt to determine the relationship of the electron fluxes with D_{st} [*Forbush et al.*, 1962] yielded rather mixed results because the other four processes caused large effects that could not be readily identified and removed as has been done here. Presumably it will be possible to interpret many of the previous outer zone electron observations in terms of the five processes.

The 40 MeV trapped protons respond predictably to the D_{st} variations [*McIlwain*, 1966], as also do the 1 MeV trapped protons [*Fillius*, 1966]. The data published by *Davis and Williamson* [1966, see their Figure 10] have shown that 140 keV protons clearly depend upon D_{st}, and can be represented by Eq. (4) with β equal to about +120 γ. The 20 to 100 keV electron time variations at $L = 3.75$ displayed in *Davis and Williamson's* Figure 9 show a clear *anti*-correlation with D_{st}, which is represented with fair accuracy by Eq. (4) with β equal to about −25 γ. The equation

$$I = 0.06\,(10 - D_{st}) \qquad (25)$$

gives an equally good fit where I is the measured energy flux in ergs cm^{-2} sr^{-1} sec^{-1} and D_{st} is in gammas. The chief deviations of the data from this equation are at times of rapid decreases in D_{st} and when D_{st} is low. The former may well be due to the local time asymmetries in the ring current particles which have been demonstrated to exist at early times during magnetic storms [*Cahill*, 1966; *Akasofu*, 1966]. The deviations when D_{st} is small may be due to the fact that D_{st} also includes the magnetic effects of the time dependent D_{cf} currents. Equation (25) would indicate that these 20 to 100 keV electrons are actually a constituent of the long sought ring current particles. If similar fluxes extend over a reasonably large volume, such as 4×10^{28} cm^3, then they would produce about 0.5% of the total magnetic field depression.

Frank [1966] has shown that electrons of still lower energies comprise an important part of the ring current particle energy density. Historically, one reason for assuming that the ring current particles are protons is that the loss of low energy protons due to charge exchange gives about the ob-

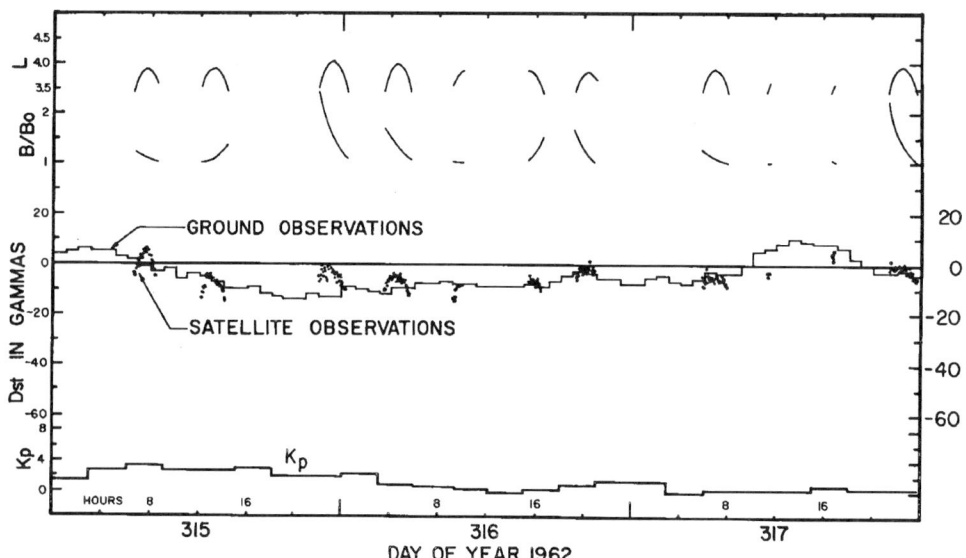

Figure 11. D_{st} values derived from ground observations and from the fluxes of 0.5 MeV electrons measured by the Explorer 15 satellite at L values greater than 3.4. The motion of the satellite in B, L space is shown in the upper part of the figure.

served decay time constant of 2 ± 1 days. It is now clear however, that the mechanisms responsible for persistent decay are capable of causing a loss of low energy electrons in comparatively short times, thus there remains little reason for the prior prejudice in favor of protons.

10. D_{st} BASED UPON PARTICLE FLUX VARIATIONS

Many fluctuations of the trapped particle intensities are clearly caused by global changes in the earth's magnetic field. This suggests the possibility of using the trapped particle measurements themselves to measure D_{st}. Since the particles respond to the changes in magnetic flux inside the magnetic shell upon which they are trapped, they are insensitive to the effects of ionospheric currents which plague ground based observations. The particle fluxes are of course also perturbed by the action of the other four processes, thus it is unlikely that they can be used to measure the variations in D_{st} over any extended period of time.

Figures 11 and 12 show that the energetic electrons in the outer zone can be used to obtain D_{st} values for a time period of at least one week where all are obtained from the Explorer 15 0.5 MeV electron data taken at L values greater than 3.4 during the week beginning November 11, 1962 by the following procedure.

First, Eq. (19) was used to fit the B dependence of the data taken early on day 318 of 1962. Noting that the ground observatories gave $D_{st} = -2\,\gamma$ at this time, this data was assumed to correspond to $D_{st} = -2$. A value for r in Eq. (20) was then obtained for each reading telemetered by the satellite using the equation

$$r = \frac{J}{g(B)} \exp\left(\frac{t - t_0}{16}\right), \qquad (26)$$

where J is the measured flux, $g(B)$ is the fit to the data on day 318, and where the exponential factor is employed to remove the decay due to Process 2. Equation (20) was then solved for K for each r value. The D_{st} values were then assumed to be equal to $-K$ and were plotted versus time as shown in Figures 11 and 12.

For a large fraction of the time during this one week interval, the ground and satellite D_{st} values agree to within $10\,\gamma$. Some of the discrepancies are undoubtedly due to errors in the normalization of the spin average data obtained at points off the equator to omnidirectional intensities on the equator. Other discrepancies may be ascribed to other processes. It is quite probable however, that many of the discrepancies are due to errors in the ground based values. One indication of this is that the data from the different ground stations often differ from each other by more than $10\,\gamma$ in a fashion which suggests contamination due to ionospheric currents. Another is the almost 12-hour gap between Hermanus and Honolulu; any asymmetry in the ring current field may cause a large error in the longitudinal average. Specifically, the magnetographs for day 319 of 1962 show the presence of a local time asymmetry of the type found by *Akasofu* [1966] and *Cahill* [1966]. Furthermore, the magnetographs indicate that at 1600 hours UT the maximum decrease in the field was located in the gap between Hermanus and Honolulu. The discrepancy at this time which can be seen in Figure 12 and which was mentioned before in connection with Figure 8 is almost certainly due to poor longitudinal averaging in the ground data and not

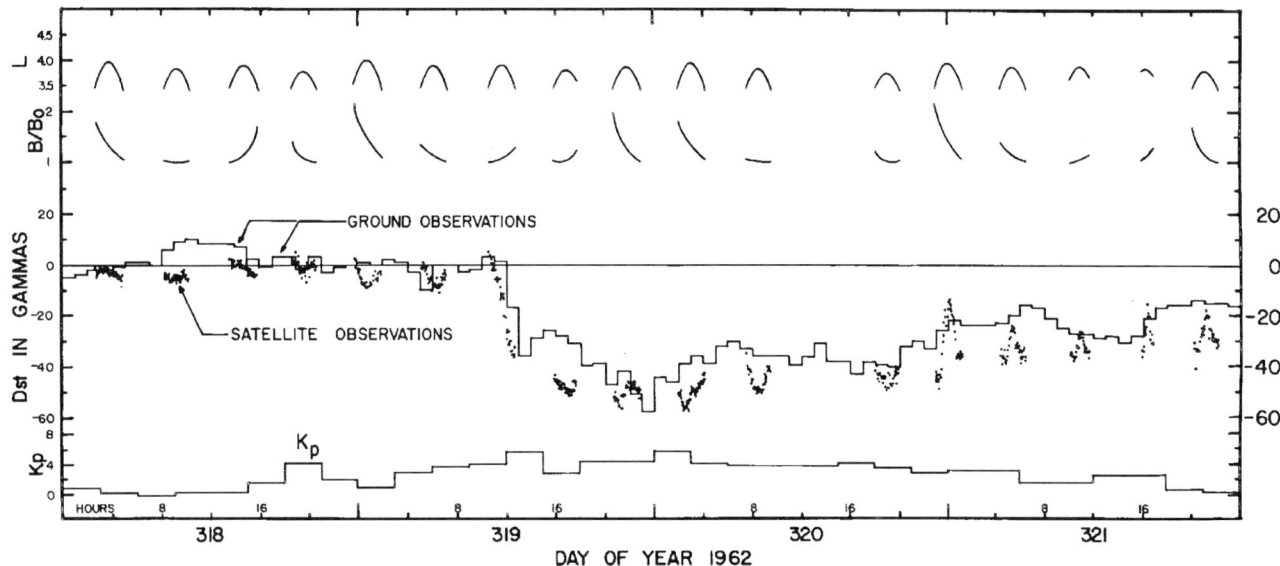

Figure 12. Continuation of Figure 11. Note that the fluctuations during a satellite pass tends to be larger at the time K_p is high. The large discrepancy (about 20 γ) at 1600 hours UT on day 319 of 1962 can be shown to be due to poor longitudinal averaging in the ground based data.

to errors in the satellite measurements.

The particle data used here are by no means the best which can be obtained for determining D_{st}. First, a major improvement would result if the satellite orbit were circular and had zero inclination so that data could be obtained continuously and could be predicted more accurately since motion in B, L space would be much smaller. Second, the proton fluxes would probably provide a better measure since they do not seem to be as radically perturbed by the other processes, or at least the effects of the other processes tend to cancel each other. Third, the primary error made in the satellite measurements of D_{st} is in assuming that the field at the satellite is $B + D_{st}$, where B is the value computed with a spherical harmonic representation of the earth's field. If a magnetometer on the satellite were to measure the field at the satellite to an accuracy of $\pm 1 \gamma$, it appears quite possible that the spatial averages of the field-changes inside the particles' orbits could be obtained with an accuracy of about $\pm 2 \gamma$.

11. RADIAL DIFFUSION

When the magnetic field perturbations are longitude dependent and occur within times comparable or short compared with the drift period of the particles, radial diffusion will occur. It can be shown that the effects of this diffusion for any given perturbation will invariably be of second order compared with the adiabatic effects. To measure the non-adiabatic effects directly therefore requires that the adiabatic effects be removed with a high accuracy. This in return requires very accurate values for D_{st}. It is important therefore, that further efforts be made to improve the determinations of D_{st}.

Another mechanism which can cause radial diffusion of electrons even in the absence of fast field fluctuations has been suggested by *Roederer* [1965]. When the field lines are distorted into nondipolar shapes in a longitude dependent fashion such as by an asymmetric ring current or by magnetospheric boundary (D_{cf}) and tail current systems, then the electrons on the same line of force at one longitude but which have different pitch angles will drift to different lines of force. Thus, if the rapid pitch angle diffusion implied by Process 2 is taking place, the drift paths of the electrons will be continuously changing as their pitch angles are changed. The net result is radial diffusion across lines of force.

12. CONCLUSION

It has been shown that the slowly changing global magnetic fields, as determined by D_{st} values, cause large and predictable changes in the outer zone particle fluxes. The D_{st} values can be used therefore to remove the effects of these adiabatic changes and thus make it possible to study the non-adiabatic effects of the other processes with far greater accuracy.

Acknowledgements. The author gratefully acknowledges the assistance of Dr. J. Valerio in the preparation of the Explorer 15 and Explorer 26 experiments. The many discussions with Drs. R.W. Fillius and L.J. Cahill were particularly helpful in the preparation of this paper. This work was supported in part by the National Aeronautical and Space Administration Grant NsG-538 and Contract NAS 5-3063.

REFERENCES

Akasofu, S.-I. and S. Chapman, On the asymmetric development of magnetic storm in low and middle latitudes, *Planet. Space Sci.,*

12, 607, 1964.

Akasofu, S.-I., Electrodynamics of the magnetosphere: Geomagnetic storms, U. of Iowa Preprint, 66-19, May 1966.

Andronov, A.A. and V.Y. Trakhtengerts, Kinetic instability of the earth's outer radiation belt, *Geomag. Aeron., IV*, 181, 1964.

Cahill, L.J., Jr., Inflation of the inner magnetosphere during a magnetic storm, UCSD Preprint SP-66-2, April 1966. (also *J. Geophys. Res., 71*, 4505, 1966.)

Chang, D.B., Some plasma instabilities of the magnetosphere, Radiation Trapped in the Earth's Magnetic Field, *Proc. NATO Adv. Study Inst., 491*, 1966.

Chang, D.B. and L.D. Pearlstein, On the effect of resonant magnetic-moment violation on trapped particles, *J. Geophys. Res., 70*, 3075, 1965.

Cornwall, J.M., Scattering of energetic trapped electrons by very-low-frequency waves, *J. Geophys. Res., 69*, 1251, 1964.

Cornwall, J.M., Micropulsations and the outer radiation zone, *J. Geophys. Res., 71*, 2185, 1966.

Davis, L.R. and J.M. Williamson, Outer zone protons, Radiation Trapped in the Earth's Magnetic Field, *Proc. NATO Adv. Study Inst., 215*, 1966.

Davis, Leverett, Jr. and D.B. Chang, On the effect of geomagnetic fluctuations on trapped particles, *J. Geophys. Res., 67*, 2169, 1962.

Dessler, A.J. and R. Karplus, Some effects of diamagnetic ring currents on Van Allen radiation, *J. Geophys. Res., 66*, 2289, 1961.

Dragt, A.J., Effect of hydromagnetic waves on the lifetime of Van Allen radiation protons, *J. Geophys. Res., 66*, 1641, 1961.

Dungey, J.W., Loss of Van Allen electrons due to whistlers, *Planet. Space Sci., 11*, 591, 1963.

Dungey, J.W., Effects of electromagnetic perturbations on particles trapped in the radiation belts, *Space Sci. Rev., IV*, 199, 1965.

Eviatar, A., The role of electrostatic plasma oscillations in electron scattering in the earth's outer magnetosphere, *J. Geophys. Res., 71*, 2715, 1966.

Falthammar, C.-G., Effects of time-dependent electric fields on geomagnetically trapped radiation, *J. Geophys. Res., 70*, 2503, 1965.

Falthammar, C.-G., On the transport of trapped particles in the outer magnetosphere, *J. Geophys. Res., 71*, 1487, 1966.

Fillius, R.W., Storm time changes in low energy trapped protons, (in preparation), 1966.

Forbush, S.E., G. Pizzella and D. Venkatesan, The morphology and temporal variations of the Van Allen Radiation Belt, October 1959 to December 1960, *J. Geophys. Res., 67*, 3651, 1962.

Frank, L.A., A survey of electrons $E > 40$ keV beyond 5 earth radii with Explorer 14, *J. Geophys. Res., 70*, 1593, 1965.

Frank, L.A., Inward radial diffusion of electrons of greater than 1.6 million electron volts in the outer radiation zone, *J. Geophys. Res. 70*, 3533, 1965.

Frank, L.A., Explorer 12 observations of the temporal variations of low energy electron intensities in the outer radiation zone during geomagnetic storms, U. of Iowa Preprint, 66-8, March 1966.

Herlofson, N., Diffusion of particles in the earth's radiation belts, *Phys. Rev. Lett., 5*, 414, 1960.

Hoffman, R.A., R.L. Arnoldy and J.R. Winckler, Observations of the Van Allen radiation regions during August and September 1959, 6. Properties of the outer region, *J. Geophys. Res., 67*, 4543, 1962.

Kellogg, P.J., Van Allen radiation of solar origin, *Nature, 183*, 1295, 1959.

Kennel, C.F. and H.E. Petschek, Limit on stably trapped particle fluxes, *J. Geophys. Res., 71*, 1, 1966.

McIlwain, C.E., The radiation belts, natural and artificial, *Science, 142*, 355, 1963.

McIlwain, C.E., Redistribution of trapped protons during a magnetic storm, *Space Research, V*, 374, 1964.

McIlwain, C.E., Ring current effects on trapped particles, *J. Geophys. Res., 71*, 3623, 1966.

O'Brien, B.J., Lifetimes of outer-zone electrons and their precipitation into the atmosphere, *J. Geophys. Res., 67*, 3687, 1962.

O'Brien, B.J., High-latitude geophysical studies with satellite Injun 3, 3. Precipitation of electrons into the atmosphere, *J. Geophys. Res., 69*, 13, 1964.

Parker, E.N., Geomagnetic fluctuations and the form of the outer zone of the Van Allen radiation belt, *J. Geophys. Res., 65*, 3117, 1960.

Paulikas, G.A. and S.C. Freden, Precipitation of Energetic Electrons into the Atmosphere, *J. Geophys. Res., 69*, 1239–1249, 1964.

Paulikas, G.A., J.B. Blake, and S.C. Freden, Precipitation of energetic electrons at middle latitudes, *J. Geophys. Res., 71*, 3165, 1966.

Pearlstein, L.D., M.N. Rosenbluth and D.B. Chang, High-frequency "L Cone" flute instabilities inherent to two-component plasmas, *General Atomics Report GA-6708*, 1965. (see also *Phys. Fluids, 9*, 953–956, 1966.)

Roberts, C.S., Electron loss from the Van Allen zones due to pitch angle scattering by electromagnetic disturbances, Radiation Trapped in the Earth's Magnetic Field, *Proc. NATO Adv. Study Inst., 403*, 1966.

Roederer, J., Private Communication, 1965.

Sugiura, M. and S.J. Hendricks, Private Communication, 1966.

Trakhtengerts, V.Y., The mechanism of generation of very low frequency electromagnetic radiation in the earth's outer radiation belt, *Geomag. Aeron., III*, 365, 1963.

Trakhtengerts, V.Y., Kinetic instability of the outer radiation zone of the earth, *Geomag. Aeron., V*, 865, 1965.

Wentzel, D.G., Hydromagnetic waves and the trapped radiation—Part 1. Breakdown of the adiabatic invariance; Part 2. Displacements of the mirror points, *J. Geophys. Res., 66*, 359, 1961.

Wentzel, D.G., Hydromagnetic waves and the trapped radiation, Part 3. Effects on protons above the proton belt, *J. Geophys. Res., 67*, 485, 1962.

Williams, D.J., A 27-day periodicity in outer zone trapped electron intensities, *J. Geophys. Res., 71*, 1815–1826, 1966.

Williams, D.J. and N.F. Ness, Simultaneous trapped electron and magnetic tail field observations, *NASA GSFC Tech. Rept. X-611-66-264*, June 1966.

C.E. McIlwain, Department of Physics, University of California at San Diego, La Jolla, California, USA.

Dynamic Physical Modelling of Trapped Particles for Satellite Survey

S. Bourdarie and D. Boscher

Departement de Technologie Spatiale, CERT-ONERA, Toulouse

T. Beutier

Centre d'Etudes Spatiales des Rayonnements, Toulouse

The measurements made on board AMPTE and CRRES are striking examples of the importance of the magnetic activity in determining particle (electron and proton) dynamics. Therefore in order to properly model radiation belts, various phenomena involved in particle transport during magnetic storms must be taken into account. We have attempted such an approach to model with a four dimension code in the phase space based on adiabatic invariants and the Boltzman equation. Both diffusive and convective particle transport are performed. Substorms are simulated by increasing convective electric field and injecting particles with keV range energies in the nightside region. These simulations produce electron and proton fluxes as a fonction of time for different energies in various configurations (meridian cut, equatorial cut, along orbits, ...).

1. INTRODUCTION

The aim of this study is the modelling of the charged particle transport in the internal magnetosphere, for energy greater than one keV. We make use of the adiabatic invariants formalism to obtain a four dimensional convection-diffusion equation in the phase space. Therefore we spread out the three dimensional model [*Salammbô Beutier*, 1993] to the fourth dimension, the local time (azimuthal variable). The theory is based on both the eccentric tilted dipolar magnetic field model [*Schulz*, 1991, p. 92] and the Volland-Stern convective electric field model [*Volland*, 1973; *Stern*, 1975]. The important point in our calculation is that we take into account the processes we had in the Fokker-Planck diffusion equation and we calculate in the whole internal magnetosphere (for all pitch angles).

Then we want to test the built tool. We carry out a magnetic storm scenario and we let evolve the system for eight hours. This simulation produces differential omnidirectional fluxes for protons and electrons in magnetic equatorial and meridian planes. We analyze flux maps at three different times, for several energies to estimate if the code works as expected. So we define a magnetic storm model and we calculate particle transport in the radiation belts. Then we produce unidirectional or omnidirectional differential fluxes a satellite can measure along its trajectory. The problem now is how can we extract all parameters we have at first from these flux measurements.

2. THEORETICAL APPROACH

First we write an equation to reproduce either diffusive or convective transport of charged particles with energies between about one keV and several tens of MeV. This equation has to govern the evolution of a particle population in the four dimension phase space. We begin with the Hamilton formalism [*Landau et al.*, 1963] and we define, as usual, the action variables by:

$$J_i = \frac{1}{2\pi} \oint \left(\vec{p}_i + q\vec{A} \right) d\vec{l}_i, \qquad (1)$$

where the integral is made over the periodic motion, when the magnetic field B is set to a constant in time and the electric

Radiation Belts: Models and Standards
Geophysical Monograph 97
Copyright 1996 by the American Geophysical Union

field is zero [*Beutier et al.*, 1995ab].

We then write the Boltzmann equation in the action variables / angle variables phase space and, assuming that all transports are made with constant J_1 and J_2 (or relativistic magnetic moment M and second invariant J), we average the Boltzmann equation over the two first phases, associated with the gyration and bounce motions. Moreover, we take the interaction term as the Fokker-Planck operator, so we can write the temporal variation of distribution function $f(J_1, J_2, J_3, \varphi_3, t)$ as:

$$\frac{\partial f}{\partial t} + \frac{dJ_3}{dt}\frac{\partial f}{\partial J_3} + \frac{d\varphi_3}{dt}\frac{\partial f}{\partial \varphi_3} = -\sum_i \frac{\partial}{\partial J_i}(D_i f)$$
$$+ \sum_{i,j} \frac{\partial}{\partial J_i}\left(D_{ij}\frac{\partial f}{\partial J_i}\right) + \text{Sources} - \text{Losses}, \quad (2)$$

where all the coefficients (the D_i friction terms, the D_{ij} diffusion coefficients, the $\partial J_3/\partial t$ radial transport term and the $\partial \varphi_3/\partial t$ azimuthal transport term) are averaged over the gyration and bounce motion.

3. PARTICLE TRANSPORT MODELLING

To reproduce particles dynamic it is necessary to recognize all the physical phenomena which participate to the transport [*Beutier et al.*, 1995ab]. The diagram in Figure 1 gathers all the characteristics of the four dimensional Salammbô model describing the proton and electron transport modelling [*Bourdarie et al.*, 1996].

The diagram on Figure 1 is composed of three levels:

- the first one, white, describes necessary physical processes to trap the particles, with the losses below, convective and diffusive transports at left and non-stochastic process at right;

- the second one, light grey, shows the properties which have an influence on all these processes plus the external source above;

- the third one, dark grey, underlines all the origins of the particle dynamics.

In the model, particles are trapped in an eccentric tilted dipole field and are carried by this magnetic field and a Volland-Stern convective electric field (plus the corotation field). They have a drift velocity with two components, one azimuthal (the $d\varphi_3/dt$ term in Eq. (2) and one radial (related to the dJ_3/dt term in Eq. (2). Protons are decelerated by friction with plasmaspheric cold electrons and atmospheric atoms (the D_i term in Eq. (2); they are lost by charge exchange with ambient hydrogen atoms (the losses term in Eq. (2), by precipitation on the Earth (loss cone) and by convection on the day-side. For electrons it is nearly the same. They are decelerated by friction with plasmaspheric cold electrons and atmospheric atoms (the D_i term in Eq. (2); they are lost by pitch angle diffusion either given by Coulomb collisions or by wave-particle interactions (the D_{ij} term in Eq. (2), by precipitation on the Earth (loss cone) and by convection on the dayside.

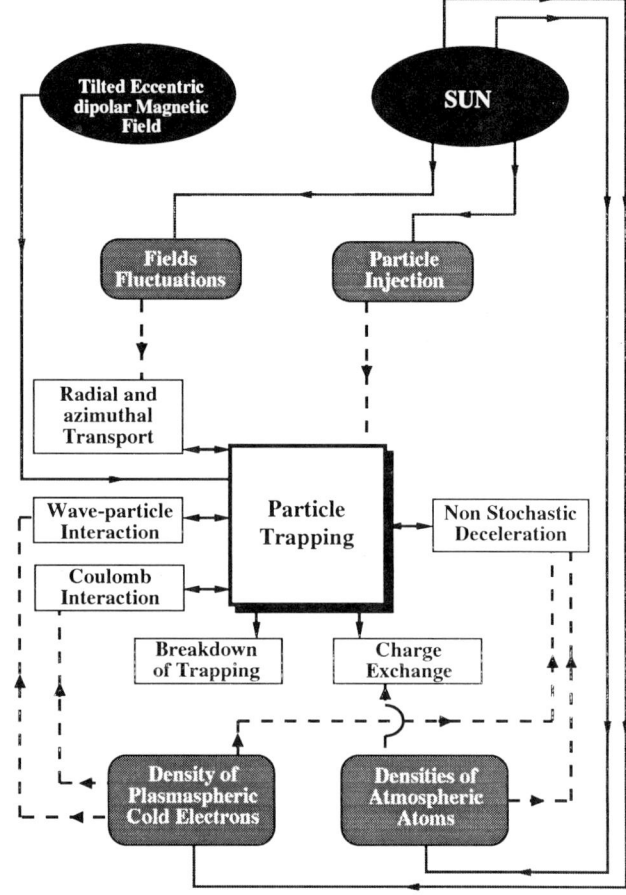

Figure 1. Radiation belt modelling

The convective access of the model allows to take into account particles with an energy range from one keV to about a hundred MeV. So it is possible, increasing the convective electric field and injecting particles on the night side with low energies, to simulate periods of strong magnetic activity.

4. SIMULATION

We verify if the code results are in good agreement with reality, particularly the transport of low energy particles (1 keV-1 MeV) injected during periods of strong magnetic activity. To do that, we elaborate a magnetic storm scenario:

1. the simulation begins at 09:00 UT;

2. we take as initial conditions the distribution functions deduced from the NASA AP-8 model for protons and an empty magnetosphere for electrons. This choice comes from CRRES satellite measurements, which show a mean steady state for the proton radiation belt, on the other hand the outer electron radiation belt is very variable in time;

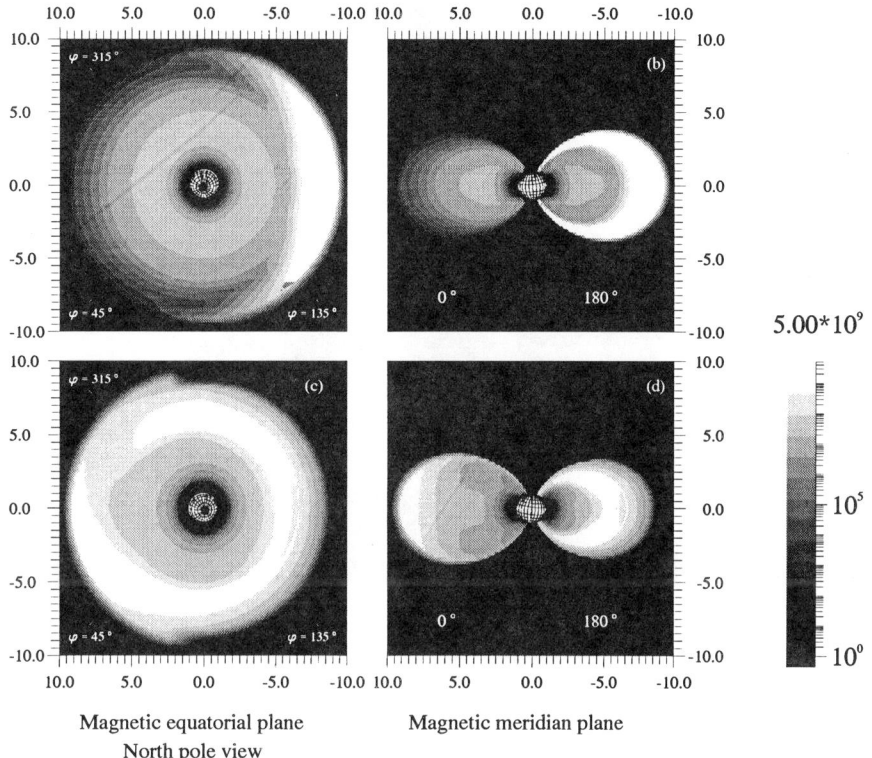

Figure 2. Omnidirectional differential flux maps (MeV^{-1} cm^{-2} s^{-1}) in L units for both magnetic equatorial and magnetic meridian planes for two times (a and b at 0902 UT, and c and d at 1300 UT) for 5 keV energy protons.

3. we increase instantaneously the convective electric field (from 0.1 mV/m for a quiet period to 1.1 mV/m for a disturbed period at 6 R_E);

4. the injection duration is fixed at 2 minutes;

5. the injection spectrum is given by

$$f = f_0 (\sin \alpha_e)^{0.646} \exp\left(-\frac{E}{kT}\right), \qquad (3)$$

where α_e is the equatorial pitch angle, E is the kinetic energy, f_0 is taken to 10^{31} for protons and 10^{33} for electrons and kT is equal to 8 keV at 9 R_E (the mean temperature of the Central Plasma Sheet (CPS) is taken to be 5 keV at 10 R_E) [*Chen et al.*, 1994], we note that the pitch angle distribution has been deduced from measurements;

6. particles are injected at $L = 9$ between 23h and 1h LT;

7. the plasmapause location is set to 4.5 R_E.

Thus we simulate particle dynamics with Eq. (2). So we produce electron and proton distribution functions in the action-angle phase space for each time step. Then we deduce omnidirectional differential fluxes.

4.1. Proton results

In Figure 2, proton results in both the noon-midnight magnetic meridian plane and the magnetic equatorial plane are given. They are presented as omnidirectional differential flux images for a kinetic energy of 5 keV. The equatorial image is seen for the North pole and the sun is at the left. Two times are given. The first one (a and b) is 2 minutes after the beginning of the simulation (just at the end of the injection) and the second (c and d) 4 hours after, so at 13h UT. After 2 minutes, protons with lower energies are convected inward and drifted by the magnetic and corotation electric fields (these two fields act in opposite side for protons). After four hours particles at high L values begin to be lost on the dayside and others are trapped. The three space dimensions are illustrated on this figure, but to have a look to the fourth one in the phase space, we have represented in Figure 3 fluxes for a 100 keV energy. At this energy, nearly all particles are trapped. 100 keV Protons are first created in the $L = 7.5$ region because particles with lower energies cannot reach lower L values after 2 minutes. But four hours after, 100 keV protons seems to be created in the $L = 4$ region; this is due to the fact that particles are transported with constant M and J. It corresponds to the ring current formation (Figure 3 c

30 DYNAMIC PHYSICAL MODELLING OF TRAPPED PARTICLES

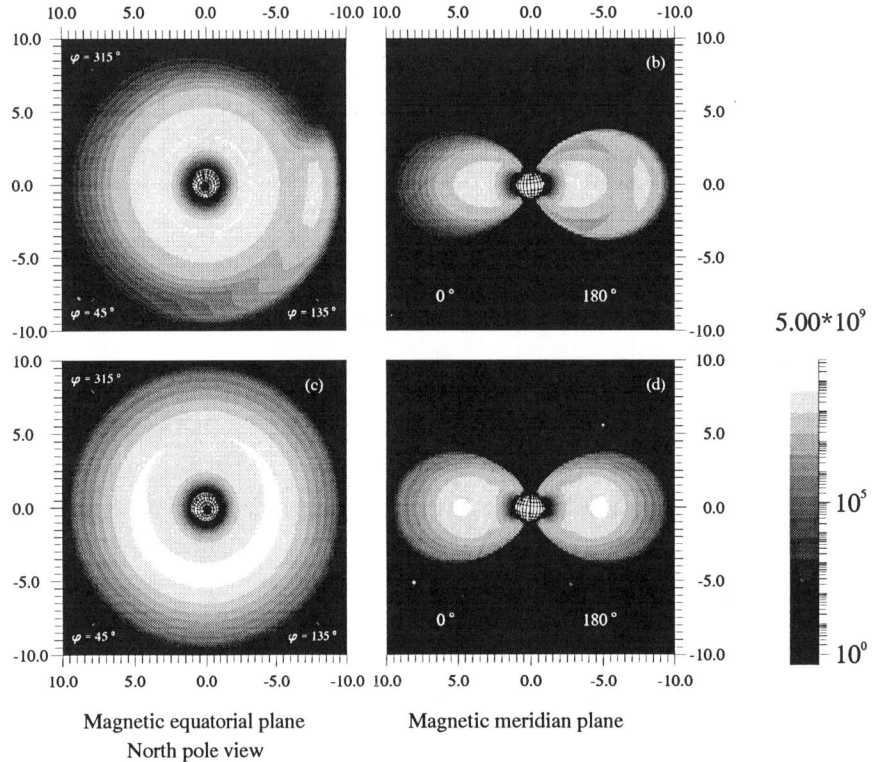

Figure 3. Omnidirectional differential flux maps $(\mathrm{MeV}^{-1}\mathrm{cm}^{-2}\mathrm{s}^{-1})$ in L units for both magnetic equatorial and magnetic meridian planes for two times (a and b at 0902 UT, and c and d at 1300 UT) for 100 keV energy protons.

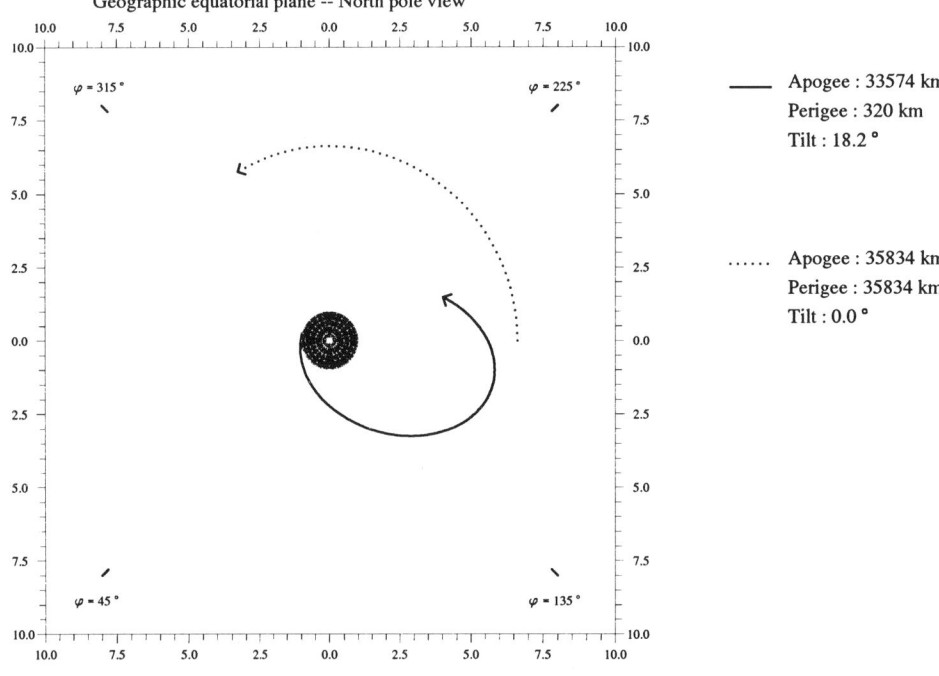

Figure 4. Equatorial projections of satellites orbit from 0900 to 1700 UT.

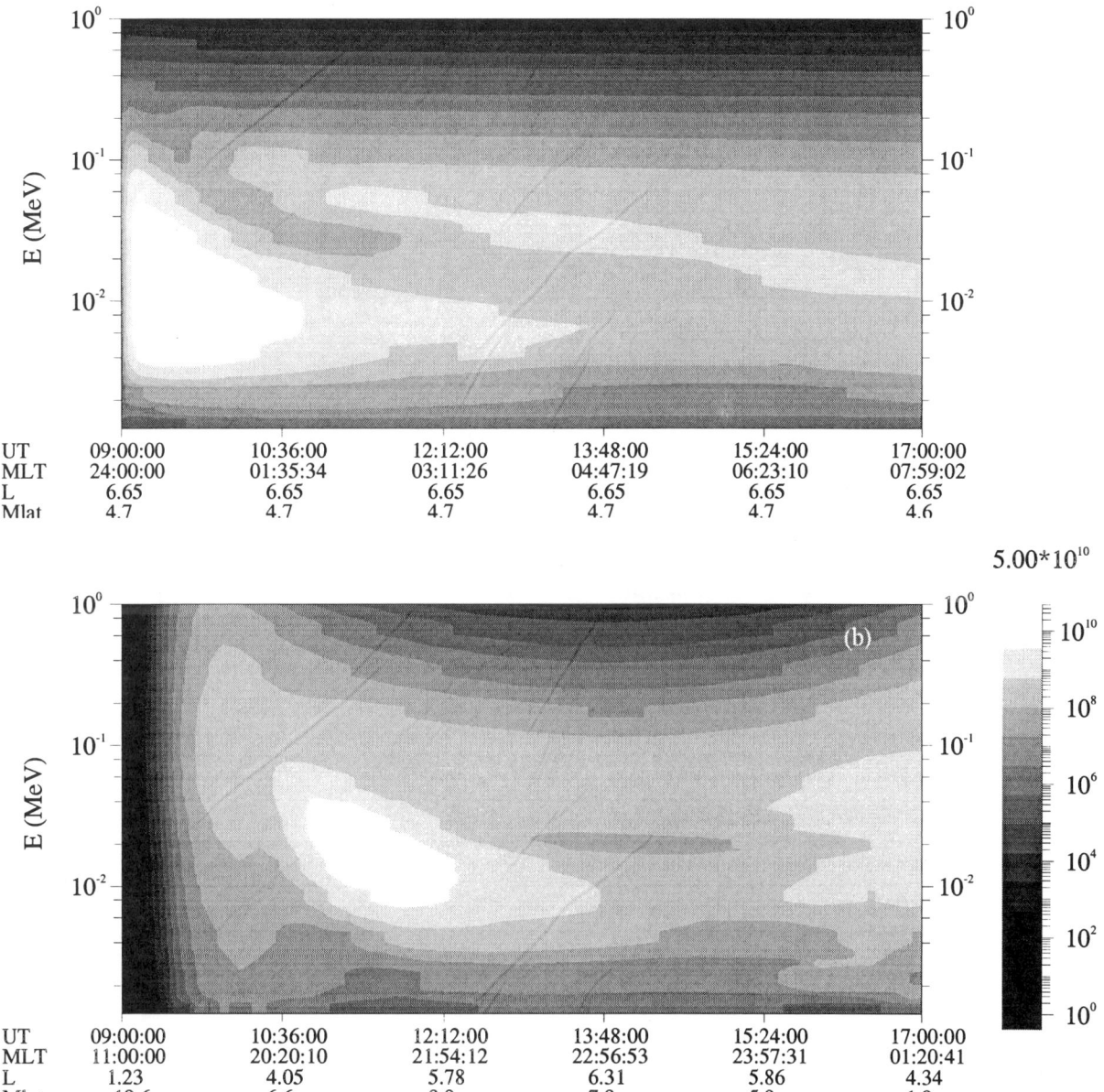

Figure 5. Omnidirectional differential flux maps (MeV^{-1}cm^{-2}s^{-1}) in energy unit versus time as seen by two satellites along their orbit: (a) geostationary orbit and (b) elliptic orbit.

and d).

So, we have a complete view of particle transport, with energies between about 1 keV and several tens of MeV, in the whole internal magnetosphere for a simple magnetic storm. Now, with this complete simulation description, we want to analyze with the code how a storm period can be interpreted with satellite measurements. Thus we simulate two satellite trajectories in the proton radiation belt. A projection of their trajectory in the geographic equatorial plane is represented in Figure 4, the first one is geostationary, at a longitude of 225°

and the second one on an elliptic orbit. The CRRES satellite orbit has been chosen to see the results.

In Figure 5 we give omnidirectional differential fluxes measured by this two satellites with an energy range between 1 keV to 1 MeV for 8 hours. The first image (a) is quite simple to analyze. At 9h UT the energy spectrum of the stationary state can be deduced. Few minutes after, fluxes of low energy particles (1-100 keV) are growing instantaneously, the injection effect is clear. Here the satellite sees all injected particles at the same time. Dynamic processes, which are en-

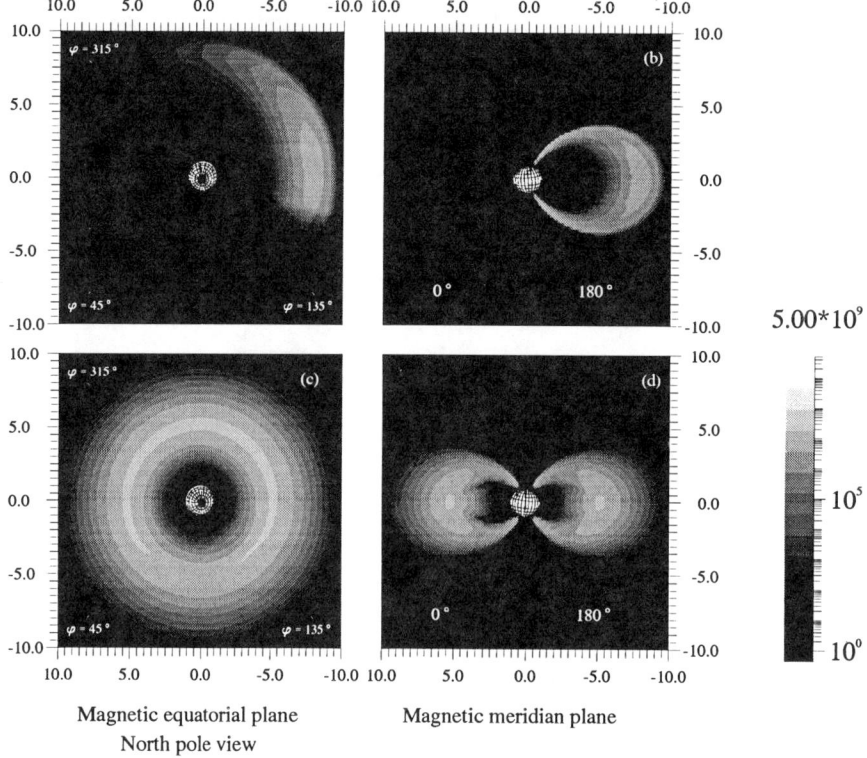

Figure 6. Omnidirectional differential flux maps (MeV^{-1}cm^{-2}s^{-1}) in L units for both magnetic equatorial and magnetic meridian planes for two times (a and b at 0902 UT, and c and d at 1300 UT) for 100 keV energy electrons.

ergy dependant, do not have enough time to spread out these new particles. This observation provides information about the injection location (near the satellite position) and the universal time of the storm beginning. Nevertheless the energy spectrum is quite difficult to extract because of the combined effect of the injection duration and the diffusive and convective processes. The drift echo is reproduced by the code and can be seen on Figure 5 (a). 100 keV protons (near 1036 UT) are observed before 10 keV ones (near 1700 UT). This is typical from drift echo. Such observation is another source of informations to characterize dynamic processes.

On the contrary, in the second case (b) all these informations are not so obvious. We can only suppose a storm period with low energy particles injected, but some difficulties can be encountered to define all the parameters of an injection.

So we can see in this simple case, where all parameters defining the injection are known at first, how it is difficult to extract them from satellite measurements. We hope our model will be useful for this analysis.

4.2. Electron results

In Figure 6, electron results are given in the same way as the proton results. They are presented as omnidirectional differential flux images at a kinetic energy of 100 keV. Two times are given. The first one (a and b) is just at the end of the injection and the second (c and d) 4 hours after the begining of the simulation. At the begining of the simulation the magnetosphere is empty. So all particles seen on Figure 6 have been created by radial convection of lower energy particles injected in $L = 9$. In Figures 6c and 6d, all 100 keV electrons are trapped. The maximum flux is near $L = 4$, it corresponds (as for protons) to the ring current formation.

The main difference with proton results is the drift which is in the opposite direction as expected. But the delay to the ring current formation is the same for both electrons and protons.

5. CONCLUSION AND PROSPECTS

A 4D calculation code has been developed, taking into account local time in the particle transport. Up to now, particles are carried by a dipole field and a Volland-Stern electric field (plus the corotation field). With this code, we can determine particle fluxes in the energy range 1 keV–10 MeV for the electrons and 1 keV–300 MeV for the protons. The simulations are performed for any pitch angle with L varying from 1 R_E to 9 R_E. In this energy range, this code is like a particle one, but with the assumption that the two first adiabatic invariants are conserved. The determination of the convective access of particles to internal regions of the magnetosphere then is possible. In the high energy range, diffusion processes are prominent.

The model enables us to simulate the evolution of the whole

of the radiation belts during storm and substorm periods. Fluxes versus time as seen by a satellite along its orbit can also be determined.

To be complete, the numerical model has to be associated with satellite measurements. So more information will be obtained for sources (particles injections) with geostationary, polar and GTO satellites, particles transport (ring current) with GTO satellites and losses (auroral precipitations) with polar satellites. Finally a global view of the radiation belts can be seen by neutral imager. Such images are easy to deduce from our results.

Acknowledgements. This work has been performed within the framework of the ARES memorandum of understanding, an agreement of cooperation between CERT-ONERA, the Midi-Pyrénées Observatory, (OMP, 14, Avenue Edouard Belin, 31400 Toulouse), and CNRS (CESR, 9, Avenue du Colonel Roche, 31029 Toulouse). We would like to thank Michel Blanc (OMP), Jean-André Sauvaud (CESR) and Manola Romero (CERT) for their valable advice and fruitful discussions.

REFERENCES

Beutier, T., Modélisation Tridimentionnelle pour l'Etude de la Dynamique des Ceintures de Radiation, *Thèse de doctorat—ENSAE*, 1993.

Beutier, T., D. Boscher, and M. France, Salammbô: A three-dimensional simulation of the proton radiation belt, *J. Geophys. Res.*, 100, 17781, 1995a.

Beutier, T., and D. Boscher, A three-dimensional analysis of the electron radiation belt by the Salammbô, *J. Geophys. Res.*, 100, 14853, 1995b.

Bourdarie, S., D. Boscher, T. Beutier, J.A. Sauvaud and M. Blanc, Electron and proton belt dynamic simulations during storm periods. A new asymmetric convective-diffusive model, *J. Geophys. Res.*, to be submitted, 1996.

Chen, M. W., et al., Simulations of phase space distributions of storm time ring current, *J. Geophys. Res.*, 99, 5745, 1994.

Landau, L., and E. Lifchitz, Mécanique, Physique Théorique - Tome 1, *Editions MIR*, 1963.

Schulz, M., The magnetosphere, in Geomagnetism, Vol. 4, Ed. P. Jacobs, pp. 87–293, 1991.

Stern, D., A Model of the Terrestrial Electric Field, *Eos Trans. AGU*, 55, 403, 1975.

Volland, H., A Semiempirical Model of Large-Scale Magnetospheric Electric Fields, *J. Geophys. Res.*, 78, 171, 1973.

S. Bourdarie, D. Boscher, CERT-ONERA/DERTS, 2, Avenue E. Belin, PO 4025, 31055 Toulouse Cedex, France

T. Beutier, CESR, 9, Avenue du Colonel Roche, 31029 Toulouse Cedex, France

DISCUSSION

Q: J. Albert. Your equation contains a 3×3 diffusion tensor for the 3 adiabatic invariants. Which terms do you actually include?

A: S. Bourdarie. Protons: no cross terms at all (convection instead). Electrons: $D11$, $D12$ (pitch angle scattering).

Q: V. Bashkirov. Have you studied the influence of the variance diffusion models on the results you obtained?

A: S. Bourdarie. Not yet.

Q: A. Korth. What kind of electron pitch angle distribution was used for the simulation?

A: S. Bourdarie. The pitch angle distribution used for the simulation was not isotropic. It was $(\sin \alpha_e)^{0.646}$ where α_e is the equatorial pitch angle. But a few minutes after the injection, the pitch angle distribution was isotropic.

Q: M. Hudson. What was the injection energy spectrum for both electrons and protons? What is the slope?

A: S. Bourdarie. Exponential, not a power law.

Q: G.V. Popov. 1. How many particles were injected? 2. Were self-consistent electric fields taken into account?

A: S. Bourdarie. 1. 1031 protons and 1033 electrons. 2. They were not taken into account.

Q: X. Li. The induction electric field during substorm onset should be included, since this electric field can be comparable or larger than the convection electric field?

A: S. Bourdarie. The right (real) magnetic field model is not available, so the implementation of the induction electric field cannot be included at this time.

Anomalous Cosmic Rays: The Principal Source of High Energy Heavy Ions in the Radiation Belts

R.A. Mewaldt, R.S. Selesnick and J.R. Cummings

California Institute of Technology, Pasadena, California

Recent observations from SAMPEX have shown that "anomalous cosmic rays" are the principal source of high energy (> 10 MeV/nuc) heavy ions trapped in the radiation belts. This component of interplanetary particles is known to originate from interstellar atoms that has been accelerated to high energies in the outer heliosphere. The mechanism by which anomalous cosmic rays with ~ 1 to ~ 50 MeV/nuc are trapped in a radiation belt at $L \approx 2$ has now been verified. We discuss models for accelerating and trapping anomalous cosmic rays and review observations of their composition, energy spectra, pitch angle distribution, and time variations. Extrapolation of the fluxes observed at ~ 600 km to higher altitude and other time periods is also discussed.

1. INTRODUCTION

Just over twenty years ago a new component of cosmic rays was discovered when unexpected enhancements were observed in the energy spectra of 1 to 50 MeV/nuc He, N, and O measured in interplanetary space during solar quiet times. Soon after the discovery of this so-called "anomalous cosmic ray" (ACR) component, *Fisk, Koslovsky, and Ramaty* [1974] proposed what has proven to be the correct explanation for its origin, suggesting that ACRs represent a sample of neutral interstellar particles that have drifted into the heliosphere, become ionized by the solar wind or UV radiation, and then accelerated to energies of tens of MeV/nuc in the outer heliosphere. Once accelerated, they can re-enter the inner heliosphere as low energy cosmic rays (see Figure 1). There is now evidence for ACR contributions to seven elements (H, He, C, N, O, Ne, and Ar) and over the past two decades the properties of ACRs have been studied extensively throughout the heliosphere (see reviews in *Simpson* [1995] and *Klecker* [1995]).

Shortly after the discovery of ACRs *Blake and Friesen* [1977] suggested a mechanism for trapping ACRs in the magnetosphere, but it was more than a decade later that the first solid evidence for this mechanism was provided [*Grigorov et al.,* 1991]. With the launch of the Solar, Anomalous, and Magnetospheric Particle Explorer (SAMPEX) in 1992 it became possible to study high-energy heavy ions in the magnetosphere in detail for the first time. SAMPEX measurements have shown that anomalous cosmic rays are the dominant source of heavy ions with > 10 MeV/nuc in the magnetosphere, and that they are located in a relatively narrow belt at $L \approx 2$ [*Cummings et al.,* 1993, 1994]. In this paper we review briefly the properties of anomalous cosmic rays in interplanetary space, describe models for their acceleration and subsequent trapping in the magnetosphere, and present observations of trapped ACRs provided by SAMPEX over the past three years.

2. ANOMALOUS COSMIC RAYS IN INTERPLANETARY SPACE

The model by Fisk, Koslovsky and Ramaty (hereinafter FK&R) predicts that there should be ACR contributions to species that are mainly or partially neutral in the interstellar medium. The observed abundances of He, C, N, O, Ne, and Ar are generally consistent with this picture, and provide a means of measuring the composition of the neutral interstellar medium [*Cummings and Stone,* 1995]. For example, the low abundance of carbon in ACRs (see Figure 1) implies that only $\sim 1\%$ of the carbon in the interstellar medium is in a neutral state. The model of FK&R also predicts that ACRs should be singly charged, in contrast to galactic cosmic rays, which are essentially fully stripped, and there is now abundant evid-

Radiation Belts: Models and Standards
Geophysical Monograph 97
Copyright 1996 by the American Geophysical Union

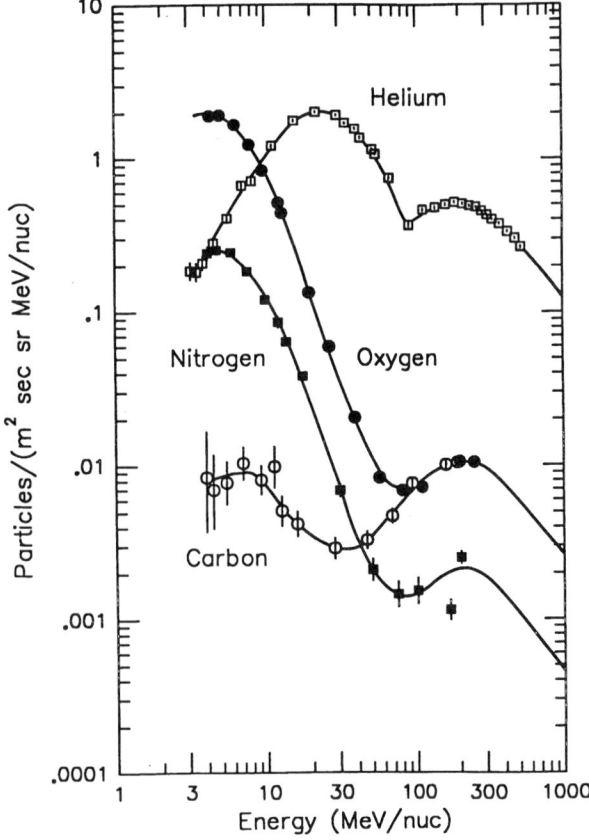

Figure 1. Energy spectra of He, C, N, and O observed by Voyager 2 at ~ 23 AU during the 1987 solar minimum illustrating the anomalous low-energy enhancements in the spectra (from *Mewaldt et al.* [1994]). At energies > 50 MeV/nuc galactic cosmic rays dominate.

ence that the bulk of ACRs with ~ 10 MeV/nuc are singly charged (see *Klecker et al.* [1995] and references therein).

As Pioneer 10 & 11, and later Voyager 1 & 2, began to explore the outer solar system they found that the intensity of ACRs increased with distance from the Sun, and the distribution of ACRs in the heliosphere has now been measured out to 60 AU, and to latitudes as high as 80° (e.g., *Cummings et al.* [1995]). Ulysses has recently measured the abundances of the "pick-up" ions that are the seed population for ACR acceleration (e.g., *Geiss et al.* [1995]). It is now believed that the bulk of ACR acceleration takes place at the solar wind termination shock [*Pesses et al.*, 1981; *Jokipii*, 1990] estimated to be at a distance of ~ 80 to 100 AU from the Sun. Because the access of low energy cosmic rays to inner solar system is strongly affected by interplanetary conditions ("solar modulation"), ACRs are detectable at 1 AU only near solar minimum. Figure 2 illustrates variations in the intensity of 8 to 27 MeV/nuc oxygen over the past 27 years. Note that the intensity at 1 AU varies by a factor > 100, and is reasonably well represented by the neutron monitor count rate taken to the 25th power.

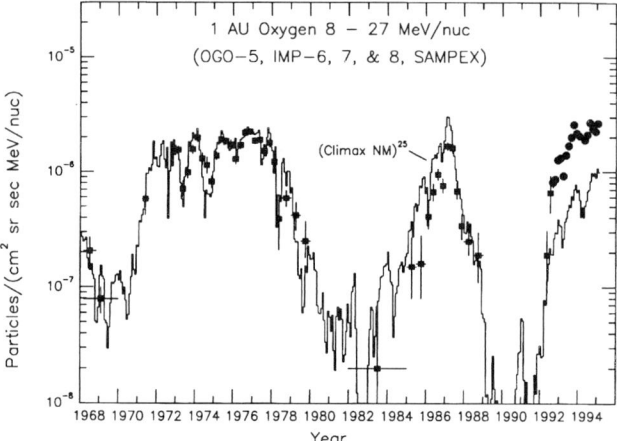

Figure 2. Quiet time measurements in the flux of 8 to 27 MeV/nucleon oxygen over the past 27 years. Data from 1972 to mid-1992 are from IMP-8, after which they are from the HILT sensor on SAMPEX (B. Klecker, private communication). For references to the earlier data see *Mewaldt et al.* [1993]. The solid curve is proportional to the Climax neutron monitor counting rate (*R. Pyle and J. Simpson*, private communication) taken to the 25th power.

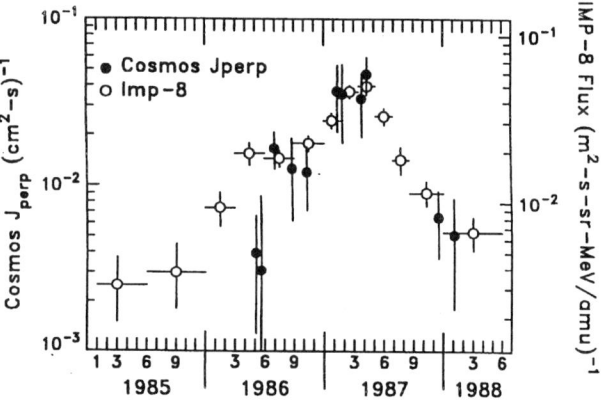

Figure 3. Measurements of the flux of 5 to 30 MeV/nuc trapped oxygen from a series of Cosmos flights at a typical altitude of ~ 300 km are compared with the interplanetary flux of 8 to 27 MeV/nuc oxygen measured on IMP-8, and with the scaled Mt. Washington neutron monitor count rate taken to the 30th power (from *Grigorov et al.* [1991]).

3. TRAPPING ANOMALOUS COSMIC RAYS

Long before there was much solid evidence in favor of the model by FK&R, *Blake and Friesen* [1977] suggested that if ACRs were indeed singly charged, then some of them would become trapped in the Earth's radiation belts. They reasoned that singly-charged ACRs can penetrate to lower geomagnetic latitude than fully-stripped galactic cosmic rays or solar particles with the same energy/nuc, because they have greater rigidity. If one of these ions were to brush against

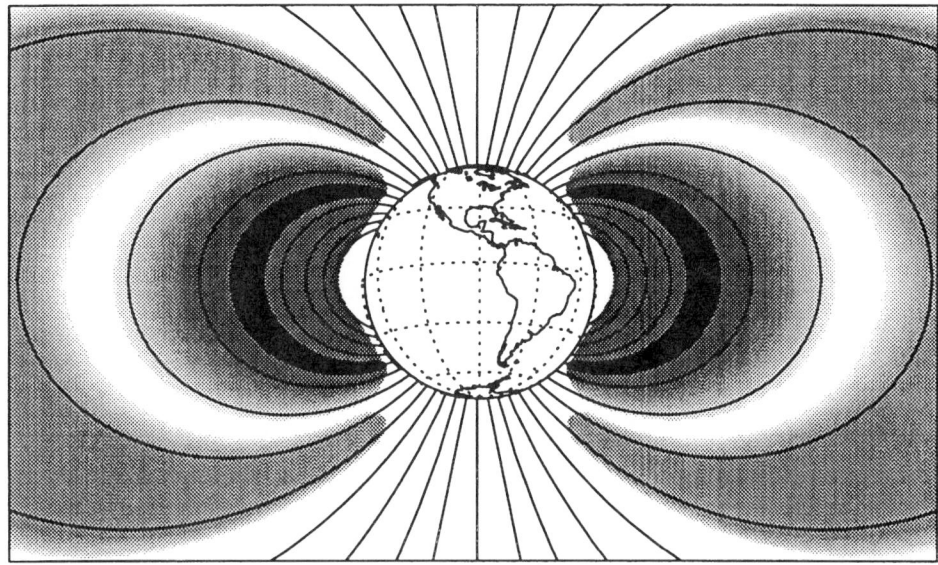

Figure 4. Illustration of the Earth's radiation belts. The narrow belt at $L = 2$ (dark shading) includes trapped anomalous cosmic rays that originate from the local interstellar medium.

Table 1. Anomalous Cosmic Ray Abundance Ratios

	SAMPEX (1992–1993)		Voyager 2, 1987 [1]
	Trapped [1] 16–45 MeV/nucleon	Interplanetary [2] >17 MeV/nucleon	Interplanetary 16–30 MeV/nucleon
C/O	~ 0.0004	0.014 ± 0.009	0.020 ± 0.004
N/O	0.09 ± 0.01	0.19 ± 0.03	0.194 ± 0.013
Ne/O	0.04 ± 0.01*	0.06 ± 0.02*	0.048 ± 0.006

*Lower energy limit for SAMPEX Ne/O is 18 MeV/nucleon.
[1] Selesnick et al. [1995a]
[2] Mewaldt et al. [1996]

the upper atmosphere, losing some or all of its remaining electrons, its rigidity would suddenly decrease by a large factor (up to eight in the case of singly-charged oxygen). Depending on its pitch angle, it might then become trapped in a stable orbit. *Blake* [1990] estimated that trapped ACRs might have lifetimes of weeks to months before losing their energy to the residual atmosphere.

While Blake and Friesen had been stimulated by observations from Skylab [*Chan and Price*, 1975; *Biswas et al.*, 1975], the composition reported by these experiments did not fully agree with that of interplanetary ACRs, and this model received little attention for more than a decade. *Grigorov et al.* [1991], using dielectric track detectors flown on a series of low-altitude Cosmos flights, observed a population of 5 to 30 MeV/nuc heavy ions whose angular distribution indicated a trapped population, and whose composition and time variations (see Figure 3) were consistent with that of interplanetary ACRs observed simultaneously by IMP-8. They correctly identified these as the trapped ACRs predicted by *Blake and Friesen* [1977].

4. SAMPEX OBSERVATIONS

Soon after this discovery, SAMPEX was launched into an 82° inclination low-Earth orbit carrying four instruments (MAST, PET, LICA, and HILT) designed to measure energetic nuclei and electrons over three decades in energy (see *Baker et al.* [1993] and associated articles in the same issue). Measurements with the MAST instrument on SAMPEX [*Cook et al.*, 1993] quickly showed that trapped ACRs with > 15 MeV/nuc are located in a narrow belt centered at $L \approx 2$ and embedded within the inner Van Allen belt [*Cummings et al.*, 1993ab], as illustrated in Figure 4. Since 1992 SAMPEX has been mapping the distribution and time variations of trapped ACRs, as well as other energetic particle components. These observations have generally confirmed the model

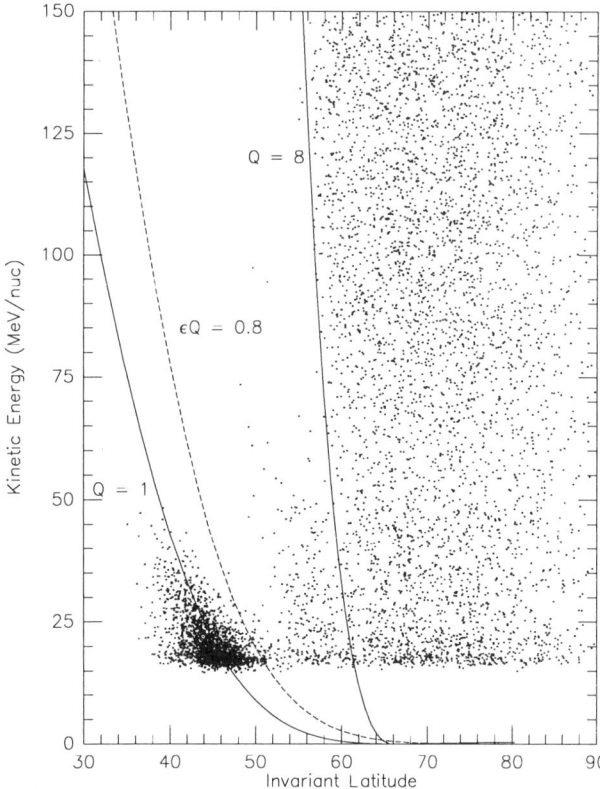

Figure 5. Measured energy/nucleon vs. invariant latitude for oxygen ions observed during solar-quiet days from 7/92 to 2/94. Calculated geomagnetic cutoffs for particles arriving from the west (estimated by Störmer dipole approximation) are shown for singly-ionized ($Q = +1$) and fully stripped oxygen ($Q = +8$). At latitudes $\Lambda > 60°$ a mixture of ACRs and GCRs is observed. At mid-latitudes, GCRs no longer have access, but singly-charged ACRs have access. At $\Lambda \approx 45°$ a population of trapped ACRs is observed. The dotted line labeled $\varepsilon Q = 0.8$ approximately bounds the trapped fluxes (see discussion in text).

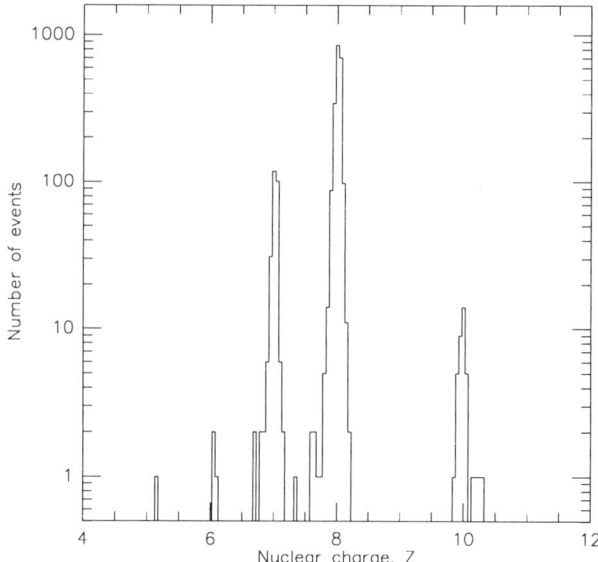

Figure 6. Composition of trapped ACRs with > 15 MeV/nuc observed by the MAST sensor on SAMPEX from 7/92 to 2/94 (see *Selesnick et al.* [1995a]).

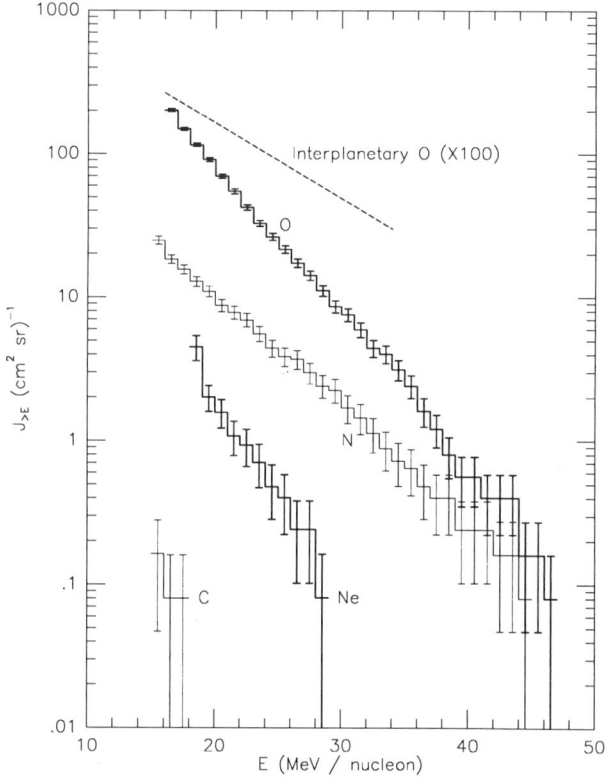

Figure 7. Integral energy spectra of trapped C, N, O, and Ne fluences observed by SAMPEX from 7/92 to 2/94 (see *Selesnick et al.* [1995a]). The relative interplanetary oxygen spectrum observed by SAMPEX is also shown [*Mewaldt et al.*, 1993].

of *Blake and Friesen* [1977], although they have modified a number of details in the model.

To illustrate the distribution of ACRs observed over the SAMPEX orbit Figure 5 shows measured kinetic energy vs. invariant latitude (Λ) for oxygen nuclei with > 15 MeV/nuc. For comparison, nominal geomagnetic cutoffs for particles arriving from the west, estimated from the Störmer dipole approximation, are shown for singly ionized and fully stripped oxygen. Three distinct particle populations are evident. At high latitudes ($> 60°$) there is a mixture of GCR and ACR oxygen. At mid-latitudes ($\sim 50°$ to $60°$), fully stripped GCRs are not allowed, but singly-charged (and possibly also multiply-charged) ACRs can be observed down to the appropriate geomagnetic cutoff. At low latitudes ($< 50°$), there is a grouping near and below the estimated cutoff for singly-charged oxygen. Figure 5 demonstrates the use of the Earth's field as a magnetic spectrometer for separating out anomalous cosmic rays from fully stripped GCRs, and this technique

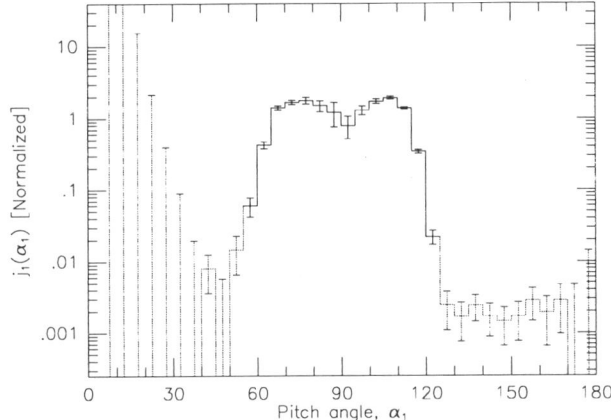

Figure 8. Normalized pitch angle distribution for trapped oxygen with $> 16\,\mathrm{MeV/nuc}$. The pitch angles were transformed to a single value of $R = L\cos^2\Lambda = 1.3$, as described in *Selesnick et al.* [1995a].

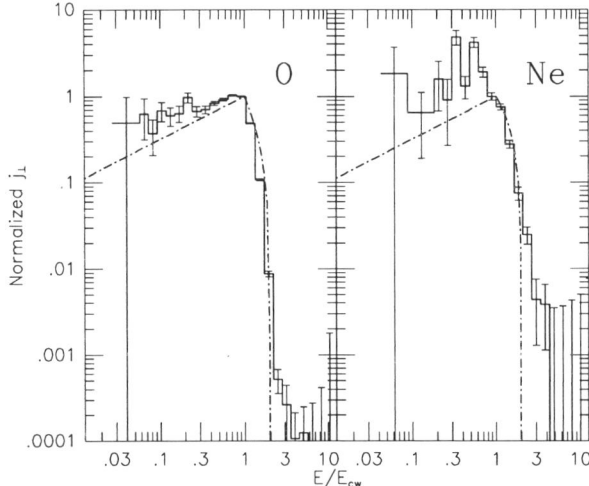

Figure 9. Spectra of trapped O and Ne based on data from the LICA, HILT, and MAST sensors on SAMPEX. The energy is normalized by the local western geomagnetic cutoff energy. Model spectra are shown by the dot-dash curves (see *Selesnick et al.* [1995b]). The differences at low energies may be due to approximations in the model or to possible systematic uncertainties in normalizing data from the three instruments.

has been used to measure the composition of a pure sample of ACRs and to extend measurements of the ACR energy spectrum $\sim 100\,\mathrm{MeV/nuc}$ [*Mewaldt et al.*, 1996].

The data in Figure 5 also illustrate *Blake and Friesen*'s [1977] trapping mechanism. Low energy galactic cosmic rays or solar particles with charge states of $Q = A/2$ do not have access to $L = 2$, but singly-charged ACRs can penetrate down to the geomagnetic cutoff for $Q = 1$. Once stripped, they can be trapped if they have the proper pitch angles, and if they undergo adiabatic motion, as characterized by the "adiabaticity" parameter

$$\varepsilon = 5.18 \times 10^{-5} \frac{A}{Q_s} L^2 [E(E+1863)]^{1/2}. \quad (1)$$

Here E is kinetic energy in MeV/nuc, Q_s is the charge state after stripping in the upper atmosphere, and particles are assumed to be stably trapped if $\varepsilon < \varepsilon_0$. *Blake and Friesen* [1977] originally suggested $\varepsilon_0 = 1/3$, in which case the trapped component in Figure 5 would extend out to $\Lambda = 60°$ ($L = 4$). Although the exact value of ε_0 is a subject of discussion, SAMPEX observations show that $\varepsilon_0 \approx 0.1$ is a more appropriate value [*Cummings et al.*, 1993a; *Tylka*, 1994; *Selesnick et al.*, 1995a]. Figure 5 illustrates that the trapping boundary for oxygen can be represented approximately by $\varepsilon Q = 0.8$ (consistent with $\varepsilon_0 = 0.1$ and $Q_s = 8$ for oxygen).

The composition of trapped nuclei with $> 15\,\mathrm{MeV/nuc}$ at $L = 2$ is shown in Figure 6. In addition to O and N, also detected by *Grigorov et al.* [1991] (see also *Bobrovskaya et al.* [1993]), a substantial amount of Ne, and a small amount of C can be observed. There is also evidence for trapped Ar [*Jonthal et al.*, 1993; *Mazur et al.*, 1993]. From a comparison of these abundances with those of interplanetary ACRs (Table 1) it appears that ACR nitrogen may be less efficiently trapped than oxygen, while trapped ACR carbon is depleted by more than a factor of 40 (see discussion in *Selesnick et al.* [1995a]). This depletion of C is expected if $\varepsilon_0 = 0.1$ because C with a rigidity just at the geomagnetic cutoff will

have $\varepsilon Q > 0.8$, even if $Q_s = 6$ after stripping, so apparently C is difficult to trap. Similarly, it is not expected that ACR helium would be trapped at all by this mechanism.

The composition of trapped heavy ions in Table 1 is unlike that of any other suggested sources of magnetospheric ions [*Cummings et al.*, 1993a]. The observed C/O ratio of ~ 0.0004 is much less than that of solar energetic particles or the solar wind (where C/O ~ 0.5), the Ne/O ratio of ~ 0.04 is much greater than that in the ionosphere (where Ne/O $\sim 5 \times 10^{-4}$). These and other sources may also contribute to the population of trapped heavy ions, especially at lower energies, where there have been a number of earlier observations of nuclei with $Z > 2$ (see e.g., the review by *Spjeldvik and Fritz* [1983]). It is clear, however, that trapped ACRs have been the dominant source of high energy ($> 10\,\mathrm{MeV/nuc}$) heavy ions with $Z > 2$ observed in the radiation belts over the past decade.

The energy spectra of the trapped nuclei, shown in Figure 7, are considerably softer than the corresponding interplanetary spectra. The spectrum for trapped 16 to 45 MeV/nuc oxygen is approximately exponential with an e-folding energy of $\sim 4\,\mathrm{MeV/nuc}$. The e-folding energy for interplanetary oxygen is $\sim 8\,\mathrm{MeV/nuc}$. The trapped spectra also appear to soften with increasing atomic number. These differences are not understood, but may be due in part to the energy dependence of the stripping cross sections.

Assuming that the observed intensity j is a separable function $j = U(E,L)\,V(\alpha)\,W(t)$ of energy/nuc E, pitch angle α, and time t, *Selesnick et al.* [1995a] were able to determine the energy spectra, pitch angle distribution, and time variations of trapped O in 1992 and 1993. Figure 8 shows that the pitch angle distribution of $> 16\,\mathrm{MeV/nuc}$ oxygen is peaked at 90°, with a broad, flat maximum. *Selesnick et al.* [1995a] argued that the nearly isotropic distribution observed outside the loss cone implies that electron stripping is a single-step process. In this case, both the source and loss rates of trapped ACRs would be proportional to the atmospheric density at the mirror point altitude, resulting in an isotropic pitch angle distribution at a given altitude.

This analysis approach has recently been extended to a broader energy interval (~ 1 to 50 MeV/nuc) by including data from the LICA and HILT sensors on SAMPEX [*Selesnick et al.*, 1995b]. An analysis of data from seven L-shells ranging from $L = 1.66$ to $L = 2.53$ shows that in each case the maximum intensity occurs at an energy corresponding to the western geomagnetic cutoff for singly-charged ions. Demonstration of this organization by geomagnetic cutoff is provided in Figure 9, in which data from all L-shells is plotted as a function of E/E_{cw}, where E_{cw} is the western cutoff energy. Trapped particles with energies below the cutoff have apparently lost energy as they brushed against the upper atmosphere at their mirror points. The trapped spectra at energies above the geomagnetic cutoff are affected by the interplanetary source spectra, the energy dependence of the stripping cross sections, and the requirement for adiabatic motion [*Selesnick et al.*, 1995ab]. Also shown in Figure 9 is the result of a simple model for trapped ACRs which is in reasonable agreement with the experimental data. Similar models have also been described by others [*Blake*, 1990; *Tylka*, 1994; *Tylka et al.*, 1996].

5. A SEMI-EMPIRICAL MODEL OF TRAPPED ANOMALOUS COSMIC RAYS

The results described above determine the trapped ACR intensity at and below the SAMPEX orbit. *Selesnick et al.* [1995a] have also placed limits on the intensity of trapped ACRs above the SAMPEX orbit (see Figures 11 and 12 of their paper). A lower limit can be derived by considering the omnidirectional intensity due only to ions observed by SAMPEX, assuming that there are none with higher mirror points. A more reasonable estimate is obtained by assuming the intensity at 90° pitch angles at points above the SAMPEX orbit is equal to the value observed at the orbit, as discussed above. Here the intensity generally increases with altitude on a given L-shell due to the narrowing of the loss cone. Note that the true intensity may differ from this if other loss processes are important, or if the fluxes have not attained a steady state because the trapping lifetimes are long compared to variations in the interplanetary source strength. It is planned to test this model with simultaneous SAMPEX/COSMOS observations at two altitudes during 1994–1996.

To estimate the intensity of trapped ACRs as a function of time we assume proportionality to the interplanetary source strength (Figure 2). Then the intensities presented in *Selesnick et al.* [1995a] can be scaled by the factor $F_{nm} = (\text{Climax}/3993)^{25}$ where "Climax" is the counting rate of the CLIMAX neutron monitor during the period of interest (=3993 for the SAMPEX 7/92 to 2/94 period of *Selesnick et al.* [1995a] and =4024 for the 7/92 to 2/95 period of *Selesnick et al.* [1995b]). Over the first two and one-half years after the SAMPEX launch the flux of trapped ACRs has increased by a factor of ~ 3 to ~ 4, in reasonable correlation with the increase in the interplanetary fluxes as solar minimum approaches. For a typical solar minimum Climax rate of ~ 4250 the expected trapped ACR fluxes are ~ 5 times higher than in *Selesnick et al.* [1995a]. The ratio of the trapped oxygen to trapped proton fluxes at 15 MeV/nuc is $\sim 10^{-6}$.

6. SUMMARY

All observations to date of trapped ACRs confirm that the basic mechanism proposed by *Blake and Friesen* [1977], and amplified by others, is responsible for the belt of trapped ACRs observed at $L = 2$. The mechanism appears to be reasonably efficient for N, O, Ne, and possibly Ar, but C is apparently trapped very inefficiently, and ACR He is not trapped at all. Although simple models of the trapping mechanism are able to account for the observations, there remain questions as to the relevant stripping cross sections, and the associated trapping efficiencies and lifetimes. The Cosmos observations suggest that trapped ACRs vary in intensity by more than two orders of magnitude over the solar cycle, in proportion to the interplanetary ACR intensity. The trapped intensity has increased by a factor of ~ 3 to 4 from mid-1992 to early 1995, and may increase by another factor of 2 or 3 by the time of solar minimum.

Acknowledgements. This work was supported by NASA under contract NAS5-30704 and grant NAGW-1919. We appreciate contributions to this work by our SAMPEX colleagues B. Blake, A. Cummings, B. Klecker, R. Leske, G. Mason, J. Mazur, E. Stone, and T. von Rosenvinge. We thank B. Klecker, R. Pyle, and J. Simpson for the use of unpublished data in Figure 2.

NOTE ADDED IN PROOF

Since the submission of this paper new measurements by SAMPEX [*Mewaldt et al.*, Ap. J. Letters, in press, 1996] have shown that the vast majority of interplanetary ACRs with > 30 MeV/nuc are multiply-charged, with charge states of $Q = 2$, $Q = 3$, and probably higher. This contrasts to energies of ~ 10 MeV/nuc, where most ACRs are singly charged. As pointed out by *Mewaldt et al.* [1996], the dominance of multiply-charged ACRs at high energy can explain why trapped ACRs have a steeper energy spectrum than interplanetary ACRs (Figure 7), and why the trapped ACR spectrum extends only to ~ 45 MeV/nuc, much less than the interplanetary spectrum (see Figure 5). To be trapped by the *Blake and Friesen* [1977] mechanism interplanetary ions must have access to invariant latitudes below the $\varepsilon Q = 0.8$ line in Figure 5. Since only singly-charged oxygen can reach this region, trapped ACR oxygen is derived from only O^{+1}, and not from multiply-charged ACR oxygen that dominates at higher energy.

REFERENCES

Baker, D.N., G.M. Mason, O. Figueroa, G. Colon, J.G. Watzin and R.M. Aleman, An overview of the Solar, Anomalous, and Magnetospheric Particle Explorer (SAMPEX) mission, *IEEE Trans. Geosci. Remote Sensing, 31*, 531–541, 1993.

Blake, J.B., Geomagnetically trapped heavy ions from anomalous cosmic rays, *Proc. 21st Internat. Cosmic Ray Conf. (Adelaide), 7*, 30–33, 1990.

Blake, J.B. and L.M. Friesen, A technique to determine the charge state of the anomalous low-energy cosmic rays, *Proc. 15th Internat. Cosmic Ray Conf. (Plovdiv), 2*, 341–346, 1977.

Bobrovskaya, V., N.L. Grigorov, M.A. Kondratyeva, M.I. Panasyuk, Ch.A. Tretyakova, D.A. Zuravlev, J.H. Adams, Jr. and A.J. Tylka, Cosmos observations of anomalous cosmic ray N and Ne in the inner magnetosphere, *Proc. 23rd Internat. Cosmic Ray Conf. (Calgary), 3*, 432–435, 1993.

Chan, J.H. and P.B. Price, Composition and energy spectra of heavy nuclei of unknown origin detected on Skylab, *Astrophys. J., 375*, L539–L542, 1975.

Cook, W.R., A.C. Cummings, J.R. Cummings, T.L. Garrard, B. Kecman, R.A. Mewaldt, R.S. Selesnick, E.C. Stone and T.T. von Rosenvinge, MAST: A mass spectrometer telescope for studies of the isotopic composition of solar, anomalous, and galactic cosmic ray nuclei, *IEEE Trans. Geosci. Remote Sensing, 31*, 557–564, 1993.

Cummings, A.C. and E.C. Stone, Elemental composition of the anomalous cosmic ray component, *Proc. 24th Internat. Cosmic Ray Conf. (Rome), 4*, 497–500, 1995.

Cummings, J.R., A.C. Cummings, R.A. Mewaldt, R.S. Selesnick, E.C. Stone and T.T. von Rosenvinge, New evidence for geomagnetically trapped anomalous cosmic rays, *Geophys. Res. Lett., 20*, 2003–2006, 1993a.

Cummings, J.R., A.C. Cummings, R.A. Mewaldt, R.S. Selesnick, E.C. Stone, T.T. von Rosenvinge and J.B. Blake, SAMPEX measurements of heavy ions trapped in the magnetosphere, *IEEE Trans. Nucl. Sci., 40*, 1459–1462, 1993b.

Cummings, J.R., A.C. Cummings, R.A. Mewaldt, R.S. Selesnick, E.C. Stone and T.T. von Rosenvinge, SAMPEX observations of geomagnetically trapped anomalous cosmic rays, in *Proceedings of the 23rd Internat. Cosmic Ray Conf.*, Invited, Rapporteur and Highlight Papers, edited by D.A. Leahy, R.B. Hicks, and D. Venkatesan, pp. 475–482, World Scientific, Singapore, 1994.

Fisk, L.A., B. Kozlovsky and R. Ramaty, An interpretation of the observed oxygen and nitrogen enhancements in low-energy cosmic rays, *Astrophys. J., 190*, L35–L38, 1974.

Geiss J., G. Gloeckler, U. Mall, R. von Steiger, A.B. Galvin and K.W. Ogilvie, Interstellar oxygen, nitrogen, and neon in the heliosphere, *Astron. Astrophys., 282*, 924–933, 1994.

Grigorov, N., M.A. Kondratyeva, M.I. Panasyuk, Ch.A. Tretyakova, J.H. Adams, Jr, J.B. Blake, M. Schulz, R.A. Mewaldt and A.J. Tylka, Evidence for anomalous cosmic ray oxygen ions in the inner magnetosphere, *Geophys. Res. Lett., 18*, 1959–1962, 1991.

Jokipii, J.R., The anomalous component of cosmic rays, in *Physics of the Outer Heliosphere*, S. Grzedzielski and D.E. Page, Pergamon Press, 1990.

Klecker, B., The anomalous component of cosmic rays in the 3-D heliosphere, *Space Science Reviews, 72*, 419–430, 1995.

Klecker, B., M.C. McNab, J.B. Blake, D. Hovestadt, H. Kastle, D.C. Hamilton, M.D. Looper, G.M. Mason, J.E. Mazur and M. Scholer, Charge state of anomalous cosmic ray nitrogen, oxygen, and neon: SAMPEX observations, *Astrophys. J., 442*, L69–L72, 1995.

Mazur, J.E. and G.M. Mason, Observations of low energy trapped anomalous cosmic rays using SAMPEX, *Trans. Am. Geophys. U., 76*, S 237, 1995.

Mewaldt, R.A., A.C. Cummings, J.R. Cummings, E.C. Stone, B. Klecker, D. Hovestadt, M. Scholer, G.M. Mason, J.E. Mazur, D.C. Hamilton, T.T. von Rosenvinge and J.B. Blake, The return of the anomalous cosmic rays to 1 AU in 1992, *Geophys. Res. Lett., 20*, 2263–2266, 1993.

Mewaldt, R.A., J.R. Cummings, R.A. Leske, R.S. Selesnick, E.C. Stone and T.T. von Rosenvinge, A study of the composition and energy spectra of anomalous cosmic rays using the geomagnetic field, *Geophys. Res. Letters*, in press, 1996.

Mewaldt, R.A., A.C. Cummings and E.C. Stone, The anomalous cosmic rays—Interstellar interlopers in the heliosphere and the magnetosphere, *EOS Trans. Am. Geophys. Un., 85*, 185 and 193, 1994.

Pesses, M.E., J.R. Jokipii and D. Eichler, Cosmic ray drift, shock wave acceleration, and the anomalous component of cosmic rays, *Astrophys. J., 246*, L85–L88, 1981.

Selesnick, R.S., A.C. Cummings, J.R. Cummings, R.A. Mewaldt, E.C. Stone and T.T. von Rosenvinge, Geomagnetically trapped anomalous cosmic rays, *J. Geophys. Res., 100*, 9503–9509, 1995a.

Selesnick, R.S., R.A. Mewaldt, E.C. Stone, G.M. Mason, J.E. Mazur, J.B. Blake, M.D. Looper, B. Klecker and D. Hovestadt, Observations of geomagnetically trapped anomalous cosmic rays, *Proc. 24th Internat. Cosmic Ray Conf. (Rome) 4*, 1013–1017, 1995b.

Simpson, J.A., The anomalous nuclear component in the three-dimensional heliosphere, *Adv. Space Res., 16*, 135–149, 1995.

Spjeldvik, W.N., and T.A. Fritz, Experimental determination of geomagnetically trapped energetic heavy ion fluxes, *Energetic Ion Composition in Earth's Magnetosphere*, R.G. Johnson, ed., Terra Scientific, Tokyo, 1983.

Tylka, A.J., Theoretical modeling and interpretation of trapped anomalous cosmic rays, in *Proceedings of the 23rd International Cosmic Ray Conference*, Invited, Rapporteur and Highlight Papers, edited by D.A. Leahy, R.B. Hicks and D. Venkatesan, pp. 475–482, World Scientific, Singapore, 1994.

Tylka, A.J., P.R. Boberg and J.H. Adams, Jr., LET spectra of trapped anomalous cosmic rays in low-earth orbit, *Adv. Space Res., 17*, 47–51, 1996.

R.A. Mewaldt, R.S. Selesnick and J.R. Cummings, California Institute of Technology, Pasadena, CA 91125, USA.

DISCUSSION

Q: A.L. Vampola. Why do you assume that the anomalous cosmic rays don't peak low on the field line? The data seems to show this and the maximum generation is down there.

A: R.A. Mewaldt. The lifetime of the particles scales with the source rate, so we expect an isotropic distribution initially. We will compare this with COSMOS data.

Q: G. Ginet. What is the ratio of the equatorial omnidirectional flux of the anomalous radiation belt to the standard electron and proton belts?

A: R.A. Mewaldt. I do not have the numbers right in front of me but I believe the ratio is at most 10^{-4}.

Formation of the Radiation Belts by Anomalous Cosmic Rays and Similar Phenomena

A.V. Dmitriev, V.D. Ilyin, S.N. Kuznetsov and B.Yu. Yushkov

Skobeltsyn Institute of Nuclear Physics, Moscow State University

A model of charged energetic particle motion in a dipole field (in the range of rigidities from applicability of the adiabatic approximation up to the threshold of cosmic ray cut-off) is considered. The concept of the particle guiding centre trajectory is refined. The characteristics of this trajectory at the geomagnetic equator induce changes in a parameter which is the analogue of the pitch angle of a particle are studied. It is shown that when the adiabatic particle motion is violated due to the large rigidity of the particle, the particle conserves the analogue of the magnetic moment during motion from the equator to the mirror point and back. Violation of the magnetic moment (changes of an analogue of the particles' pitch angle) occurs when the particle intersects the equatorial plane. The current model is used for the analysis of motion of particles with large rigidity. It is applied to study phenomena in the magnetosphere such as penetration of incompletely ionised nuclei into the magnetosphere and their stripping in the residual atmosphere. For incompletely ionised nuclei penetrating into the magnetosphere, conditions of particle motion and their penetration into the atmosphere for different stages of stripping are considered. A region of stable motion for repeated stripping of ions was found. The lifetimes of these particles is estimated.

1. INTRODUCTION

The range of charged particle motion in the geomagnetic dipole trap is described by the expression [*Störmer*, 1955]

$$R = \frac{C_{St} \cos^2 \lambda}{\gamma + \sqrt{\gamma^2 - \sin\alpha \sin\varphi \cos^3 \lambda}}, \quad (1)$$

where C_{St} is the Störmer length unit in Earth radii, $C_{St}^2 = 57.7\, Q/p$. Here p is the particle momentum in GeV/c, Q is the particle charge in proton (electron) units, λ is the geomagnetic latitude, γ is half the component of the particle moment in the direction of the dipole axis, when the particle with a given p is located at infinity, α is the particle pitch angle and φ is the particle gyration phase, measured from the meridional plane.

Radiation Belts: Models and Standards
Geophysical Monograph 97
Copyright 1996 by the American Geophysical Union

C_{St} and γ define the field line $L_0 = C_{St}/2\gamma$, along which the particle approaches the dipole.

In order to describe the motion of particles which are trapped or move periodically the adiabatic (drift) theory is used. According to this theory the particle motion is considered to be a combination of three independent motions: Larmor gyration about the guiding center, bounce motion of the guiding center and drift of the guiding center around the dipole. A specific role in conservation of stable particle motion is played by conservation of the first adiabatic invariant, the magnetic moment μ. In a coordinate system rotating around the dipole with a velocity equal to the particle drift velocity,

$$\mu = \frac{p^2 \sin^2 \alpha}{2mB}, \quad (2)$$

where B is measured at the particle's location. It should be mentioned that in an immobile coordinate system μ will undergo reversible variations, associated with the particle gyration phase φ.

The conditions in which the drift theory is applicable require that the value of the adiabaticity parameter $\chi = \rho/R_c$ be small. Here ρ is the Larmor radius of the particle corres-

44 FORMATION OF THE RADIATION BELTS

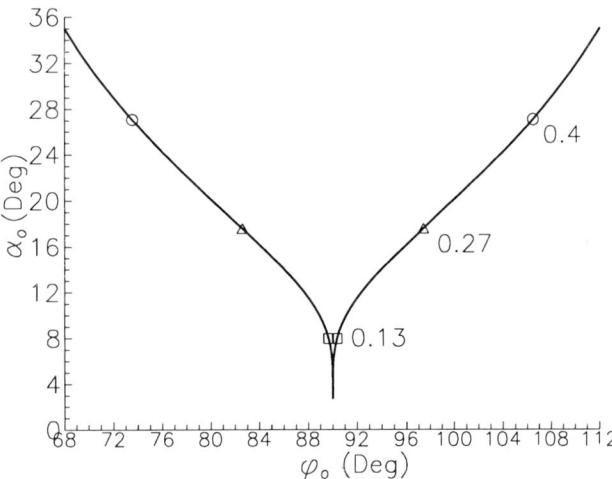

Figure 1. Connection between α_0 and φ_0 for central trajectories

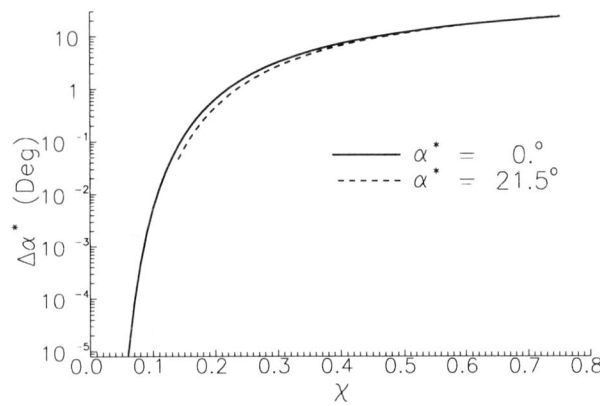

Figure 2. Dependence of the mean angle of particles $\overline{\Delta\alpha(0°)}$ and $\overline{\Delta\alpha(21.5°)}$

ponding to pitch angle $\alpha = \pi/2$ at the equator and R_c is the curvature radius of the field line at the equator. χ may also be described as

$$\begin{aligned} \chi &= 5.18 \times 10^{-5} pc \frac{L_0^2}{Q} \\ &= 2.24 \times 10^{-3} \sqrt{E} \frac{L_0^2}{Q} \\ &= 0.75 \gamma^{-2}, \end{aligned} \quad (3)$$

where pc and E are in MeV.

Numerical integration [*Amirkhanov et al.*, 1988] of the equations of motion shows that for $\chi \geq 0.13$ noticeable irreversible fluctuations of μ appear. In *Amirkhanov et al.* [1991] and *Ilyina et al.* [1993] it is shown that there exists a coordinate system in which particle motion between two successive crossings of the equatorial plane occurs with conservation of

an analogue of the magnetic moment μ^*

$$\mu^* = \frac{p^2 \sin^2 \alpha^*}{2mB}, \quad (4)$$

where α^* is the angle between the particle velocity vector and the tangent to the central trajectory (CT)—the trajectory of a particle passing through the dipole center. At an arbitrary latitude we can calculate α^* using the expression

$$\cos \alpha^* = \cos \alpha_0 \cos \alpha + \sin \alpha_0 \sin \alpha \cos(\varphi_0 - \varphi), \quad (5)$$

where the index '0' indicates the CT. The CT characteristics are functions of the adiabaticity parameter χ. During the equator crossing the CT undergoes a kink, whereas α_0 does not change, and φ_0 changes to $(\pi - \varphi_0)$. Therefore, during the equator crossing α^* (and the quasi-moment μ^*) undergoes a sharp change

$$\begin{aligned} \sin \Delta\alpha^* &= \sin \alpha_0 \cos \varphi_0 \\ &\simeq 1.4 \exp(-0.96/\chi), \end{aligned} \quad (6)$$

$$\sin \alpha_0 = 1.225 \chi^{1.07}. \quad (7)$$

It is possible to determine φ_0 from α_0 and $\Delta\alpha^*$ using Eqs. (6) and (7) or vice versa.

Figure 1 shows the connection between α_0 and φ_0 for particles moving away from the dipole towards the equator (left branch), and for particles moving from the equator towards the dipole (right branch). Certain selected values of χ are shown on the right branch.

Figure 2 shows the dependences of $\overline{\Delta\alpha^*}$ on χ. The maximum value of $\Delta\alpha^*$ may exceed $\overline{\Delta\alpha^*}$ by a factor of 2.

2. GUIDING CENTER MODEL OF NONADIABATIC MOTION FOR ARBITRARY PITCH ANGLES

For several values of χ and α^* ranging from 5° to 60° we also studied the characteristics of the guiding center trajectories $\alpha_0(\alpha^*)$ and $\varphi_0(\alpha^*)$. For $\alpha^* < 60°$ and $\chi < 0.4$ we have approximately:

$$\sin \alpha_0(\alpha^*) = \frac{1.18 \chi^{1.07}}{\psi(\alpha^*)}, \quad (8)$$

$$\begin{aligned} \sin \overline{\Delta\alpha^*}(\alpha^*) &= \sin \alpha_0(\alpha^*) \cos \varphi_0(\alpha^*) \\ &\simeq 1.4 \exp[-\psi(\alpha^*)/\chi], \end{aligned} \quad (9)$$

where

$$\psi(\alpha^*) = \frac{1}{\sqrt{2} \sin^2 \alpha^*} \left(\frac{1+\sin^2 \alpha^*}{\sin \alpha^*} \ln \frac{1+\sin \alpha^*}{\cos \alpha^*} - 1 \right). \quad (10)$$

From these expressions it is easy to obtain that with increasing α^*, $\overline{\Delta\alpha^*}$ decreases.

In order to develop the model it is necessary to resolve the following problems:

- obtain a more accurate dependence of α_0 and φ_0 on χ and α^*;
- define more accurately than in [*Ilyina et al.*, 1993] the phase accumulation during particle motion along the field line from the equator to the mirror point and back.

Accounting for particle gyration around the guiding center trajectory (and not around the field line) leads to the following expression for the phase accumulation $\Delta\Phi_0$:

$$\Delta\Phi_0 = \frac{3\left(\sin^{-1.348}\alpha^* - 0.255\right)}{\pi\chi\left[1 + \frac{\chi}{3}\sin\alpha_0(\alpha^*)\right]^2}. \quad (11)$$

Numerical calculation showed that $\Delta\Phi_0$ depends on the initial phase of the particle. The main harmonic has the form

$$\Delta\Phi_1 = A\sin 2\Phi, \quad (12)$$

where

$$A = 3.35\,\chi^{1.857}(\sin^{-1.09}\alpha^* - 1). \quad (13)$$

We have not yet studied fluctuations of $\mu^*(<10\%)$ during 1/2 bounce period, associated with the value of the initial phase.

According to the theory of adiabatic invariant violation [*Chirikov*, 1987] in the vicinity of the stability boundary the conservation of μ is determined by the existence of resonances $r = \tau_2\overline{\omega}/4\pi$, where r are integer numbers. If $\delta\alpha^*$—the distance between neighbouring resonances—is greater than $\overline{\Delta\alpha^*}$, reversible variations of μ^* will be observed. If $\delta\alpha^*$ is of the same order, or less, than $\overline{\Delta\alpha^*}$, this will lead to irreversible variations. For the geomagnetic dipole we can find the nonadiabatic motion boundary [*Ilyin et al.*, 1993].

3. ANALYSIS OF TRAPPING CONDITIONS OF ANOMALOUS COSMIC RAYS COMPONENTS

An analytic version of this theory for irreversible variations $\Delta\mu \ll \mu$ was used for describing the dynamics of the outer boundary of the proton radiation belt in quiet magnetospheric conditions [*Ilyin et al.*, 1986] as well as during magnetic storms [*Ilyin et al.*, 1988]. In this paper we will give further development to the basic ideas, formulated in [*Ilyin et al.*, 1993]. In the latter paper it was pointed out that the changes in the rigidity of fully stripped oxygen ions of the anomalous component at $L < 3$ are quite sufficient to make the particle trapped.

Here we will consider this problem in detail. We will analyse the particle dynamics for any degree of stripping. We assume that the non-dipolity of the magnetic field is first of all revealed in the variations of the geomagnetic field with longitude at low altitudes. We neglected the non-dipolity of the field at $L = 2.15$ at the equator. Figure 3 shows an example of equatorial pitch angle α_e dependence on the geomagnetic longitude at $L = 2.15 \pm 0.02$ in the northern and southern hemispheres. Here

$$\alpha_e = \arcsin\sqrt{\frac{B_e}{B}}. \quad (14)$$

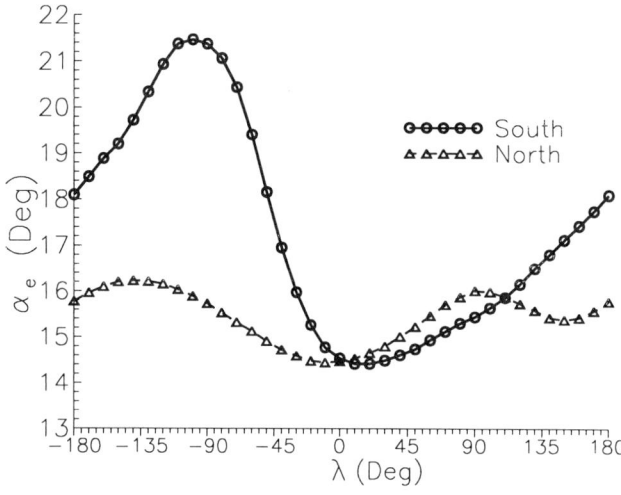

Figure 3. Dependence of equatorial pitch angles on geomagnetic longitude in the northern and southern hemisphere for particles mirroring at ~ 350 km

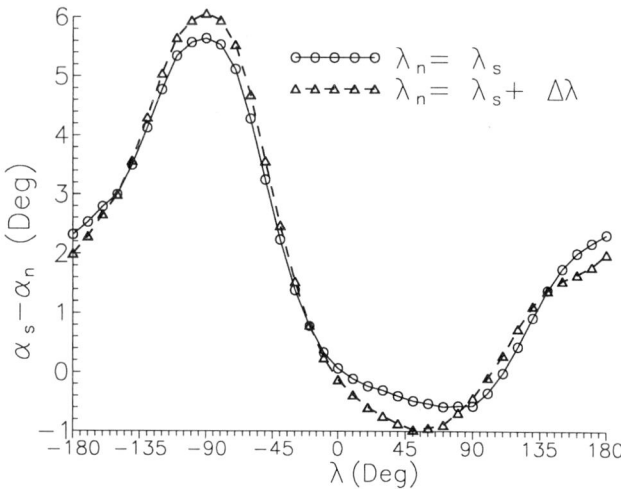

Figure 4. Difference in equatorial pitch angle in Northern and Southern hemisphere: solid line for the same geomagnetic meridian, dashed line with account for drift

Particles mirroring under these conditions in the Brasil magnetic anomaly (BMA) have equatorial pitch angle $21.5°$. Figure 2 shows that the difference between $\overline{\Delta\alpha^*}(0°)$ and $\overline{\Delta\alpha^*}(21.5°)$ is insignificant. If one electron is stripped at ~ 350 km, then 7 electrons will be stripped at ~ 200–220 km. We can see that there may be a difference in the equatorial pitch angle of the particle for mirror point altitude $h = $ const on the same L shell.

The dependence of pitch angle on longitude at the same meridian is shown in Figure 4. We need to compare B and the particle's pitch angle not at the different ends of one field line, but with a shift in longitude, corresponding to the drift

Table 1. Connection of lifetimes (n) and sizes of injection regions ($\Delta\lambda$) of nonadiabatic particles with their charge states.

Q	χ	n	$\Delta\lambda$
1	0.75–1.04	—	360°N,360°S
2	0.375–0.52	34	360°N,360°S
3	0.25–0.347	192	360°S
4	0.188–0.26	650	120°S
5	0.15–0.208	2400	10°S
6	0.125–0.173	140	< 7°S
7	0.107–0.148	—	20°S
8	0.094–0.13	—	20°S

during half of a bounce period:

$$\Delta\lambda = 38.8°\chi\left(1 - 0.185\sin^2\alpha\right). \quad (15)$$

This longitude interval is basically a function of χ and is weakly dependent on the pitch angle. In Figure 4 we also show the difference of pitch angles for particles with $\chi = 0.4$ in the northern and southern hemispheres. It can be seen that qualitatively the difference in pitch angles is the same as for the solid line.

4. PARTICLE DYNAMICS IN THE NONADIABATIC MOTION RANGE

Analysing the data, shown in Figures 2 and 4, we can see that for at all longitudes (including northern longitudes conjugated with the BMA), due to pitch angle fluctuations particles may increase their pitch angle to $> 21.5°$. For smaller χ values the probability of the pitch angle increasing to values $> 21.5°$ for recharged particles decreases and the northern hemisphere starts playing a less important role as a source of recharged particles, and at $\chi < 0.21$ we can neglect the northern hemisphere as a source of radiation belt particles. Particles with $\chi < 0.13$ at $L = 2.15$ are trapped according to *Ilyin et al.* [1993] and can be generated only in the Southern hemisphere at altitudes where the necessary stripping of anomalous component (AC) ions occurs, i.e. in the longitude interval 20–25°. With increasing χ the longitude range in the Southern hemisphere, from which the particles may reach the BMA, increases and at $\chi \geq 0.26$ covers the whole surface of the Earth. From Figure 1 we can see that ACR in the $(14.9-20.8)/L^2$ rigidity range may produce on $L \sim 2.15$ particles with the values of χ, given in Table 1, depending on the charge of oxygen.

It can be seen that the anomaly particles with $Q \sim 6-8$ (trapped) are generated. Particles with charge $Q \leq 4$ may undergo the first stage of stripping in the northern as well as southern hemisphere. Particles moving in the stochastic mode with $Q \sim 4-7$ may be produced only in the Southern hemisphere. If particles are in the stochastic region, then during the first half-bounce for more than half of particles the mirror points will be raised. Further on the effects of mirror point raising is observed until the range of stochastic motion contains a practically isotropic distribution of particles, after this dynamic equilibrium is achieved. Using the calculation from *Ilyin et al.* [1993] we can determine how many bounce periods it takes to reach dynamic equilibrium, i.e. the lifetimes of particles in bounce periods.

Table 1 shows an estimate of the lifetimes for oxygen ions (in a bounce period number n) with various degrees of stripping with initial rigidity $16/L^2$GV. The first column Q is the stripping stage of oxygen ions, the second column χ is the interval of nonadiabatic number χ, the third column n is the lifetime of oxygen ions in bounce period number n, the fourth column $\Delta\lambda$ is the longitude interval after stripping in which particles can reach pitch angle 21.5° in the BMA. For $Q = 2-6$ values of n are calculated for $\chi = 0.4, 0.267, 0.2, 0.167, 0.133$ corresponding. The Northern hemisphere is marked by N, the Southern one by S.

The trapping efficiency is ~ 0.5. Particles with $Q = 2$ are generated in the Northern and the Southern hemisphere at all longitudes (this longitude range exceeds the BMA longitude range by a factor of ~ 36), particles with $Q = 3$ are generated at all longitudes in the Southern hemisphere. The Northern hemisphere gives a small contribution, therefore the region where such particles are generated exceeds the width of the BMA by a factor of ~ 18. Particles with $Q = 4$ are generated only in the Southern hemisphere in the longitude range $\sim 120°$, i.e. the region of particle generation is ~ 6 times greater than the BMA. Particles with $Q = 5$ are generated practically only in the BMA, and so are particles with $Q = 6$. Particles which have $Q = 5$ at the eastern boundary of the BMA will encounter it once again, having changed their pitch angle by $\Delta \sim 0.28°$, i.e. they will undergo further stripping. And only particles which were stripped in the western part will attain charge state $Q = 5$. Therefore, for this particles the source will have a longitude interval less than half of the BMA longitude size. For particles with $Q \leq 4$ the probability of undergoing further stripping in the BMA region is negligibly small. Therefore, their life time is defined only by the time of diffusion of these particles in the trapping region. For particles with $Q \sim 8$ the life times are defined by ionisation losses.

5. CONCLUSION

At present we can list a broad range of cases in which it is necessary to account for non-adiabaticity of motion in a dipole field. These cases are:

1. formation of particle belts due to sharp changes in the charge of the particles (neutron decay, ion stripping);
2. changes of the magnetic field during magnetic storms, affecting the particle trapping conditions;
3. dynamics of charged albedo particles.

REFERENCES

Amirkhanov I.V., Zhidkov E.P., Ilyina A.N., Ilyin V.D., Kuznetsov S.N. and Yushkov B.Yu., Nonadiabatic motion of energetic protons motion in the geomagnetic dipole field, *Bull. Acad. Sci. USSR, Phys. Ser.*, 52, 129–131, 1988.

Amirkhanov I.V., Ilyina A.N., Ilyin V.D. and Yushkov B.Yu., On the nonadiabatic theory of charged particles motion in the geomagnetic dipole field, *Cosmic Research, 29*, 243–249, 1991.

Chirikov B.V., Particle dynamics in magnetic traps, *Reviews of Plasma Physics, 13*, edited by B.P. Kadomtsev, Consultant Bureau, New York, 3–73, 1987.

Ilyin V.D., Ilyin I.V. and Kuznetsov S.N., Stochastic instability of charged particles in a geomagnetic trap, *Cosmic Research, 24*, 75–83, 1986.

Ilyin V.D., Ilyin I.V. and Kuznetsov S.N., Stochastic instability of charged particles in a geomagnetic trap during magnetic storms, *Cosmic Research, 26*, 362–370, 1988.

Ilyin V.D., Ilyin I.V. and Kuznetsov S.N., Accumulation of oxygen ions in the geomagnetic trap, *Cosmic Research, 31*, 687–690, 1993.

Ilyina A.N., Ilyin V.D., Kuznetsov S.N., Yushkov B.Yu., Amirkhanov I.V. and Ilyin I.V., Model of nonadiabatic charged-particle motion in the field of a magnetic dipole *JETP, 77*, 246–252, 1993.

Störmer C., *The Polar Aurora*, Clarendon Press, 1955.

A.V. Dmitriev, V.D. Ilyin, S.N. Kuznetsov and B. Yu. Yushkov, Skobeltsyn Institute of Nuclear Physics, Moscow State Universty, 119899, Moscow, Russia.

Jovian, Solar, and other Possible Sources of Radiation Belt Particles

D.N. Baker[1], S.G. Kanekal[2], M.D. Looper[3], J.B. Blake[3], and R.A. Mewaldt[4]

It is well known that electrons, protons, and heavier ions can be accelerated to high energies ($\gtrsim 1$ MeV) throughout the solar system by a variety of mechanisms. We review several of the sources of energetic ions and electrons that can produce enhanced fluxes of particles near the Earth's orbit. Solar energetic particles and particles accelerated at interplanetary shock waves are considered. We also review the properties and potential terrestrial influence of Jovian electrons. Recent measurements from the SAMPEX spacecraft in low-Earth orbit are examined to look for extraterrestrial sources of electrons and ions. We find clear evidence of both solar and Jovian electrons at high latitudes and at high altitudes around the Earth, but the durably trapped outer zone electron population seems best and most completely explained by an internal acceleration mechanism.

1. INTRODUCTION

It is widely recognized that the plasma populating the Earth's magnetosphere may originate both from the ionosphere (e.g., *Gloeckler and Hamilton* [1987]) and from the solar wind (e.g., *Cowley* [1980]). High-energy particles within the Earth's radiation belts can, in principle, be accelerated *in situ* in the magnetosphere [*Schulz and Lanzerotti*, 1974] or they can originate from sources beyond the Earth's immediate influence [*Scholer*, 1979]. Notable possible source regions include solar flare sites in the sun's atmosphere [*Mason et al.*, 1994], interplanetary shock waves driven by solar disturbances [*Fisk*, 1971], and corotating interplanetary stream interaction regions [*Gloeckler*, 1984]. It is also known that Jupiter's magnetosphere is a copious source of relativistic electrons which fills the inner solar system (e.g., *Teegarden et al.* [1974]) and these electrons have been suggested as a possible source of high-energy electrons in Earth's outer radiation belt [*Baker et al.*, 1979].

Our purpose in this paper is to examine sources of radiation belt particles external to the Earth's magnetosphere. We will particularly focus on sources that may be responsible for the highly dynamic and (often) regularly variable relativistic electron population in Earth's outer radiation zone.

2. ENERGETIC ION SOURCES IN THE HELIOSPHERE

Figure 1 shows in a schematic way the energetic ion populations of the solar system. As is evident, the region near 1 astronomical unit (AU) is awash with energetic ions which have been accelerated in a wide variety of locations. Among the particle populations of interest are solar energetic particles (SEPs) which are produced by explosive dissipation events (e.g., flares) in the sun's corona [*Mason et al.*, 1994] and which can represent intense heavy ion-rich particle enhancements at Earth. Also of interest are energetic storm particle (ESP) events which are due to strong acceleration caused by traveling interplanetary shock waves moving outward from the sun. These ESP events can represent the most intense high-energy ion events that are observed near Earth (e.g., *Zwickl and Kunches* [1989]).

As seen in Figure 1, there can also be strong acceleration of ions at the forward-reverse shock pairs that form typically beyond 1 AU in association with corotating solar wind stream interactions regions [*Gloeckler*, 1984]. These "corotating ion" events can be frequent and regular sources of high-energy ions as the solar wind streams emerging from solar coronal holes propagate outward through the heliosphere. These corotating events often exhibit ~27-day periodicity if the solar wind streams are long-lived (as they often are during sunspot minimum conditions).

Another shock-related source of relatively energetic ions is the diffuse upstream ions accelerated upstream of the Earth's

[1]LASP, University of Colorado, Boulder, Colorado
[2]NASA/Goddard Space Flight Center, Greenbelt, Maryland
[3]Aerospace Corp., Los Angeles, California
[4]California Inst. of Tech., Pasadena, California

Radiation Belts: Models and Standards
Geophysical Monograph 97
Copyright 1996 by the American Geophysical Union

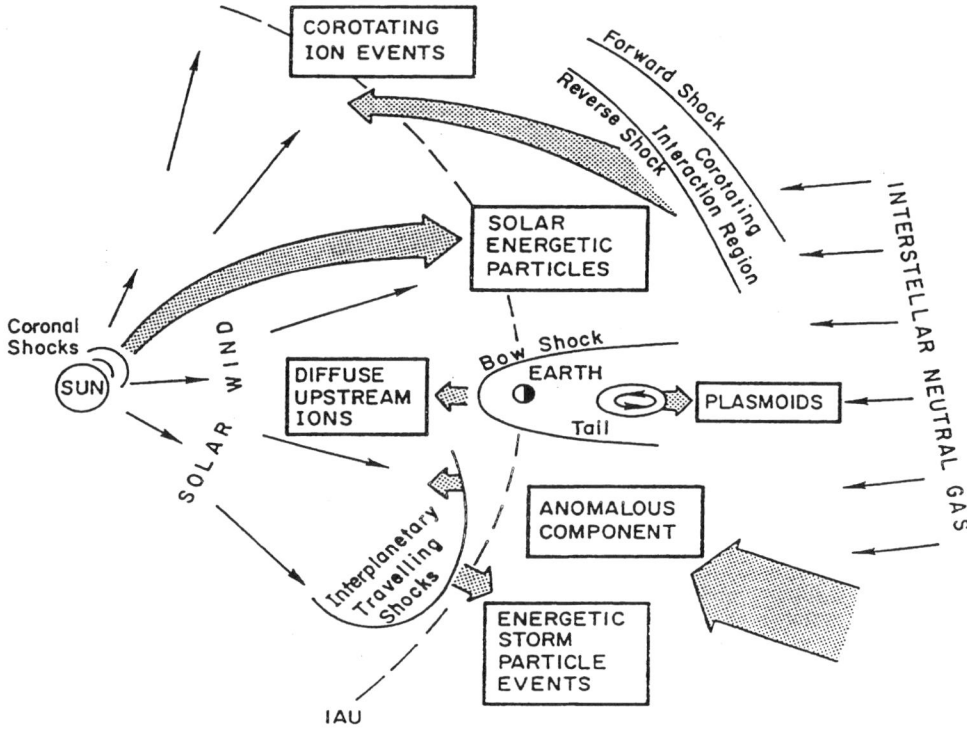

Figure 1. A schematic diagram illustrating the many types of energetic particle populations and associated acceleration mechanisms in the heliosphere (from *Gloeckler* [1984]).

bow shock [*Scholer*, 1985]. The present view is that solar wind ions initially "reflected" by the bow shock are further accelerated by Fermi processes in the quasi-parallel region of the bow shock. Solar protons and alpha particles may be accelerated to energies $> 1\,\mathrm{MeV}$ as they scatter back and forth between magnetic irregularities in the foreshock region upstream of the Earth.

A final population of interest is the anomalous cosmic ray component. The Solar, Anomalous, and Magnetospheric Particle Explorer (SAMPEX) spacecraft has detected a radiation belt surrounding the Earth that traps material from the nearby interstellar medium. This recently discovered belt is embedded in the lower of the two previously-known Van Allen belts. SAMPEX pinpointed the new belt whose existence was first predicted nearly 20 years ago [*Blake and Friesen*, 1977].

The new belt consists of trapped heavy ions, including the nuclei of atoms of nitrogen, oxygen, and neon which are part of the ACR (anomalous cosmic ray) component [*Cummings et al.*, 1993]. These gases make up the tenuous ($\lesssim 1\,\mathrm{cm}^{-3}$) interstellar gas which, if electrically neutral, can penetrate the heliosphere. Some of these neutral interstellar atoms are singly ionized by solar UV radiation, and are then accelerated to cosmic ray energies at the solar wind "termination shock". If one of these singly-charged cosmic rays encounters the Earth's atmosphere and loses its remaining electrons, it may become trapped in the Earth's magnetic field. Once inside the new belt, these atoms may be trapped for an extended period before leaking out into space or into the atmosphere. At present, the intensity of ACR oxygen inside the belt is about 400 times greater than in interplanetary space. The rate of trapping varies with the interplanetary ACR intensity, and as a result, the intensity of ions trapped in the belt varies by perhaps a factor of 100 to 1000 over the solar cycle.

As noted by *Scholer* [1979], any high energy particles which impinge upon the magnetosphere can, with finite probability, penetrate the magnetopause. Thus, it is likely that the magnetosphere is populated to some extent by all of the above sources. The relevant question is not whether external sources populate the radiation belts, but rather the question is whether a given source is significant—or even dominant—at a particular time in a given energy range.

3. JOVIAN ELECTRONS AS A POSSIBLE RADIATION BELT SOURCE

Electrons from Jupiter were first recognized by *Teegarden et al.* [1974] as the source of "quiet-time" cosmic ray electron increases. Various groups have, since the first suggestion of Teegarden et al., reported on the source characteristics, the interplanetary propagation, and the observed properties at 1 AU of Jovian electrons outside the terrestrial magnetosphere (e.g., *Krimigis et al.* [1975]; *Mewaldt et al.* [1976]; *Chenette et al.* [1977]).

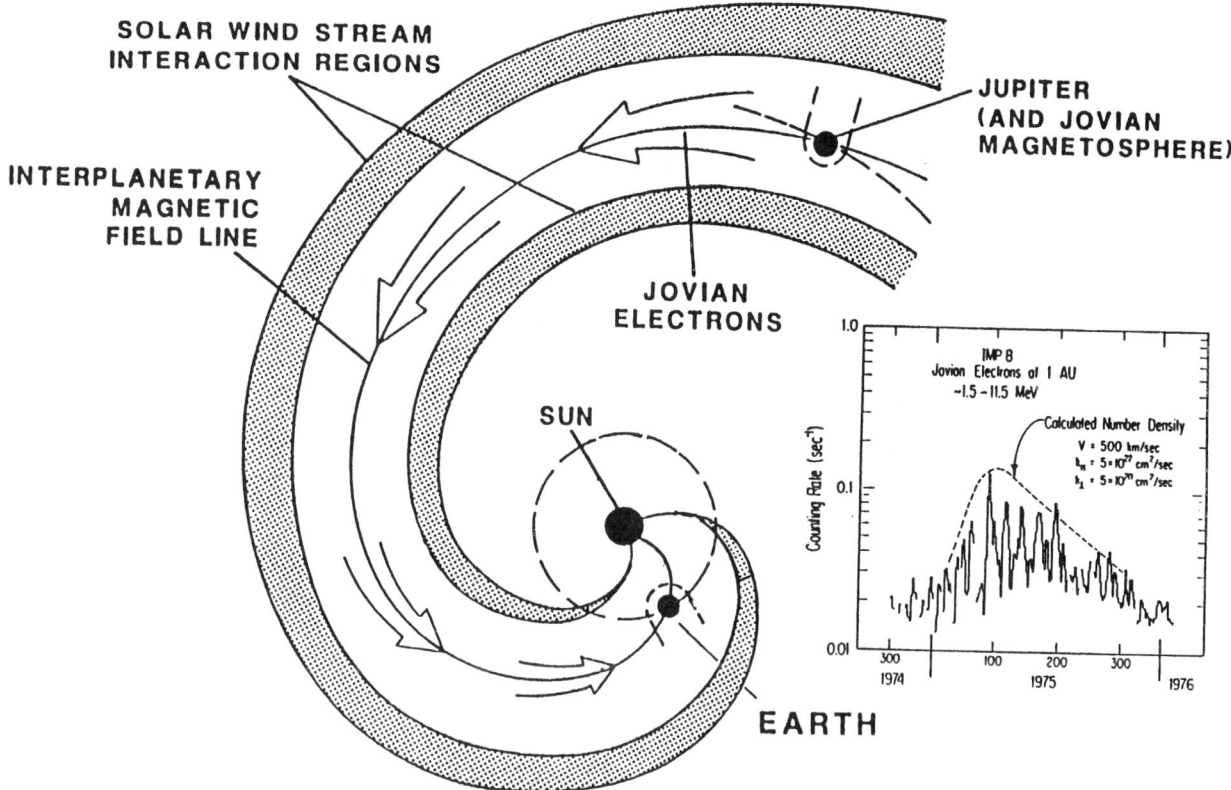

Figure 2. Diagram viewing downward from the north pole of the sun showing solar wind stream interaction regions and the possible path followed by Jovian electrons along interplanetary magnetic field lines toward Earth. Inset: 48-hr averages of IMP-8 1.5–11.5 MeV electron counting rates for late 1974 to early 1976. The dashed curve is the predicted shape of peak Jovian electron intensities in the convection-diffusion model (after *Conlon* [1978]).

Jovian electrons ($E \lesssim 10$ MeV) both at Jupiter and in the interplanetary medium near Earth have a spectrum with a power law spectral index $\gamma < 2$ [*Baker and Van Allen*, 1976; *Mewaldt et al.*, 1976]; which is remarkably hard; in fact this spectral feature has been used by most experimenters to identify Jovian electrons and to separate out terrestrial and solar backgrounds from the Jovian signal [*Krimigis et al.*, 1975; *Mewaldt et al.*, 1976; *Chenette et al.*, 1977]. By way of contrast, in most regions of the terrestrial magnetosphere (as at 6.6 R_E) and in solar energetic particle events, electron spectra have $\gamma \gtrsim 3$.

Also, Jovian electrons tend to consist of flux increases of several days duration which then recur with 27-day periodicities (e.g., *Teegarden et al.* [1974]; *Mewaldt et al.* [1976]). Furthermore, these electrons appear after high speed solar wind streams have gone past the Earth and, therefore, Jovian electrons appear at 1 AU during periods of declining solar wind speed and low K_p. The 27-day periodicity of the Jovian increases has been attributed by *Conlon* [1978] to the effects of recurrent, fast solar wind streams overtaking slower plasma which form corotating interaction regions (CIRs). The CIRs then form barriers to the cross-field diffusion of Jovian electrons, and thus the Jovian particles are constrained to propagate within the region between CIRs. The entire heliospheric CIR pattern rotates about the sun with the 27-day solar rotation period; when Jupiter and Earth are both within a given CIR "cavity", a Jovian increase is seen at Earth (see Figure 2).

A third point is that Jovian electron increases at 1 AU exhibit a long-term periodicity of ~13 months which is the synodic Jovian period as viewed from the Earth [*Chenette et al.*, 1977; *Teegarden et al.*, 1974]. The modulation arises from the fact that once every 13 months Earth and Jupiter are directly connected along the ideal average IMF spiral field line and fluxes will be high during this period; low fluxes result when Earth is far from nominal connection since Jovian electrons must then diffuse large distances across interplanetary field lines to reach the Earth.

In Figure 3 we show a portion of CPA (Charged-Particle Analyzer) 1.4–2.0 MeV electron data at geostationary orbit for 1977–78. These data are from *Baker et al.* [1979] showing magnetospheric trapped particles. In eight of nine months during this period, Baker et al. identified the time of peak Jovian electron flux increases outside the magnetosphere (from Caltech IMP-8 data). Each of these peak Jovian electron times is shown by the dot in Figure 3. The missing dot (September 1977) corresponds to a time of a large solar

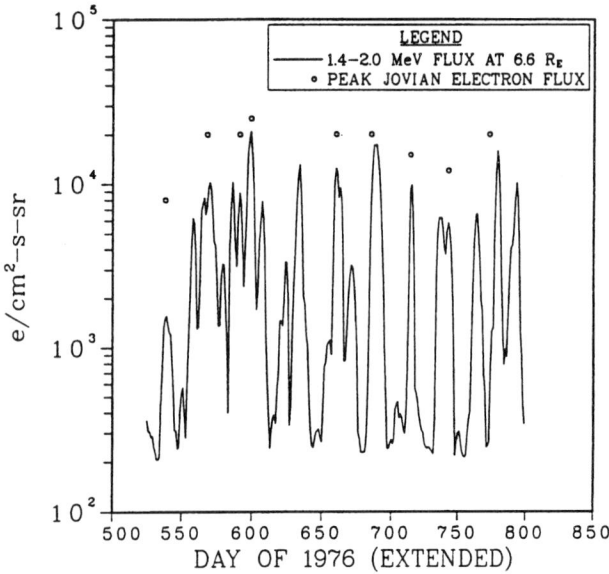

Figure 3. Charged-particle analyzer electron data (1.4–2.0 MeV) for 1977–78 at geostationary orbit. The dots show times that Jovian electron fluxes peaked outside the Earth's magnetosphere (from *Baker et al.* [1979]).

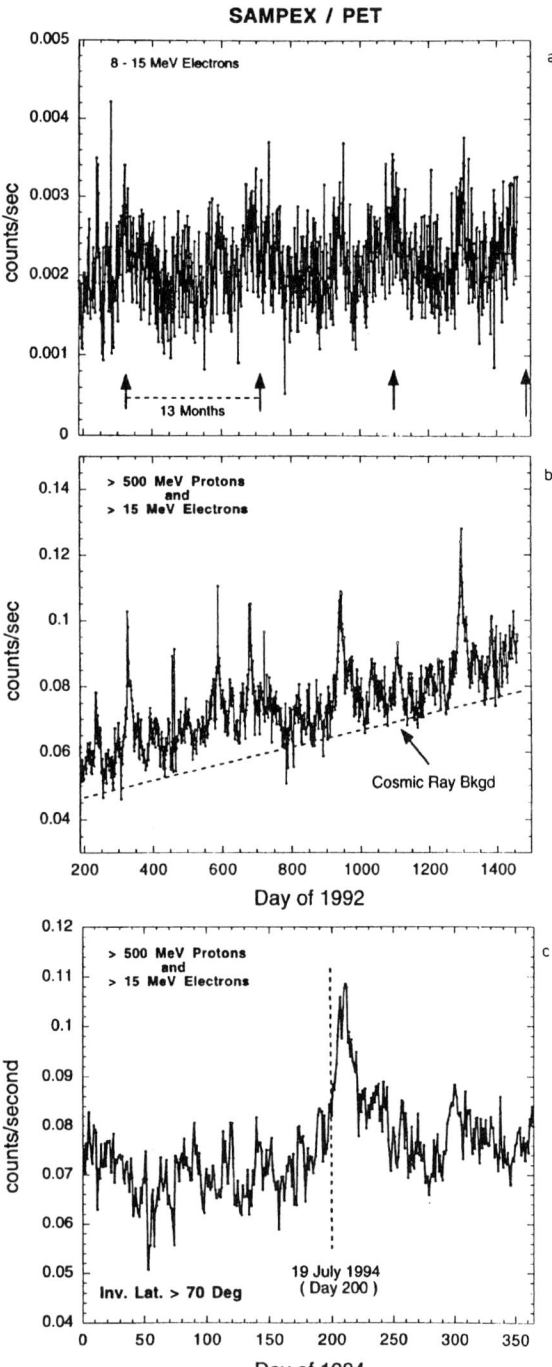

Figure 4. High energy electron fluxes measured by the SAMPEX spacecraft over the polar caps. (a) This figure shows daily averages of 8–15 MeV electrons from July 1992 through 1995 and suggests a 13-month modulation. (b) This figure shows higher energy data that include galactic protons plus > 15 MeV electrons. (c) This figure shows a detail of Figure 4b for 1994. A large flux increase on ~Day 200 (19 July) of 1994 may have been caused by Comet Shoemaker-Levy 9 impact at Jupiter.

flare electron increase.

It is seen by Figure 3 that Jovian electron increases occur, in close temporal association with most of the major high-energy flux peaks at $6.6\,R_E$. When examined on a more detailed basis, it is observed that the Jovian electrons outside the magnetosphere commonly peak 1–3 days prior to the peak seen in electrons at $6.6\,R_E$ inside the magnetosphere. As noted, Jovian electrons have very hard energy spectra ($\gamma \lesssim 2$) by magnetospheric standards and similar spectral hardening occurs during each of the flux peaks in Figure 3 as well.

Based on results such as those shown in Figure 3, *Baker et al.* [1979] sketched an external source scenario. In this model Jovian electrons, controlled by interplanetary solar wind stream structure, appear regularly in the vicinity of Earth. Such Jovian electrons are observed down to energies as low as ~200 keV and they can readily enter the distant plasma sheet where the magnetic field is weak. The Jovian population (distinct because of its very hard spectrum) then becomes part of the plasma sheet population and begins to participate in the overall magnetospheric dynamics. During sunward convection in the plasma sheet, Jovian electrons are moved nearer the Earth; during substorms the Jovian population is convected strongly with plasma sheet particles and is "injected" into the outer radiation region. Inward radial diffusion would also be significant for such particles.

A question addressed by *Baker et al.* [1979] is whether Jovian electron absolute intensities at a given energy outside the magnetosphere are similar to those at geostationary orbit. It was found that $\gtrsim 1$ MeV fluxes at geostationary orbit are of the order of 10^3 times higher than in the interplanetary medium. However, the first adiabatic invariant ($\mu \sim E^2/B$) should be conserved [*Schulz and Lanzerotti*, 1974] as electrons are transported from interplanetary space (or deep in the tail) to $6.6\,R_E$. Thus, the ambient magnetic field will be increased from ~ 5 nT to over 100 nT. This factor of ~ 20 increase in B and conservation of μ (plus the hard spectrum) can account for higher fluxes at geostationary orbit by a factor of about 20. Also, if the characteristic source time is short compared to the characteristic loss time at $6.6\,R_E$, then Jovian fluxes could build up further in the outer zone. Source time scales should be associated primarily with substorm occurrence frequency. Loss time scales should be associated with radial diffusion times and, also, lifetimes against pitch angle scattering into the loss cone. Since several substorms occur, on average, each day, it is not unreasonable to suppose that a buildup could occur. Furthermore, Jovian electrons appear for several days outside the magnetosphere and 10–20 substorms can occur in this period, thus causing a buildup of fluxes in the outer zone.

Based on early work such as described above, it seemed possible that Jovian electrons would play an important role in populating the electron radiation belts. Subsequent work [*Christon et al.*, 1989] suggested that Jovian electrons are not the dominant cause of low-to-moderate energy (0.2–5.0 MeV) electron enhancements at geostationary orbit. Recent SAMPEX data [*Baker et al.*, 1994] also suggests that internal magnetospheric acceleration is the dominant source of typical relativistic electron enhancements in the outer radiation zone. However, it may still be the case that the highest energy trapped electrons are introduced by the Jovian source [*Baker et al.*, 1986].

An interesting illustration of very high-energy electron influence from an external source is shown in Figure 4. Figure 4a shows measurements over the period 1992–95 from the P4–P7 sensors of the PET telescope of SAMPEX (*Looper et al.* [1994] and references therein). The data are daily averages for times when SAMPEX was at invariant latitudes above $70°$. These data correspond, therefore, to polar cap fluxes and are a measure of ~ 8–15 MeV electrons essentially in the interplanetary medium. Although statistical fluctuations are large, the data suggest both by the highest maximum count rates and by the trends in the minimum count rates that there is a systematic modulation of the fluxes with something over a 1-year period. We show by the vertical arrows periods separated by 13 months. This 13-month period looks very consistent with the modulation seen in Figure 4a and supports the view that Jovian electrons are seen by SAMPEX over the polar caps.

An even higher PET energy range for electrons (which also includes galactic cosmic ray protons >500 MeV) is shown in Figure 4b. The period covered is the same as Figure 4a and is again for latitudes $>70°$. The solar cycle trend in the background level (indicated by the dashed line) is consistent with galactic cosmic ray modulation. Of interest are the long-lasting events extending well above the background level (for example, around Day 325, Day 590, Day 925, and Day 1275). Other SAMPEX data suggest that these are not proton enhancements and thus they are most likely very energetic electron events. During the period 16–22 July 1994, fragments of Comet Shoemaker-Levy 9 plunged into the upper atmosphere of Jupiter. Numerous scientific reports have described the visible, IR, and UV wavelength observations of these huge impacts. Recent papers have also discussed extreme ultraviolet, X-ray, and radio signatures detected at Earth in association with the S/L 9 impacts (e.g., *dePater et al.* [1995]; *Waite et al.* [1995]). The radio emissions, in particular, suggest that magnetospheric electrons at Jupiter in the energy range 1–300 MeV were significantly perturbed by the comet fragments [*dePater et al.*, 1995].

In Figure 4c we expand a portion of the SAMPEX record and we examine the high energy interplanetary electron flux variations. We find evidence that energetic electrons increased substantially in flux beginning about 19 July 1994 ($\sim$Day 200). The clearest increase was for $\gtrsim 5$ MeV electrons which remained elevated in flux for ~ 2 months. The spectrum of these electrons was apparently quite hard since, as seen in Figure 4c, there was a very clear flux enhancement in electrons with $E>15$ MeV. This is consistent with the hard energy spectrum seen previously for Jovian electrons. Jupiter and Earth were not ideally connected via the nominal interplanetary magnetic field in July 1994. Therefore, an even stronger signal might have been seen if ideal magnetic connection had obtained.

We continue to examine high-energy electron data with SAMPEX and other magnetospheric spacecraft in order to understand the acceleration and transport processes that ultimately populate the highly variable outer zone. The large peak in July 1994 is suggestive of a S/L 9 effect, but the question remains as to what caused the other large peaks in Figure 4b.

4. SOLAR ENERGETIC ELECTRON SOURCE

As noted above, it has been well established that energetic protons and other ions from the sun can penetrate into the terrestrial magnetosphere [*Scholer*, 1979; *Fennell*, 1973]. Some portion of these ions can be trapped and can constitute a reasonably persistent (days to weeks) component of the magnetospheric particle environment. A question arises as to whether or not solar energetic electrons are also a significant source of outer zone electron flux enhancements.

In an earlier study, *Baker et al.* [1986] examined data from IMP-8 and other available upstream spacecraft and compared energetic electron measurements with concurrent data of similar energy at geostationary orbit. It was concluded that for electrons with $E \gtrsim 0.1$ MeV, there was not a one-to-one relationship between solar electron events and geostationary electron enhancements. In fact, in most cases there were no large solar electron events that could be associated with geostationary recurrent flux events.

New measurements from SAMPEX allow a further examination of this question of solar electron entry. In Figure 5a, we show SAMPEX data from January–February 1994. The measurements are for the northern polar cap region with a selection criterion applied such that magnetic latitude is greater than $70°$. In the open magnetosphere model, this criterion would suggest that at such high latitudes SAMPEX was sampling essentially interplanetary-connected field lines. Thus, solar particles would have rather direct access to this region (e.g., *Fennell* [1973]).

A large solar energetic particle event was observed on 20–26 February 1994 (see *Baker et al.* [1995]). As seen in Figure 5a, the 2–6 MeV electron flux measured by the Proton-Electron Telescope (PET) onboard SAMPEX rose rapidly above background levels on ~ 21 February and intensities

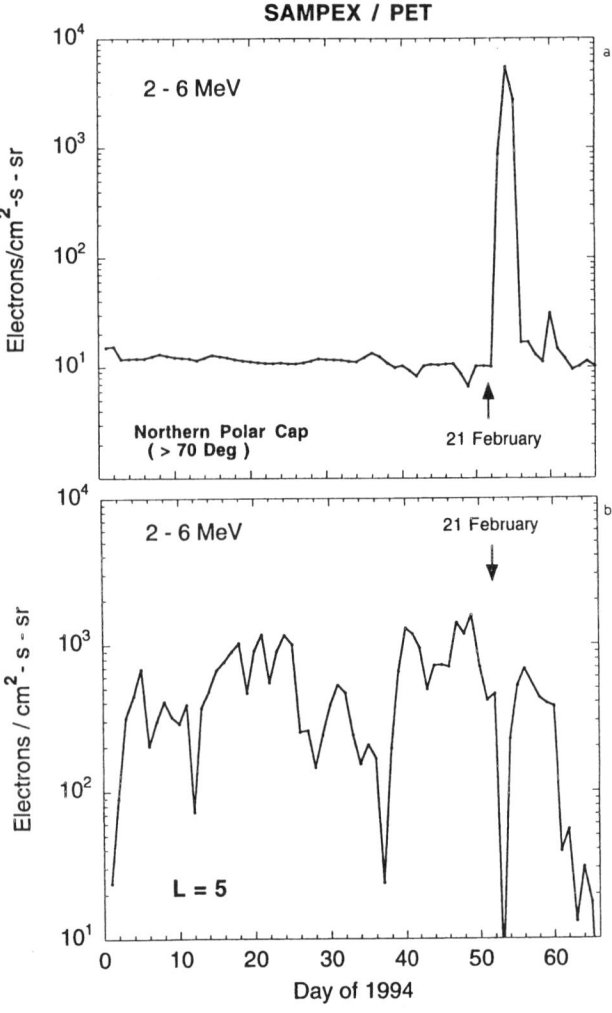

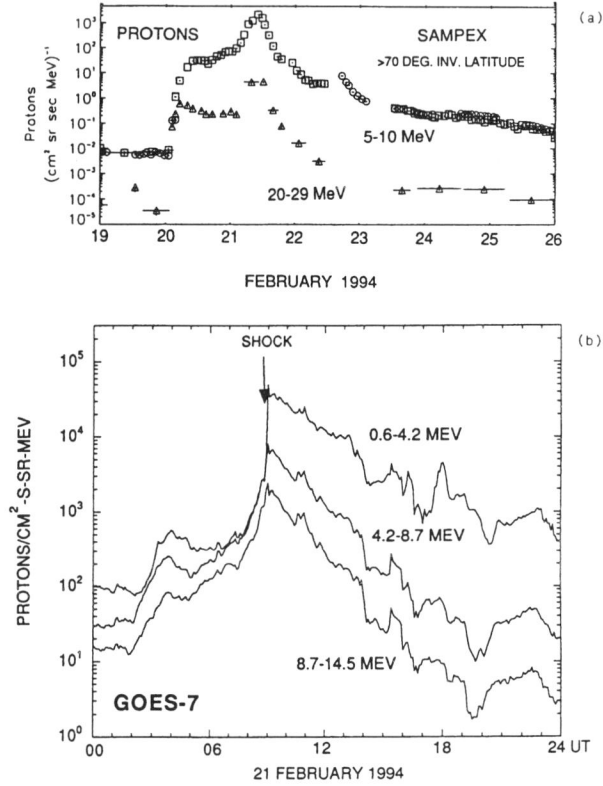

Figure 5. (a) Polar cap electron fluxes (2–6 MeV) measured by SAMPEX during Jan.–Feb. 1994. A solar energetic electron event commenced on 21 February 1994. (b) Same as (a) but showing trapped electron fluxes measured at lower latitudes for $L = 5$. Electron intensities decrease rapidly on 21 February.

Figure 6. (a) Summary of proton fluxes for the period 19–25 February 1994 as measured over the polar caps by SAMPEX sensors. (b) Data from GOES-7 at geostationary orbit for the period 0000-2400 UT on 21 February 1994. (From *Baker et al.* [1995]).

remained high for several days. These data suggest a strong solar electron event in this case.

In Figure 5b, we show measurements of electron fluxes again in the 2-6 MeV channel of SAMPEX for the same period of time as in Figure 5a. However, in Figure 5b we show data for the trapping region at $L = 5$. In January and early in February the figure shows that there were several large increases (and decreases) in the relativistic electron flux in the outer zone. Previous papers [e.g., *Baker et al.* [1994]] showed that these electron enhancements were driven by high speed solar wind streams hitting the magnetosphere. However, on 21 February—when the polar cap fluxes were increasing dramatically—the 2–6 MeV electron flux at $L = 5$ actually dropped precipitously. Thus, despite there being strong and rapid access of solar electrons to the polar cap in this case, such electrons did not (at least not immediately) enter into the outer zone trapping region. Thus, it appears that solar energetic electrons do not constitute a prompt or dominant source of outer zone electrons. By the same token, during all of January and early February 1994, it appears from Figure 5 that there were numerous large changes in the outer zone ($L = 5$) electron fluxes (Day 14–24 and Day 38–49) that had no counterpart in the polar cap flux. These results suggest that at these times neither solar nor Jovian electrons were the dominant cause of magnetospheric trapped electron events.

In some contrast we see clear evidence that, for the February 1994 event, solar protons had rather ready access to the outer portion of the Earth's radiation belt. In Figure 6a we show high latitude proton measurements from SAMPEX in the energy ranges 5–10 MeV and 20–29 MeV. The period shown is 19–25 February. As described by *Baker et al.* [1995], the proton flux seen by SAMPEX above magnetic latitude 70° jumped up early on 20 February and then a further large increase occurred on 21 February when a strong shock wave struck the magnetosphere. Figure 6b shows data from the GOES-7 spacecraft at geostationary orbit for 21 February. Energetic proton fluxes were elevated on 20 February at GOES-7 [R. Zwickl, private communication], but the flux at geostationary orbit increased very sharply and substantially at ~0900 UT as the interplanetary shock wave passed through the magnetosphere [*Baker et al.*, 1995]. Thus, both SAMPEX and GOES-7 saw similar proton signatures at about the same time. Hence, solar and interplanetary ions can constitute a prompt outer magnetosphere contribution, but solar electrons seem not to be such a strong source.

5. SUMMARY

Long-term data show a high coherence of relativistic electron flux variations throughout the entire outer radiation zone (*Baker et al.* [1994] and references therein). It is seen that strong high-energy electron modulation occurs on 27-day time scales. Some influence of Jovian electrons is possible (especially at the highest energies). There is little evidence for a direct solar energetic electron source within the radiation belts. Rather, there is ample evidence that outer zone relativistic electrons are acclerated within the magnetosphere on relatively short time scales ($\lesssim 1$ day). Such acceleration is driven by the impact of high-speed solar wind streams on the magnetosphere. On the other hand, there is a direct evidence that solar energetic ions can penetrate deeply into the Earth's outer magnetospheric regions, thus allowing flare and shock-generated ions to be present for many days after solar ion events commence.

Acknowledgements. This work was supported by NASA grants through the SAMPEX program. We thank X. Li for useful discussions.

REFERENCES

Baker, D.N. and J.A. Van Allen, Energetic electrons in the Jovian magnetosphere, *J. Geophys. Res., 81*, 617, 1976.

Baker, D.N., P.R. Higbie, R.D. Belian and E.W. Hones, Jr., Do Jovian electrons influence the terrestrial outer radiation zone? *Geophys. Res. Lett., 6*, 531, 1979.

Baker, D.N., J.B. Blake, R.W. Klebesadel and P.R. Higbie, Highly relativistic electrons in the Earth's outer magnetosphere, I. Lifetimes and temporal history 1979–1984, *J. Geophys. Res., 91*, 4265, 1986.

Baker, D.N., J.B. Blake, S. Kanekal, B. Klecker and G. Rostoker, Satellite anomalies linked to electron increase in the magnetosphere, *EOS, 75*, 401, 1994.

Baker, D.N., S. Kanekal, J.B. Blake and J.H. Adams, Jr., Charged-Particle Telescope on Clementine, *J. Spacecraft Rockets, 32*, 1060, 1995.

Blake, J.B. and L.M. Friesen, A technique to determine the charge state of the anomalous low energy cosmic rays, 15 *Int. Cosm. Ray Conf., 2*, 341, 1977.

Chenette, D.L., T.F. Conlon, K.R. Pyle and J.A. Simpson, Observations of Jovian electrons at 1 AU throughout the 13-month Jovian synodic year, *Ap. J. (Letters), 215*, L 95, 1977.

Christon, S.P., D.L. Chenette, D.N. Baker and D. Moses, Relativistic electrons at geosynchronous orbit, interplanetary electron flux, and the 13-month Jovian synodic year, *Geophys. Res. Lett., 16*, 1129, 1989.

Conlon, T.F., The interplanetary modulation and transport of Jovian electrons, *J. Geophys. Res., 83*, 541, 1978.

Cowley, S.W.H., Plasma populations in the simple open model magnetosphere, *Space Sci. Rev., 26*, 217, 1980.

Cummings, J.R., A.C. Cummings, R.A. Mewaldt, R.S. Selesnick, E.C. Stone and T.T. vonRosenvinge, New evidence for anomalous cosmic rays trapped in the magnetosphere, *Geophys. Res. Lett., 20*, 2003–2006, 1993.

dePater, I., et al., Outburst of Jupiter's synchrotron radiation after the impact of comet Shoemaker-Levy 9, *Science, 268*, 1879, 1995.

Fennell, J.F., Access of solar protons to the Earth's polar cap, *J. Geophys. Res., 78*, 1036, 1973.

Fisk, L.A., Increases in the low-energy cosmic ray intensities at the front of propagating interplanetary shock waves, *J. Geophys. Res., 76*, 1662, 1971.

Gloeckler, G., Characteristics of Solar and Heliospheric Ion Populations Observed Near Earth, *Adv. Space Res., 4*, 127–137, 1984.

Gloeckler, G. and D.C. Hamilton, AMPTE ion composition results, *Phys. Scr., T17*, 73, 1987.

Krimigis, S.M., E.T. Sarris and T.P. Armstrong, Observations of Jovian electron events in the vicinity of Earth, *Geophys. Res. Lett., 2*, 561, 1975.

Looper, M.D., J.B. Blake, R.A. Mewaldt, J.R. Cummings and D.N. Baker, Observations of the remnants of the ultrarelativistic electrons injected by the strong SSC of 24 March 1991, *Geophys. Res. Lett., 21*, 2079, 1994.

Mason, G.M., J.E. Mazur and D.C. Hamilton, Heavy ion isotopic anomalies in 3He-rich solar particle events, *Astrophys. J., 425*, 843–848, 1994.

Mewaldt, R.A., E.C. Stone and R.E. Vogt, Observations of Jovian electrons at 1 AU, *J. Geophys. Res., 81*, 2397, 1976.

Scholer, M., Energetic solar particle behavior in the magnetosphere, in *Solar-Terrestrial Predictions Proceedings, 2*, 446, edited by R.F. Donnelly, NOAA, Boulder, CO, 1979.

Scholer, M., Diffuse acceleration, in Collisionless Shocks in the Heliosphere: Reviews of Current Research, *Geophys. Monogr. Ser., 35*, edited by B.T. Tsurutani and R.G. Stone, pp. 287–301, AGU, Washington, DC, 1985.

Schulz, M. and L.J. Lanzerotti, *Particle Diffusion in the Radiation Belts*, Springer, New York, 1974.

Teegarden, B.J., F.B. McDonald, J.H. Trainor, W.R. Webber and E.C. Roelof, Interplanetary MeV electrons of Jovian origin, *J. Geophys. Res., 79*, 3615, 1974.

Waite, J.H., Jr., et al., ROSAT observations of x-ray emissions from Jupiter during the impact of Comet Shoemaker-Levy 9, *Science, 268*, 1598, 1995.

Zwickl, R.D. and J. Kunches, *EOS, 70*, 1258, 1989 (abstract).

D.N. Baker, LASP, University of Colorado, Campus Box 392, Boulder, CO 80309-0392

S.G. Kanekal, NASA/Goddard Space Flight Center, Greenbelt, MD, USA

M.D. Looper and J.B. Blake, Aerospace Corp., Los Angeles, CA, USA

R.A. Mewaldt, California Inst. of Tech., Pasadena, CA, USA

MHD/Particle Simulations of Radiation Belt Formation During a Storm Sudden Commencement

M.K. Hudson[1], S.R. Elkington[1], J.G. Lyon[1], V.A. Marchenko[1], I. Roth[2], M. Temerin[2], M.S. Gussenhoven[3]

MHD fields from a global 3D simulation of the March 24, 1991 storm sudden commencement are used to push particles in a guiding center test particle simulation of radiation belt formation during this event. This simulation extends our previous model to include oscillations evident in the electric and magnetic field data from the CRRES satellite following the SSC. Formation of a double (in L) proton belt results, and the new electron belt extends earthward of the new proton belt, consistent with CRRES observations.

1. INTRODUCTION

The rapid formation of new radiation belts on a particle drift time scale due to storm sudden commencement (SSC) induced electric fields was well documented by observations from the CRRES satellite during the March 24, 1991 geomagnetic storm [*Blake et al.*, 1992a; *Vampola and Korth*, 1993]. A model for this rapid acceleration has been applied to electrons [*Li et al.*, 1993] and protons [*Hudson et al.*, 1995], reproducing many features of the new radiation belts formed around $L \simeq 2.5$ during that event, e.g. drift echoes associated with acceleration over a limited range of longitude, a peaked energy spectrum for electrons around 15 MeV and a broad flat spectrum of protons from 20–40 MeV at the observed L values, within a few drift periods of the injection. CRRES was fortuitously located at the inner edge of the new belt as it formed, then it proceeded inward to perigee and back out through a drastically altered radiation belt environment on the next orbit, see Figure 1, which shows flux vs. energy and L shell during two consecutive half-orbits, accumulated over five hours each, during which the new radiation belts were observed to have formed. Figure 1a shows the primary source population for the new proton belt [*Hudson et al.*, 1995], namely solar protons which penetrate into $L = 4$ prior to

[1] Physics and Astronomy Department, Dartmouth College, Hanover, New Hampshire
[2] Space Sciences Laboratory, University of California, Berkeley, California
[3] Phillips Laboratory, Hanscom AFB, Massachusetts

Radiation Belts: Models and Standards
Geophysical Monograph 97
Copyright 1996 by the American Geophysical Union

the arrival of the interplanetary shock at the magnetopause [*Gussenhoven et al.*, 1994], as well as the inner edge of the new proton belt as it forms, seen at $L \simeq 2.5$ with a broadly peaked energy flux around 25 MeV. A double belted structure, in addition to the inner zone, is seen in Figure 1b on the outbound pass, as well as movement of the solar proton boundary to lower L values relative to Figure 1a. The new proton and electron belts which formed during this event persisted past the end of the CRRES mission six months later [*Blake et al.*, 1992b; *Looper et al.*, 1994].

The induction electric field associated with shock compression of the dayside magnetosphere is responsible for the rise in fluxes by several orders of magnitude on a drift time scale. The analytic form assumed for that electric field in the model of *Li et al.* [1993] is :

$$\mathbf{E}_w = \hat{e}_\phi E_0 \left[1 + c_1 \cos(\phi - \phi_0)\right] \left[e^{-\xi^2} - c_2 e^{-\eta^2}\right] \quad (1)$$

where $\hat{e}_\phi$ points westward in the convention used by *Lyon et al.* [1995]. Faraday's law determines $\mathbf{B}_w(\xi)$, which is added to a background dipole magnetic field in the analytic model. The two exponentials in Eq. (1) describe inbound and outbound pulses which superimpose to approximate compression and relaxation of the magnetosphere. In Eq. (1) the phases of the inbound and outbound pulses are $\xi = [r + v_0(t - t_{\rm ph})]/d$ and $\eta = [r - v_0(t - t_{\rm ph} + t_{\rm d})]/d$, respectively, where v_0 denotes the pulse propagation speed and d determines its width. $c_1(>0)$ describes the local time dependence of the electric field amplitude, which is largest at azimuthal impact angle ϕ_0; $t_{\rm ph} = t_{\rm i} + (c_3 R_E/v_0)[1 - \cos(\phi - \phi_0)]$ represents the longitudinal time delay of the pulse from ϕ_0 to other local times where c_3 determines the magnitude of the delay; c_2 determines the partial reflection of the pulse, $t_{\rm d} = 2.1\,R_{\rm E}/v_0$ indicates that the reflection occurs at $r = 1.05\,R_{\rm E}$ in earth radii and $t_{\rm i}$ is the initial reference time. At $t = 0$ the pulse

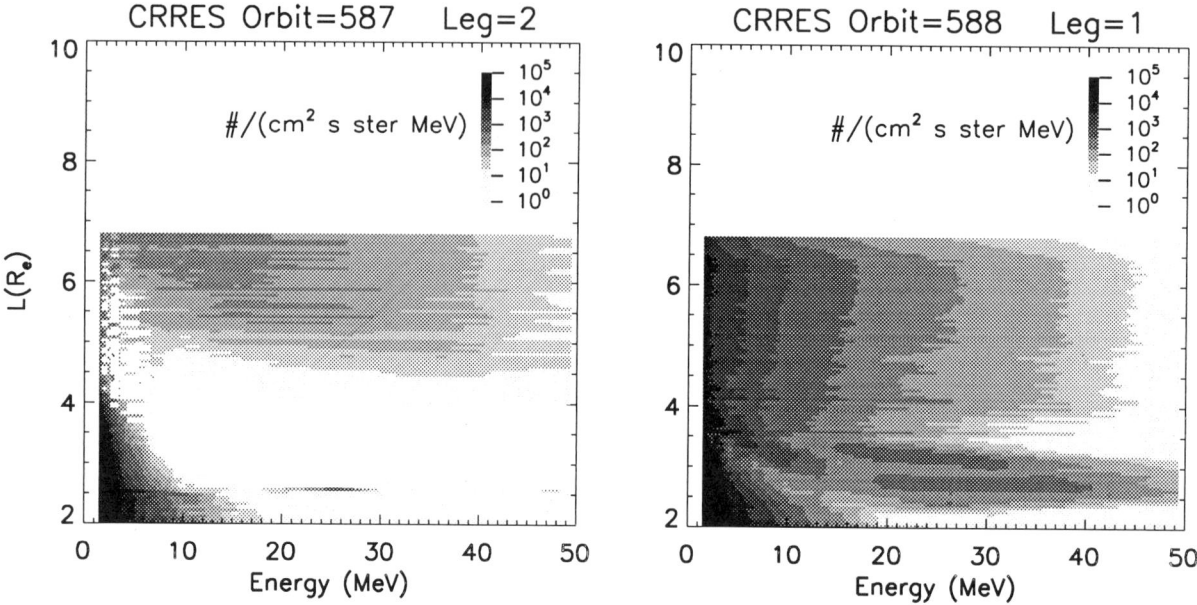

Figure 1. a) Proton differential number flux at 90° pitch angle, measured by Protel on CRRES [*Violet et al.*, 1993] and interpolated between discrete energy channels, for half-orbit in which SSC occurred. b) Same format as a) for outbound half-orbit of CRRES following March 24, 1991 SSC.

maximum is at about 25 R_E at ϕ_0, and it reaches 10, 1.05 and 10 R_E again at 48, 76 and 105 seconds, respectively. The pulse is assumed to strike the magnetopause at 1500 MLT, consistent with the arrival times of newly accelerated electrons and protons observed at CRRES [*Blake et al.*, 1992a]. Pulse parameters used by *Li et al.* [1993] for electrons and *Hudson et al.* [1995] for protons are: $E_0 = 240\,\text{mV/m}$, $c_1 = 0.8$, $c_2 = 0.8$, $c_3 = 8.0$, $v_0 = 2000\,\text{km/s}$, $t_i = 80\,\text{s}$, $\phi_0 = 45°$ and $d = 30,000\,\text{km}$.

A maximum amplitude $E_0 = 240\,\text{mV/m}$ was assumed, based on extrapolation of the measured electric field on the nightside at the location of the CRRES satellite [*Wygant et al.*, 1994], and their suggestion that this magnitude was required to explain particle energy gain by tens of MeV based on the path integral of $q\mathbf{E} \cdot \mathbf{l}$ from the day to nightside, of outer zone electrons, e.g. initially in the 1–9 MeV range. This model, including $\mathbf{B}_w$, reproduced the electron drift echoes reported by *Blake et al.* [1992a] in remarkable detail, while *Hudson et al.* [1995] reproduced observed proton drift echoes [*Blake et al.*, [1992a] using the same field model and a combination of solar and inner zone proton source populations.

It is the purpose of this paper to present first results for protons and electrons using output from a global MHD simulation of this shock event [*Lyon et al.*, 1994]. The discontinuity in the solar wind was modelled as a shock propagating from a direction 20° duskward of noon with a postshock solar wind speed of 1400 km/s [*Shea and Smart*, 1993], density of 20 cm^{-3} and IMF $B_z = -20\,\text{nT}$ assumed. No in situ upstream measurement of solar wind parameters was available for this event, however the assumed speed is bracketed by the delay time between energetic particle and shock arrival at the earth [*Blake et al.*, 1992a], 2000 km/s, and shock speed observed at Ulysses at 2.5 AU during this event, 1000 km/s [*Phillips et al.*, 1992]. A $K_p = 3$ model plasmasphere was assumed [*Moore et al.*, 1987], and the inner boundary of the MHD simulation domain was taken to be at $L = 1.8$. Fields from the MHD simulation are interpolated to particle locations in each time step, and used to push particles in the equatorial plane, as in the analytic model. The particle pushing code is otherwise the same: a relativistic guiding center code which assumes conservation of the first adiabatic invariant.

Unlike the analytic model, the MHD code extends to a magnetopause where **B** reverses for southward IMF. However, the particle acceleration occurs in the magnetosonic pulse which moves inward ahead of the solar wind shock compression of the magnetopause. This results from the greater magnetosonic speed within the outer dayside magnetosphere ($\sim 2000\,\text{km/s}$) than shock propagation speed in the solar wind ($\sim 1400\,\text{km/s}$). Thus, there is a sharp rise in **B** and associated induction **E** where most of the acceleration occurs, and particle behavior remains adiabatic. Those particles which find themselves in a $\mathbf{B} \sim 0$ region where their behavior would be nonadiabatic are removed from the simulation. These particles would make no significant contribution to the flux at low L values where the new belts form.

2. MHD AND PARTICLE DRIFT SIMULATIONS

Figure 2 shows the azimuthal electric field (positive eastward) from MHD simulations a) at 14 MLT at $L = 4.5$ and b) at the location of CRRES, at 0300 MLT and $L = 2.5$ [*Lyon et al.*, 1994]. A smaller radial component of the electric field

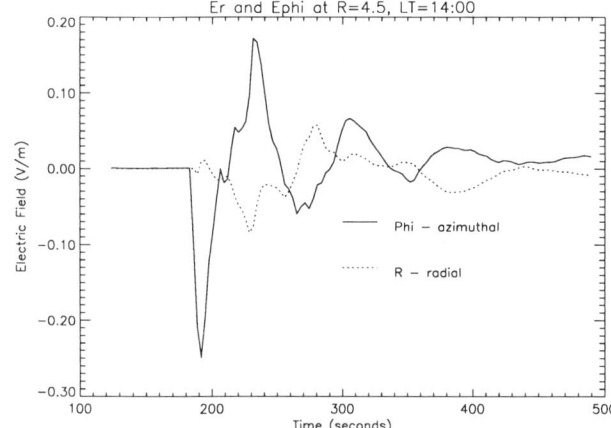

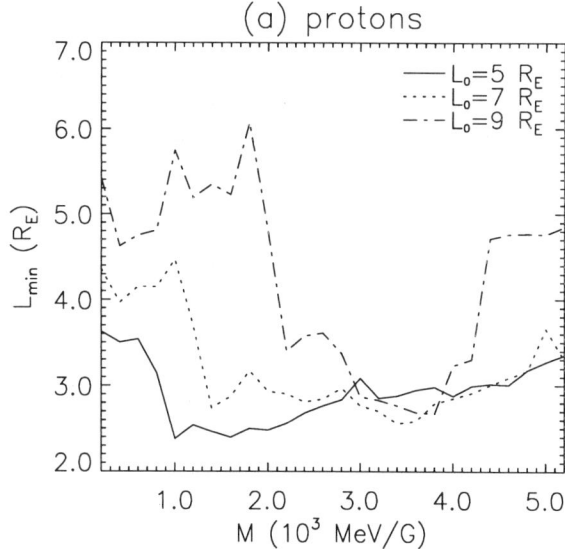

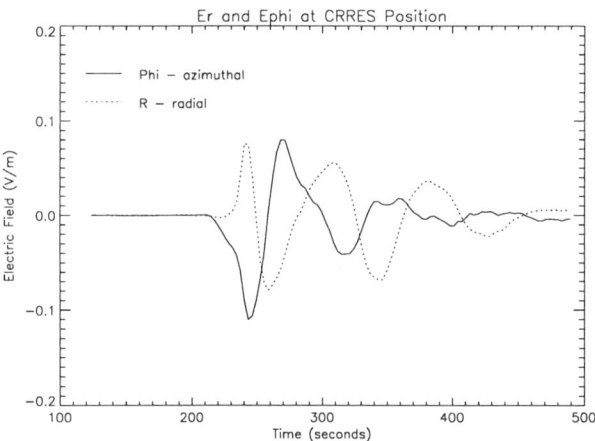

Figure 2. Electric field azimuthal and radial components (positive eastward) from MHD simulations [*Lyon et al.*, 1994] a) at 1400 MLT at $L = 4.5$ and b) at the location of CRRES, at 0300 MLT and $L = 2.5$, when the SSC occurred

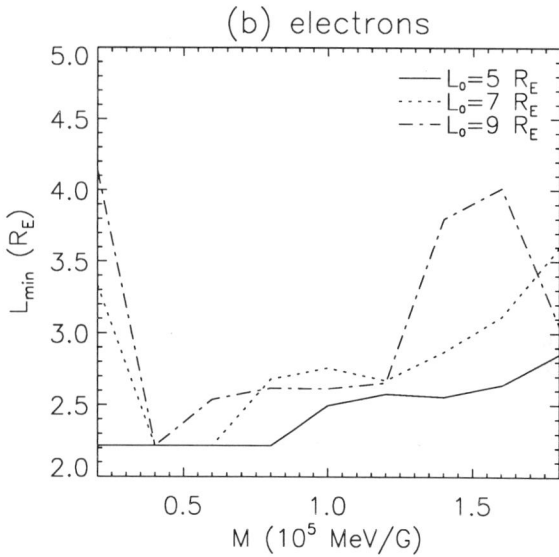

Figure 3. Plot of minimum L value reached by any (a) proton and (b) electron in a ring initially distributed uniformly in longitude, all at the same initial L shell and magnetic moment M.

E_r, not included in the analytic model, is also shown. The dayside electric field amplitude is consistent with the estimate by *Wygant et al.* [1994], while the amplitude at CRRES (and change in magnetic pressure) is large by roughly a factor of two for the MHD simulation parameters used here, relative to CRRES measurement, agreeing with CRRES fields when solar wind velocity is reduced to 1000 km/s. In addition to the bipolar electric field signature included in the analytic model, which corresponds to a compression and partial relaxation of the magnetosphere as seen in the magnetic SSC pulse [*Araki et al.*, 1995], there are subsequent ringing oscillations with a period of ~ 2 minutes as seen by the CRRES magnetometer and electric field instrument [*Wygant et al.*, 1994]. Using the MHD field data to push particles incorporates the additional effects of E_r and the ringing oscillations.

Figure 3 is a plot of the minimum L value reached by any (a) proton or (b) electron in a ring distribution in longitude, all at the same initial L shell and magnetic moment M. There is a single range of optimum M for minimum L (maximum energization) in the electron plot, while the protons have a secondary minimum at lower M and higher L value, besides the primary one around $M = 2$–4×10^3 MeV/G and $L = 2.5$. The minimum L value for electrons occurs at about $L = 2.3$ for $M = 5 \times 10^4$ MeV/G, slightly lower in L than for protons, consistent with the CRRES measurements [*Blake et al.*, 1992a]. The double minimum structure for the protons produces a secondary outer belt, as seen in Figure 1b in the

CRRES data.

The existence of an optimum M or energy range for acceleration demonstrates the resonant nature of the acceleration mechanism. Protons (electrons) which are drifting westward (eastward) at a speed which optimizes the interaction time with the pulse near maximum amplitude undergo the greatest acceleration. The pulse simply pushes cold plasma in and then out again, while very energetic particles which spend little time where the pulse amplitude is large are less affected than intermediate energies [*Li et al.*, 1993; *Ginet et al.*, 1994]. The particle drift includes both gradient **B** and **E** × **B** contributions in the equatorial plane. The electric field component E_ϕ produces inward (and outward) radial motion, while the generally smaller E_r component, which is neglected entirely in the analytic field model, affects drift phase bunching. The gradient **B** drift actually reverses direction where the pulse magnetic field gradient exceeds that of the ambient dipole, and allows more energetic particles, since this is an energy dependent drift, to spend longer in the region where the pulse amplitude is large [*Hudson et al.*, 1996].

A significant point to note in Figure 3 is the collapse of breadth in initial L distribution down to a concentrated range of final L for both protons and electrons undergoing maximum acceleration. This produces the greatly enhanced fluxes observed immediately following the SSC [*Blake et al.*, 1992a], which cannot be accounted for simply by accelerating particles on nearby drift shells, e.g. inside $L = 5$ [*Li et al.*, 1993].

3. FLUX WEIGHTING

Figure 4 shows the (a) input and (b) output flux of protons for the analytic field model simulation after 300 seconds. The relative flux vs. energy and L shell is plotted for 440,000 protons, with a flux accumulation time in a given energy-L bin which exceeds one full drift period for the minimum energy plotted (1.5 MeV), normalized by the drift period at that energy. Thus flux is drift-averaged in this plot. A different source population weighting scheme is used than in *Hudson et al.* [1995], to insure a smoother variation in flux with L. Energy values in the simulation are discretely binned in intervals of 1.5×1.05^n MeV, where n runs from 1 to 60. The weighting scheme is described by three cutoff energies. The inner zone protons are simulated with energy $W < (4.5/L)^5$ in MeV. Inside that energy range, the flux is a W^{-5} power law in energy with a $10^{-.9L}$ radial envelope. The solar protons are characterized by two cutoff energies used to show that a relatively smaller portion of them penetrates to lower L values, while the bulk is concentrated above $L = 5.5$ for energies > 10 MeV. The bulk of solar protons is initialized with $W > (8.5/L)^7$, and the remaining (lower L-penetrating) protons are simulated with $W > (6.5/L)^6$. The difference between the two solar proton fluxes is one order of magnitude. The flux of both solar proton populations depends on W as $W^{-.3}$, as in [*Hudson et al.* [1995]. There is a numerical factor to insure that the inner zone flux is stronger in absolute numbers than the solar proton flux. The total source population, as modelled in Figure 4a, compares well with measured protons prior to the SSC in Figure 1a. In weighting particle flux j (cm^{-2}/s) based on initial energy,

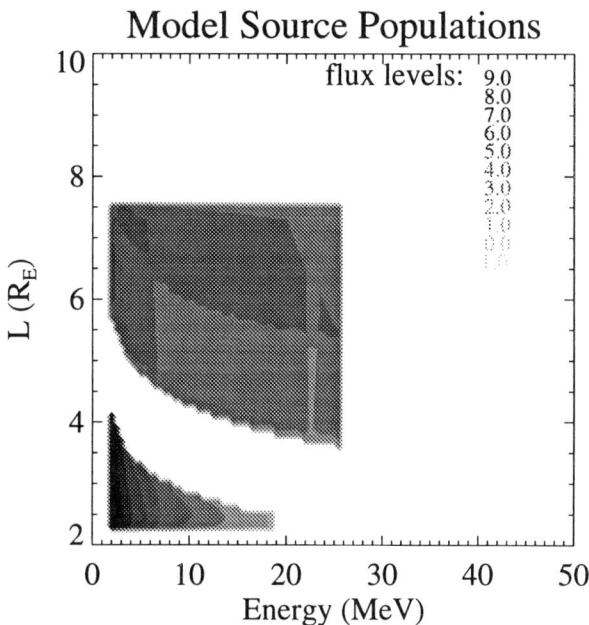

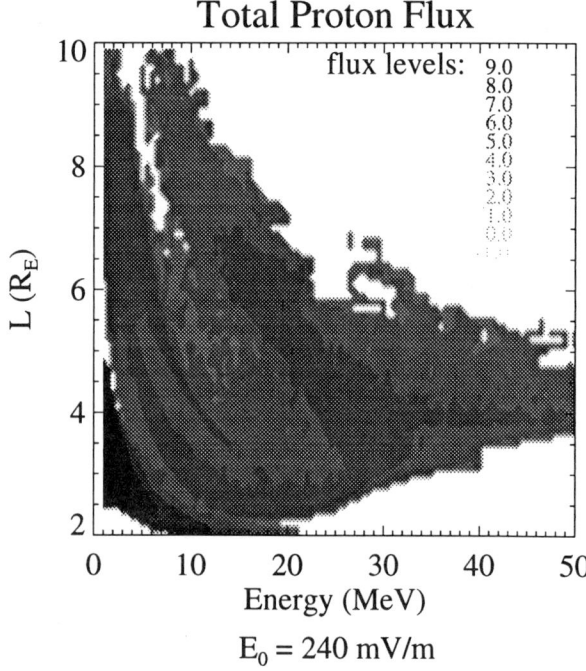

Figure 4. Relative proton flux vs. energy and L shell a) for input source population, compare with Figure 1a, and b) after 300 seconds, averaged over $\Delta L = 0.2$ and $\Delta W = 0.5$ MeV, using analytic field model to push particles.

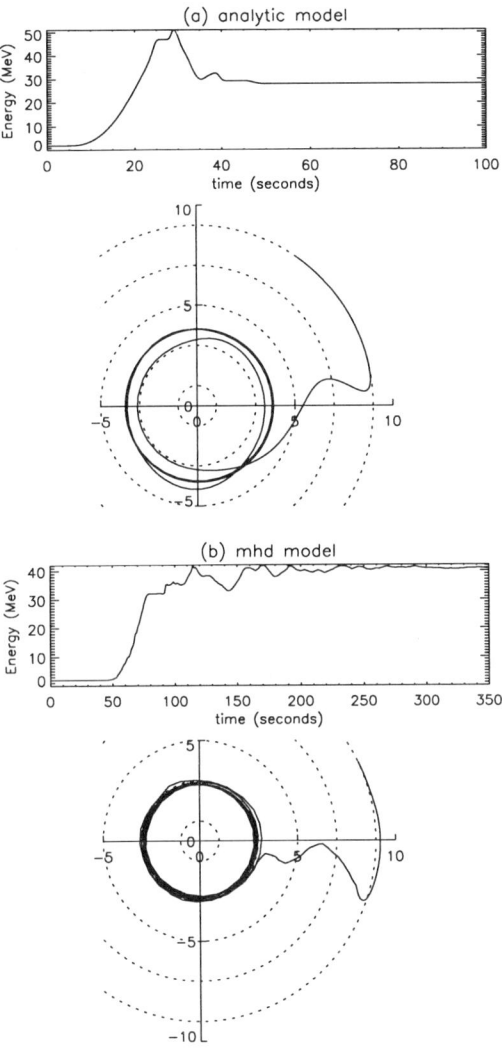

Figure 5. a) Plot of proton energy W and trajectory in the equatorial plane for analytic field model [Hudson et al., 1996], with $M = 4.7 \times 10^3$ MeV/G. b) Same plots for proton with $M = 5.3 \times 10^3$ MeV/G, using MHD fields (same initial energy and L, but equatorial dipole field strength of .275 G vs. .31 G is used in MHD model).

with the preceeding envelope in L included, we assume that the distribution function is conserved along the trajectory, therefore that flux divided by energy is conserved according to Liouville's theorem. The structure in Figure 4b shows a peak in the proton energy spectrum around 25–30 MeV, which is in good agreement with Figure 1a.

4. DISCUSSION

Both analytic and MHD simulation fields have been used to model the formation of new radiation belts during the March 24, 1991 SSC, and each reproduce important features of the CRRES measurements. The energy peak around 25–30 MeV at $L = 2.5$ in Figures 1a and 4b are in good agreement. Protons are transported inward conserving the magnetic moment M, and the downward sweeping arc of constant M is evident in both plots. In the analytic model one sees the reduction in energy of protons first accelerated by the incident pulse, then partially decelerated by the reflected pulse with a different drift phase relationship in the interaction with the two pulses. This deceleration appears as an upward rising plume from $L = 4$–6 and from 18–30 MeV in Figure 4b. For comparison with MHD field results, Figure 5a, from Hudson et al. [1996], follows the energy history of a single proton initialized at $L = 9$ with $W = 2$ MeV, arriving finally at $L = 3.67$ with $W = 28$ MeV after acceleration by the analytic field model. Figure 5b shows the energy history of a proton initialized at $L = 9$ with $W = 2$ MeV using the MHD fields, with very similar behavior to that seen in Figure 5a. However the proton in Figure 5b appears to lose less energy interacting with the oscillatory components of the MHD pulse than the proton in Figure 5a, which loses energy in its interaction with the reflected pulse in the analytic model. A comparison of Figure 4b with the corresponding plot using MHD fields (not shown), indicates that this behavior is seen in aggregate. More energy is retained in the inner magnetosphere in the MHD simulation, which includes a model plasmasphere and, in principle, a better description of partial reflection at gradients in the Alfvén speed than the analytic superposition of an incident and reflected soliton [Eq. (1)].

5. CONCLUSION

The formation of new radiation belts in less than a drift time scale has been simulated in the equatorial plane using field output from a global MHD simulation of the March 24, 1991 SSC event. While the magnitude of this event was remarkable, it was not unique during the lifetime of CRRES (July 1990–October 1991). The techniques described in this paper can be applied to smaller events with similar particle morphology but less radial transport and energization, for which both upstream solar wind and in situ CRRES measurements are available. Simulating global magnetospheric response to input solar wind data on the MHD scale, and use of the resulting fields to push particles in the interior of the magnetosphere, will extend our ability to forecast radiation belt transient response to changing solar wind conditions, and facilitate evaluating the relative importance of the mechanism described here to the average properties of the radiation belt environment.

Acknowledgements. Work at Dartmouth was supported by AFOSR grant F49620-93-1-0101, and at Dartmouth and Berkeley by NASA grant NAG 5-1098, also LANL Contract 6858V0016-3A. Work at the Aerospace Corporation was supported by the Air Force under contract Fo4701-88-C-0089. Computations were performed on the SDSC and PSC Crays.

REFERENCES

Araki, T., S. Fujitani, K. Yumoto, K. Shiokawa, T. Ichinose, H. Luehr, D. Orr, D.K. Milling, H. Singer, G. Rostoker, S. Tsunomura, Y. Yamada and C.F. Liu, The anomalous sudden commencement on March 24, 1991, *J. Geophys. Res.*, in press, 1995.

Blake, J.B., W.A. Kolasinski, R.W. Fillius, and E.G. Mullen, Injection of electrons and protons with energies of tens of MeV into $L < 3$ on March 24, 1991, *Geophys. Res. Lett., 19*, 821, 1992.

Blake, J.B., M.S. Gussenhoven, E.G. Mullen and R.W. Fillius, Identification of an unexpected space radiation hazard, *IEEE Trans. Nuc. Sci., 39*, 1761, 1992.

Ginet, G.P., W.J. Burke, and J. Albert, An analysis of electron energization seen in simulations of the March 24, 1991 SSC, *EOS Trans. Am. Geophys. Union, 75*, 305, 1994.

Gussenhoven, M.S. Gussenhoven, E.G. Mullen and M.D. Violet, Solar particle events as seen on CRRES, *Adv. Space Res., 14*, (10)619, 1994.

Hudson, M.K., A.D. Kotelnikov, X. Li, I. Roth, M. Temerin, J. Wygant, J.B. Blake, and M.S. Gussenhoven, Simulation of proton radiation belt formation during the March 24, 1991 SSC, *Geophys. Res. Lett., 22*, 291, 1995.

Hudson, M.K., A.D. Kotelnikov, X. Li, J.G. Lyon, I. Roth, M. Temerin, J.R. Wygant, J.B. Blake, M.S. Gussenhoven, K. Yumoto and K. Shiokawa, Modelling formation of new radiation belts and response to ULF oscillations following March 24, 1991 SSC, *Taos Workshop, AIP Conference Proceedings*, in press, 1996.

Li, X., I. Roth, I., M. Temerin, J.R. Wygant, M.K. Hudson, and J.B. Blake, Simulation of the prompt energization and transport of radiation belt particles during the March 24, 1991 SSC, *Geophys. Res. Lett., 20*, 2423, 1993.

Lyon, J.G., M.K. Hudson, J.A. Fedder and C.C. Goodrich, Global MHD simulation of the March 24, 1991 SSC, *EOS Trans. Am. Geophys. Union, 75*, 539, 1994.

Moore, T.E., D.L. Gallagher, J.L. Horwitz, and R.H. Comfort, MHD wave breaking in the outer plasmasphere, *Geophys. Res. Lett., 14*, 1007, 1987.

Phillips, J.L., S.J. Bame, J.T. Gosling, D.J. McComas, B.E. Goldstein, E.J. Smith, A. Balogh and R.J. Forsyth, Ulysses plasma observations of coronal mass ejections near 2.5 AU, *Geophys. Res. Lett., 19*, 1239, 1992.

Shea, M.A. and D.F. Smart, March 1991 Solar-terrestrial phenomena and related technolgical consequences, Proceedings of 23rd International Cosmic Ray Conference, Calgary, 1993.

Vampola, A.K., A. Korth, Electron drift echoes in the inner magnetosphere, *Geophys. Res. Lett., 19*, 625, 1993.

Violet, M.D., K. Lynch, R. Redus, K. Riehl, E. Boughan, and C. Hein, Proton telescope (PROTEL on the CRRES spacecraft, *IEEE Trans. Nuc. Sci., 40*, 242, 1993.

Wygant, J.R., F. Mozer, M. Temerin, J.B. Blake, N. Maynard, H. Singer, and M. Smiddy, Large amplitude electric and magnetic field signatures in the inner magnetosphere during injection of 15 Mev electron drift echoes, *Geophys. Res. Lett., 21* 1739, 1994.

M.K. Hudson, S.R. Elkington, A.D. Kotelnikov, J.G. Lyon and V.A. Marchenko, Physics and Astronomy Department, Dartmouth College, Hanover NH 03755

I. Roth and M. Temerin, Space Sciences Laboratory, University of California, Berkeley, CA 94720

M.S. Gussenhoven, Phillips Laboratory, Hanscom AFB, MA 01731

Cross Field Entry of High Charge State Energetic Heavy Ions into the Earth's Magnetosphere

W.N. Spjeldvik

Department of Physics, Weber State University, Ogden, Utah

The sun emits time variable fluxes of energetic ions and electrons over a wide range of energies, charge states and elemental abundance ratios. Generally, the solar particle fluxes come in high charge states corresponding to solar coronal pseudo equilibrium temperatures of millions of Kelvin. Typically helium ions come as He^{2+}, carbon ions as C^{5+}, oxygen ions as O^{6+}, and iron ions as Fe^{12+}. At keV and lower MeV energies, their geomagnetic rigidity is generally unsuitable for direct deep penetration into the magnetosphere, but a very small fraction of these solar energetic particle fluxes stochastically penetrate the magnetosphere in processes describable as diffusive entry. Two conditions favor the effectiveness of this penetration process: (1) enhanced flux intensities in the energetic solar particle emission; and (2) enhanced geomagnetic activity perturbing the outer magnetosphere. Whenever the ambient magnetospheric radial distribution gradient is positive for a given particle species, the stochastic process favors net diffusive penetration from the magnetospheric exterior, but when that gradient is negative, net particle expulsion and outward leakage is expected. This paper presents a computation of the predicted equilibrium structure of radiation belt carbon and iron ions at the geomagnetic equator assuming a specified solar energetic particle flux source acting as a boundary condition and a specified geomagnetic activity as reflected in the cross-field diffusive transport coefficients.

1. AN INTRODUCTORY THOUGHT EXPERIMENT

Let us assume that a certain fraction of the solar particle emission (solar wind and solar energetic particle fluxes) consists of heavy ions. For simplicity, assume that by number density the C/H-ratio is 0.001 and the Fe/H-ratio is 0.000001. In agreement with hydrogen distributions observed by *Christon et al.* [1988], let us further assume that all solar ions entering the Earth's magnetosphere through the magnetopause are phase space distributed in velocity according to a κ function, given by:

$$f_i(v) = f_{i0}\left[1 + v^2/(v_t^2 \kappa)\right]^{-\kappa-1} \quad (1)$$

where i is the particle elemental species, v is the particle velocity, v_t is the representative thermal characteristic velocity, and κ is a characteristic parameter. Following the work of *Christon et al.* [1988] we take $\kappa = 5.5$ and $v_t = 833 \, \text{km/s}$. For i=hydrogen we take f_{i0} to be $10^{-24} \, \text{cm}^{-6}\text{s}^3$, and for i=carbon and i=iron this number is multiplied by the above noted factors. The upper left panel in Figure 1 depicts the assumed distributions for these elemental species, and the panel below shows how these distributions appear as ion number flux versus velocity. Apart from the different shapes of the flux spectra, the relative abundance ratios are preserved.

Most spacecraft instruments sensitive to energetic ions measure the ion energy, and so these distributions and fluxes have been converted to total ion energy distributions in the

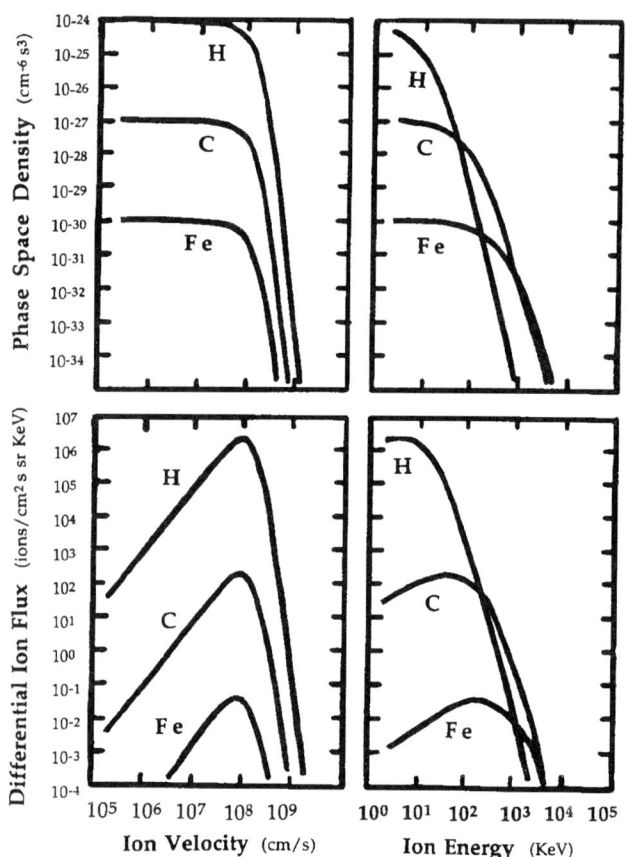

Figure 1. Result of a thought experiment. Left panels show velocity distributions of phase space density and ion fluxes under assumed κ function velocity dependence, and right scales show energy distributions of the same quantities. The results show that even with small relative abundances of heavy ions versus velocity there can be high relative abundances versus total ion energy of the heavy ion component.

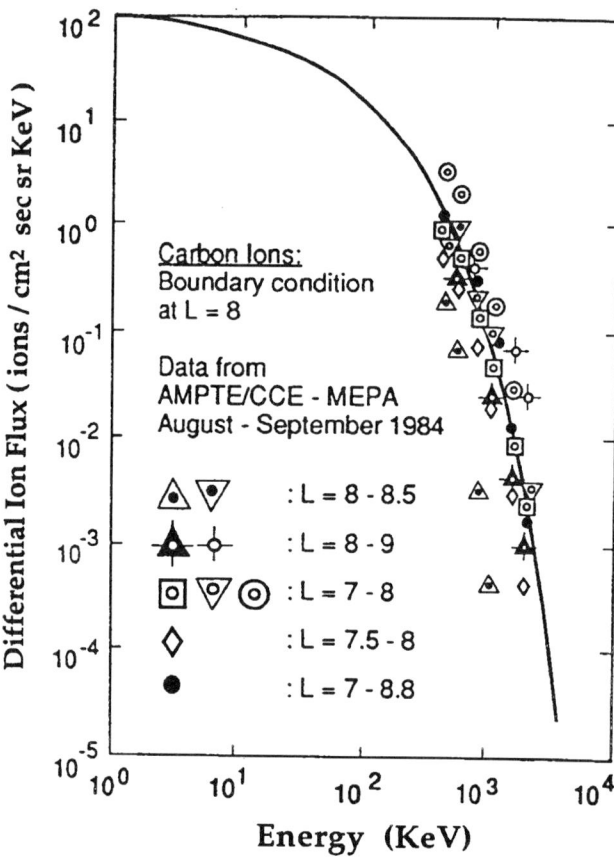

Figure 2. Boundary condition spectra adopted at $L = 8$ based on data from the AMPTE/CCE spacecraft. The carbon ion spectrum has been extrapolated towards lower energies as indicated here.

corresponding right side panels in this figure. Here we note that when compared at equal total ion energy, the phase space densities of carbon and iron ions exceed the phase space density of hydrogen ions at energies in the hundreds of keV per ion and above. Furthermore, the fluxes of the heavy ions are also seen to exceed the fluxes of hydrogen at energies above about 1 MeV (total ion energy).

This is in general agreement with observations, as reported with mass discriminating ion detectors flown on several spacecraft (e.g. *Fritz and Wilken* [1976]; *Panasyuk et al.* [1978]; *Spjeldvik* [1979]; and others). Thus, even if heavy ions are in low relative abundance at the source, by virtue of the higher mass these ions can appear more abundant than hydrogen ions at some energies when injected into the Earth's magnetosphere based on similar velocity distributions in the solar energetic particle emission.

2. APPLICATION OF CLASSICAL DIFFUSION THEORY TO HEAVY IONS

Stochastic processes related to the turbulent magnetic field in the magnetopause and magnetosheath region allow a limited exchange of energetic charged particles between the Earth's magnetosphere and the heliosphere. Outward leakage most likely occurs after the magnetospheric particle content has been temporarily enhanced (e.g. following a magnetic storm or similar event). Inward diffusive leakage should take place at other times when the radial profile of the energetic phase space distribution in the outer magnetosphere is positive.

Once injected into the geomagnetic cavity, these energetic ions distribute themselves according to cross-field diffusion mechanisms and according to the phase space redistribution and loss mechanisms acting in the radiation belt region. The theoretical formalism for the interior magnetospheric transport has been summarized in general by *Schulz and Lanzerotti* [1974] and for oxygen ions in particular by *Spjeldvik and Fritz* [1978]. The reader is referred to these works for mathematical details, and it here suffices to indicate parameters used:

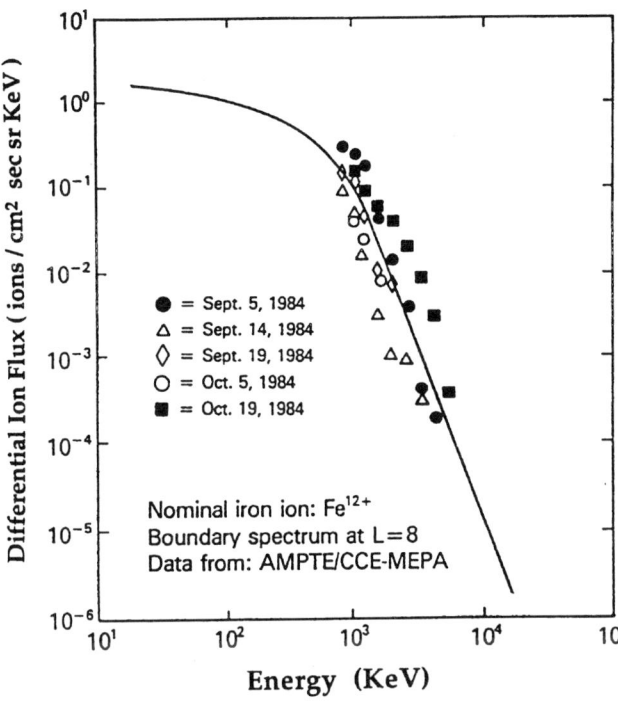

Figure 3. Boundary condition spectra adopted at $L = 8$ based on data from the AMPTE/CCE spacecraft. The iron ion spectrum has been extrapolated towards lower energies.

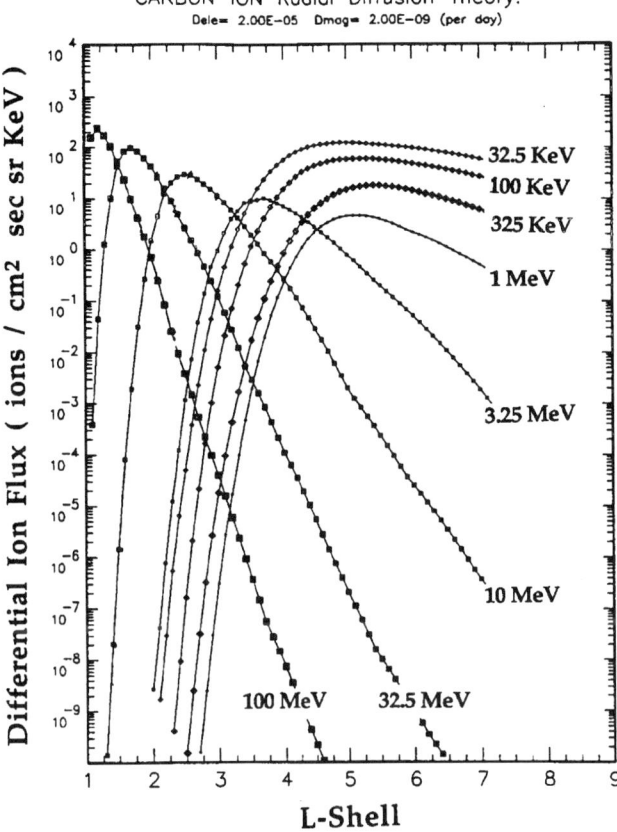

Figure 4. Predicted differential fluxes of radiation belt carbon ions displayed versus geomagnetic L shell for a range of total carbon ion energies from 32.5 keV to 100 MeV.

(a) Effective radial diffusion coefficients are specified as the result of a combination of geoelectric and geomagnetic field fluctuations as specified by *Cornwall* [1972]:

$$D_{LLe} = D_{ele}L^{10} / \left[L^6 + (\mu/\mu_0)^2 Q^{-2}\right] \quad (2)$$

$$D_{LLm} = D_{mag}L^{10} \quad (3)$$

where L = geomagnetic L-shell, μ is the ion magnetic moment, $\mu_0 = 1$ MeV/gauss, Q is the net ionic charge state, and D_{ele} and D_{mag} are the radial diffusion sub-coefficients for electric and magnetic fluctuations respectively. This simple form is arrived at by assuming field fluctuation spectra following an f^{-2} relation [*Falthammar*, 1968], where f is the fluctuation frequency. In this work we have assumed "moderate" geophysical conditions corresponding to the application of "medium" radial diffusion coefficients with $D_{ele} = 2 \times 10^{-5}$ day^{-1} and $D_{mag} = 2 \times 10^{-9}$ day^{-1}.

(b) Outer radiation zone differential flux boundary conditions on carbon and iron ions have been estimated from ion flux data obtained with the AMPTE/CCE spacecraft at $L = 8$, and these are depicted in Figures 2 and 3.

(c) The exospheric environment specification is identical to the plasmaspheric free electron densities and exospheric neutral atomic hydrogen densities utilized by *Spjeldvik* [1979] and correspond to a mean exobase temperature of 950 K and a plasmapause located at $L = 4.1$.

(d) Charge exchange cross sections for collisions between (atomic) carbon ions and neutral atomic hydrogen, and between (atomic) iron ions and neutral atomic hydrogen have been compiled by the author from a survey of the experimental literature. This survey began with a summary by *Dehmel et al.* [1973], it included data given by *Berkner et al.* [1978, 1981], and it added more recent cross section data. Space limitation does not permit an exposition of these cross sections in this paper, but the author aims for a separate paper of the atomic collision aspects. The interested reader may obtain a tabulation of the herein adopted cross sections (extrapolated to cover the range 1 keV to 100 MeV) by direct communication with the author.

3. PREDICTED RADIAL DISTRIBUTIONS OF ATOMIC CARBON AND IRON IONS

Based on these parameters the descriptive radial diffusion equation (e.g. *Spjeldvik and Fritz* [1978]) was solved numerically for carbon and iron ions. Figures 4 and 5 show overall predicted radial distributions of carbon and iron ions computed with the indicated parameters for a range of ion

Table 1. Tabulation of charge state distributions of atomic carbon ions given as percent abundance (by flux) for L shells from 2 through 8 and ionic charge states 1 through 6 at fixed total ion energies: 100 keV, 1 MeV, and 10 MeV.

100 keV L shell	\multicolumn{6}{c}{Carbon Ion Charge State}					
	1	2	3	4	5	6
2	86.2	13.6	0.2	0.0	0.0	0.0
3	82.4	17.4	0.2	0.0	0.0	0.0
4	62.2	30.6	6.2	0.9	0.1	0.0
5	35.1	29.5	18.6	9.6	5.5	1.7
6	16.2	17.4	16.7	16.0	18.7	15.0
7	5.4	6.7	7.4	10.0	21.1	49.4
8	0.0	0.0	0.0	0.0	0.0	100.0

1 MeV L shell	Carbon Ion Charge State					
	1	2	3	4	5	6
2	23.1	58.6	15.6	1.9	0.8	0.0
3	28.5	58.8	11.4	1.3	0.0	0.0
4	30.5	57.8	10.9	0.8	0.0	0.0
5	30.2	56.8	11.9	0.8	0.2	0.1
6	15.7	43.7	22.1	9.4	5.2	3.9
7	19.1	11.4	14.7	15.1	19.3	20.4
8	0.0	0.0	0.0	0.0	0.0	100.0

10 MeV L shell	Carbon Ion Charge State					
	1	2	3	4	5	6
2	0.0	0.0	0.0	0.0	0.9	99.1
3	0.0	0.0	0.0	0.3	11.8	87.9
4	0.0	0.0	0.2	3.9	28.7	67.2
5	0.0	0.0	0.0	0.2	9.9	89.9
6	0.0	0.0	0.0	0.0	0.7	99.3
7	0.0	0.0	0.0	0.0	0.0	100.0
8	0.0	0.0	0.0	0.0	0.0	100.0

Table 2. Tabulation of charge state distributions of atomic iron ions given as percent abundance (by flux) for L shells from 2 through 8 and ionic charge states 1 through 12 at fixed total ion energies: 100 keV, 1 MeV, and 10 MeV.

100 keV L shell	Iron Ion Charge State											
	1	2	3	4	5	6	7	8	9	10	11	12
2	76.3	21.7	1.8	0.2	0.0	0.0	0.0	0.0	0.0	0.0	0.0	0.0
3	69.5	27.3	2.9	0.3	0.0	0.0	0.0	0.0	0.0	0.0	0.0	0.0
4	35.3	28.2	16.4	9.8	5.8	3.1	1.0	0.3	0.1	0.0	0.0	0.0
5	13.2	13.4	11.6	11.1	11.4	11.9	10.0	7.8	5.1	2.9	1.3	0.3
6	4.8	5.3	5.1	5.6	6.6	8.3	9.4	10.7	11.6	12.8	12.5	7.4
7	1.3	1.7	1.7	1.9	2.4	3.2	3.9	5.1	6.9	11.2	21.8	38.9
8	0.0	0.0	0.0	0.0	0.0	0.0	0.0	0.0	0.0	0.0	0.0	100.0

1 MeV L shell	Iron Ion Charge State											
	1	2	3	4	5	6	7	8	9	10	11	12
2	26.1	32.7	25.3	10.5	4.2	1.0	0.2	0.0	0.0	0.0	0.0	0.0
3	26.1	32.4	25.2	10.7	4.4	1.0	0.2	0.0	0.0	0.0	0.0	0.0
4	26.7	32.4	24.7	10.4	4.4	1.1	0.2	0.1	0.0	0.0	0.0	0.0
5	19.8	26.2	23.5	13.9	8.7	4.5	2.2	0.8	0.3	0.1	0.0	0.0
6	8.1	12.8	14.8	12.8	12.1	11.0	10.0	7.7	5.5	3.2	1.6	0.4
7	2.0	3.8	5.1	5.2	5.7	6.6	8.3	9.7	11.8	13.7	16.3	11.8
8	0.0	0.0	0.0	0.0	0.0	0.0	0.0	0.0	0.0	0.0	0.0	100.0

10 MeV L shell	Iron Ion Charge State											
	1	2	3	4	5	6	7	8	9	10	11	12
2	0.0	0.0	0.2	0.9	9.8	14.6	22.3	27.0	17.8	4.1	1.2	1.1
3	0.1	1.9	11.8	25.3	34.6	13.7	5.9	3.4	1.9	1.1	0.3	0.0
4	0.2	2.4	12.7	27.0	34.2	12.5	5.7	2.7	1.5	0.8	0.3	0.0
5	0.2	2.3	12.7	26.2	33.9	13.2	5.5	3.1	1.7	0.9	0.3	0.0
6	0.0	0.7	3.9	8.9	18.2	14.3	13.2	13.8	12.9	9.2	4.1	0.8
7	0.0	0.0	0.0	0.2	0.8	1.6	3.0	6.7	10.5	18.4	29.0	29.8
8	0.0	0.0	0.0	0.0	0.0	0.0	0.0	0.0	0.0	0.0	0.0	100.0

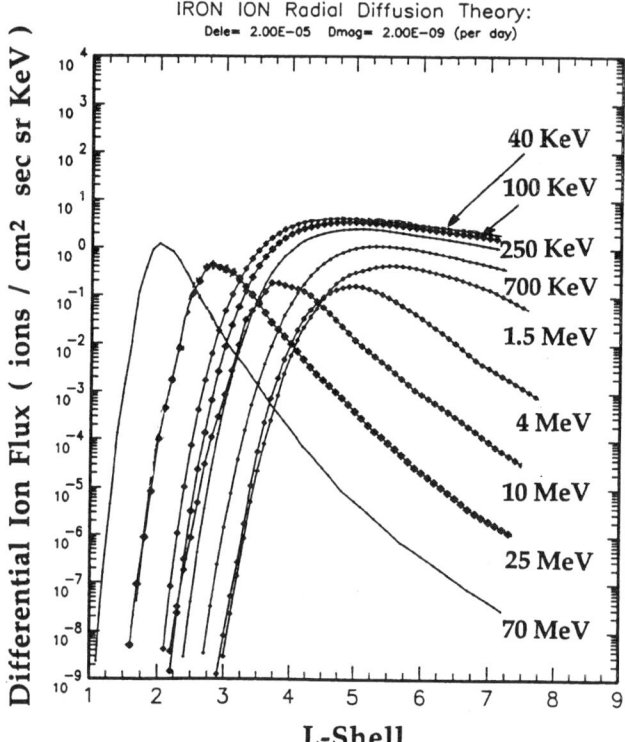

Figure 5. Predicted differential fluxes of radiation belt iron ions displayed versus geomagnetic L shell for a range of total iron ion energies from 40 keV to 70 MeV.

energies. These fluxes are summed over all available ionic charge states, for carbon: $Q = 1$ through 6 and for iron: $Q = 1$ through 12 (with iron ion charge states 13 through 26 disregarded).

Detailed ion charge state distributions were also computed, but present space limitations permit only an example to be presented here (it is intended to give a detailed exposition of these features in a separate paper). Table 1 shows the carbon ion charge state distributions as a display of relative abundance (in percent of the total carbon ion flux at a given energy and L shell) versus L shell for three fixed total ion energies: 100 keV, 1 MeV, and 10 MeV. Table 2 shows the corresponding result for iron ions. As expected, the higher charge states are favored at the higher energies.

4. COMPARISON WITH OBSERVATIONS

There are no currently available charge state distribution observations for radiation belt carbon and iron ions in the hundreds of keV through MeV energy range, so these predictions can not yet be verified or refuted on observational grounds.

It is, however, possible to compare the overall fluxes of iron ions (summed over all available ionic charge states) with available observations made with the MEPA-instrument flown on the AMPTE/CCE spacecraft [McEntire, personal communication, 1992] on L shells well below the outer

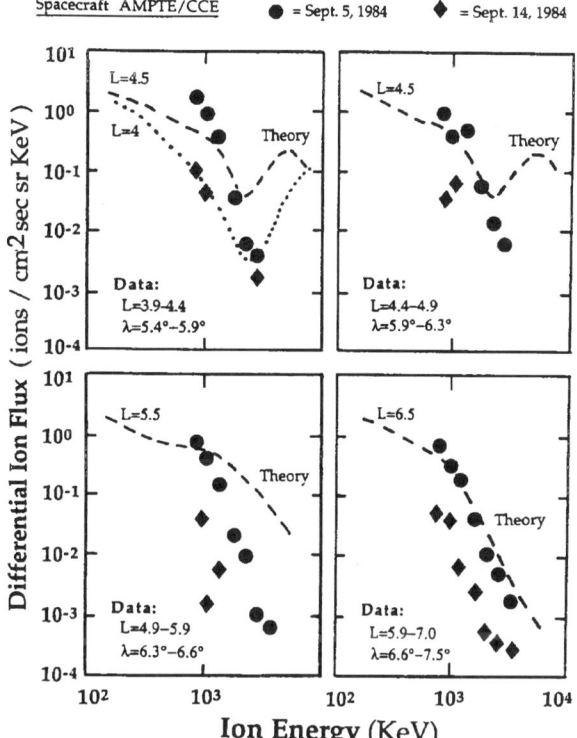

Figure 6. Comparison between equilibrium transport theory and time evolving observation for energetic iron ions, using data from the MEPA instrument on the AMPTE/CCE spacecraft.

carbon ions and around 1 MeV for iron ions. These features are dependent on the model parameters adopted here, particularly the energy dependence of the radial diffusion coefficients and the ion charge exchange cross sections. Further work comparing these predictions with observations extending over a greater energy range is warranted, and in particular charge state discriminating observations are particularly needed.

Acknowledgements. The author expresses his gratitude to Dr. Richard McEntire and the staff of the Space Department at the Applied Physics Laboratory in Maryland for support and access to the AMPTE/CCE spacecraft data, and for many enlightening discussions. The work was supported in part by a travel allocation from Weber State University and in part by NASA research award NAS 5-32915.

boundary (which for computational purposes was placed at $L = 8$ in this modeling effort). Such a comparison is given in Figure 6 and covers two spacecraft passes through the central radiation belt region close to the geomagnetic equator (at geomagnetic latitudes generally less than $7°$) on Sept. 5 and on Sept. 14, 1984. It should be noted that the theory pertains to steady state equilibrium while the data (which is charaterized by low count statistics) are for a time actual period that appeared to consist of a sequence of ion flux injections and subsequent relaxations. Nevertheless, the correspondance is not unfavorable, although of course very restricted in energy coverage.

5. SUMMARY AND CONCLUSIONS

In this paper computed fluxes of carbon and iron ions in the outer radiation zone of the Earth have been presented and compared with available spacecraft observations. Since there appears to be a reasonable agreement between the predicted carbon and iron ion flux distribution and direct observations from the AMPTE/CCE spacecraft, it seems reasonable to assume that the diffusive entry and radial redistribution processes do indeed in some measure describe the high charge state heavy ion component in the Earth's outer radiation zone. The predictions indicate interesting spectral features, including spectral depressions centered at a few hundred KeV for

REFERENCES

Berkner, K.H., W.G. Graham, R.V. Pyle, A.S. Schlarchter, J.W. Stearns and R.E. Olson, Electron-Capture and Impact-Ionization Cross Sections for Partially Stripped Iron ions Colliding with Atomic and Molecular Hydrogen, *J. Phys., B11*, 875–885, 1978.

Berkner, K.H., W.G. Graham, R.V. Pyle, A.S. Schlarchter and J.W. Stearns, Electron Capture, Electron Loss and Impact Ionization Cross Sections for 103- to 3400 KeV/AMU Multicharged Iron Ions Colliding with Molecular Hydrogen, *Phys. Rev., A23*, 2891–2904, 1981.

Christon, S.P., D.G. Mitchell, D.J. Williams, L.A. Frank, C.Y. Huang and T.E. Eastman, Energy Spectra of Plasma Sheet Ions and Electrons from 50 eV/e to 1 MeV During Plasma Temperature Transitions, *J. Geophys. Res., 93*, 2562–2572, 1988.

Cornwall, J.M., Radial Diffusion of Ionized Helium and Protons: A Probe for Magnetospheric Dynamics, *J. Geophys. Res., 77*, 1756, 1972.

Dehmel, R.C., H.C. Chau and H.H. Fleichmann, Experimental Stripping Cross Sections for Atoms and Ions in Gases, 1950–1970, *Atomic Data, 5*, 231–289, 1973.

Falthammer, C.-G., Radial Diffusion by Violation of the Third Adiabatic Invariant, in Earth's Particles and Fields, B.M. McCormac, editor, p. 157, Reinhold, New York, 1968.

Fritz, T.A. and B. Wilken, Substorm Generated Fluxes of Heavy Ions at the Geostationary Orbit, in Magnetospheric Particles and Fields, B.M. McCormac, editor, p. 171, D. Reidel Publishers, Dordrecht-Holland, 1976.

Panasyuk, M.I., T.A. Fritz and W.N. Spjeldvik, Equatorial Measurements of Protons and Helium Ions in the Radiation Belts: Comparison of Soviet and American Experiments, International Symposium on Nuclear Physics, Leningrad, Russia (in Russian), Conference Publication, 1978.

Schulz, M. and L.J. Lanzerotti, *Particle Diffusion in the Radiation Belts*, Springer Verlag, New York, 1974.

Spjeldvik, W.N. and T.A. Fritz, Theory for Charge States of Energetic Oxygen Ions in the Earth's Radiation Belts, *J. Geophys. Res., 83*, 1583–1594, 1978.

Spjeldvik, W.N., Expected Charge States of Energetic Ions in the Magnetosphere, *Space Science Rev., 23*, 499–538, 1979.

W.N. Spjeldvik, Department of Physics, Weber State University, Ogden, Utah

CRRES observations and radial diffusion theory of radiation belt protons

J.M. Albert

Center for Electromagnetics Research, Northeastern University, Boston, Massachusetts

The phase space density of high-energy, equatorially mirroring radiation belt protons, as measured by CRRES, is analyzed in terms of radial diffusion. Only the period prior to the magnetic storm of 24 March 1991 is considered. The observed profiles differ drastically from the NASA AP-8 models, and also from steady state solutions of the radial diffusion equation, indicating a non-steady configuration. Rates of change of the observed profiles are calculated according to the diffusion equation and compared to the observed time development. Good agreement is obtained for $L \leq 2.5$ and E up to about 20 MeV.

1. INTRODUCTION

The Proton Telescope (PROTEL) instrument on the Combined Release and Radiation Effects Satellite (CRRES) provided measurements of proton differential flux for energies between 1 and 100 MeV throughout the inner magnetosphere. The data have been binned by energy, L, and equatorial pitch angle, and time averaged over the periods before and after the Sudden Storm Commencement of 24 March 1991 [*Gussenhoven et al.*, 1993]. Here the pre-storm observations of equatorially mirroring protons are analyzed in terms of phase space density f as a function of L and the first adiabatic invariant μ.

The current understanding of the long-term structure of the proton radiation belt involves radial diffusion, driven by electromagnetic field perturbations resonant with the azimuthal drift frequency [*Fälthammar*, 1965; *Cornwall*, 1968]. Sources include both proton flux incident on an outer boundary and protons due to cosmic ray albedo neutron decay (CRAND), which predominates in the innermost region of the magnetosphere. These sources are balanced by the loss processes of Coulomb collisions with plasmaspheric electrons, resulting in energy drag, and charge exchange mostly with background hydrogen atoms, resulting in neutralization, as well as absorption by the atmosphere.

Radiation Belts: Models and Standards
Geophysical Monograph 97
Copyright 1996 by the American Geophysical Union

The proton radial diffusion equation is written as

$$\frac{\partial f}{\partial t} = L^2 \frac{\partial}{\partial L}\left[\frac{D_{LL}}{L^2}\frac{\partial f}{\partial L}\right] + \frac{G(L)}{\mu^{1/2}}\frac{\partial f}{\partial \mu} - \Lambda f + S, \quad (1)$$

where the terms on the right hand side represent radial diffusion, Coulomb collisions, charge exchange, and CRAND respectively. The CRAND model used is that of *Claflin and White* [1974], which is based on direct measurements of the neutron source. The Coulomb collision term is proportional to the cold electron density, and the charge exchange term depends on the neutral hydrogen density [*Spjeldvik*, 1977]. Radial diffusion coefficients have been estimated by many workers, in a variety of ways. *Schulz* [1991] gives as typical values $D_m = 7 \times 10^{-9} L^{10} R_e^2/$day and $D_e = 10^{-4} L^{10}/(L^4 + \mu^2) R_e^2/$day, where μ is in MeV/G.

Steady state solutions of the radial diffusion equation with $L < 3$ agree qualitatively quite well with the NASA model AP-8 MAX, an empirical model based on data from many different satellites. However, compared to AP-8, the CRRES observations display large discrepancies, as shown in Figures 1 and 2 (AP-8 values of unidirectional integral flux have been converted to directional differential flux and then to phase space density).

The observed phase space density has a large depression around $L = 2$, with f increasing sharply at smaller L to peak around $L = 1.7$. The depleted region widens rapidly in L as μ increases from about 200 MeV/G, extending beyond $L = 3$ for $\mu = 1500$ MeV/G. Adjusting the various parameters in the diffusion equation, such as the diffusion coefficients, to try to improve the agreement of the steady state solutions with the CRRES profiles is in general unsuccessful.

70 RADIAL DIFFUSION OF PROTONS

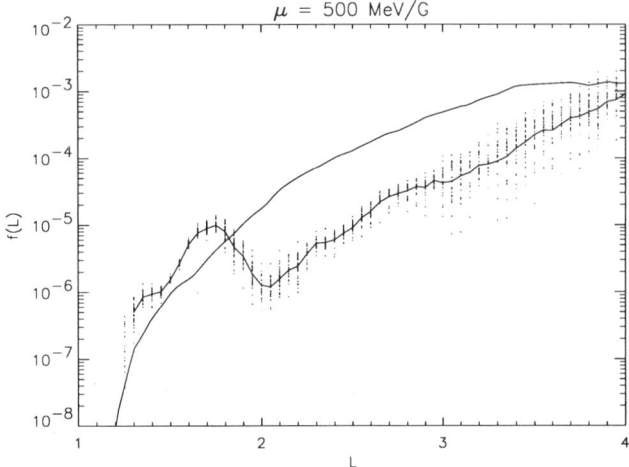

Figure 1. Phase space density $f(L)$ from CRRES observations (individual data and average) and from AP-8 MAX, for μ=500 MeV/G.

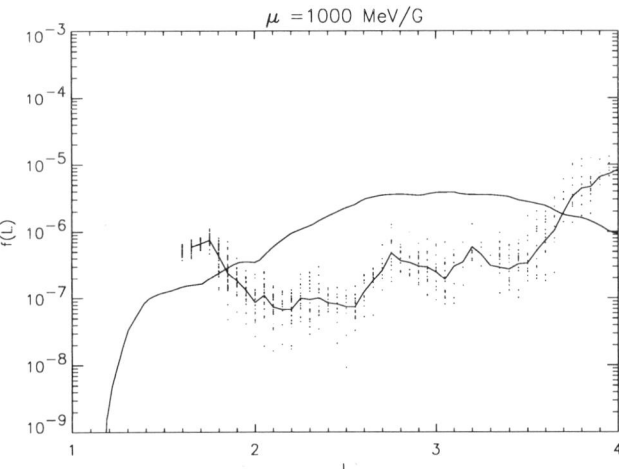

Figure 2. Same as Figure 1 for μ=1000 MeV/G.

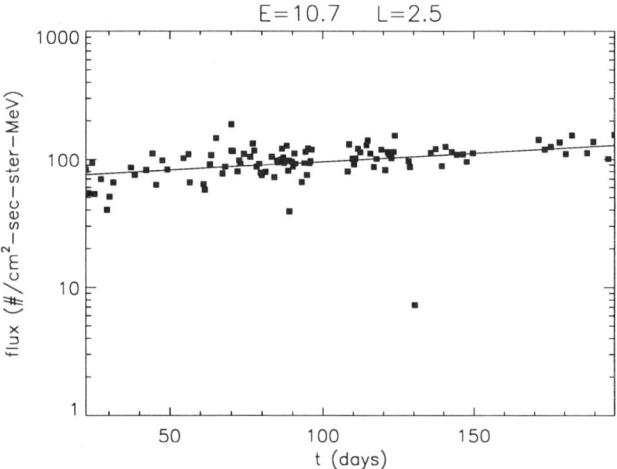

Figure 3. Time series of CRRES flux data for $L = 2.5$, $E = 10.7$ MeV (prior to the 24 March 1991 SSC), and the least-squares fit.

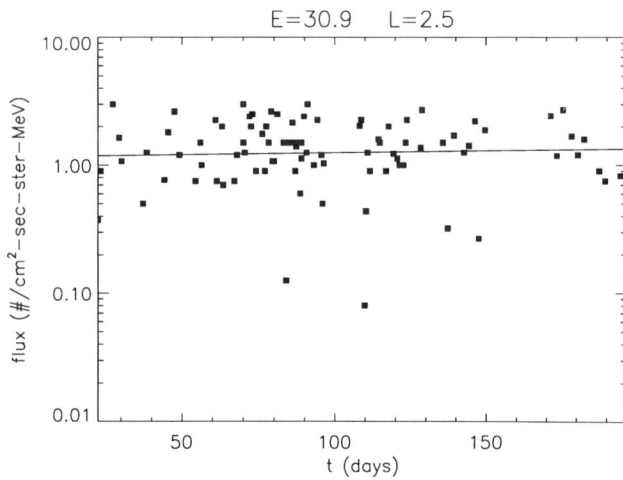

Figure 4. Same as Figure 3 for $L = 2.5$, $E = 30.9$ MeV.

2. OBSERVED GROWTH RATES

The previous section indicates that the observed configuration is far from steady state. In fact, systematic time variation of the fluxes can be seen. Figure 3 shows the individual flux measurements from CRRES orbits 50 to 550 (essentially the pre-storm period), for $L = 2.5$, $E = 10.7$ MeV. In this case it is possible to assign a convincing value to $\gamma \equiv (\partial j/\partial t)/j$ by fitting a least-squares straight line to $\log j(t)$. Plotting a histogram of the deviations of $\log f$ from this line shows that they are small and distributed roughly according to a Gaussian, indicating that the best fit is acceptably good. Figure 4 shows the fluxes for $L = 2.5$, $E = 30.9$ MeV. For this large value of energy, the fluxes are too scattered to determine a meaningful rate of change. Figure 5 shows the fluxes for $L = 3$, $E = 5.7$ MeV. At this larger value of L, there are clear signs of abrupt, nondiffusive changes in the measured fluxes, requiring a different approach to physical modeling.

3. COMPUTED GROWTH RATES

Because the data is sufficient to represent f over a wide range of L and μ, it is possible to explicitly evaluate the terms on the right-hand side of the diffusion equation. If nonzero, this total rate should indicate the rate of evolution from the oberved state towards steady state. This evaluation requires an expression for the diffusion coefficient and other parameters; the same standard values were used as in the steady state simulations. Also required is a reliable method for calculating derivatives of f from its values, which for this purpose are noisy and only fairly well resolved (23 energy channels

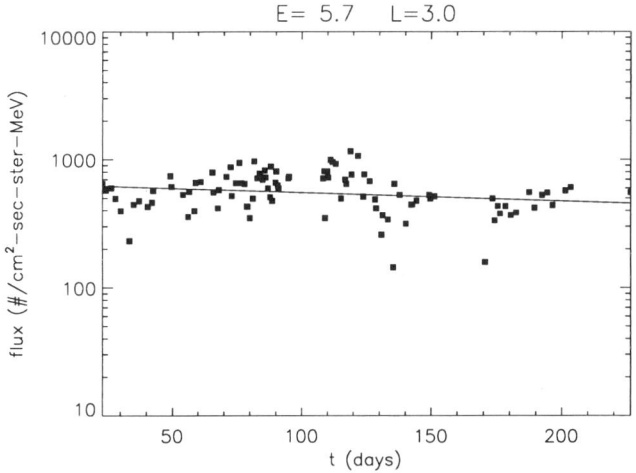

Figure 5. Same as Figure 3 for $L = 3$, $E = 5.7\,\text{MeV}$.

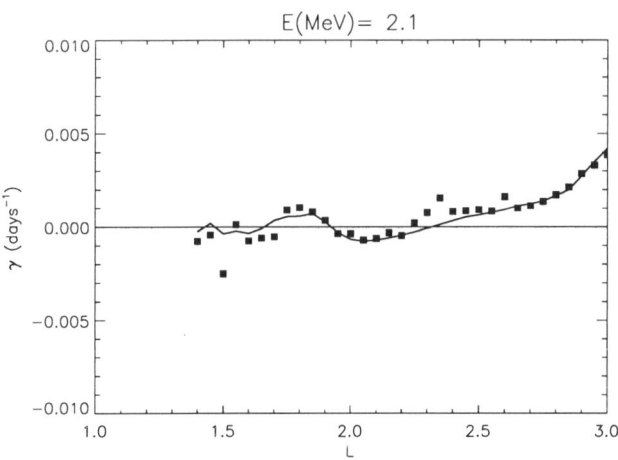

Figure 6. Observed (points) and computed (curves) growth rates, in days^{-1}, vs. L for $E = 2.1\,\text{MeV}$.

over nearly 1–100 MeV, and 20 bins per R_E). Savitzky-Golay low-pass filters c_n were used [Press et al., 1992], which are defined so that

$$\frac{d^k F}{dx^k} \approx \sum_{n=-N}^{N} c_n^{(k,p,N)} F(x + n\Delta x) \quad (2)$$

gives the kth derivative of a polynomial of degree p fit through $2N+1$ values of $F(x)$. First, the (L, E) grid of time-averaged flux values was interpolated, and extrapolated where necessary to obtain f on a regularly spaced grid of (L, μ). The coefficients were then applied to $\log f$ to compute $\partial f/\partial(\log \mu)$, and again to μ-smoothed values of $\log f$ to obtain $\partial f/\partial L$ and $\partial^2 f/\partial L^2$. Parameters used were such that each set of derivatives involved biquartic fits through 13×13 values of (L, μ).

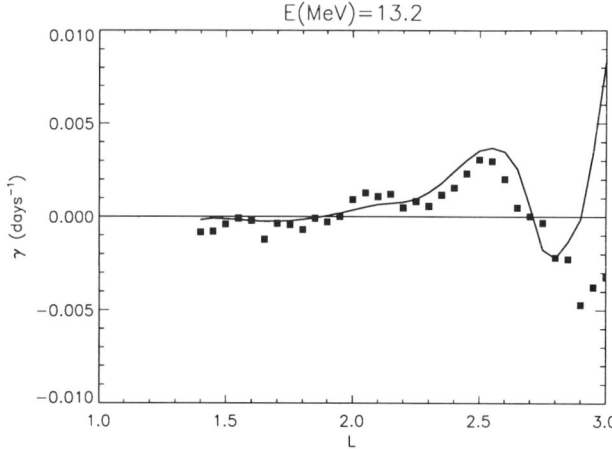

Figure 7. Same as Figure 6 for $E = 10.7\,\text{MeV}$.

Figures 6 and 7 show typical results, in terms of growth rates: $(\partial j/\partial t)/j$ from the time series is compared to $1/f$ times the right hand side of Eq. (1). In some cases terms nearly balance, so the total rate is sensitive to small errors. Nevertheless, qualitative agreement is apparent for energy below 20 MeV and L up to about 2.5. This L range of agreement overlaps the depletion discussed above. Whether the depletion was formed by a large storm prior to the CRRES mission, or in some other way, the flux seems to evolve according to the diffusion equation so as to eventually smooth it away.

As discussed above, the scatter in the measurements is too great for energy greater than 20 MeV, even at $L = 2.5$, to extract growth rates from the time series. At higher L, even for energy less than 20 MeV, the time series clearly do not display smooth evolution in accordance with the assumption of a single diffusive time scale.

REFERENCES

Claflin, E.S. and R.S. White, A study of equatorial inner belt protons from 2 to 200 MeV, *J. Geophys. Res., 79*, 959, 1974.

Cornwall, J.M., Diffusion processes influenced by conjugate-point wave phenomena, *Radio Sci., 3*, 740, 1968.

Fälthammar, C.-G., Effects of time-dependent electric fields on geomagnetically trapped radiation, *J. Geophys. Res., 70*, 2503, 1965.

Gussenhoven, M.S., E.G. Mullen, M.D. Violet, C. Hein, J. Bass and D. Madden, CRRES High energy proton flux maps, *IEEE Trans. Nucl. Sci., 40*, 1450, 1993.

Press, W.H., S.A. Teukolsky, W.T. Vetterling and B.P. Flannery, *Numerical Recipes*, Cambridge University Press, New York, 1992.

Schulz, M., The Magnetosphere, in *Geomagnetism, 4*, Academic Press, New York, 1991.

Spjeldvik, W.N., Equilibrium structure of equatorially mirroring radiation belt protons, *J. Geophys. Res., 82*, 2801, 1977.

J.M. Albert, Center for Electromagnetics Research, Northeastern University, Boston, MA 02115

Dynamic models of the energetic ions in the Earth's radiation belts

V.F. Bashkirov

Skobeltsyn Institute of Nuclear Physics, Moscow State University

A model of proton, helium and oxygen ion with arbitrary pitch angle in the energy range from 0.01 to 1.0 MeV/e transport in the Earth's magnetosphere during geomagnetic storms was developed. Using the results of the transport of ions in the Earth's magnetosphere we map phase space distributions from the stationary pre-storm ion distributions. The pre-storm ion distributions were obtained from the steady-state solution of the radial diffusion equation with the losses. The pre-storm distributions using the Olsen-Pfitzer geomagnetic field model take into account geomagnetic field asymmetry in the external region of the Earth's magnetosphere. The dynamic models of the the ring current ions pitch angle distributions reflect and explain isotropic, "butterfly" and "head-and-shoulders" pitch angle distribution cases experimentally measured by the AMPTE and CRRES satellites during geomagnetic storms.

1. INTRODUCTION

Spatial-energy distributions of the energetic ions in the Earth's magnetosphere during a geomagnetic storm are both experimentally and theoretically scantily known and understood. So far theoretical works on stormtime ion distributions in the Earth's magnetosphere used only the simplest geomagnetic and electric field models and/or examined only particles with equatorial pitch angle $\alpha_0 \sim 90°$ (e.g. *Lyons* [1977]; *Chen et al.*, [1994]). The experimental data available are still lacking to develop empirical models of ion spatial-energy distributions in the Earth's radiation belts during geomagnetic storms and substorms. This problem has great both theoretical and applied importance.

The goal of our work is development of the dynamic models of the proton, helium and oxygen ion fluxes with $\alpha_1 \le \alpha_0 \le 90°$ (α_1 is the atmosphere loss cone) in the energy range $10 \, \text{keV/e} \le E \le 10 \, \text{MeV/e}$ in the Earth's magnetosphere during a geomagnetic storm. This work is the first step where we present results of numerical simulations of the temporal evolution of the ion pitch-angle distributions and explained the isotropic, "butterfly" and "head-and-shoulders" pitch angle distributions cases experimentally measured in the Earth's radiation belts during a geomagnetic storms (e.g. *Sibeck et al.* [1987]).

2. SIMULATION OF THE ION STORM TIME DISTRIBUTIONS

The stormtime phase-averaged distribution $f \sim j(E)/E$ of charged particles in the Earth's magnetosphere could be obtained from the diffusion equation derived from a Fokker-Planck equation (e.g. *Bourdarie et al.* [1995]) or mapped from pre-storm distributions using results of the guiding-center transport of charged particle distributions (e.g. *Chen et al.* [1994]). A comparison made between the radial diffusion and guiding-center drift by *Riley and Wolf* [1992] has shown that the diffusion formalism gives only an approximately correct answer for a single real storm but does much better for an average over a statistical ensemble of storms.

Using results of the stormtime transport of ions with arbitrary pitch angle in the energy range from 0.01 to 1.0 MeV/e in the Earth's magnetosphere, phase space distributions are mapped (interpolated) from the stationary pre-storm ion distributions in accordance with Liouville's theorem. It is well known that the $2 < R/R_E < 9$ and $0.01 < E < 10.0$ MeV life-time of the particles due to charge exchange and Coulomb collision far exceeds the typical evolution time considered in this work (< 3–4 hours) (e.g. *Fok et al.* [1991]). In the present work only "old" particles pre-existing in the $2 < R/R_E < 9$ region are considered.

The ion trajectory is traced through magnetospheric magnetic B (B in nT) and electric $\nabla\Psi$ (Ψ is the electric field potential in V) fields during a model geomagnetic storms having a main phase duration ~2–6 hr ($|D_{st}| \sim 100$–200 nT) both by using the guiding-center formalism and by solving the equation of particle motion:

$$\frac{\partial^2 \mathbf{r}}{\partial t^2} = \frac{Q}{M}\left(\nabla\Psi + \frac{\partial \mathbf{r}}{\partial t} \times \mathbf{B}\right), \quad (1)$$

where M is the mass (in proton mass units m_p) and Q is the charge (in electron's charge units e) of the particle.

The conventional expressions for magnetic:

$$\begin{aligned}\mathbf{V}_{DM} &= \frac{M}{2Q}(v_\perp^2 + 2v_\parallel^2)\frac{\mathbf{B}\times\nabla_\perp \mathbf{B}}{B^3} \\ &= 10^{11} E(1+\cos^2\alpha)\frac{\mathbf{B}\times\nabla_\perp \mathbf{B}}{B^3}\,[m/s], \quad (2)\end{aligned}$$

and electric:

$$\mathbf{V}_{DE} = 10^4 \frac{\nabla\Psi \times \mathbf{B}}{B^2}\,[m/s] \quad (3)$$

drift velocity averaged over the magnetic field line between mirror points l_m and l_m^*

$$\mathbf{V}_D = \frac{\int_{l_m}^{l_m^*}\frac{\mathbf{V}_{DM}(l)+\mathbf{V}_{DE}(l)}{\cos\alpha(l)}dl}{\int_{l_m}^{l_m^*}\frac{dl}{\cos\alpha(l)}} \quad (4)$$

are used. The dynamic Olson-Pfitzer model [*Pfitzer et al.*, 1988], which describes the geomagnetic field at storm better than the Tsyganenko models [*Jordan et al.*, 1992], is used to determine the magnetic field during a geomagnetic storm. The stormtime electric field models are not developped yet and the precise influence of complex stormtime magnetic and electric field fluctuations in energetic ion distributions in the Earth's magnetosphere will be studied in future papers. So only simple field models were used. It is assumed that the electric field $\mathbf{E} = -\nabla\Psi$ corresponds to the Volland-Stern model [*Volland*, 1973; *Stern*, 1973] and the time dependent storm-associated enhancement:

$$\Psi(t) = -\frac{V_\Omega}{L} + \frac{V_0}{2}\left(\frac{L}{L^*}\right)^2 \sin\phi + \frac{\Delta V(t)}{2}\frac{L}{L^*}\sin\phi. \quad (5)$$

Here L is McIlwain's parameter, ϕ is longitude, $V_\Omega = 90$ kV, $V_0 = 50$ kV, $L^* = 8.547$ and $\Delta V(t)$ is the storm-accociated enhancement in the cross-tail potential (for more detail see *Chen et al.* [1994]).

During a geomagnetic storm the "wave-particle" interaction usually negligible during geomagnetically quiet conditions could affect ion distributions like charge-exchange and Coulomb collision loss processes, so for a correct solution of this problem it is necessary to take into account radial

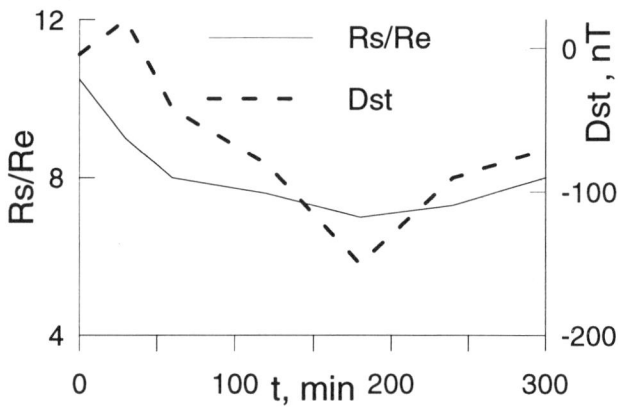

Figure 1. D_{st} index and standoff distance R_S for the model geomagnetic storm studied

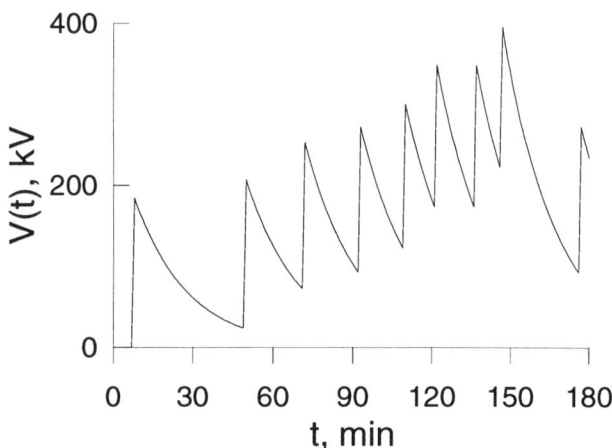

Figure 2. The $\Delta V(t)$ potential in our model storm (more detailed in *Chen et al.* [1994]).

and pitch-angle diffusion and loss processes. But the "wave-particle" interaction is a difficult and unresolved problem for the ions, therefore it was not considered in this paper.

The steady-state structure of the ion spatial-energy distributions in the Earth's radiation belts principally confirmed with quiet time experimental data could be found from the radial diffusion equation with losses [*Spjeldvik et al.*, 1990]:

$$\frac{\partial f}{\partial t} = L^2\frac{\partial}{\partial L}\left(\frac{D_{LL}}{L^2}\frac{\partial f}{\partial L}\right) + \text{Source} - \text{Losses} = 0, \quad (6)$$

where D_{LL} is the radial diffusion coefficient. We use the steady-state models of the spatial-energetic distributions of proton, helium and oxygen ion fluxes $j_0(R, LT, E)$ in the energy range $0.001 < E < 10.0$ MeV/e at $2 < R/R_E < 9$

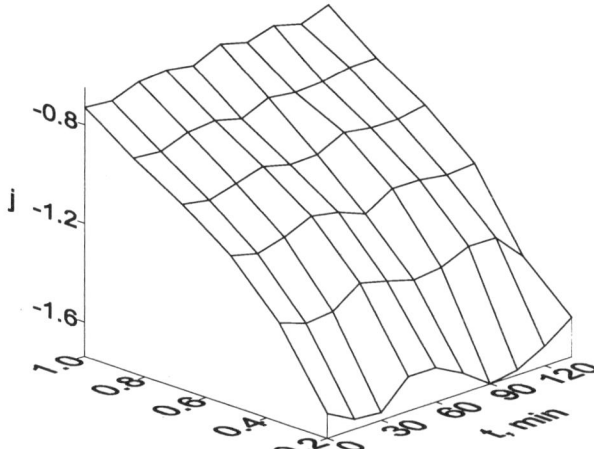

Figure 3. PADs of 300 keV protons at $R = 3 R_E$ and LT=0000 during a model storm

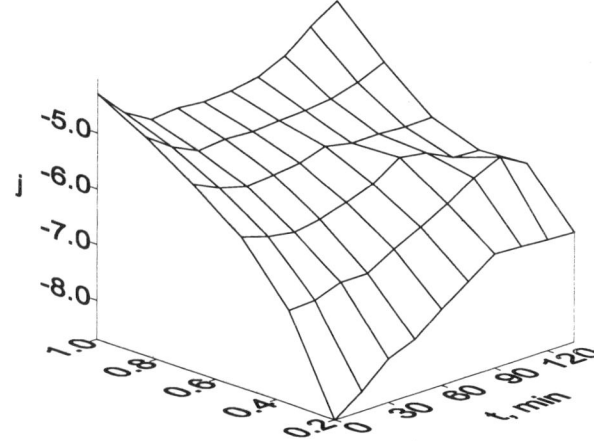

Figure 4. PADs of the 50 keV protons at $R = 3 R_E$ and LT=1200 during a model storm

taking into account both the real longitude asymmetry of the Earth's magnetosphere and the main loss processes (charge exchange and Coulomb collisions) [*Bashkirov et al.*, 1996; *Beliaev et al.*, 1996] as the pre-storm distributions. These models are consistent with experimental data for geomagnetically quiet and weakly disturbed conditions [*Bashkirov et al.*, 1994].

3. THE NUMERICAL SIMULATION RESULTS

Using the guiding-center formalism is a more convenient but less accurate method than solving the equation of particle motion for tracing ions in the Earth's magnetosphere. We found that for ions with $10° \leq \alpha_0 \leq 90°$ for $0.01 \leq E \leq 1.0$ MeV/e both methods yield the same results even during a geomagnetic storm. That is not correct for the particles having higher energy and smaller equatorial pitch angle. So the results presented below were obtained using the guiding-center formalism.

It was found that the pitch angle distributions (PADs) of the ions in the Earth's radiation belts vary considerably during a main phase of a storm. Not only the anisotropy value but also the form of PADs essentially changes. A number of local features appear. Both character and value of these variations strongly depend on E and R.

The results of numerical simulation of the temporal evolution of proton PADs are presented below (for other ion species the results are mainly similar to the proton ones). Figures 3–6 show some typical samples of the proton PADs evolution during the main phase of the geomagnetic storm (Figures 1–2). In the three-dimensional plots the ion differential flux (in arbitrary units) is plotted as a function of $\sin \alpha_0$ and time (in minutes) after storm onset, for some values of E, R and LT in the geomagnetic equator plane.

Figure 3 shows PADs of the 300 keV protons at $R = 3 R_E$ and LT=0000. The time scale of the variations corresponds to a period of the particle drift around the Earth ("drift echo"). For ions having $E > 100$ keV/e we observe weak temporal variation of PADs in the inner magnetosphere at $R < 4.5 R_E$.

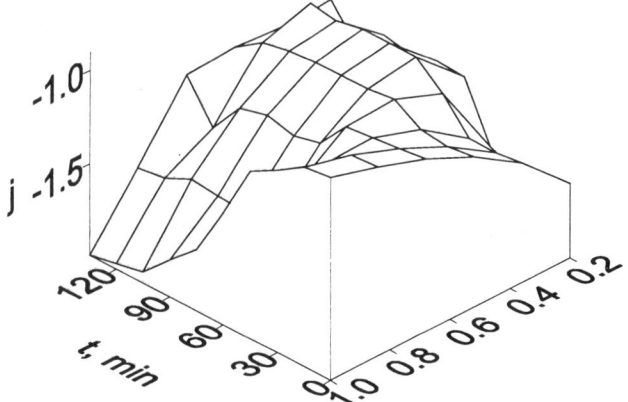

Figure 5. PADs of the 100 keV protons at $R = 4.5 R_E$ and LT=1800 during a model storm

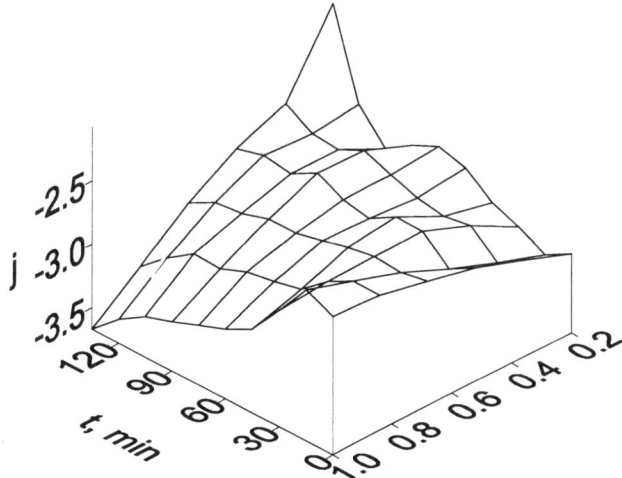

Figure 6. PADs of the 500 keV protons at $R = 4.5 R_E$ and LT=0600 during a model storm

These variations are periodic and reflect the initial PAD asymmetry with LT which decreases with decreasing R [*Bashkirov et al.*, 1996] or stormtime ion acceleration by a storm sudden commencement [*Hudson et al.*, 1995].

The effects of the appearance of "butterfly" and "head-and-shoulders" ion pitch angle distributions are shown in Figures 4–6. These effects are caused by differential particle drift. Indeed, the drift velocity strongly depends on the equatorial pitch angle of the particle (decreasing with decreasing α_0). Ions with different equatorial pitch angles arrive at the observation point (R, LT) from different points (R_0, LT_0). The pre-storm $|dj_0/dR|$ is very high in the Earth's inner radiation belts at $R < 7\,R_E$ [*Beliaev et al.*, 1996] which leads to the formation of "butterfly" and "head-and-shoulders" PADs during a storm.

4. CONCLUSION

It has been found that it is possible to use the guiding center drift assumption for a numerical simulation of the storm-time evolution of the ion spatial-energetic distribution for $0.01 < E < 1.0\,\text{MeV}/e$ and equatorial pitch angle $\alpha_0 > 10°$.

It has been shown that PADs of the ions in the Earth's radiation belts considerably vary during a main phase of a geomagnetic storm. Not only the anisotropy value but also the form of PADs essentially changes. A number of local features appears. Both character and value of these variations strongly depend on E, R and LT. The asymmetry in the PAD variation increases with decreasing E and increasing R.

The low energy ion PAD mainly depends on the electric field while the more energetic ion PAD ($E > 100\,\text{keV}$) depends on magnetic field variations in the inner Earth's magnetosphere.

Acknowledgements. The research described in this publication was made possible in part by grant MCU 000 from the International Science Foundation.

REFERENCES

Bashkirov, V.F., A.S. Kovtyukh and M.I. Panasyuk, Numerical simulation of the proton pitch angle distributions in the Earth's radiation belts, Preprint INP MSU, 94-5/327, 1994.

Bashkirov, V.F., A.S. Kovtyukh and M.I. Panasyuk, Influence of the Charge Exchange and Coulomb Collisions of the proton Pitch Angle Distributions From in the Earth's Radiation Belts, *Adv. Space Res., 17,* 25–28, 1996.

Beliaev, A.A., E.G. Koroteyeva and M.I. Panasyuk, Sensitivity of the model of the heavy ions diffusion in the Earth's radiation belts to the ionosphere ions fluxes, *Adv. Space Res., 17,* 169–172, 1996.

Bourdarie, S., D. Boscher and T. Beutier, Dynamical physical modeling of trapped particles for satellite survey, these proceedings.

Chen, M.W., L.R. Lyons and M. Schulz, Simulation of phase space distributions of storm time proton ring current, *J. Geophys. Res., 99,* 5745–5759, 1994.

Fok, M.-C., J.U. Kozyra, A.F. Nagy and T.E. Cravens, Lifetime of ring current particles due to Coulomb collisions in the plasmasphere, *J. Geophys. Res., 96,* 7861–7867, 1991.

Hudson, M.K., A.D. Kotelnikov, X. Li, I. Roth, M. Temerin, J. Wygant, J.B. Blake and M.G. Gussenhoven, Simulation of proton radiation belt formation during the March 24, 1991 SSC, *Geophys. Res. Lett., 22,* 291–294, 1995.

Jordan, C.E., J.N. Bass, M.S. Gussenhoven, H.J. Singer and R.V. Hilmer, Comparison of magnetospheric magnetic field models with CRRES observation during the August 26, 1990 Storm, *J. Geophys. Res., 97,* 16,907–16,9210, 1992.

Lyons, L.R., Adiabatic evolution of trapped particle pitch angle distributions during a storm main phase, *J. Geophys. Res., 82,* 2428–2432, 1977.

Pfitzer, K.A., W.P. Olson and T. Mogstad, A Time dependent, source driven magnetospheric field model, *EOS, 69,* 426, 1988.

Riley, P. and R.A. Wolf, Comparison of diffusion and particle drift descriptions of radial transport in the Earth's inner magnetosphere, *J. Geophys. Res., 97,* 16,865–16,876, 1992.

Sibeck, D.J., R.W. McEntire, A.T.Y. Lui, R.E. Lopez and S.M. Krimigis, Magnetic field drift shell splitting: Cause of unusual dayside particle pitch angle distributions during storms and substorms, *J. Geophys. Res., 92,* 13,485–13,497, 1987.

Spjeldvik, W.N., Equilibrium charge state distribution of geomagnetically trapped ions: Analytic consideration and a useful algorithm, *Ann. Geophys., 8,* 59, 1990.

Stern, D.P., A study of the electric field in an open magnetospheric model, *J. Geophys. Res., 78,* 7292–7305, 1973.

Volland, H., A semiempirical model of large-scale magnetospheric electric field, *J. Geophys. Res., 78,* 171–180, 1973.

V.F. Bashkirov, Skobeltsyn Institute of Nuclear Physics, Moscow State University, Moscow, 119899, Russia

Physical Radiation Belt Models
Report of Discussion Group A

Reporter: T. Beutier

Centre d'Etudes Spatiales des Rayonnements, Toulouse

Participants: W. Spjeldvik (chair), M.K. Hudson (co-chair), T. Beutier (reporter), J. Albert, V. Bashkirov, S. Bourdarie, M. Walt

The aim of this group has been to answer the essential question: What do we want from physical radiation belt models? And so to answer the following questions: Which kind of models can be used? Are these models accurate? What are the principal sources of the radiation belts? What data are necessary to build good models?

The first purpose of physical models is the understanding of physical processes which allow to create, support and modify the structure and dynamics of the radiation belts. From the point of view of experiments they also must be able to extrapolate beyond existing measurements in all dimensions of the problem (in space, energy, angle, etc.) and into the future in order to make predictions. This forecasting requires to find simple scaling laws and to determine which are the better variables to use to describe the radiation belt behaviour. Finally, the physical models can be useful to organise data.

To answer all these requirements the best approach is to use a set of codes able to describe all the features of radiation belts. These codes are essentially of four types:

1. MHD models, two fluids, hybrid or kinetic;

2. equatorial or 3D (symmetric in azimuth) diffusive models, the time scale of this kind of codes being of some drift period;

3. 4D (space coordinates and energy) diffusive-convective models, the time scale of this kind of codes being of some bounce period;

4. Test particle codes.

All these codes being strongly dependent on magnetic field models, a new question must be asked:

> *What is the best magnetic field to use (knowing that in function on our purpose it is not necessary to use the same field...)?*

The sources to use are also a crucial issue. It is possible to find two kinds of sources. First there are the "direct" sources such as CRAND (essentially for very high trapped particle energies) and much lower sources as anomalous cosmic rays, solar cosmic rays, jovian electrons and ions coming from the ionosphere. Secondly, the "physical" sources such as adiabatic acceleration (radial diffusion), wave acceleration (by whistler, ULF and VLF waves) and shock acceleration (during storms) can be considered.

However, another problem subsists: the current radiation belt models are not adequate for space science needs. In particular, it is necessary to review existing models at low altitude (essentially for protons) and to elaborate a new model for electrons (especially in the outer zone, cf. CRRES results).

To accomplish this, more data are needed. In particular, low altitude data (in particular in the slot region) will be useful. It is also necessary to have more magnetic and electric field measurements (in particular for waves).

In conclusion, it is necessary to answer the fundamental question asked by engineers who need radiation belt models: Can physical models or elements of them be used for "empirical" or "engineering" models?

The answer of the community is yes. If the physical models conform to the observed data then the physical models may be used to extend the models for engineering purposes beyond the range of the data compilation. Moreover, such extension may be spatially, temporally or event/epoch related to simulate conditions infrequently observed or not yet observed. This is the present challenge of radiation belt modelling.

T. Beutier, CESR, 9, Avenue du Colonel Roche, 31029 Toulouse Cedex, France

DISCUSSION

Q: J.F. Lemaire. In addition to MHD models, I believe you should consider kinetic models and thermo-electric charge separation electric fields whose intensity can be very high at the interface between hot and cool plasma regions. These types of electric field are ignored in ideal MHD models but need to be included in comprehensive models of the magnetosphere.

A: T. Beutier. The existing models must be extended to bi-fluid, kinetic models. These tools have to be improved wish test-cases.

Q: D.N. Baker. We must understand how the radiation belts can be "dumped" very rapidly (e.g., 3–4 November 1993). How can this happen? What are the loss processes?

A: M.K. Hudson. Outer zone electrons can be dumped very rapidly by inward motion of the magnetopause and by wave-induced pitch angle diffusion. On the D_{st} minimum time scale of ring current buildup, the adiabatic trapping boundary for protons moves inward as they are subjected to an increasing curvature gradient.

Recent Development in the NASA Trapped Radiation Models

S.F. Fung

Space Physics Data Facility, Code 632, NASA Goddard Space Flight Center, Greenbelt, Maryland

The NASA omnidirectional trapped radiation models are static, empirical models developed at the NASA National Space Science Data Center (NSSDC) using an extensive compilation of spacecraft measurements taken between 1958 and 1978. The latest models, AP-8 for protons (0.1–400 MeV) and AE-8 for electrons (0.04–7 MeV), which provide integral fluxes of the trapped particles, are the last in a series of earlier electron (AE-1–7) and proton (AP-1–7) models. These models have been widely distributed and used in assessing the near-Earth trapped radiation environment in many applications. However, because the models are static and statistical in nature, the "errors" associated with the models are application-dependent. Systematic errors in these models are also difficult to be quantified because of the large number of data sets used. Different processing techniques were needed to process the different data sets in order to remove temporal and spatial variations, while preserving the energy dependence in the data sets. Thus, the resultant models are not suited for emulating the various geomagnetic conditions. After a brief review of the current AP-8 and AE-8 models and their comparisons with recent energetic particle observations, we will describe an effort at the NASA Goddard Space Flight Center to develop a new generation of trapped radiation models.

1. INTRODUCTION

The NASA Trapped Radiation Environment Modeling Program (TREMP) began in the mid-1960's when Dr. J.I. Vette became the head of the NASA National Space Science Data Center (NSSDC) and has since produced a series of empirical radiation belt models (see Table 1; adapted from *Gaffey and Bilitza* [1994]) based on many past energetic particle measurements. A review of TREMP and the various models can be found in *Vette* [1991a] and *Gaffey and Bilitza* [1994]. The latest models in the series are the AP-8 model [*Sawyer and Vette*, 1976] for the trapped energetic protons and the AE-8 [*Vette*, 1991b] model for energetic electrons.

The NASA trapped radiation models, due to their comprehensiveness in data coverages, have been the de facto standards against which other trapped radiation models are compared [*Lemaire et al.*, 1995; *Beutier and Boscher*, 1995; *Beutier et al.*, 1995; *Panasyuk*, 1996]. They have been widely distributed and used in many studies for spacecraft engineering designs, space instrument design and development and space mission planning. Figures 1 and 2 show the time period and L-shell coverages of the various data sets used for constructing the AP-8 and AE-8 models.

Of all the NASA trapped radiation models listed in Table 1, only the AP-8 and AE-8 models, the culminations of their earlier counterparts, are being distributed by the NASA NSSDC. These models and documentations are available from the NSSDC's Request Office and electronically from its anonymous-FTP directory or its world wide web site at: **http://nssdc.gsfc.nasa.gov**.

Since a lot of work has been devoted to addressing the accuracies of the AP-8 and AE-8 models [*Gussenhoven et al.*, 1991; *Vette*, 1991a; *Gaffey and Bilitza*, 1994; *Heynderickx and Beliaev*, 1994], we will not discuss them in detail but only point out some results from comparing the models with recent observations in order to deliberate on the model deficiencies.

In the following sections, we will provide a brief overview of the characteristics of the existing NASA trapped radiation models. They will then be compared with long-term aver-

Radiation Belts: Models and Standards
Geophysical Monograph 97
This paper is not subject to U.S. copyright.
Published in 1996 by the American Geophysical Union

Table 1. NASA Trapped radiation models for electrons and protons (adapted from *Gaffey and Bilitza* [1994]).

Name	Energy (MeV)	L Range	Epoch	Comments
Electrons				
AE-1	0.3–7	1.2–3	7/63	Starfish not cons.
AE-2	0.04–7	1.1–6.3	8/64	
AE-3	0.01–5	6.6		geostationary
AE-4	0.04–4.85	3–11	1964/67	solar max and min
AE-5	0.04–4	1.2–2.8	10/67	Starfish removed
AE-5P	0.04–4	1.2–2.8	1964	AE-5, solar min
AE-6	0.04–4	1.2–2.8	1967	AE-5, solar max
AE-I7	—	—	—	withdrawn from circ.
AE-8	0.04–7	1.2–11	1964/67	
Protons				
AP-1	30–50	1.17–4.6	9/63	
AP-2	15–30	1.17–3.5	9/63	
AP-3	> 50	1.17–3.15	9/63	
AP-4	4–15	1.17–2.9	9/63	
AP-5	0.1–4	1.2–6.6		geost. inc.
AP-6	4–30	1.2–4	12/64	
AP-7	> 50	1.15–3	1/69	as AP-3; solar max
AP-8	0.1–400	1.15–6.6	1964/70	solar min and max

aged trapped particle measurements obtained by the NOAA-10 and the Japanese OHZORA (EXOS-C) satellites. After identifying certain deficiencies in the existing models, we will describe a renewed effort at the Space Physics Data Facility (SPDF), a sister organization of the NSSDC within the Space Science Data Operations Office (SSDOO), NASA Goddard Space Flight Center, to develop a new generation of trapped radiation models by using modern data management and analysis tools.

2. THE NASA TRAPPED RADIATION MODELS: AP-8 & AE-8

2.1. *Model Overview*

Details of the models are given in *Vette* [1991a] and the other references listed above. Only a brief overview will be given here. All of the models developed in the TREMP (Table 1) are static, empirical models. In constructing those models, energetic particle observations obtained by various instruments (including Geiger-Mueller tubes, scintillator/photomultiplier tubes (S/PMT), ionization chambers, solid-state detectors (SSD), beta ray spectrometer, solid-state telescopes, S/PMT spectrometers and SSD clusters) on the various spacecraft (see Figures 1 and 2) are collected, combined and processed to statistically compute the long-term and large (spatial) scale averages of the differential, integral, and omni-directional fluxes as functions of L and B/B_0, where L is the McIlwain parameter, B is the local magnetic field strength and B_0 is the minimum B magnitude for the given L value. It is important to note however that the basic architecture and the output of the models rest upon the specific magnetic field models used to compute L and B/B_0. Although both the proton (AP-8) and electron (AE-8) models are static, two versions (SOLMIN and SOLMAX) of each model were constructed to reflect solar cycle variations associated with the solar minimum and solar maximum conditions. Differences between these versions occur only for $L < 3$ and at low altitudes (< 1000 km). Local-time or longitudinal variations have been averaged out in the distributed model software, with corrections given in the documentation.

Table 2. Characteristics of the AP-8 and AE-8 models.

Characteristics	AP-8	AE-8
Issue date	12/76	12/83
No. of satellites	24	24
No. of instruments	29	26
(channels)	(101)	(94)
L Range	1.15–6.6	1.2–11
Energy range (MeV)	0.1–400	0.04–4.5 (inner zone)
		0.04–7.0 (outer zone)
Epoch		
SOLMIN	1964	1964
SOLMAX	1970	1967

2.2. *Data Coverages*

As shown in Figure 1, the collection of data sets used to construct the NASA trapped radiation models cover almost two solar cycles. Though solar proton, galactic cosmic ray

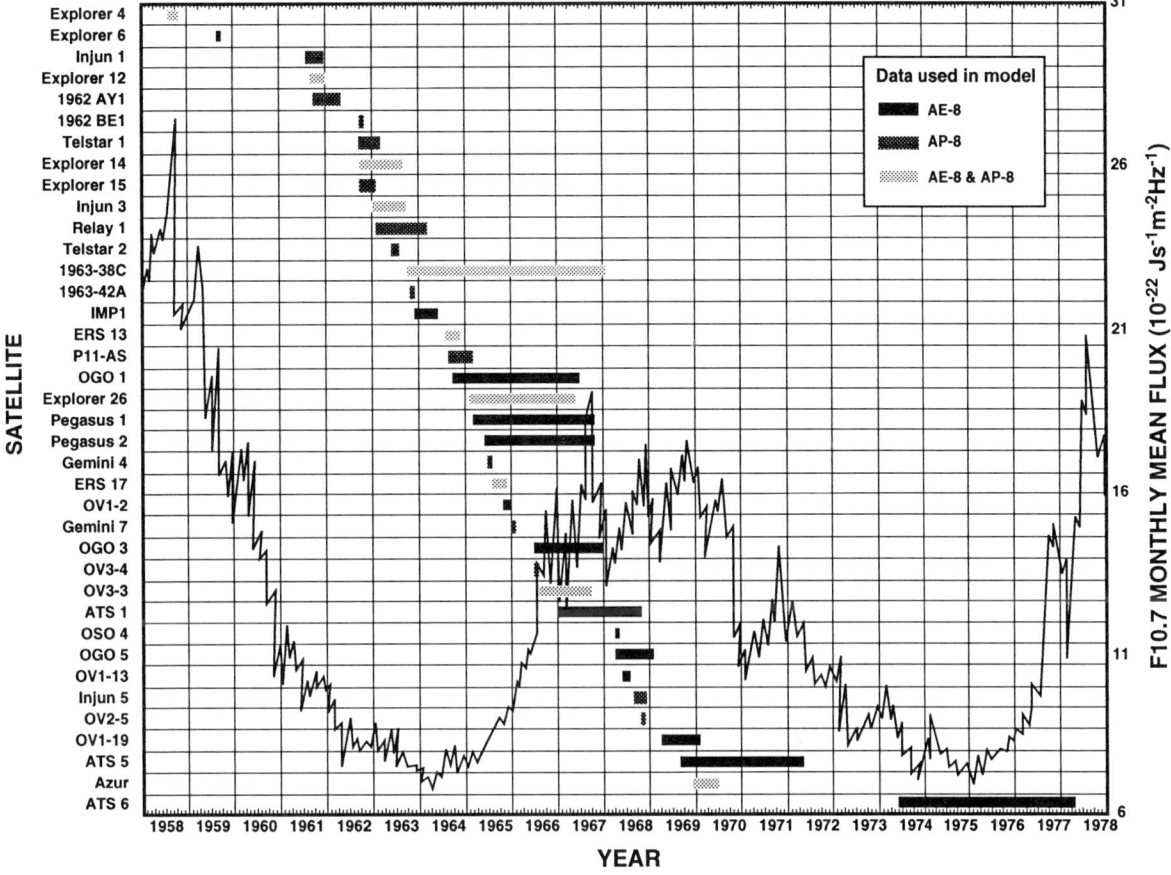

Figure 1. Time coverages of data used in the NASA AE/AP-8 trapped radiaton models.

and high charge-state particle data were excluded from the trapped radiation data base, the earlier electron models were biased by the residual energetic electrons from the Starfish experiment in 1962. In constructing the later electron models, particularly for the solar minimum period, a Starfish decay model was implemented in conjunction with more recent data from the ATS-6 satellite, covering 1974.5–1978.3, to re-calibrate the data and produce the much improved AE-6–8 models.

Figure 2 indicates the L-shell coverages of the data sets used in the AP-8 and AE-8 models. It by no means suggests that sufficient spatial coverages were attained. Figures 3a and 4b illustrate the energy coverages of the various data sets used. The dashed and solid lines indicates discrete and integral energy measurements, respectively. What seem to be missing are long-term spectral data sets to provide adequate spatial and spectral data for proper construction of empirical models. Table 2 summarizes the characteristics of the AP-8 and AE-8 models. In both models, the SOLMIN and SOLMAX versions differ mostly only for altitudes < 1000 km and $L < 3$.

3. COMPARISONS WITH RECENT OBSERVATIONS

Although the NASA trapped radiation models were constructed 2 decades ago by using some even older data sets, much care was exercised in treating the data such that the models have been regarded as good to "about a factor of 2" [*Vette*, 1991a]. This statement, of course, cannot be taken literally without considerations as we will explain below. Here we present a comparison between the NASA models with some recent trapped radiation measurements taken by the NOAA-10 (10/86–8/91) and the Japanese OHZORA (EXOS-C) (2/84–3/87) satellites.

Both the NOAA-10 [*Raben et al.*, 1995] and OHZORA [*Nagata et al.*, 1985] satellites have two orthogonal telescopes for detecting protons and electrons. To analyze the data, we have assumed that the mirroring particle pitch angle distributions can be modeled by the traditional form,

$$J(\theta) = J_\perp \sin^N(\theta), \qquad (1)$$

where the perpendicular flux $J_\perp$ and the anisotropy index N can be solved from the measurements of the particle fluxes $J(\theta)$ at pitch angles θ by the two orthogonal detectors.

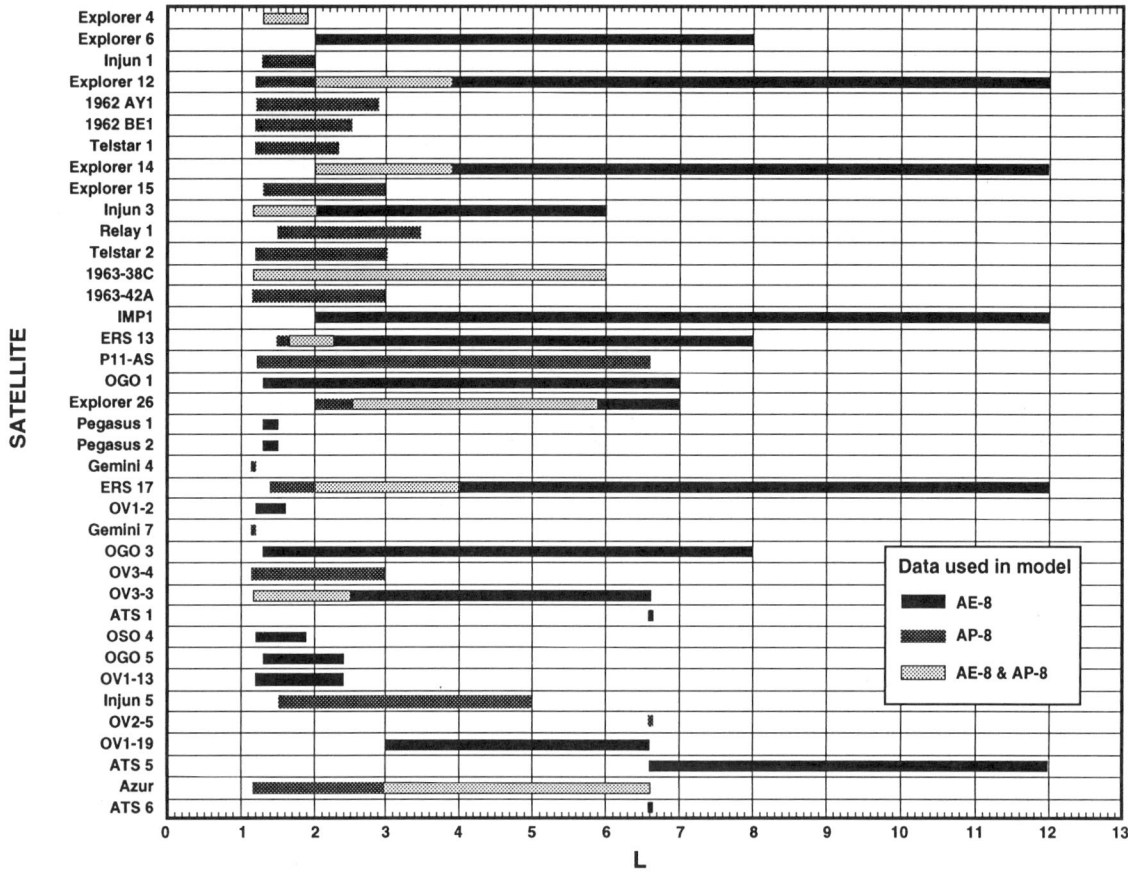

Figure 2. L coverages of data used in the NASA AE/AP-8 trapped radiaton models.

Figures 5a and 5b show the preliminary results of omnidirectional fluxes of 80–250 keV trapped protons observed from the NOAA-10 satellite at about 850 km in September and October, 1989, respectively, as a function of L-shell. The satellite orbit planes were roughly fixed in local times for the two monthly periods; and the almost circular satellite orbit led to crossings of a given L-shell by the satellite at points having slightly different magnetic field strengths (with an average given by B). Thus, from the satellite positions at the times of observations and a given magnetic field model, a relationship between the parameters B/B_0 and L can be determined by a second-order polynomial fit, $Y = M_0 + M_1 X + M_2 X^2$ with $Y = B/B_0$ and $X = L$. Because of longitudinal variations in the geomagnetic field, different fits are obtained for the two periods as the NOAA-10 orbit plane changed. The best-fit magnetic coordinates B/B_0 and L of the satellite can then be used to obtain the corresponding AP-8 model fluxes for comparison with observations.

It is apparent that the trapped radiation fluxes predicted by the AE-8 and AP-8 models (based on the coordinates B/B_0 and L) for a given geographic location will be different when different magnetic field models are used. This has led to some confusion and recommendations for the proper applications of the models [*Gaffey and Bilitza*, 1994; *Heynderickx et al.*, 1996].

In order to evaluate how well the models predict the radiation environment, the AP-8 MAX models are superposed as light dots on the NOAA-10 observations (dark dots) in Figures 5a and 5b. The solid curves are fits to the model fluxes of 80–250 keV protons (light dots) along the fits of B/B_0 as a function of L, computed by using the contemporary IGRF field model and the 1970 IGRF model, respectively. As we will discuss in the next section, substantial differences in the trapped proton fluxes along the satellite trajectory are predicted by the models based on the different magnetic field models in the two widely separated epochs.

It is apparent that the average trend in the data shown in Figure 5a for the month of September 1989 agrees remarkably well with the model (1989 curve) at $L > 2.2$, even though individual measurements may depart significantly from the model. In October 1989 (Figure 5b), there occurred several great solar energetic particle events [*Reeves, et al.*, 1992; *Boberg et al.*, 1995] which might account for the overall enhanced level of average proton fluxes as compared to the model.

At $L < 2$, however, the observed fluxes appeared to have exceeded that of the model predictions by as much as two

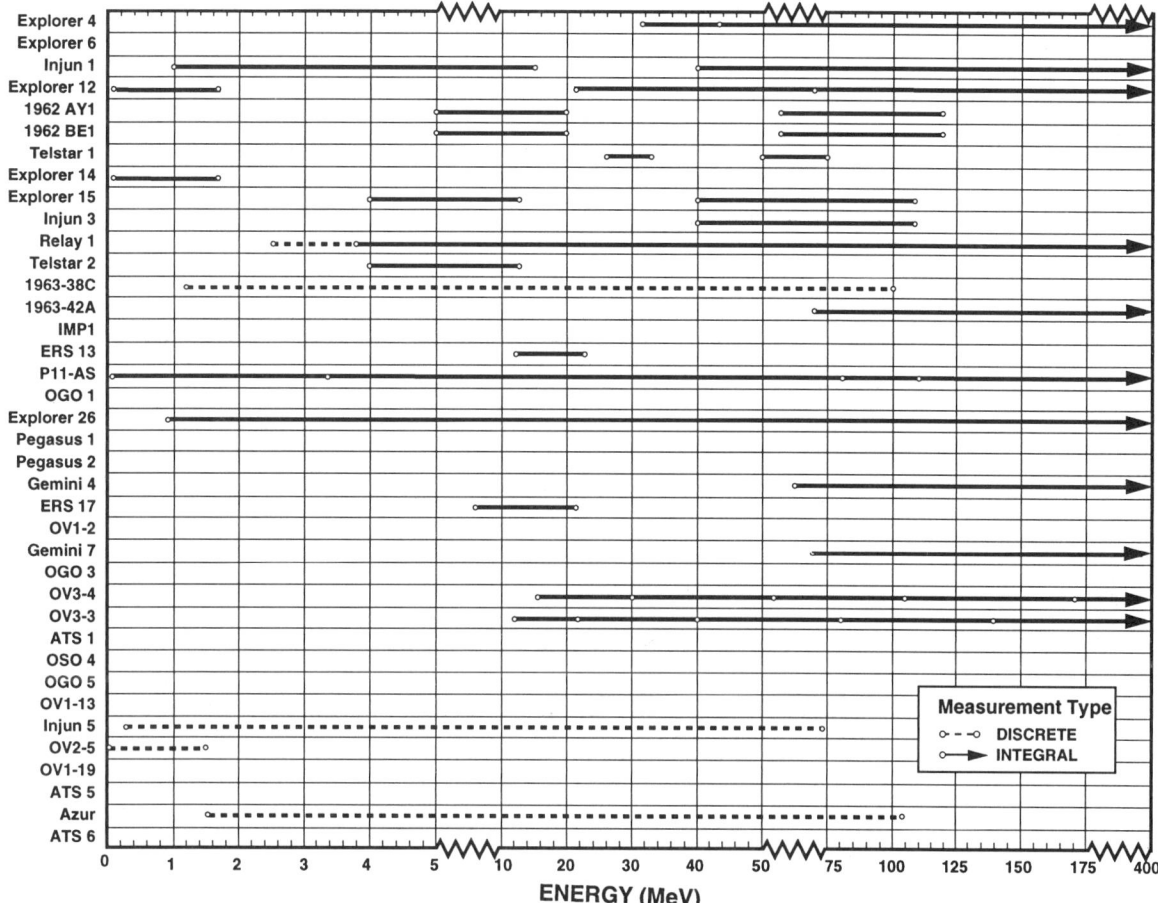

Figure 3. Energy coverages of data used in the NASA AP-8 trapped proton model.

orders of magnitude or more. An examination of the geographic locations of all the data points indicated that they were all taken within the South Atlantic Anomaly (SAA) region. It is important to note that the pitch angle distributions for the events observed at $L < 1.7$ also tend toward isotropy (N becoming smaller, see lower panels of Figures 5a and 5b), indicating the possibility of contamination by the much higher energy trapped protons (> 20 MeV) in the inner radiation zone. Further investigations will be needed to resolve these discrepancies.

Figures 6a and 6b display the compilations of almost three years' (1984–87) observations by the Japanese OHZORA (EXOS-C) satellite in two different local time zones, 5.5–6.5 hrs. and 11.5–12.5 hrs, respectively, near the solar minimum period. The figures show the L-shell profiles of quiet-time ($|D_{st}| < 30$ nT) omni-directional fluxes of trapped electrons between 0.2 and 3.2 MeV. Although many of the individual electron measurements in the outer radiation zone ($3 < L < 7$) deviate from the AE-8 MIN flux levels (superposed 1964 curves), analogous to the proton distributions shown in Figures 5a and 5b, their long-term averages remain consistent with the model predictions.

Considerable local time variations are also evident in the long-term OHZORA data, particularly in the "slot" and inner radiation regions ($L \leq 3$). For example, Figures 6a and 6b show that the trapped electron fluxes in the "slot" region near $L = 3$ tend to average to a higher level in the earlier local time zone (5.5–6.5 hrs) than those near local noon (11.5–12.5 hrs). On the other hand, the electron distributions in the noon sector appear to be more isotropic, with the anisotropy indices N tending to smaller values than the ones observed in the earlier sector. These variations are in constrast with the digital versions of the existing trapped radiation models (although local-time corrections are included in the construction and documentation of the electron model [Vette, 1991a]). We should note here that the apparent local-time variations described here have not been discussed previously and deserve further study.

The AE-8 MIN models for the epochs of 1986 and 1964 are also shown in Figures 6a and 6b for comparisons with the OHZORA data. Unlike the proton results shown in Figure 5, both AE-8 models in the different epochs compare favorably with the observations in the outer radiation zone ($3 < L < 9$). On the other hand, both models appear to perform poorly in

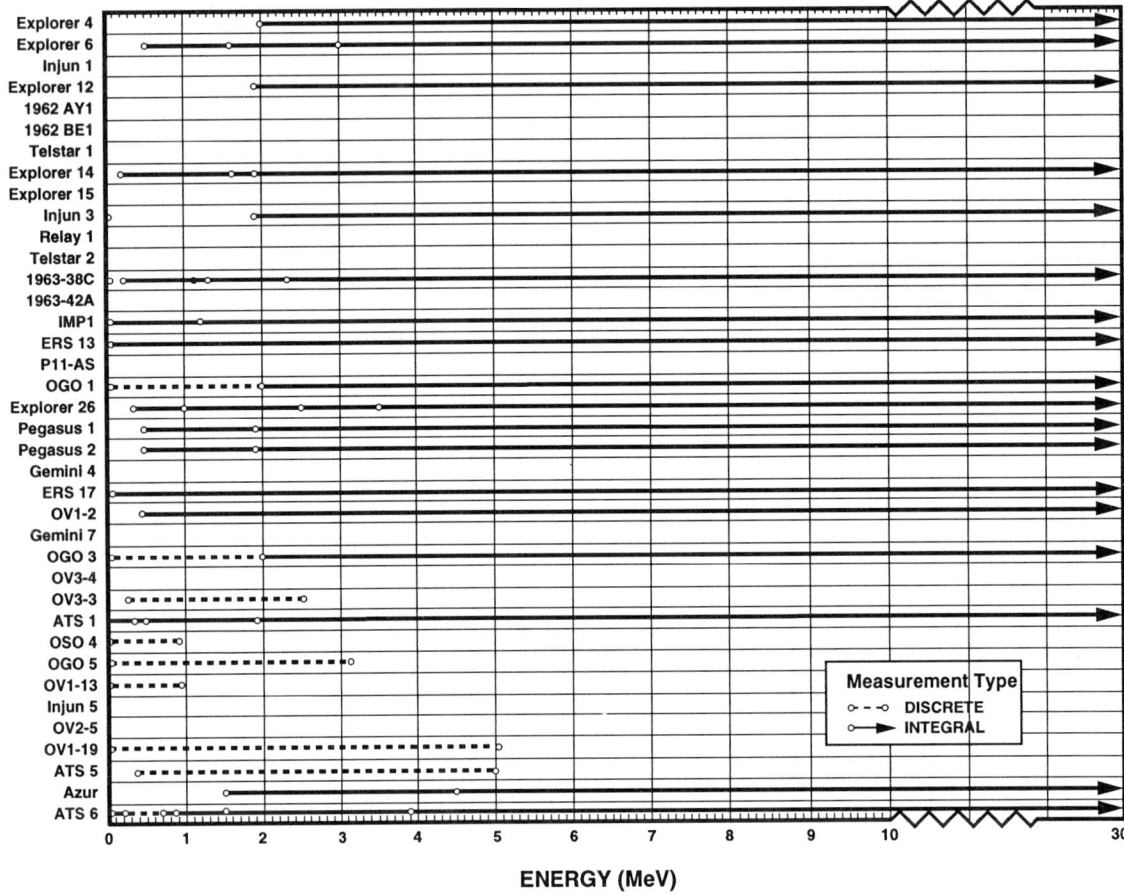

Figure 4. Energy coverages of data used in the NASA AE-8 trapped electron model.

the lower L-shells, though effects of possible contaminations by high energy proton events cannot be ruled out.

Unlike the AP-8 model, there appears to be less disagreement in the AE-8 models when the model predictions from the two epochs are compared. However, the magnetic cutoff conditions (the relationship between B/B_0 and L) seen by the OZHORA spacecraft in the different local time zones are slightly different. Because of the steepness of the cutoffs, which in turn are also different from the fixed magnetic cutoff implemented in the AE-8 model [Vette, 1991b], the trapped electron fluxes predicted by the AE-8 models at $L < 3$ are different for the different local time zones as shown in Figures 6a and 6b.

3.1. "Errors" in the Existing Models

In general, the "errors" in the NASA empirical trapped radiation models are difficult to assess and quantify precisely [Vette, 1991a]. We almost have to evaluate the situations on a case-by-case basis.

For example, as mentioned above in comparing the trapped proton (Figures 5a and 5b) and trapped electron (Figures 6a and 6b) observations with the NASA AP-8 MAX and AE-8 MIN models, we have intentionally used a contemporary, interpolated-IGRF magnetic field model to examine the performance of the models under the circumstances. This of course is inconsistent with the recommendations that proper comparisons with the trapped radiation models should adhere to using a field model appropriate to their respective epochs (see Tables 1 and 2) [Vette, 1991a; Gaffey and Bilitza, 1994; Heynderickx et al., 1996]. Nevertheless, such procedure appears to be a common practice among users of the models and has been regarded as a major contributor to the discrepancies between the models and observations.

Both the AP-8 and AE-8 model curves in the Figures 5 and 6 show significant departure from individual observations and predict much lower fluxes in the inner radiation zone ($L < 3$). The most dramatic effect is seen in the electron profile near local noon (Figure 6b) in which the AE-8 MIN model is practically absent. This latter effect may be attributed to the incongruence between the contemporary magnetic field used to produce the model curves in Figures 5 and 6 and the original magnetic field models (B and L) used to construct the model flux lookup tables.

On the other hand, it would appear that both the AP-8 MAX and AE-8 MIN models based on an updated mag-

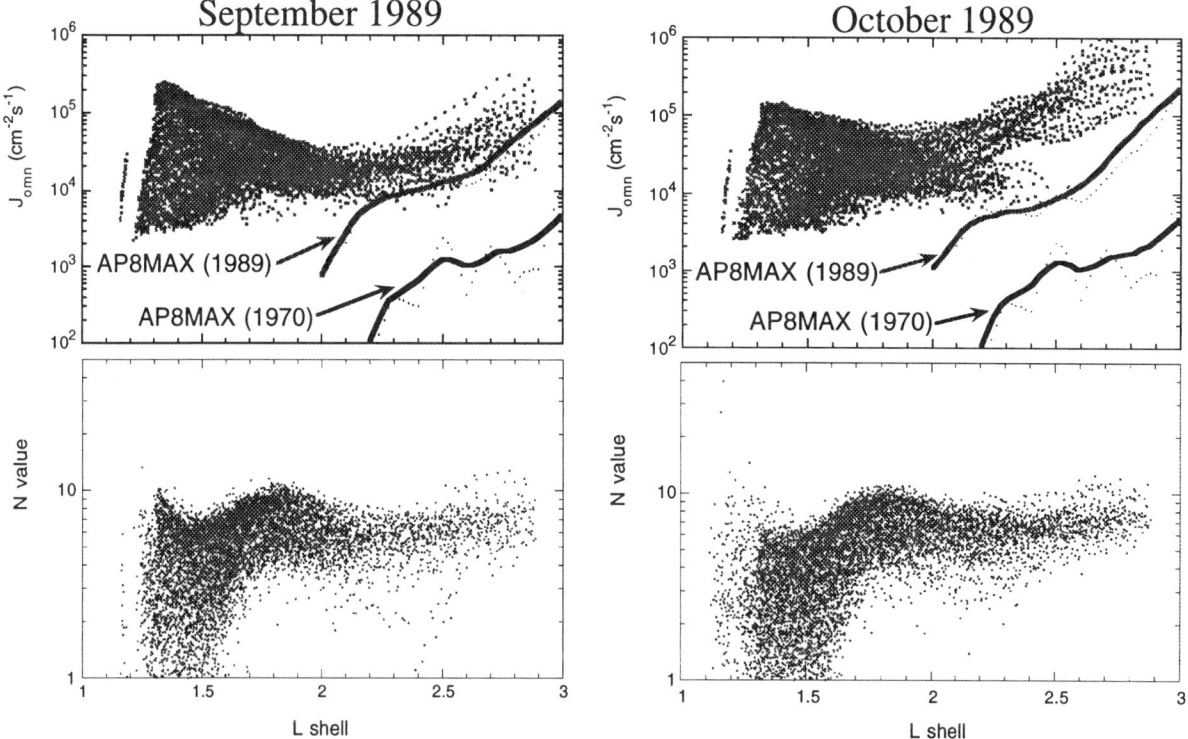

Figure 5. L-shell distributions of omni-directional fluxes and anisotropy indices N of 80–250 keV protons calculated from NOAA-10 measurements in (a, left) September (with $M_0 = -0.3962$, $M_1 = 3.9538$, and $M_2 = -0.5671$) and (b, right) October (with $M_0 = -0.81$, $M_1 = 6.1723$, and $M_2 = -3.4656$) of 1989. The high fluxes with small N values at $L < 1.7$ suggest the presence of contamination by energetic (> 20 MeV) protons in the inner radiation zone.

netic field perform reasonably well in the outer radiation zone ($3 < L < 7$) in that the steady-state model flux levels are consistent with the (logarithmic) average observed fluxes. The apparent agreement may be due to the relative insensitivity of the geomagnetic field in this region to changes in the high order magnetic moments and the use of B/B_0 (instead of B) and L as the basic coordinate system effectively nullifies the changes caused by the variations in the dipole moment [*Gaffey and Bilitza*, 1994; *Vette and Sawyer*, 1986].

In order to assess the model performance in the presence of secular changes of the geomagnetic field, the AP-8 MAX (1970) and AE-8 MIN (1964) models are also displayed in Figures 5 and 6. The models predict dramatic increases in the trapped proton fluxes at < 1 MeV as a result of the secular changes in the magnetic field, similar to those reported earlier for much higher energy particles ($\gg 10$–100 MeV) (e.g. *Gaffey and Bilitza* [1994], *Heynderickx et al.* [1996]). Such increases have generally been regarded as unrealistic [*Heynderickx et al.*, 1996], particularly in view of the apparent overestimation of > 100 MeV proton dosages by the AP-8 MIN model [*Gussenhoven et al.*, 1991]. Though the earlier conclusion was obtained only by adapting different magnetic field models to the AP-8 model, as we have also done, the consistency between the models and observations shown in Figures 5 and 6 tend to support an actual increase in the < 1 MeV trapped proton fluxes over the course of twenty years; and that the trapped radiation model based on an updated magnetic field acutally performs better. More in-depth studies are therefore needed before a best approach to applying the AE-8 and AP-8 models to predict long-term radiation exposure can be determined.

As we have seen, a fair comparison between the trapped radiation models and observations is difficult to ascertain. Apparent discrepancies between the models and observations may result when:

1. different quantities, such as unidirectional fluxes, omni-directional fluxes and dosages, are compared as each of these comparisons will involved different sets of added assumptions (i.e., the results of comparisons may be application dependent.); or

2. long-term averages represented by the models are compared to individual measurements which are more susceptible to short-term geomagnetic variations and are likely to differ from the models, sometimes by a large margin; or

3. there is a lack of commonality in the time, space, and energy coverages and in the magnetic field models used as in the above example.

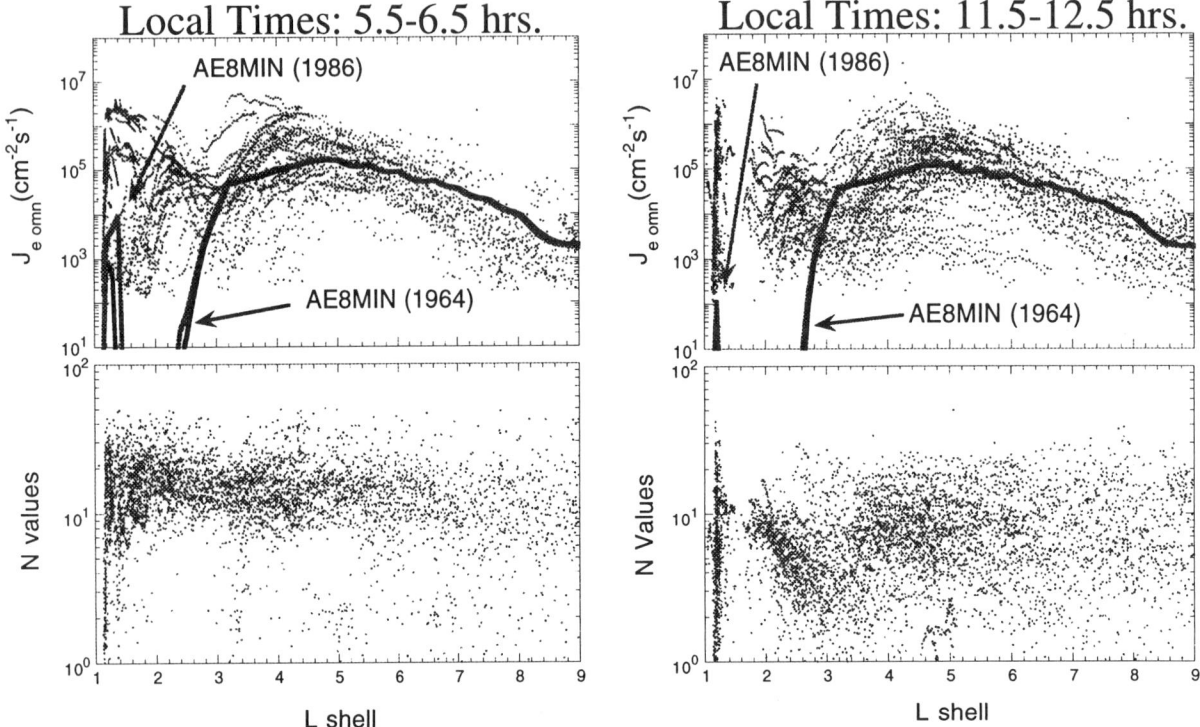

Figure 6. L-shell distributions of omni-directional fluxes and anisotropy indices N of 0.19–3.2 MeV electrons calculated from OHZORA measurements taken in 1984–87 during quiet periods ($|D_{st}| < 30$ nT in the (a, left) 5.5–6.5 hrs (with $M_0 = -0.2453$, $M_1 = 3.8966$, and $M_2 = -0.5277$) and (b, right) 11.5–12.5 hrs (with $M_0 = -0.2342$, $M_1 = 3.984$, and $M_2 = -0.6346$) local time sectors.

In addition, large discrepancies will occur also in regions of steep gradients such as in low altitude and low L (< 3) regions where the trapped particles encounter sharp atmospheric and magnetic cutoffs, at high L-values (> 3) where the electrons fluxes are highly modulated by geomagnetic activities, and at high energies at which the particle spectra are steep and inadequately sampled.

3.2. Deficiencies in Existing Models

A number of deficiencies in the NASA trapped radiation models can readily be identified. They are:

1. lack of long-term data for adequate spatial coverages;
2. old data sets did not uniformly provide spectral and pitch angle distributions;
3. only solar minimum and maximum, no other dynamical variations such that short-term radiation environment prediction is unreliable, nor possible;
4. inaccurate at low altitudes (< 1000 km);
5. cannot be easily updated with new or additional data; and
6. no explicit local-time dependence in AP-8 and in AE-8 at $L < 5$.

4. A NEW GENERATION OF TRAPPED RADIATION MODELS

4.1. Requirements on Next Generation Models

The greatest deficiency in the existing models may be their inability to predict the dynamical behavior of radiation environment. It has been known that the radiation belt electron fluxes can enhance by a few orders of magnitude during geomagnetic storms. The slot region can be filled in quite rapidly and would take days to recover (see e.g., *Baker et al.* [1994]). Generally speaking, temporal or dynamic variations of the trapped radiation can be affected by geomagnetic activities as well as the solar cycle variations.

Unlike physical models in which the time evolution of a system is simulated, empirical models can only provide a snapshot of a system under a given set of average conditions approximated by the data sets used to construct the models. Therefore, discrepancies are likely to result from comparing observations to an empirical model, unless the physical conditions under which the observations were made exactly match the average conditions which defined the models.

The existing trapped radiation models are ineffective at low altitudes (< 1000 km) where low earth-orbiting spacecraft,

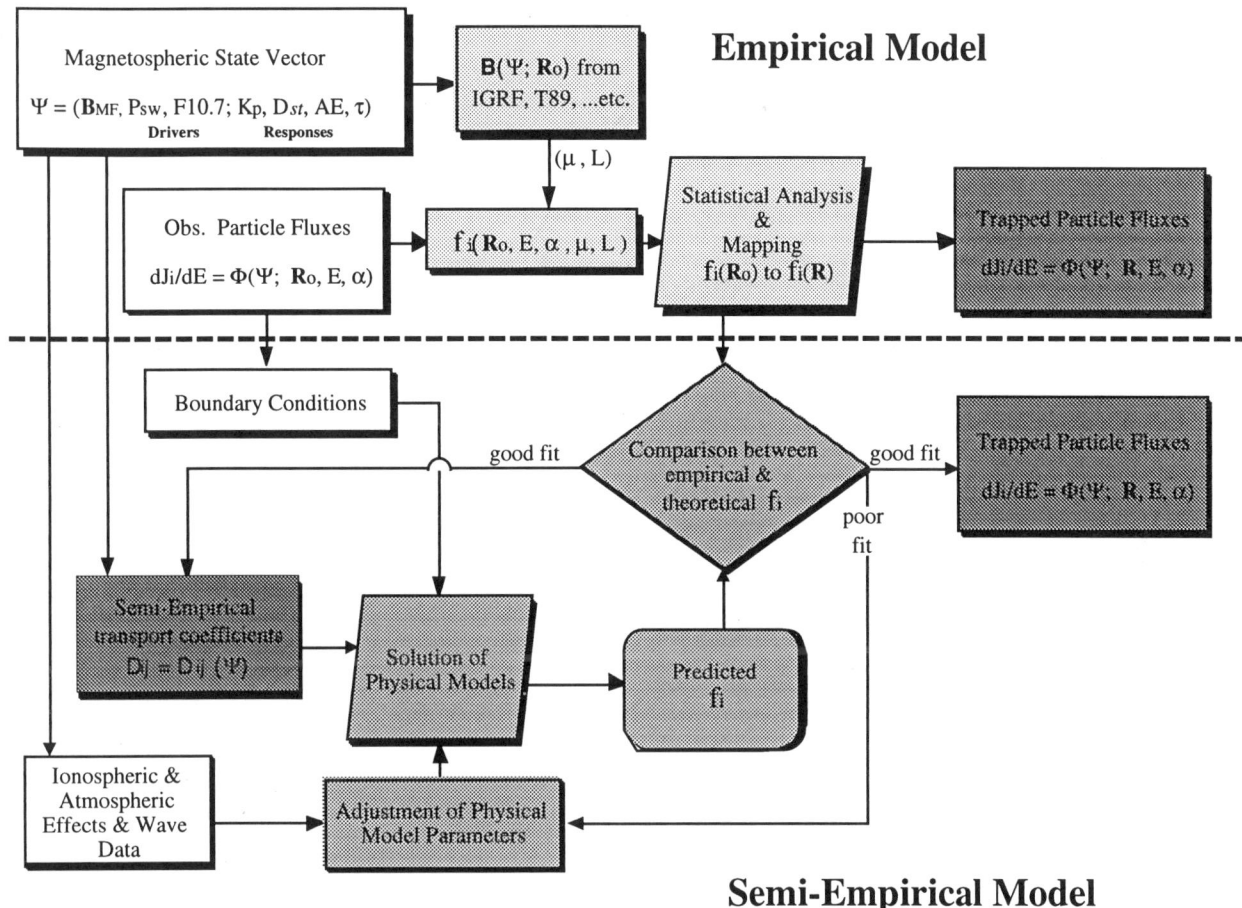

Figure 7. Flow chart of a new generation semi-empirical trapped radiation model.

including the space shuttles, the space stations, and many scientific missions are flown. Thus, extensions of the models to low altitudes are needed by incorporating atmospheric precipitation and absorption and the associated geophysical and geomagnetic processes.

Other enhancements to the models may include a better definition of the South Atlantic Anomaly (SAA) region, in spatial location vs. altitude and particle energy, and in the presence of dynamic and temporal changes (e.g., new proton belt, and westward drift). As routine, long-term data on heavy ions become available, they should also be incorporated.

4.2. Model Architecture

Unlike the purely empirical models, the new generation trapped radiation models will be semi-empirical, combining recent advances in physical modeling techniques and state-of-the-art data and information management technology. A schematic of the modular architecture of the model is shown in Figure 7. Although there are two parts to the full model, separated by the dashed line, the purely empirical and semi-empirical portions complement each other and can function quite independently.

The quantities in the white boxes in Figure 7 are input to the models while those in the dark gray boxes are the nominal output. The light grey parts of the chart represent the core of the empirical model where the particle data associated with a given magnetospheric state (see below) and observed at various locations **R** are suitably mapped onto their corresponding magnetic coordinates (e.g., μ and L in two dimensions), based on a magnetic field model (e.g., IGRF, T89, etc.). The projected particle distribution functions can then be statistically analyzed to produce the model output fluxes.

The medium grey parts of Figure 7 constitute the physical modeling portion where a physics-based model is implemented. Currently, a three-dimensional diffusive model after those of *Beutier et al.* [1995], *Beutier and Boscher* [1995] and *Bourdarie et al.* [1996] is being implemented. As seen from the flow chart, the physical model depends on the magnetospheric conditions and will have observations as boundary conditions. Work is being done on investigating the empirical dependence of the transport coefficients on the geomagnetic conditions. It is envisioned that the semi-physical model can be used to fill the data gaps in the empirical trapped radiation data base.

4.3. *The Data Base and the New Empirical Models based on "Magnetospheric States"*

Since all empirical models are constructed by statistically analyzing a collection of data that are "representative" or "typical" of some conditions, any variations of the model must be brought about by choosing different data sets which correspond to discernibly different geomagnetic conditions. Our ability to identify a data sample which corresponds to a well-defined set of conditions is therefore central to the validity of the models.

Since the motions and dynamics of charged particles are largely controlled by magnetic fields, it is therefore reasonable to assume that the trapped radiation environment in the magnetosphere evolves with the geomagnetic field as well. *Hoffman et al.* [1988; 1994] and *Fujii et al.* [1994] have shown that the large scale magnetospheric electrodynamic patterns as seen in the field-aligned currents, ionospheric convection patterns, and particle precipitation boundaries in the polar region can more or less be defined generically during quiescent and substorms periods. This suggests that the state of the particle environment can be specified if the configuration of the geomagnetic field is known.

A trapped energetic particle data base $\mathbb{D}_i$ for the ith species particles (electrons, protons, and ions) can be written symbolically as

$$\mathbb{D}_i = \{\boldsymbol{\Psi}; \boldsymbol{\Phi}_i\}, \qquad (2)$$

where $\boldsymbol{\Psi}$ is a "magnetospheric state vector" which has all the necessary and sufficient parameters for identifying the "state" of the magnetosphere. An example of $\boldsymbol{\Psi}$ may be given by

$$\boldsymbol{\Psi} = [\mathbf{B}_{\text{IMF}}, P_{\text{SW}}, F_{10.7}; K_{\text{p}}, D_{\text{st}}, AE; \boldsymbol{\tau}], \qquad (3)$$

in which the interplanetary magnetic field vector $\mathbf{B}_{\text{IMF}}$, the solar wind pressure P_{SW}, and the solar $F_{10.7}$ radio flux are the driver parameters, while the geomagnetic indices, K_{p}, D_{st} and AE, are the magnetospheric response parameters; the vector $\boldsymbol{\tau}$ contains the relative time delays (correlation or response times) associated with the various magnetospheric state elements (e.g., *Bargatze et al.* [1985], *Blanchard and McPherron* [1995]).

The example of $\boldsymbol{\Psi}$ given here by no means suggests a unique set of all possible parameters. The temporal gradient of these and other geomagnetic parameters may also be important in defining a magnetospheric state [*Ahn et al.*, 1992; *Goertz et al.*, 1993]. On the other hand, the apparent correlations among some of the geomagnetic indices [*Cade et al.*, 1995; *Blanchard and McPherron*, 1995] may eventually decrease the number of necessary parameters. We should note in any case that the magnetospheric states as defined here are solely for operational purposes (such as selection of subset of data pertinent to a given set of geomagnetic conditions) and need not be discrete. They are, however, analogous to phases of storms and substorms in that they are qualitatively different parts of a continuous process.

The "particle flux vector" $\boldsymbol{\Phi}_i = \boldsymbol{\Phi}$ contains all the information associated with the particle measurements. Because of the multi-dimensional nature of particle measurements, $\boldsymbol{\Phi}$ will be a collection of data files in common data format (CDF) (see below), i.e., $\boldsymbol{\Phi} = \{\boldsymbol{\Phi}_{\boldsymbol{\Psi}}\}$. Each $\boldsymbol{\Phi}_{\boldsymbol{\Psi}}$ file contains a subset of data (a single set of measurements) associated with $\boldsymbol{\Psi}$. A typical $\boldsymbol{\Phi}_{\boldsymbol{\Psi}}$ may then be given by

$$\boldsymbol{\Phi}_{\boldsymbol{\Psi}} = [\mathbf{R}; B/B_0, L; E, \theta, J_\theta, J_{\text{omn}}], \qquad (4)$$

where $\mathbf{R}$ is the position of observation with corresponding magnetic coordinates, e.g., B/B_0 and L based on a magnetic field model, E is the particle energy, θ is the pitch angle, J_θ is the directional flux, and J_{omn} is the corresponding omni-directional flux. As many of the details about the individual data sets are preserved by the data files in CDF (see below), very little pre-processing, such as averaging of data sets, will be needed. Finally, we should recognize that for any given $\boldsymbol{\Psi}$, there will be portions of phase space where no particle measurements are available. In such cases, we expect that the physical model can be employed to extrapolate between point observations.

Since the particle measurements taken at different times but having similar associated $\boldsymbol{\Psi}$ are combined to form different $\boldsymbol{\Phi}_{\boldsymbol{\Psi}}$ for different locations $\mathbf{R}$, the trapped radiation data $\boldsymbol{\Phi} = \{\boldsymbol{\Phi}_{\boldsymbol{\Psi}}\}$ in $\mathbb{D}_i$ are identifiable only by their associated state vector $\boldsymbol{\Psi}$. Thus, $\mathbb{D}_i$ is an "object-oriented" data base in which the data are selectable by the parameters (elements of $\boldsymbol{\Psi}$) which characterize the state of the magnetosphere (the object).

4.4. *Construction of Empirical Models*

With the data base $\mathbb{D}_i$ sorted by magnetospheric states, a trapped radiation model suitable for a given magnetospheric state $\boldsymbol{\Psi}_0$ can be constructed by selecting all $\boldsymbol{\Phi}_{\boldsymbol{\Psi}}$ with $\boldsymbol{\Psi} \approx \boldsymbol{\Psi}_0$ and averaging the particle fluxes (J_θ or J_{omn}) for each location $\mathbf{R}$. A separate model for each $\boldsymbol{\Psi}_0$ can then be constructed. Since there is no explicit time dependence in $\mathbb{D}_i$, therefore all the models constructed will be static.

Recently, work has been done in investigating the predictions of magnetospheric activities based on their correlations with solar wind parameters [*Baker et al.*, 1990; *Goertz et al.*, 1993; *Blanchard and McPherron*, 1995]. Simple electromechanical models have also been devised to model the magnetospheric responses to various solar wind input [*Vassiliadis et al.*, 1993; *Klimas et al.*, 1992; 1994]. It is envisioned that these works can provide the basis for predicting the time evolution of the geomagnetic indices and thus effectively provide the means to enable the otherwise static, empirical models to follow the temporal variations of the trapped radiation environment.

4.5. *Common Data Format (CDF)*

The NSSDC's CDF [*Goucher and Mathews*, 1994; *Goucher et al.*, 1994] is a data file structure designed for storing multi-dimensional data, including both the data and information about the data (meta-data) in one binary file. Because of the self-documenting and self-describing nature of the CDF, all the information of and about a given data set is retained and is readily available. As every piece of data within a CDF data file can be addressed and recalled independently, the file format is particularly useful for managing large volumes of multi-dimensional data sets. A CDF file is also suited for data manipulations and analysis, and is transportable among various computer platforms (UNIX, VAX, PC, Macintosh). As a binary file, a CDF data file requires much less computer storage capacity than the equivalent ASCII file, and is thus better suited for archival purposes.

Because of its uniformity and versatility, the NSSDC's Common Data Format has widely been adopted as the data format by many recent space missions, including those of the International Solar-Terrestrial Program (ISTP). The ISTP Guidelines for CDF [*Kessel et al.*, 1993], designed for space physics implementations, have served to form the unifying data standards. Consequently, they have also been adopted by various science campaigns of the Interagency Consultative Group (IACG), formed by the multi-national space agencies (ESA, ISAS, IKI, and NASA) to promote understanding of the complex space plasma environment surrounding Earth, other planets, and in the interplanetary space. As the successes of these programs rely on extensive sharing and analysis of data from many different data sources, which is also a key to the successful construction of a comprehensive trapped radiation data base $\mathbb{D}_i$ as described above, CDF data files written according to the ISTP guidelines will provide the commonality and basis for inter-comparisons of different data sets.

A set of data retrieval and analysis tools based on CDF data files are being developed at the NSSDC. The goal is to make the CDF data products useful, understandable, and available in standardized and automated forms that are convenient for research and analysis purposes. We anticipate that the CDF and the associated data management and analysis tools will also aid the development of the next generation trapped radiation models. Further information on CDF and tools can be obtained via the internet on the world-wide web at **http://nssdc.gsfc.nasa.gov/space/spdf/sp_use_of_cdf.html**.

5. SUMMARY AND CONCLUSIONS

We have reviewed and compared the NASA AP-8 (MAX) and AE-8 (MIN) trapped radiation models with the trapped particle measurements taken by the NOAA-10 and the Japanese OHZORA satellites. These efforts have led to the discussion and identification of a number of deficiencies in the existing models. As a renewed and collaborative effort between the NSSDC and the SPDF of the NASA/Goddard Space Flight Center to construct a new generation of trapped radiation models, the new approach will be semi-empirical, departing from the purely empirical formulations.

The technique for data base construction, outlined in Sect. 4., will take advantage of the state-of-the-art data management and analysis tools that are only available in recent years, and will effectively allow the empirical models to reflect geomagnetic variations. It is envisioned that the current effort will result in the development of a set of trapped radiation models which will have the following advantages:

1. Data sets, remaining distinct, are not pre-processed by gross statistical averaging;
2. "easily" updated and extended by new data sets;
3. Multi-dimensional data structure allows the retention of physical details of data (e.g., pitch angle distributions);
4. Semi-empirical approach is more physical than pure statistical extrapolations; and
5. Models vary with geomagnetic conditions.

In addition, we anticipate that our effort will promote national and international collaborations for exchange of data and technical expertise.

Acknowledgements. We thank T. Kohno (IPCR, Saitama, Japan) and K. Nagata (Tamagawa University, Tokyo, Japan) for providing the OHZORA energetic particle data to the NSSDC, and the National Geophysical Data Center (NGDC) for making the NOAA-10 data available. The author would also like to thank L.C. Tan for his assistance in data analysis. Discussions with D.M. Sawyer, J.H. King, M.J. Teague, D. Bilitza, J.F. Cooper and D. Boscher are also appreciated.

REFERENCES

Ahn, B.-H., Y. Kamide, H.W. Kroehl, and D.J. Gorney, Cross-polar cap potential difference, auroral electrojet indices, and solar wind parameters, *J. Geophys. Res., 97*, 1345, 1992.

Baker, D.N., R.L. McPherron, T.E. Cayton, and R.W. Klebesadel, *J. Geophys. Res., 95*, 15133–15140, 1990.

Baker, D.N., J.B. Blake, L.B. Callis, J.R. Cummings, D. Hovestadt, S. Kanekal, B. Klecker, R.A. Mewaldt, and R.D. Zwickl, Relativistic electron acceleration and decay time scales in the inner and outer radiation belts: SAMPEX, *Geophys. Res. Lett., 21*, 409–412, 1994.

Bargatze L.F., D.N. Baker, R.L. McPherron, and E.W. Hones, Magnetospheric impulse response for many levels of geomagnetic activity, *J. Geophys. Res., 90*, 6387–6394, 1985.

Beutier, T., and D. Boscher, A three-dimensional analysis of the electron radiation belt by the Salammbo code, *J. Geophys. Res., 100*, 14853–14861, 1995.

Beutier, T., D. Boscher, and M. France, SALAMMBO: A three-dimensional simulation of the proton radiation belt, *J. Geophys. Res., 100*, 17181–17188, 1995.

Blanchard, G.T., and R.L. McPherron, Analysis of the linear response function relating Al to VBs for individual substorms, *J. Geophys. Res., 100*, 19155–19165, 1995.

Boberg, P.R., A.J. Tylka, J.H. Adams Jr., E.O. Fluckiger, and E. Kobel, Geomagnetic transmission of solar energetic protons during the geomagnetic disturbances of October 1989, *Geophys. Res. Lett., 22*, 1133–1136, 1995.

Bourdarie, S., D. Boscher, and T. Beutier, Dynamic physical modeling of trapped particles for satellite survey, Proceedings of the Workshop on Radiation Belts: Models and Standards, this volume, 1996.

Cade III, W.B., J.J. Sojka, and L. Zhu, A correlative comparison of the ring current and auroral electrojets using geomagnetic indices, *J. Geophys. Res., 100*, 97–105, 1995.

Fujii, R., R.A. Hoffman, P.C. Anderson, J.D. Craven, M. Sugiura, L.A. Frank, and N.C. Maynard, Electrodynamic parameters in the nighttime sector during auroral substorms, *J. Geophys. Res., 99*, 6093–6112, 1994.

Gaffey, J.D., and D. Bilitza, NASA/National Space Science Data Center Trapped Radiation Models, *J. Spacecraft and Rockets, 31*, 172–176, 1994.

Goertz, C.K., L.-H. Shan, and R.A. Smith, Prediction of geomagnetic activity, *J. Geophys. Res., 98*, 7673–7684, 1993.

Goucher, G.W., and G.J. Mathews, A Comprehensive Look at CDF, NSSDC/WDC-A-R&S 94-07, NASA/Goddard Space Flight Center, August 1994.

Goucher G.W., J. Love, and H. Leckner, A Discipline-Independent Scientific Data Management Package—The National Space Science Data Center's (NSSDC) Common Data Format (CDF), Proceedings of the 1992 STEP Symposium/5th COSPAR Colloquium, Applied Physics Laboratory, Laurel, Maryland, February 1994.

Gussenhoven, M.S., E.G. Mullen, D.H. Brautigam, E. Holeman, C. Jordan, F. Hanser, and B. Ditcher, Preliminary comparison of dose measurements on CRRES to NASA model predictions, *IEEE Trans. Nucl. Sci., 38*, 1655–1662, 1991.

Hoffman, R.A., M. Sugiura, N.C. Maynard, R.M. Candey, J.D. Craven, and L.A. Frank, Electrodynamic patterns in the polar region during periods of extreme magnetic quiescence, *J. Geophys. Res., 93*, 14515–14541, 1988.

Hoffman, R.A., R. Fujii, and M. Sugiura, Characteristics of the field-aligned current system in the nighttime sector during auroral substorms, *J. Geophys. Res., 99*, 21303–21325, 1994.

Heynderickx D., and A. Beliaev, Identification of an error in the distribution of the NASA model AP-8 MIN, *J. Spacecraft and Rockets, 32*, 190–192, 1994.

Heynderickx, D., J. Lemaire, and E. Daly, Historical review of the different procedures used to compute the *L*-parameter, *Radiat. Meas.*, in press, 1996.

Kessel, R.L., R.E. McGuire, H.K. Hill, N.J. Schofield, Jr., and J. Love, ISTP Guidelines for CDF, in International Solar-Terrestrial Physics (ISTP) Key Parameter Generation Software (KPGS) Standards and Conventions, Version 1.2, W. H. Mish, NASA, 1993. (Request copy through the ISTP project at NASA/Goddard Space Flight Center)

Klimas, A.J., D.N. Baker, D.A. Roberts, D.H. Fairfield, and J. Buchner, A nonlinear dynamics analogue model of geomagnetic activity, *J. Geophys. Res., 97*, 12253, 1992.

Klimas, A.J., D.N. Baker, D. Vassiliadis, and D.A. Robertss, Substorm recurrence during steady and variable solar wind driving: evidence for a normal mode in the unloading dynamics of the magnetosphere, *J. Geophys. Res., 99*, 14855, 1994.

Lemaire, J., A.D. Johnstone, D. Heynderickx, D.J. Rogers, S. Szita, and V. Pierrard, Trapped Radiation Environment Model Development, TREND 2 Final Report, *Aeronomica Acta, A 393*, Belgian Institute for Space Aeronomy, 1995.

Nagata, K., T. Kohno, H. Murakami, A. Nakamoto, N. Hasebe, T. Takenaka, J. Kikuchi, and T. Doke, OHZORA high energy particle observations, *J. Geomag. Geoelectr., 37*, 329–345, 1985.

Panasyuk, M.I., Empirical models of terrestrial trapped radiation, *Adv. Space Res., 17*, (2)137–(2)145, 1996.

Raben, V.J., D.S. Evans, H.H. Sauer, S.R. Sahm, and M. Huynh, TIROS/NOAA satellite space environment monitor data archive documentation: 1995 update, NOAA Tech. Memo. ERL SEL-86, Space Environment Lab, Boulder, CO, February, 1995.

Reeves, G.D., T.E. Cayton, S.P. Gasry, and R.D. Belian, The great solar energetic particle events of 1989 observed from geosynchronous orbit, *J. Geophys. Res., 97*, 6219–6226, 1992.

Sawyer, D.M., and J.I. Vette, AP-8 Trapped Proton Environment for Solar Maximum and Solar Minimum, NSSDC/SDC-A-R&S 76-06, NASA Goddard Space Flight Center, Greenbelt, Maryland, 1976.

Vassiliadis, D., A.S. Sharma, K. Papadopoulos, An empirical model relating the auroral geomagnetic activity to the interplanetary magnetic field, *Geophys. Res. Lett., 20*, 1731–1734, 1993.

Vette, J.I., The NASA/National Space Science Data Center Trapped Radiation Environment Model Program (TREMP) (1964–1991), NSSDC/WDC-A-R&S 91-29, NASA Goddard Space Flight Center, Greenbelt, Maryland, November, 1991a.

Vette, J.I., The AE-8 Trapped Electron Model Environment, NSSDC WDC-A-R&S 91-24, NASA Goddard Space Flight Center, Greenbelt, Maryland, November, 1991b.

Vette, J.I., and D.M. Sawyer, Short report on radiation belt calculations, unpublished, 1986.

S.F. Fung, Space Physics Data Facility, Code 632, NASA Goddard Space Flight Center, Greenbelt, Maryland

DISCUSSION

Q: G.V. Popov. I suppose that your definition of the vector Ψ is a draft one only? I mean that the parameters or indices which must be included in Ψ need further investigation, don't they?

A: S.F. Fung. Yes, the listed state vector elements are given as examples of parameters which are driven by magnetospheric processes operating at different time scales. In general, one might find it necessary to over-determine the system to have a better prescription of the magnetospheric state.

Q: A.A. Beliaev. Is NASA going to include "old" datasets already in the AE/AP models in "new generation" models?

A: S.F. Fung. Yes, to the extent that they remain available and are of good quality.

Q: G.D. Reeves. The limitations of the AP-8 model in the South Atlantic Anomaly might be over emphasized by March, September and October 1989. In addition to being storm times these were 3 of the largest solar energetic particle events of the last solar cycle. Backgrounds may make the NOAA data unreliable?

A: S.F. Fung. Those periods were chosen because they were large storm times.

Q: D.N. Baker. Do you consider the recent history of how the magnetosphere arrived at a given "state"? That is, do you consider the dynamics of various parameters prior to the specified time and thus allow for a kind of "hysteresis" in the state variables?

A: S.F. Fung. Yes, the various magnetospheric state parameters are to be taken at some appropriate time constants, represented by τ collectively, prior to the time of interst. The time constants associated with different parameters will likely be different due to the varying characteristic time scales of different magnetospheric processes.

Q: D.N. Baker. Could you please elaborate upon the fast search method that you use to find the "nearest neighbours" of a given state of the magnetosphere?

A: S.F. Fung. The time it takes to perform a near-neighbour search in a database depends on both the architecture of the database and the algorithm used. By presorting and binning the data into appropriate ranges or windows, the search time can be shortened dramatically.

Q: D. Heynderickx. The magnetic field model used to build AP-8 MAX is not Jensen & Cain but GSFC 12/66, updated to

1970. This latter magnetic field model should be used when accessing AP-8 MAX?

A: S.F. Fung. Thank you for pointing that out. When comparing observations to the NASA trapped radiation models (AE-8 and AP-8), it is important to bear in mind that serious discrepancies may result from the fact that a different magnetic field model may have been used to process the data.

Q: J.F. Lemaire. The state vector should contain independent solar wind and geomagnetic parameters or indices, not interdependent ones.

A: S.F. Fung. Yes, the example of a magnetospheric state vector given in my presentation does contain the solar wind "driver" parameters and the magnetospheric "response" parameters. A set of time delays, represented by τ, is also included to anticipate the "hysteresis" of the magnetosphere. The specific state parameters used are only examples of parameters which may reflect magnetospheric processes operating at various time scales.

Phillips Laboratory Space Physics Division Radiation Models

M.S. Gussenhoven, E.G. Mullen and D.H. Brautigam

Space Physics Division, Phillips Laboratory, Hanscom AFB, Massachusetts

A summary is given of empirical models constructed from data taken onboard the Combined Release and Radiation Effects Satellite (CRRES). CRRES, in a low-inclination, geosynchronous-transfer orbit, returned data from July 1990 to October 1991. This encompassed a very interesting period of solar activity, just post-solar maximum. In particular, CRRES measured the effects on the inner magnetosphere of the solar particle and shock events that occurred in March 1991. These effects were major and long-lasting for high energy protons (10–100 MeV) and electrons (> 10 MeV). CRRES models for proton flux and high LET and low LET dose were created for time periods before and after the event. Data from the DMSP and APEX satellites extend the low altitude portion of the dose maps to other times in the solar cycle. A quasi-static outer zone electron flux model was also constructed using the CRRES data. The outer zone relativistic electron population is extremely variable. It is reasonably well-ordered by the 15-day running average of the magnetic index A_p. Six A_{p15} ranges were used to make the electron models. Use of these models with A_p provide a reasonably accurate time-varying representation of the outer zone. Comparison of the electron and proton flux models to the NASA AP-8 MAX and AE-8 MAX models show substantive differences. These lead directly to questions about replacement of models which are standards in the industry. Issues which need to be considered in model replacement are discussed.

1. CRRES RADIATION MODELS

As part of the U.S. Air Force Space Radiation Effects Program (SPACERAD) [*Gussenhoven et al.,* 1985; *Gussenhoven and Mullen,* 1993] some 23 particle and field experiments were flown through the inner magnetosphere on the Combined Release and Radiation Effects Satellite (CRRES) in a 350 by 33,500 km, 18°-inclined orbit. The instruments collected data specifying the radiation belts for 14 months from July, 1990 to October, 1991, near the maximum of Solar Cycle 22. A major goal of the SPACERAD Program is an evaluation of the radiation models currently in widespread use, namely the NASA AE-8 and AP-8 models [*Vette,* 1991; *Sawyer and Vette,* 1976]. Since these models are specified for either solar maximum or solar minimum conditions, the CRRES data can only be used to evaluate the solar maximum models. What we saw in the data from CRRES, however, were dynamical variations in the high energy populations that were much greater than those between the solar maximum and solar minimum averages in the NASA models. For outer zone electrons the large dynamical variations occur on the order of tens of days. For protons and electrons below the outer zone only one major change occurred during the CRRES lifetime, namely following the shock that led to a great magnetic storm in March 1991.

The models created with the CRRES data are attempts to define the dynamical variations that occurred during the satellite lifetime and are averages of data over time periods considered appropriate to the variations. The best indicator of the shape and intensity of the outer zone electron profile was found to be a 15-day running average of the linear magnetic activity index, A_p, offset by one day. Only two states were found for the high energy protons and > 10 MeV electrons: the state existing before the March storm (the

Radiation Belts: Models and Standards
Geophysical Monograph 97
This paper is not subject to U.S. copyright.
Published in 1996 by the American Geophysical Union

quiet state) and the state existing after the storm (the active state). Models of the trapped electron and proton fluxes using CRRES data are averages, in L-bins of 1/20th R_E, of the pitch-angle-dependent data mapped to the magnetic equator ($B/B_0 = 1$) using a combined IGRF 1985 internal [*IAGA Division I Working Group*, 1986] and Olson-Pfitzer quiet external [*Olson and Pfitzer*, 1974] magnetic field models. With knowledge of pitch angle dependence, the model particle flux on the equator can be mapped back down the field lines to specify flux for all B/B_0 values. In addition to the electron and proton flux models, dose and cosmic ray models were created with the CRRES data. The dose models are in-place averages of dose data taken behind four hemispherical thicknesses of aluminum. The model grid in which the averages are constructed is every 1/20th R_E in L and every increment in B/B_0 equivalent to 2 degrees in magnetic latitude. The cosmic ray model uses the CRRES data to extrapolate heavy ion flux models of the galactic cosmic rays into the present epoch. The CRRES models were encapsulated in Personal Computer (PC) utilities and distributed to the community. The available PC utilities with a brief summary of their output and references to the models and the documentation are listed below.

1.0.1. CRRESRAD. This utility for the CRRES Space Radiation dose models calculates expected satellite dose accumulation behind aluminum hemispherical shielding for four different thicknesses, for user-specified orbits, and for quiet (before the March storm), active (after the March storm) or average geophysical activity levels [*Gussenhoven et al.*, 1992; *Kerns and Gussenhoven*, 1992]. The dose is separated into low LET and high LET contributions.

1.0.2. CRRESPRO. This utility for the CRRES proton flux models calculates proton omnidirectional fluence and integral omnidirectional fluence over the energy range 1–100 MeV for user-specified orbits and quiet (before the March storm) or active (after the March storm) geophysical activity levels [*Gussenhoven et al.*, 1993; *Meffert and Gussenhoven*, 1994].

1.0.3. PROSPEC. This utility for the CRRES proton flux models calculates proton differential flux or distribution function as a function of pitch angle over the energy range 1–100 MeV for user-specified positions or fractions of orbits and for quiet or active activity levels [*Gussenhoven et al.*, 1993; *Meffert et al.*, 1996].

1.0.4. CRRESELE. This utility for the CRRES electron models calculates outer zone electron omnidirectional fluence (differential and integral) over the energy range 0.5–6.6 MeV for six ranges of geomagnetic activity specified by a 15-day running average of A_p ($A_{p_{15}}$), and for user-specified orbits [*Brautigam et al.*, 1992; *Brautigam and Bell*, 1995a,b].

1.0.5. CHIME. This utility for the CRRES heavy ion model of the environment gives differential energy flux and LET spectra for all stable elements over the energy range 10 MeV/Nucleon–60 GeV/Nucleon from major cosmic ray sources for any specified period of time during the years 1970–2010 as a function of solar modulation [*Chenette et al.*, 1994].

1.0.6. PL-GEOSpace. In addition to the CRRES models and utilities, a 3-dimensional display capability, called PL-GEOSpace, is being developed at PL/GP which has the ability to accept certain space environment models and show, using that model, the space environment at and surrounding a given satellite as it proceeds in orbit [*Ginet et al.*, 1995]. Figure 1 is a screen print (reduced to gray-scale from the color original) from PL-GEOSpace showing the 2.0 MeV outer zone electron environment experienced by a GPS satellite for a quiet period ($A_{p_{15}} < 7.5$) and an active period ($A_{p_{15}} > 25$). The line plot on the left shows the flux history and future for the satellite using the two models. The gray dot shows the present position of the satellite both in the 3-D displays and in the line plot. The electron flux difference between the two models is almost two orders of magnitude. PL-GEOSpace will allow users tracking satellite operations to see the environment they have just experienced and the environment they may anticipate as predicted by a wide range of models of the environment, including the most up-to-date versions of the models.

2. MAJOR DIFFERENCES BETWEEN THE CRRES AND NASA TRAPPED RADIATION MODELS

The CRRES trapped proton and electron models have been compared to the NASA models in a variety of ways [*Gussenhoven et al.*, 1991, 1992, 1993; *Brautigam et al.*, 1992; *Mullen et al.*, 1992]. In addition, we have recently completed a review of the major differences found between the CRRES electron and proton flux models, and those of AE8MAX and AP8MAX [*Gussenhoven et al.*, 1996]. There are substantial differences in construction and output between the CRRES and NASA models for both species. These differences are summarized here.

2.1. *Period of Data Accumulation*

The CRRES models are constructed from data taken between July, 1990 and October, 1991, a period just following Cycle 22 Solar Maximum (July, 1989). The NASA models are constructed from data taken in the 1960's and early 1970's, a period occurring mainly in Cycle 20. From a monitor of solar activity (e.g., sunspot number, solar proton fluence at Earth, etc.) there are significant differences between the two cycles [*Preliminary Reports*, 1993, 1994], but the relevance of these to the near-Earth radiation environment has not been established.

2.2. *Instrumentation*

The NASA models were constructed from a collection of instruments on a number of satellites. The earliest instruments did not distinguish electron and proton counts, and had stated uncertainties in response that were large, e.g., 40% or greater. Many of the outer zone electron measurements used in the models, and even some of the proton measurements, were made on satellites in low altitude orbit and extrapolated up the field line, requiring an assumption about the near-equatorial pitch angle distribution of the particles. Nuclear detonations in the early 1960's significantly changed the in-

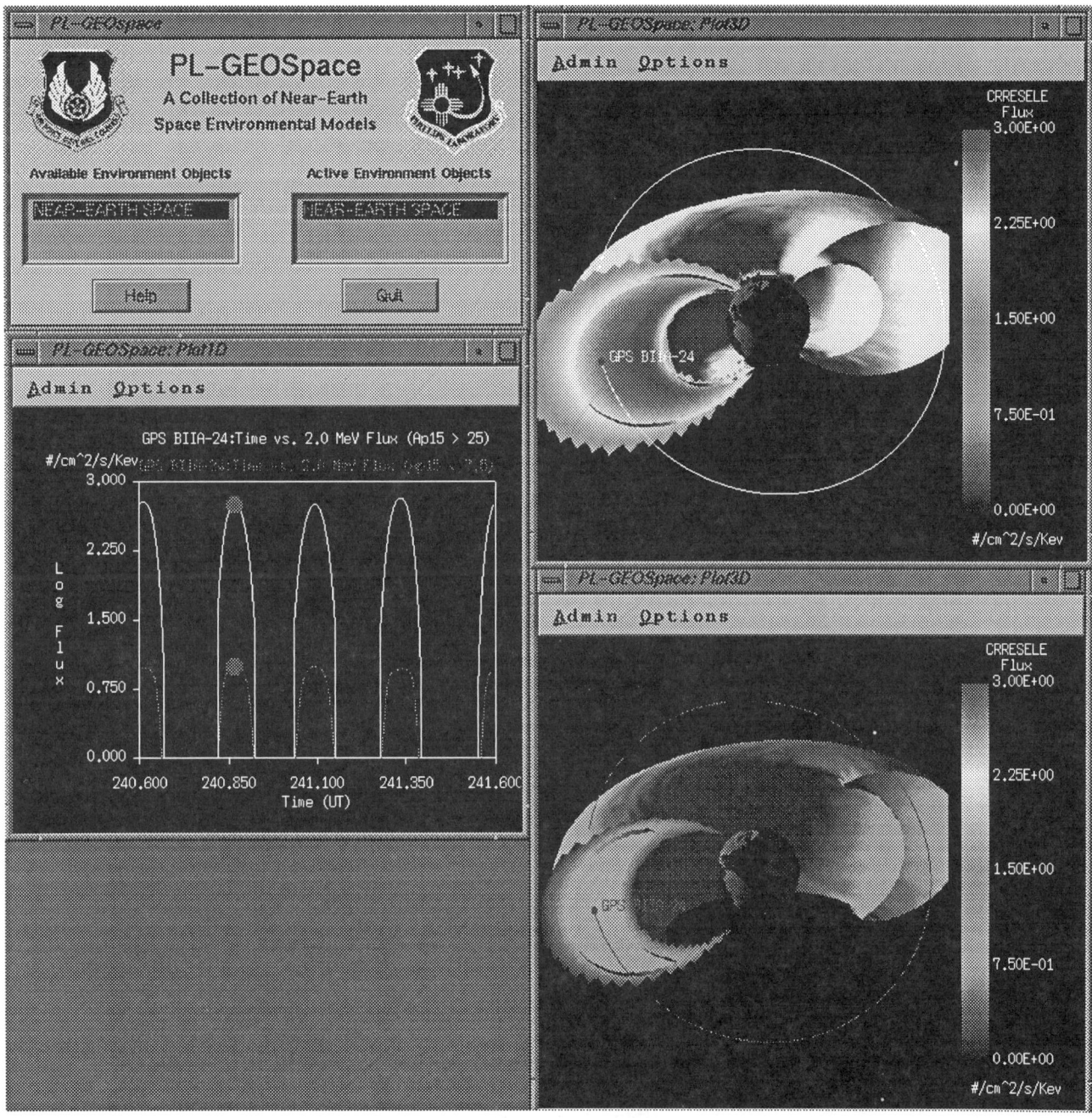

Figure 1. A screen print (reduced to gray-scale from the color original) from PL-GEOSpace showing the 2.0 MeV outer zone electron environment experienced by a GPS satellite for a quiet ($A_{p_{15}} < 7.5$) and an active period ($A_{p_{15}} > 25$).

ner belt electron environment for a very long time. *Vette et al.* [1978] summarized concerns over the data bases in the NASA models.

The CRRES models are constructed from four instruments that flew on the CRRES satellite. The instruments were all calibrated extensively prior to flight, and for two instruments sophisticated computer models were made of the instruments to simulate their response to environments that could not be created in the laboratory. The high energy electron detector showed no sensitivity to protons below 100 MeV, but its immunity to > 100 MeV protons, and > 10 MeV electrons could not be demonstrated. The high energy proton detector showed no sensitivity to electrons, including the > 10 MeV electrons found in the slot region after the March 1991 storm, but did show contamination from > 100 MeV protons. A correction was made for this latter contamination using the expected response in the loss cone and the computer model of the detector.

2.3. *Energy Coverage in the Flux Models*

For proton flux the CRRES models cover energies from 1–100 MeV. For proton flux the NASA models cover energies from 100 keV–400 MeV. For electron flux the CRRES models cover energies from 500 keV–7 MeV, while the NASA models cover energies from 40 keV–7 MeV. It should be noted that the extensive energy range of the NASA models is achieved, in part by extrapolation.

2.4. *Spatial Coverage in the Flux Models*

The B/B_0 coverage in both sets of models is from the magnetic equator to the loss cone. In L, the CRRES proton flux models extend from 1.15 to 5.5; the NASA proton flux models extend from 1.15 to 6.6. In L, the CRRES electron flux models extend from $L = 2.5$ to $L = 6.5$; the NASA electron flux models extend from $L = 1.15$ to $L = 11.5$. And again, a considerable part of the range in the NASA models is achieved by extrapolation.

2.5. *Model Dependencies*

The NASA models were developed for conditions of solar maximum and solar minimum. The CRRES proton and dose models were developed for conditions during solar maximum, before and after an extraordinarily large shock struck the magnetosphere. The CRRES electron outer zone models were developed for conditions during solar maximum specified by a fifteen day running average of the A_p magnetic index offset by one day (designated $A_{p_{15}}$). There are six CRRES electron Outer Zone models for six ranges of $A_{p_{15}}$.

2.6. *Omnidirectional Fluences on the Magnetic Equator*

Because of the large number of variables in the radiation belt models (e.g., L, energy, pitch angle, model variables) it is difficult, even when discussing omnidirectional models, to briefly summarize differences. In the generalizations we make here, comparing the NASA models and the CRRES models on the magnetic equator, we use 'approximately' with considerable latitude.

2.6.1. Proton models.
For L-values less than 1.8, that is, in the heart of the 'inner' radiation belt, all proton models are in good agreement (approximately within a factor of 2). The trend is for AP-8 MAX to be higher for energies less than 20 MeV, and for the CRRES models to be higher for energies greater than 20 MeV. For $L = 1.8$–4 there are significant differences in all three models (AP-8 MAX and the two CRRES models). The differences are organized in two groups. The first is for energies < 20 MeV. AP-8 MAX gives fluxes that can be up to an order of magnitude greater than the fluxes found in either CRRES model over a substantial portion of this region. The second is for energies > 20 MeV. The March 1991 storm created a second, stable high energy belt

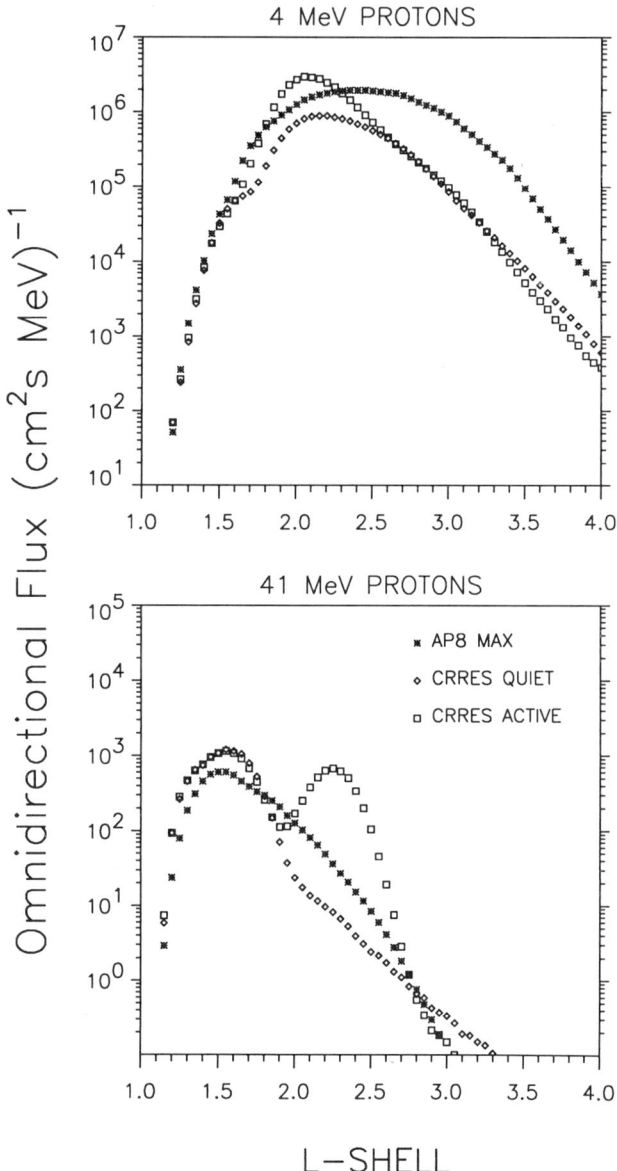

Figure 2. Profiles of 4 MeV (top) and 41 MeV (bottom) omnidirectional proton flux on the magnetic equator, as a function of L, taken from the CRRES quiet and active proton models and from AP-8 MAX.

in this region, which differentiates the CRRES active from the CRRES quiet models, the former reaching peak values more than an order of magnitude higher than the latter. But for energies > 20 MeV the CRRES flux in the active model can also greatly exceed that of AP8MAX. At 50 MeV the difference is approximately 2 orders of magnitude where the CRRES second belt peaks ($L \sim 2.3$). These differences are illustrated in Figure 2 which shows omnidirectional flux profiles on the magnetic equator taken from the two CRRES models and from AP-8 MAX for energy 4 and 41 MeV, respectively.

To summarize the differences in proton models in terms

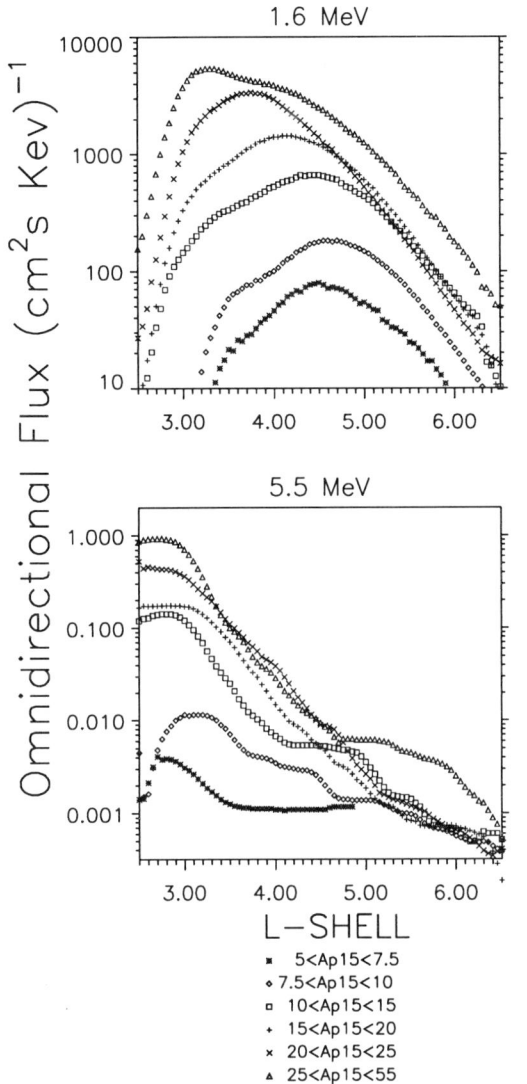

Figure 3. Profiles of 1.6 MeV (top) and 5.5 MeV (bottom) omni-directional electron flux on the magnetic equator, as a function of L, taken from the CRRES electron models specified for six ranges of $A_{P_{15}}$.

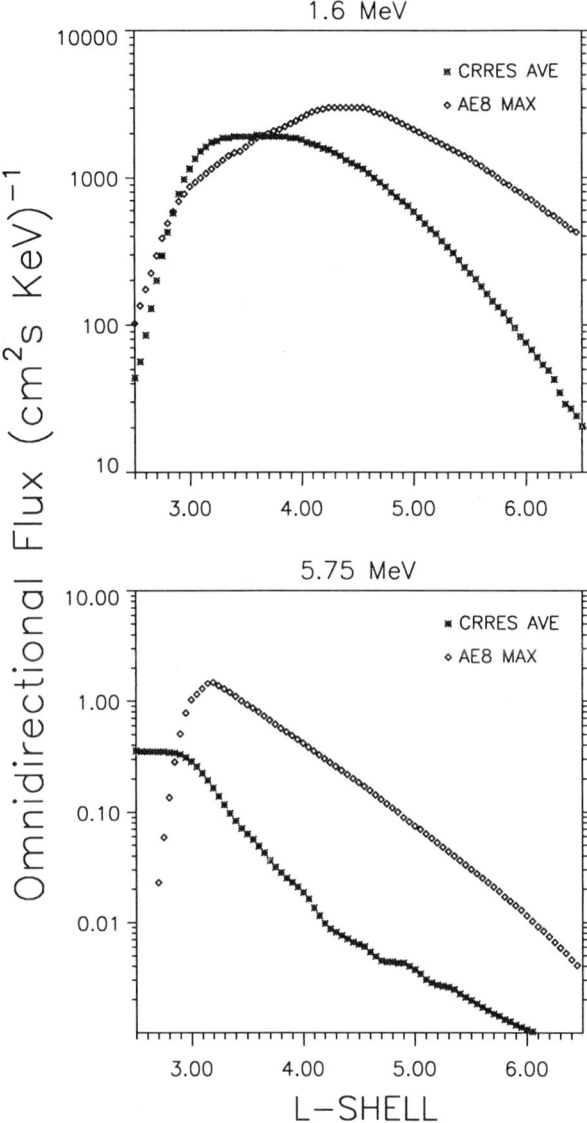

Figure 4. Profiles of 1.6 MeV (top) and 5.75 MeV (bottom) omni-directional electron flux on the magnetic equator, as a function of L, taken from the CRRES average electron model and from AE-8 MAX.

of engineering needs, for satellite orbits passing through the inner radiation belt and/or the 'slot' region, we find: At solar cell degradation energies (~ 5 MeV) AP-8 MAX predicts proton fluences higher than either of the CRRES proton models. For dose degradation within the satellite (> 20 MeV) or single event effects (> 50 MeV) the CRRES active model and, unless confined to the slot region, the CRRES quiet model, predict higher fluences than AP-8 MAX. In both cases, for given orbits and given activity levels, the differences can be significant, exceeding an order of magnitude.

2.6.2. Electron models. The CRRES electron models and AE-8 MAX differ much more substantially than the CRRES and NASA proton models. In the first place, for solar maximum there are six activity-dependent CRRES models to one average NASA model. It has been known since early in the U.S. space program that outer zone electrons can vary in intensity by orders of magnitude. The CRRES models demonstrate that there are also major changes in the spatial distribution of the outer zone electrons, and that these, as well as the flux variation, can be ordered by level of magnetospheric activity. The magnetic activity indicator that best ordered the electron models was found to be the fifteen day running average of A_p, offset by one day ($A_{P_{15}}$). $A_{P_{15}}$ was divided into six ranges and an outer zone electron model constructed for the data in each range and for the data in each of ten differential energy channels from .5 to 7 MeV. As for the proton models the electron differential number flux was mapped to the magnetic equator, and there binned by L and pitch angle, and averaged. Omnidirectional flux profiles on the magnetic equator as a function of L are shown for the six models in Figure 3. Two energies are represented: 1.6 MeV and 5.6 MeV. It is clear from this figure that electron measurements made near geosynchronous altitude are a poor monitor for the bulk of the outer zone electron population. One also sees how different is the physics of formation of the inner edge of the outer zone compared to that for protons in the inner belt (Figure 2). There is no common inner edge across the energy spectrum for the electrons, even for a restricted range in magnetospheric activity, presumably because there is no common loss mechanism such as is provided to the protons by the atmosphere. The six CRRES models can be used, with A_p, to create a time-dependent representation of the electron population in the inner magnetosphere. These have already been useful for evaluating the cause of spacecraft anomalies for certain spacecraft.

In addition to the six A_p-dependent outer zone electron models, two additional electron outer zone models were constructed from the CRRES data, the average and the "worst case" models. These are more readily compared with AE8MAX than the $A_{p_{15}}$-dependent models. They are constructed in a manner similar to that of the $A_{p_{15}}$-dependent models except that for the average model, all the data was used in a single model, and for the "worst case" model a composite of the highest flux value measured by CRRES, at each energy, L-bin and pitch angle, was created. The "worst case" model obviously does not represent a real state of the outer zone. Figure 4 is a comparison of the CRRES average outer zone model and AE-8 MAX, for two energies, 1.6 and 5.75 MeV. Omnidirectional flux profiles are plotted as a function of L on the magnetic equator. The differences in the two models are extremely large over most of the outer zone, the region of major difference increasing with increasing energy. Generally speaking the NASA model greatly exceeds (by an order of magnitude or more) the CRRES model for the larger L-values, and, for highest energies, greatly underestimates the CRRES measurements near the slot region. When we run geosynchronous, molniya, geosynchronous-transfer and GPS satellite orbits through the two models, the resulting fluence spectra are all sharply falling (for power law fit in energy, E^{-N}, N is between 6.6 and 9.5 for the CRRES average model). For large portions of the energy range from 1–7 MeV, AE-8 MAX fluence is 1–2 orders of magnitude greater than the CRRES average model.

3. ISSUES TO BE ADDRESSED IN ADVOCATING REPLACEMENT OF THE NASA TRAPPED RADIATION MODELS

The CRRES modeling efforts and comparison of these to the NASA models, which are the current standards in the industry, lead us directly to questions concerning the need to make replacement models. Three factors concerning the production of new models should be recognized up front. One, models are extremely costly to make, particularly if they require gathering new data. Two, to convince a sponsor to provide such moneys he must be assured that the new model will be of the desired quality and delivered within the time and budget constraints. Current contracting practices at NASA and ESA are poorly suited for producing specific deliverables other than hardware. Three, an investment in new models must give a tangible return to the sponsor. This probably means that the models need to be designed, in the first instance, for engineering purposes, with better understanding of the physics of the inner magnetosphere only a secondary goal. Focusing on ease of application and certainty of predictive results within the set error limits may not hold much appeal for most research scientists. Keeping these in mind we turn to issues which we are forced to consider when mission planners ask our advice about using the CRRES models to replace the NASA models.

3.1. *Sampling Time of the Radiation Belt Data*

The question to be answered here is: Did CRRES sample widely enough to be used as a replacement model? There clearly are temporal variations that are not represented during the CRRES lifetime. Very long-term changes in the trapped populations such as changes associated with changes in the magnetic moment of the Earth's dipole magnetic field or changes from solar cycle to solar cycle, are not represented in any data set, and will not be, until a long-term program to monitor various radiation populations is put into place. The 11-year cycle between successive solar maximums is represented in the NASA models by combining data into either a solar minimum or a solar maximum model. For proton fluxes the difference in intensity is about a factor of two greater in the solar minimum model. How one uses the NASA models in between solar maximum and solar minimum is not specified.

The CRRES data were taken for 14 months near solar maximum (solar maximum for the 22nd solar cycle was July, 1989, and CRRES was launched in July, 1990). CRRES measured a significant increase in the trapped proton and > 10 MeV electron population for L-values greater than 1.8 following the sudden commencement that led to the March, 1991 storm. This increase was, for selected L-ranges and energies, far greater than the factor of two increase in going from solar maximum to solar minimum. Does the magnetospheric state following the March storm represent a worst cast condition for these populations? Again, without constantly monitoring the radiation populations we cannot answer the question. We have attempted to use the contamination from high energy particles in the electrostatic analyzers on the Defense Meteorological Satellite Program (DMSP) spacecraft to monitor the trapped radiation populations from 1984 to the present [*Mullen et al.,* 1994]. No other event as large as the March event appears in that data. Solar cycle variations in the outer zone electron populations are not so easily characterized. Dosimeter data from the DMSP F7 satellite (1983–1987) showed that as one approaches solar minimum the outer zone electron 'episodes' (large increases in flux intensity) are fewer in number and decay over longer periods to lower levels than during solar maximum [*Gussenhoven et al.,* 1989]. If these are driven solely by magnetic activity as represented by A_p then the CRRES models can be used during all phases of the solar cycle. However, from the DMSP contamination data [*Gussenhoven et al.*, 1994], and from the GPS data [*T. Cayton, private communication*, 1994], it appears that the whole outer zone electron population systematically moves outward as solar minimum is approached and inward as solar maximum is approached. If this proves to be the case at least one additional time-dependency will be needed to make the CRRES electron models accurate throughout the solar cycle.

3.2. Dynamic Models

A question closely related to that concerning sampling time, is whether or not CRRES data sufficiently cover the dynamic range in the variability of the trapped radiation populations. Here, the CRRES models probably do a better job than the NASA models for two reasons. The first is, of course, that the March 1991 storm occurred during the CRRES lifetime, and the second is that the electron models are developed in a quasi-dynamic way. The dynamic range of the inner belt is not determined by the CRRES data. However, the hazard posed by a, heretofore, unmeasured 10 MeV electron population in the slot region is characterized and documented nowhere else.

3.3. Spatial Coverage

What deficiencies were there in the CRRES spatial coverage of the radiation belts? The main deficiency here is on the inner edge of the radiation belts where CRRES spent very little time and was traveling too fast to get good spatial resolution. We are using dosimeter data from the APEX satellite to evaluate the CRRES dose models, improve their spatial resolution, and, with DMSP dosimeter data, characterize the solar cycle dependence of the inner belt at low altitude [*Gussenhoven et al.*, 1996].

3.4. Instrumentation

What deficiencies in instrumentation were there in the CRRES complement of experiments and how important are they to satellite operations? The measurements that were not made on the CRRES satellite that would have helped to better characterize the trapped radiation environment are a) differential flux of protons > 100 MeV, b) differential flux of electrons > 10 MeV, and c) a sure differentiation between > 1 MeV electrons and > 50 MeV protons in the inner belt. It is not certain that any of these measurements have been satisfactorily accomplished for radiation belt flux intensities. A technology-development program to design new instrumentation for these purposes needs to be put in place if the measurements are to be made. All of these measurements could prove important to spacecraft operations and design. The > 100 MeV proton population can, in the inner belt, provide more than 50% of the integral flux above 50 MeV (the population that causes single event upsets in proton-sensitive devices), depending on how hard the spectrum is. The NASA electron models show a substantial inner zone electron population. The CRRES high energy electron detector showed substantially smaller flux, but uncertainty in the detector's response to > 100 MeV protons is the reason only outer zone electron models were released. The large number of > 10 MeV electrons in the slot region after the March storm sudden commencement was quite a surprise, but the Cerenkov detectors on CRRES could only give the barest of details of the spectral characteristics, and no pitch angle information.

3.5. Deficiencies in our Understanding of High Energy Particle Dynamics in a Real Magnetosphere

Analysis of the CRRES data and comparison with existing theory of particle energization and transport have shown that many of the physical processes thought to be well-understood do not give quantitative results that are good enough to allow us to extend empirical models outside of the measurement region, either in time, space or energy. Specific processes that are poorly understood are:

3.5.1. Entry, residence and exit of solar particles. Solar protons were found to have highly variable access into the magnetosphere which is not explained by Störmer theory in a dipole magnetic field. During the March, 1991 storm, solar protons with energies > 10 MeV penetrated below $L = 3$, while Störmer theory places their cutoff well beyond geosynchronous altitude. What controls high energy solar particle penetration? How long do they reside in the magnetosphere once inside? What, if any, processes can turn them into a trapped population?

3.5.2. Dynamics of the outer zone electrons. A problem similar to that of solar particle transport, is the question of outer zone electron variability. Outer zone electron 'episodes' (sharp increases in flux followed, after tens of days, by a rather rapid drop-off) have no satisfactory explanation. There are many magnetospheric indices and solar wind parameters that are often associated with outer zone onsets and decays, but not to the degree that they can be used to reliably say what is the state of the outer zone electrons. Since these electrons are probably the culprit for most spacecraft anomalies and since no real source has been unambiguously identified for them, we have dubbed the outer zone electron dynamics one of the greatest unsolved problems in magnetospheric physics.

3.5.3. Response of particle populations to solar wind shocks. A great deal of progress is being made in this area in order to understand how the additional belts in the magnetosphere were formed during the March 1991 storm sudden commencement [*Hudson et al.; Li et al.*, these proceedings]. The advances made here are to show how important induction electric fields can be in transporting and energizing high energy particles. To date, however, realistic magnetic field models have not been used and the inclusion of these is bound to open up even more possibilities.

3.5.4. Time dependence in diffusion. The new CRRES radiation belt models and their data bases provide a new opportunity to test the accuracy of diffusion theory along with source populations, such as CRAND [*Albert*, these proceedings]. Again, significant differences are found which have no immediate explanation, but may require that explicit time dependencies be included in the diffusion coefficients. Additional sources and losses may also be required.

3.5.5. Low altitude models. With the reality of Space Station approaching, the deficiencies in modeling the inner edge of the radiation belts and understanding its variability becomes ever more apparent. Many suggestions have been made for new coordinates that specifically include the atmospheric loss mechanism, but so far no models that incorporate these have been forthcoming. It may be the case that the development of a low altitude model may be the most attainable and useful new product we can create with existing data.

4. CONCLUSION

The CRRES radiation models and their data bases provided, in the 1990's, a truth-test for using the NASA radiation models and 1970's theoretical understanding of the inner magnetosphere. The results of the test demonstrate that we can do much better by the engineering community if we update the models on a regular basis, move toward fully dynamic models with quasi-dynamical empirical models, and make models that meet special needs, such as a low altitude model that incorporates atmospheric effects. We need to face the probability that there will not be fully instrumented satellites launched at regular intervals to provide data for model updates and find more creative ways to fill in data gaps. One way to achieve constant monitoring is through inexpensive, small, low power, low telemetry detectors that are robust for a single, highly specified measurement, say, 2 MeV electrons in the outer zone. These could be flown, as housekeeping instruments, on every satellite that traverses the outer zone. As a trade-off the satellite operator would have a measurement of the environment that can create anomalies on his spacecraft due to deep dielectric charging. Measurements, in a given region of space, of a single species, at a specific energy, could be used as the scaling factor for the rest of the radiation model for that region. To extend models to energy ranges or regions where the current models are inadequate, sophisticated instruments will have to be designed and appropriate flights found for them. Finally, we need to make better use of existing data to make special models, such as a low altitude model of the inner edges of the radiation belts. Once the dynamical variation of the region is established a monitoring program can be set up to make periodic updates of the model. We are following this type of program at Phillips Laboratory to create the best possible specification of the radiation environment for the U.S. Air Force and its space assets. Our program can only be improved by comparison of its results with those of other programs. We encourage you to use and criticize our models, to work with us and our data at Phillips Laboratory, and to share your findings with us.

REFERENCES

Blake, J.B., W.A. Kolasinski, R.W. Fillius and E.G. Mullen, Injection of electrons and protons with energies of tens of MeV into $L < 3$ on March 24, 1991, *Geophys. Res. Lett., 19*, 821, 1992.

Brautigam, D.H., M.S. Gussenhoven and E.G. Mullen, Quasi-static model of outer zone electrons, *IEEE Trans. Nucl. Sci., 39*, 1797, 1992.

Brautigam, D.H. and J.T. Bell, CRRESELE Documentation, PL-TR-95-2128, Phillips Laboratory, AFMC, Hanscom AFB, MA, 31 July 1995a.

Brautigam, D.H. and J.T. Bell, CRRES electron omnidirectional flux models and CRRESELE utility, 1995 IEEE Radiation Effects Data Workshop Record, IEEE Nuclear and Plasma Sciences Society, Piscataway N.J., 90, 1995b.

Chenette, D.L., J. Chen, T.G. Guzik, J.P. Wefel, M. Garcia-Munoz, C. Lopate, K.R. Pyle, K.P. Ray, E.G. Mullen and D.A. Hardy, The CRRES/SPACERAD heavy ion model of the environment (CHIME) for cosmic ray and solar particle effects on electronic and biological systems, *IEEE Trans. Nucl. Sci., 41*, 2332, 1994.

Ginet, G.P., R. Biasca and M. Tautz, PL-GEOSpace: Three-dimensional visualization of the dynamic space environment, 1995 IEEE Radiation Effects Data Workshop Record, IEEE Nuclear and Plasma Sciences Society, Piscataway N.J., 91, 1995.

Gussenhoven, M.S., E.G. Mullen and R.C. Sagalyn, CRRES/SPACERAD Experiment Descriptions, AFGL-TR-85-0017, Air Force Geophysics Laboratory, Hanscom AFB, MA, 24 January, 1985.

Gussenhoven, M.S., E.G. Mullen and E. Holeman, Radiation belt dynamics during solar minimum, *IEEE Trans. Nucl. Sci., 36*, 2008, 1989.

Gussenhoven, M.S., E.G. Mullen, D.H. Brautigam, E. Holeman, C. Jordan, F. Hanser and B. Dichter, Preliminary comparison of dose measurements on CRRES to NASA model predictions, *IEEE Trans. Nucl. Sci., 38*, 1655, 1991.

Gussenhoven, M.S., E.G. Mullen, M. Sperry and K.J. Kerns, The Effect of the March 1991 storm on accumulated dose for selected satellite orbits: CRRES dose models, *IEEE Trans. Nucl. Sci., 39*, 1765, 1992.

Gussenhoven, M.S., and E.G. Mullen, Space Radiation Effects Program: An overview, *IEEE Trans. Nucl. Sci., 40*, 221, 1993.

Gussenhoven, M.S., E.G. Mullen, M.D. Violet, C. Hein, J. Bass and D. Madden, CRRES high energy proton flux maps, *IEEE Trans. Nucl. Sci., 40*, 1450, 1993.

Gussenhoven, M.S., E.G. Mullen and E. Holeman, MeV electrons as measured by the DMSP J4 detector. Part II: Variations and dynamics over a solar cycle, *EOS, Trans. Am. Geophys. U., 75*(44), Fall Mtg Supplement, 541, 1994.

Gussenhoven, M.S., E.G. Mullen, D.A. Hardy, D. Madden, E. Holeman, D. Delorey and F. Hanser, Low altitude edge of the inner radiation belt: Dose models from the APEX satellite, *IEEE Trans. Nucl. Sci., 42*, 2035, 1995.

Gussenhoven, M.S., E.G. Mullen and D.H. Brautigam, Improved understanding of the Earth's radiation belts from the CRRES satellite, *IEEE Trans. Nucl. Sci.*, accepted for publication, Special Issue, April, 1996.

IAGA Division I Working Group, International Geomagnetic Reference Revision 1985, *EOS, Trans. Am. Geophys. U., 67*, No. 24, 1986.

Kerns, K.J. and M.S. Gussenhoven, CRRESRAD Documentation, PL-TR-92-2201, Phillips Laboratory, AFMC, Hanscom AFB, MA, 6 August 1992.

Meffert, J.D. and M.S. Gussenhoven, CRRESPRO Documentation, PL-TR-94-2218, Phillips Labortory, AFMC, Hanscom AFB, MA, 28 July, 1994.

Meffert, J.D., J. Bell and M.S. Gussenhoven, PROSPEC Documentation, to be published, Phillips Laboratory, AFMC, Hanscom AFB, MA, 1996.

Mullen, E.G., M.S. Gussenhoven, K. Ray and M. Violet, A double-peaked inner radiation belt: Cause and effect as seen on CRRES, *IEEE Trans. Nucl. Sci., 38*, 1713, 1991.

Mullen, E.G., M.S. Gussenhoven, D.H. Brautigam and A.R. Frederickson, Review of CRRES radiation belt measurements and engineering experiment results, ISSN 0148-7191, Society of Automotive Engineers, Inc., Salem, MA, 921373, 1992.

Mullen, E.G. and E. Holeman, MeV electrons as measured by the DMSP J4 detector. Part I: Particle identification and verification, *EOS, Trans. Am. Geophys. U.*, 75(44), Fall Mtg Supplement, 541, 1994.

Olson, W.P., and K.A. Pfitzer, A quantitative model of the magnetospheric magnetic field, *J. Geophys. Res.*, 79, 3739, 1974.

Preliminary Report and Forecast of Solar Geophysical Data, U.S. Department of Commerce, Space Environment Services Center, Boulder SESC PRF 936, 941, 945, 946, 950, 962, 1993–1994.

Sawyer, D.M., and J.I. Vette, AP8 Trapped Proton Environment for Solar Maximum and Solar Minimum, NSSCD 76-06, NASA-GSFC, Greenbelt, MD, 1976.

Vette, J.I., The AE-8 Trapped Electron Model Environment, NSSDC 91-24, NASA GSFC, Greenbelt, MD, 1991.

Vette, J.I., K.W. Chan and M.J. Teague, Problems in Modelling the Earth's Trapped Radiation Environment, AFGL-TR-78-0130, Air Force Geophysics Laboratory, Hanscom AFB, MA 01731, 1978.

D.H. Brautigam, M.S. Gussenhoven, E.G. Mullen, Space Physics Division, Phillips Laboratory, Hanscom AFB, MA 01731.

DISCUSSION

Q: J.B. Blake. Will Phillips Laboratory generate an APEX-RAD model?

A: M.S. Gussenhoven. Yes. Only dose measurements are made on APEX, and we are constructing a dose model with the data. We will incorporate the model into a utility to calculate dose for a given orbit, but it may or may not be used to extrapolate CRRESRAD to higher latitudes. We are still discussing this. The APEX/PASP+ model is discussed in *IEEE Trans. Nucl. Science*, Dec. 1995.

Q: D.N. Baker. How does one reconcile the assertion that CRRES saw very "high" relativistic electron flux with the apparent fact that overall electron fluxes would be relatively low near sunspot maximum? SAMPEX would suggest fluxes have been getting much higher as we have gone from the early 1990s to present.

A: M.S. Gussenhoven. First, I don't understand why you would expect relativistic electron fluxes to be low during solar maximum, particularly for $L > 3$. Second, we have used the contamination by 3 MeV electrons in the DMSP satellites to obtain a long baseline for electron observations over $L = 2$–6. This shows that there is a distinct solar cycle variation in the centroid of flux (analogous to center of mass) in the outer zone electrons; it moves to higher L values in the decline from solar maximum to solar minimum. This would account for a steady increase in the baseline of outer zone fluxes at geosynchronous orbit from 1991–1994, as reported by Tom Cayton, using Los Alamos satellite measurements.

Q: A.L. Vampola. You posed the question of how representative the CRRES data set is. Our preliminary results using a neural network study (reported in these proceedings) indicate that at $L = 6$, electron fluxes of 1.5 MeV were 30%–40% higher averaged over the 16-month CRRES mission than the 61-year average from 1932 to 1993.

A: M.S. Gussenhoven. If what you say is true then the CRRES period had higher than "average" 1.5 MeV electrons at geosynchronous altitude, and AE-8 gives even more excessive electron fluxes than we show.

Q: G.D. Reeves. How extensible is the CRRESRAD model to allow for inclusions of other datasets so that the models can evolve as more data become available?

A: M.S. Gussenhoven. This is a very desirable feature for models but CRRESRAD is a fixed (i.e. unextensible) model. Data quality is an important issue in including new datasets in extensible models.

A New Empirical Electron Model

D.J. Rodgers

Mullard Space Science Laboratory, University College London

New CRRES data have permitted the creation of new radiation belt models. The creation and performance of a new empirical model based on CRRES MEA data is described. From a number of candidate coordinate systems, the (L, α_0) system was chosen, where α_0 is the equatorial pitch angle and where L is defined at the mirror point. The angular resolution of the instrument is used to project fluxes from the near-equatorial spacecraft location to the whole magnetosphere. Because pitch-angle dependent L values are used both in summing the data and in using the summed data to generate total fluxes at an arbitrary spacecraft location, the model is immune to shell-splitting effects.

CRRES [*Johnson and Kierein,* 1992], launched in July 1990 into a geosynchronous transfer orbit, was ideally suited to studies of the radiation belts. An electron model has already been published, based on the CRRES HEEF instrument [*Brautigam et al.,* 1992] and a total dose model from the CRRES Space Radiation Dosimeter [*Gussenhoven et al.,* 1992]. These have used a pitch-angle independent approximation to McIlwain's L parameter and have stored fluxes in the model as omnidirectional fluxes. This limits their coverage to the near-equatorial region where the data originated. The popular AE-8 model [*Vette,* 1991] uses a similar approach but covers the whole magnetosphere by using data from a number of satellites. This paper describes the creation of a model based on data from the CRRES Medium Energy Analyzer. This model, called MEA3MSSL has been developed as part of an ongoing study for ESTEC. The aim has been to use data from one instrument to create a statistical model that is valid both at the equator and far away from it. The emphasis has been on the outer belt because this is where electron fluxes are most significant.

1. THE CHARACTERISTICS OF THE DATA

CRRES' orbit took it from 350 to 33,600 km altitude. The inclination was 18.1° and so measured fluxes near the equator. The lifetime of the mission was 18 months.

Radiation Belts: Models and Standards
Geophysical Monograph 97
Copyright 1996 by the American Geophysical Union

The Medium Energy Analyser [*Vampola et al.,* 1992] used magnetic deflection to focus entering electrons onto an array of detectors. The position of the detector determined the radius of curvature of the electron trajectory and thus the electron energy. 17 detectors measured 17 energy bins from 153 to 1534 keV. The field of view of the instrument pointed perpendicular to the satellite spin axis. Because the spin axis was usually nearly perpendicular to the magnetic field vector this gave MEA good coverage in pitch angle. The field of view of the instrument varied between 1.5° and 8° depending on energy but when the 0.5 s accumulation time is considered, the pitch angle resolution was 8°–18°. Although this is the angular resolution used in this study, the pitch angle distribution can, with more effort, be determined to about 0.5° through a deconvolution procedure.

The MEA used two methods to remove contamination by energetic penetrating radiation. The detectors were solid state and produced a pulse proportional to the energy of the particle. Discriminators counted only particles above 50% of the minimum and below 110% of the maximum energy of particles focussed onto the detector. The low discrimination threshold allowed even electrons backscattered out of the detector to be counted. Further contamination removal was achieved by subtracting counts from a shielded detector which only measured background counts. Nevertheless, there were periods, in the inner belt, when unsubtracted contamination by protons became significant.

2. CHOICE OF COORDINATE SYSTEM

Before deciding on the best coordinate system for this model, a number of possible alternatives were investigated—a study

Table 1. Format of the MEA3MSSL model file

	Parameter	Meaning
1	MODEL character*8	Name, i.e. MEA3MSSL
2	NL integer*2	No. of L_m bins
3	LS(NL) real*4	Centre values of L_m bins
4	NA integer*2	No. of α_0 bins
5	AS(NL) real*4	Centre values of α_0 bins
6	NE integer*2	No. of energy bins
7	ES real*4	Centre values of energy bins
8	FLUX(NA,NL,NE) real*4	Mean flux array

described in the final report of the TREND-2 study for ESTEC [*Lemaire et al., 1995*]. The principles used were that a good coordinate system:

1. should be as simple as possible;
2. should enable the CRRES data to cover the magnetosphere as fully as possible from the equator to the atmosphere;
3. should produce mean fluxes that vary systematically from one part of the magnetosphere to another;
4. and should have a low standard deviation meaning that similar fluxes are binned together.

300 orbits of CRRES MEA data were summed in each of five test coordinate systems:

- L and Local Time: this was used to see if we could ignore most of the detailed physics. Bins of 0.5 in L and 1 hour in local time were used. This coordinate system scored well by being very simple. From $L = 1$ to 7 about 50% of the magnetosphere was not covered. Fluxes did not show systematic variations but instead varied almost randomly due to strong aliasing between position and time in the mission.

- B/B_0 and L: this is comparable to the system used in AE-8 but using an external field model. Bins of 0.5 in L and 1 in B/B_0 were used. Fluxes varied systematically but except for the inner belt, only a small band near the equator was covered. Hence the CRRES data could only produce an equatorial model in this coordinate system.

- B_m/B_0 and L_m: by using the MEA pitch angle resolution, the values of B and L at the mirror points, far from the spacecraft, were found. Bins of 0.5 in L_m and 2 in B_m/B_0 were used. The resulting distribution showed smoothly varying fluxes with the two belts clearly visible and the decrease in fluxes moving from the equator towards the top of the atmosphere (corresponding to the local loss-cone). However, the number of bins needed to go from the equator increased steeply with L and in the outer belt not all the bins were populated. Hence coverage was incomplete.

- $\text{Log}(B_m/B_0)$ and L_m: bins of 0.5 in L_m and 2 in B_m/B_0 were used. This had the same properties of systematically varying flux as B_m/B_0 and L_m and restricted the number of bins in the outer belt so that all were adequately filled. The number of bins covering the inner belt was small however. Clearly another function of B_m/B_0 could be found to give more equal coverage over the two belts.

- α_0 and L_m: α_0 is such a function of only B_m/B_0 and has the advantage of a clear physical meaning, i.e. it is the pitch angle at the magnetic equator. Bins of 0.5 in L_m and 5° in α_0 were used. This gave an adequate number of bins in the inner and outer belts, completely filled from $L_m = 1$ to 7.

3. PROPERTIES OF THE (α_0, L_m) COORDINATE SYSTEM

The (α_0, L_m) coordinate system was chosen for use in the new model. The characteristics which make this particularly suitable are:

- α_0 and L_m are invariants of particle motion;
- an equatorial satellite with good pitch-angle resolution can cover the entire magnetosphere;
- equal bins of α_0 can be used without exaggerating the pitch-angle resolution of an equatorial satellite;
- α_0 and L_m are easy to visualise;
- adequate coverage of both inner and outer belts can be achieved without changing bin sizes between the two zones.

4. CREATION OF THE MODEL

The model was created by binning data from approximately 900 CRRES orbits in α_0 and L_m bins of 5° and 0.2 respectively. The model ranges from $L = 1$ to $L = 9$, from $\alpha_0 = 0°$ to $\alpha_0 = 90°$ and from 153–1534 keV. The Tsyganenko 1989 [*Tsyganenko, 1989*] K_p-dependent external B-field model was used along with an internal DGRF model field. The BLXTRA software [*Lemaire et al., 1995*] created by BISA, Brussels, was used to access the field models and calculate B and L_m. Five versions of the model were created, one for all K_p, and one each for K_p ranges 0 to 1+, 2 to 3+, 4 to 5+ and 6 to 7+.

The model consists of a simple text file containing a header and a 3-d array of fluxes. The text file is described in Table 1.

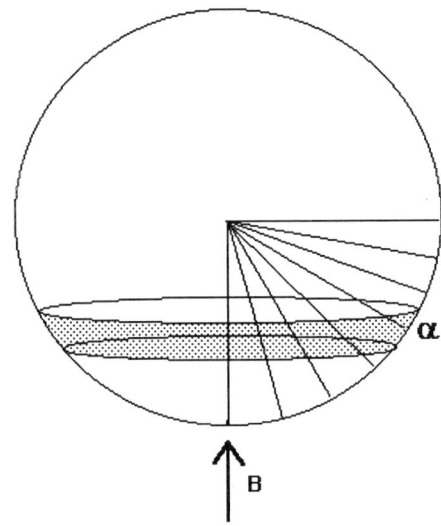

Figure 1. Schematic diagram showing the band of solid angle occupied by a pitch angle bin

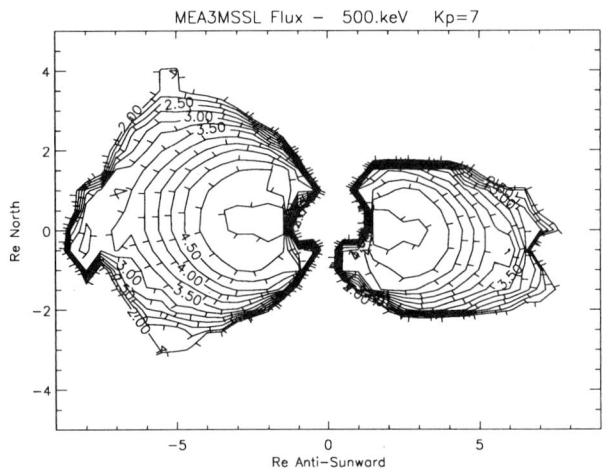

Figure 3. Contour plot of log flux in a noon-midnight cut through the magnetosphere. $K_p = 7$.

ponds to an equatorial pitch angle value, as follows:

$$\alpha_0 = \sin^{-1}\left(\sqrt{\frac{B_0}{B}} \sin \alpha\right). \quad (1)$$

Each (α_0, L_m) pair corresponds to one flux value in the model. Hence the flux is found for each pitch angle. The fluxes are weighted according to the solid angle of that pitch angle (see Figure 1), before being summed:

$$\text{Omnidirectional flux} = \\ 2 \sum_\alpha \text{flux}(\alpha) \, 2\pi[\cos(\alpha + 2.5°) - \cos(\alpha - 2.5°)]. \quad (2)$$

The final result is flux summed over 4π steradians. This calculation is performed for the two bracketing energies in the model and linearly interpolated to find the flux at the required energy.

6. RESULTS

Figure 2 shows contours of Log flux in a noon-midnight slice through the magnetosphere for $K_p = 0$ at 500 keV. The sampling grid has $0.5\,R_E$ resolution. The inner and outer belts are clearly visible. The sunward and tailward sides are fairly symmetric.

In Figure 3 are shown contours of log flux for the same slice with $K_p = 7$. The tailward side of the magnetosphere is now severely compressed, an effect of using a realistic magnetic field model. The peak of the outer belt has moved earthwards and the 'slot' between them has disappeared. There are some irregularities at the edges of the model due to the poorer statistics for high K_p values. Overall flux levels in the inner and outer belt are much higher than for low K_p.

Figure 4 shows fluxes at local midnight along the tailward equator for a range of geocentric radii. Six energies are shown. In the outer belt, both the flux and the radius of the

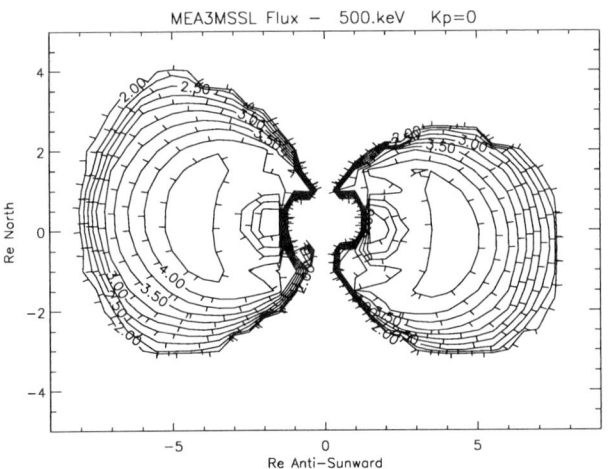

Figure 2. Contour plot of log flux in a noon-midnight cut through the magnetosphere. Tick marks indicate the downward side of each contour. $K_p = 0$. The sunward direction is to the left.

5. RECONSTRUCTION OF OMNIDIRECTIONAL FLUXES

The omnidirectional flux is calculated from the model by a subroutine. This is seen as integral to the model since it is important that the same magnetic field models are used to access the data as were used to create it. The user inputs the spacecraft longitude, latitude, geocentric distance and modified julian day, K_p and the required energy.

The code creates 18 bins of local pitch angle α. Using BLXTRA, the local magnetic field B and 18 corresponding values of L_m are found. The local pitch angle also corres-

106 A NEW EMPIRICAL ELECTRON MODEL

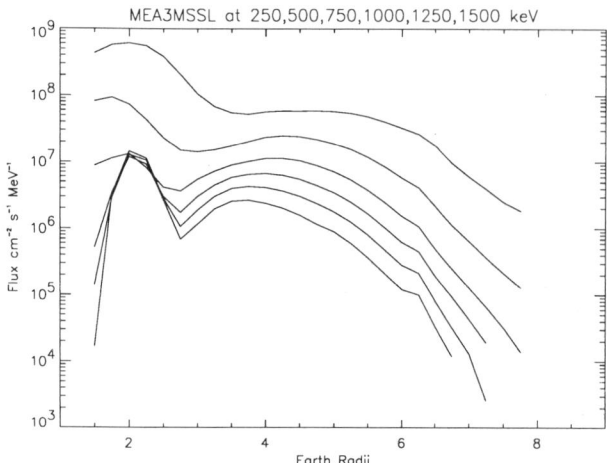

Figure 4. Flux profiles along the midnight equator for 6 energies. Lines from top to bottom on the plot correspond to energies going from lowest to highest.

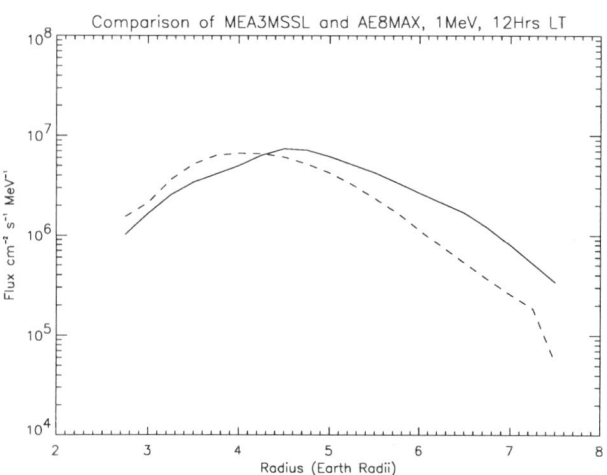

Figure 5. 1 MeV flux along the noon equator. The solid line is AE-8 MAX and the dotted line is MEA3MSSL.

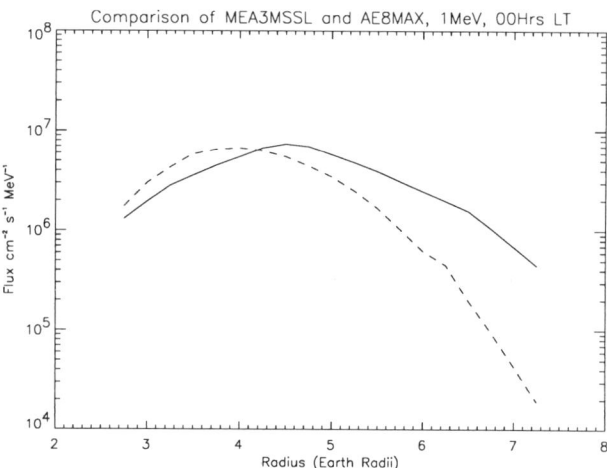

Figure 6. 1 MeV flux along the midnight equator. The solid line is AE-8 MAX and the dotted line is MEA3MSSL.

peak decrease with increasing energy. In the inner belt there is constant flux for energies above 500 keV. This is evidence that the fluxes here are dominated by noise from the proton belt, which was particularly active during the CRRES mission.

Figure 5 shows a flux profile at local noon along the sunward equator at 1 MeV for NASA's AE-8 model [*Vette*, 1991] and the MEA3MSSL model. The peak flux in the two models is similar but the MEA model drops off faster with radius. This difference becomes more pronounced in the tailward direction (Figure 6). The AE-8 model predicts almost no difference from the sunward fluxes but MEA3MSSL drops more rapidly with radius. The difference around geostationary orbit is an order of magnitude. Part of this difference is due to the presence of an external component to the magnetic field model and part is due to 'shell-splitting' [*Roederer*, 1970], i.e. the fact that a single point in space is on a range of L-shells depending on pitch angle. Both these effects are ignored by AE-8 and treated realistically by the new model. Away from the equator, MEA3MSSL and AE-8 latitudinal flux profiles are different for the same reasons.

7. DEFICIENCIES IN THE MODEL

The model is presently contaminated by high energy protons for energies above 500 keV up to $L_\mathrm{m} = 3$. Hence its use is currently restricted to the outer belt.

In the outer belt the loss cone has a width of only a few degrees and so the MEA angular resolution does not characterise it accurately. This has no effect on fluxes near the equator but results in the model becoming less accurate at low altitudes. This contrasts strongly with AE-8 which mapped the loss cone carefully.

The sudden change in the inner belt fluxes that occurred in March 1991 have shown that the 18 month CRRES mission is not a long enough database to average out infrequent large events.

8. SUMMARY AND CONCLUSIONS

The CRRES MEA instrument has been a source of high-quality data which have been used to create a new empirical electron model valid beyond $L_\mathrm{m} = 3$ and from 150 to 1500 keV. Using a single data source to cover the magnetosphere from the equator to low altitudes has been possible thanks to good pitch angle information. It removes problems of cross-calibration of data from different satellites which have had to be combined in earlier models. The new model is K_p dependent and handles 'shell-splitting' implicitly. Further improvements to the model are planned.

Acknowledgements. I am grateful to A.L. Vampola for allowing the use of the MEA data. This work was carried out under ESA contract no. 10725/94/NL/JG(SC) as part of the 'TREND' Radiation belt study directed by E.J. Daly.

REFERENCES

Brautigam, D.H., M.S. Gussenhoven and E.G. Mullen, Quasi-static Model of Outer Zone Electrons, *IEEE Trans. Nucl. Sci., 39*, 1797–1803, 1992.

Gussenhoven M.S., E.G. Mullen, M. Sperry, K.J. Kerns and J.B. Blake, The Effect of the March 1991 Storm in Accumulated Dose for Selected Satellite Orbits: CRRES Dose Models, *IEEE Trans. Nucl. Sci., 39*, 1765–1772, 1992.

Johnson M.H. and J. Kierein, Combined Release and Radiation Effects Satellite (CRRES): Spacecraft and Mission, *J. Spacecraft and Rockets, 29*, 556–563, 1992.

Lemaire J., A.D. Johnstone, D. Heynderickx, D.J. Rodgers, S. Szita and V. Pierrard, Trapped Radiation Environment Model Development: TREND-2 Final Report, *Aeronomica Acta, A-393*, 1995.

Roederer J.G., *Dynamics of Geomagnetically Trapped Radiation*, Springer-Verlag, New York, 1970.

Tsyganenko, N.A., A Magnetospheric Magnetic Field Model with a Warped Tail Current Sheet, *Planet. Space Sci., 37*, 5–20, 1989.

Vampola A.L., J.V. Osborn and B.M. Johnson, CRRES Magnetic Electron Spectrometer AFGL-701-5A (MEA), *J. Spacecraft and Rockets, 29*, 592–595, 1992.

Vette J.I., The AE-8 Trapped Electron Model Environment, NSSDC/WDC-A-R&S 91-24, 1991.

D.J. Rodgers, Mullard Space Science Laboratory, Holmbury St. Mary, Dorking, Surrey RH5 6NT, England.

DISCUSSION

Q: A.A. Beliaev. For a selected binning, was the statistics for the most and the less populated bins different?

A: D.J. Rodgers. Yes, significantly.

Q: M.S. Gussenhoven. What was the pitch-angle distribution of the electron belt centred around $L = 2$, and, in particular, was the loss cone empty?

A: D.J. Rodgers. The electrons were peaked around 90° and fell toward the loss cone. The loss cone was not empty however.

C: A.L. Vampola. I consider the $L = 2$ fluxes to be contaminated. This would explain the loss cone electrons at this L value.

Q: M. Hudson. Have you considered binning into before and after March 91 storm: it was such a step function change in the electron population.

A: D.J. Rodgers. It is hard to know which would then be "right" to compare with AE-8.

Q: S.F. Fung. The statement that the AE-8 model did not incorporate shell-splitting is puzzling, especially when the model is based on particle observations, which effectively incorporated the shell-splitting effect.

A: D.J. Rodgers. Although the particles include shell splitting, the fact that the organizing coordinate system did not use a realistic field with pitch angle dependent L sums together fluxes on the day and night side, that according to the internal field (without shell-splitting) have the same L. In fact they have different Ls. The result will be a smoothing with fluxes too high on the dayside and too low on the nightside.

C: A.L. Vampola. CRRES electron measurements at low L are subject to contamination which needs further investigation.

Q: E. Daly. When you mentioned the reconstruction of the isotropic flux, shouldn't have said "omnidirectional" flux?

A: D.J. Rodgers. Yes, that's what I meant.

Experimental Validation of South Atlantic Anomaly Motion Using a Two-Dimensional Cross-Correlation Technique

M. Lauriente

NASA Goddard Space Flight Center, Greenbelt, Maryland

A.L. Vampola and K. Gosier

University Research Foundation, Greenbelt, Maryland

Using a two-dimensional longitude-latitude cross-correlation technique, a data base consisting of Single Event Effects (SEEs), which were produced in the TOPEX satellite by energetic protons, is correlated with the location of the South Atlantic Anomaly (SAA) as predicted by the AP-8 MAX model. The results show that a current-epoch magnetic field model used with the AP-8 particle model accurately predicts the location of the proton SAA. Furthermore, the cross-correlation technique provides a means of determining the energy threshold needed to produce the SEEs. Using this technique at lower altitude on background effects in the COBE/DIRBE infrared detector system produces results which can be explained only by an error in the AP-8 model.

1. INTRODUCTION

There has been concern about the accuracy of energetic proton flux predictions at low altitude, such as for Shuttle or Space Station, which are obtained by using the standard NASA magnetospheric proton model, AP8 [*Sawyer and Vette*, 1976]. AP-8 was generated from data obtained in the late 1960's and is specified in terms of integral flux intensity above various energy thresholds at various L values and B/B_0 values. Changes in the magnetic field which have occurred since the AP-8 model was generated result in prediction errors in location if the original magnetic field model is used and in errors in flux intensity as a function of altitude if a current-epoch magnetic field model is used [*Konradi and Hardy*, 1987]. The error, in predicting present or future particle distributions, which is produced by using a trapped particle model defined by an earlier magnetic field epoch (e.g., IGRF 65, Epoch 1970 [*IAGA*, 1987] for AP-8 MAX), has been analyzed theoretically (e.g., *Lemaire et al.* [1990]). Experimental verification of the change in location, which is referred to as a "drift", has also been attempted (e.g., *Konradi et al.* [1994]).

There is a local intensity minimum in the magnetic field known as the South Atlantic Anomaly (SAA). The minimum is not confined to the Earth's surface—it maps upward to high altitude, also. The minimum is due to the fact that the geomagnetic field is not a simple dipole: it has significant high order terms (in a spherical harmonic expansion formalism); and furthermore, the magnetic axis is both offset and tilted with respect to the geographic rotational axis. The minimum is of interest because a particle which is trapped in the Earth's magnetic field and is performing bounce motion mirrors at a constant magnetic field intensity. The particle's lowest mirror altitude coincides with the location at which the geomagnetic field intensity is a minimum along the particle's drift path.

There is a secular decrease in the dipole term (and other low order terms) of the Earth's magnetic field. As the dipole term diminishes, the centroid of the field intensity minimum at the Earth's surface changes location, or "drifts". This change in the magnetic field contributes to error in trapped particle predictions, as noted above. This secular change in the field is referred to as a "drift", but is not a true drift in the sense

Radiation Belts: Models and Standards
Geophysical Monograph 97
Copyright 1996 by the American Geophysical Union

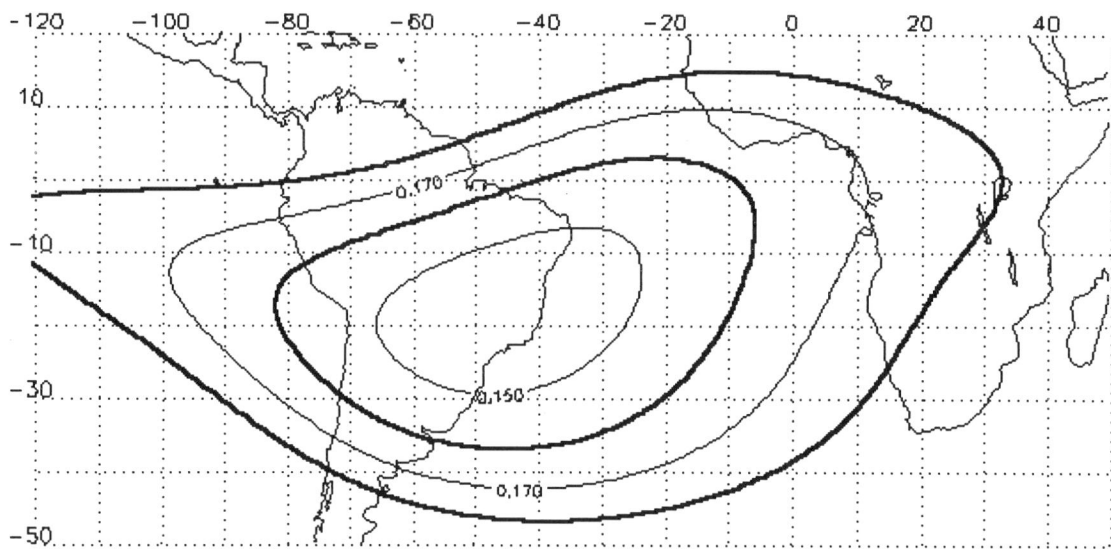

Figure 1. DGRF 65, Epoch 1970, 1336 km magnetic field contours.

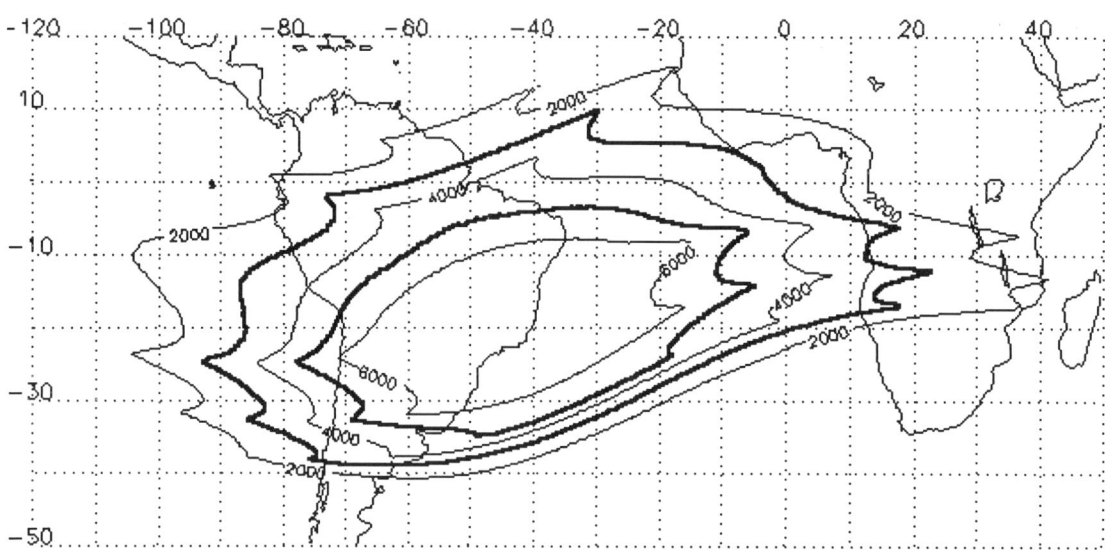

Figure 2. AP-8 MAX Epoch 1970 > 80 MeV protons at 1336 km.

that the field configuration remains constant while rotating relative to the geographic axis of the Earth. It is a change in the location of the centroid of the field intensity minimum. In 1970 at the Earth's surface, the field intensity minimum was located at $-26.2°$ latitude, $-49.9°$ longitude. By mid 1993, the minimum was located at $-27.4°$ latitude, $-54.1°$ longitude. The total sea-level "drift" during this period was $4.4°$, or $0.19°$/yr. Equivalent numbers at 1336 km were $-18.8°/-45.1°$ in 1970 and $-18.7°/-50.1°$ in 1993.5, for a drift of $0.21°$/yr. Note that neither the location of the minimum nor the "drift" rate was the same as at the surface of the Earth. At 0 km, the "drift" was primarily westward, by

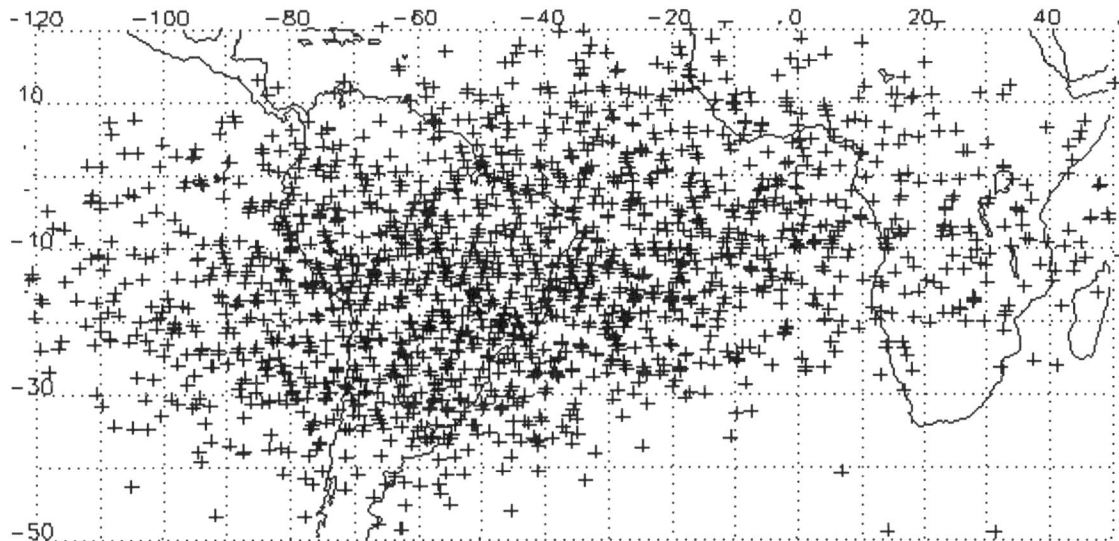

Figure 3. TOPEX SEE geographical distribution.

4.2°, and southward, by 1.2°. At 1336 km, the "drift" was almost entirely westward, by 5.0°. Thus, when discussing the "drift," and when attempting to confirm it, it is necessary to specify and be consistent in the use of altitude.

The SAA minimum in the field intensity is not a sharp, or well-defined, location. It actually covers a large area, with a complex shape. Figure 1 is a contour map of the field intensity at 1336 km (the TOPEX altitude) in the region of the SAA as modeled by DGRF 65, Epoch 1970. Epoch 1970 is used because that is the epoch that corresponds to the field model used to generate the AP-8 MAX particle model which we are addressing in this work. One may make various definitions of drift. In this work, we define drift as the change in location between the centroids of the field intensities at the two epochs. The "drift" of the centroid has been similar to the drift of the absolute minimum.

The maximum in the energetic proton flux at low altitude due to the SAA does not coincide in location to the exact region of the minimum in the magnetic field, since the particle maximum is a convolution of the magnetic SAA and the Energy-L-B/B_0 dependence of the particle ensemble. Figure 2 shows iso-intensity contours for > 80 MeV protons as predicted by the AP-8 MAX model, epoch 1970, for the 1336 km orbit altitude of TOPEX. Note that the particle contours do not correspond in detail to the magnetic field contours in Figure 1. Other energies will produce slightly different contours (and centroid locations) for two reasons. First, higher energy protons peak at lower L in the inner zone. This translates to a lower latitude (and closer to the geographic equator in the vicinity of the SAA). The second effect is due to differences in flux intensity gradients as a function of altitude (or atmospheric density) and energy. This effect translates into a westward trend with increasing energy for the location of the centroid of the proton flux intensity. The interdigitization which is seen in the contours of Figure 2 are an artifact of the AP-8 MAX model and its access using the GSFC-furnished interpolation subroutine TRARA2, which extracts flux values from the NASA particle environment models. The AP-8 models are not smoothed between L values at large B/B_0. A modified version of TRARA2 which corrects this problem has been developed, but has not yet been accepted as a standard [E. Daly, ESTEC, private communication, 1994].

Because the location of the peak in low altitude energetic protons is a convolution of the proton ensemble and the magnetic field configuration, there is no requirement a priori that the proton fluxes move in unison with the magnetic field minimum. Furthermore, the change in location both in longitude and latitude will be energy dependent for the reasons just mentioned. In this study, we use a large body of SEEs data from the TOPEX satellite as a proxy for the present position of the particle SAA and compare it with the position of the SAA as presently predicted by an updated magnetic field model in order to experimentally verify the change in the location of the SAA. Additionally, we make use of the dependence on energy of the latitude and longitude of the particle SAA to estimate the energy of the particles which produce the TOPEX SEEs.

2. TOPEX SEES

The TOPEX/Poseidon satellite, which is the orbital component of a three-year mission to constantly map the world's oceans, was launched August 10, 1992 into a circular Earth orbit at an inclination of 66° and an altitude of 1336 km. In this orbit, the satellite traverses the SAA region approximately

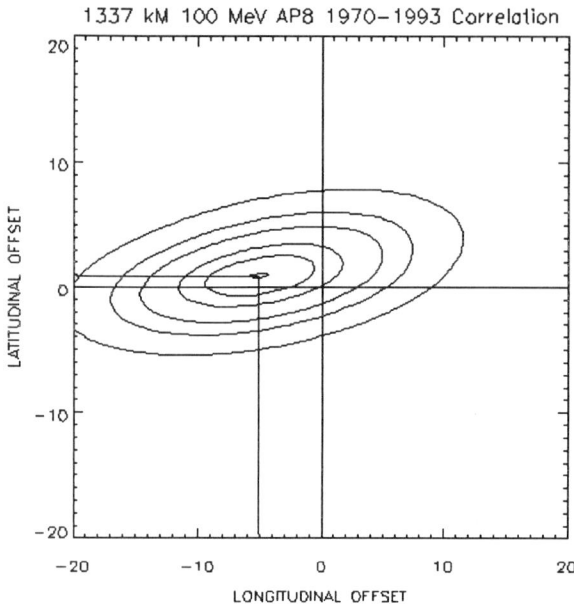

Figure 4. Epochs 1970 and 1993.5 cross-correlation contours for > 100 MeV protons at 1336 km.

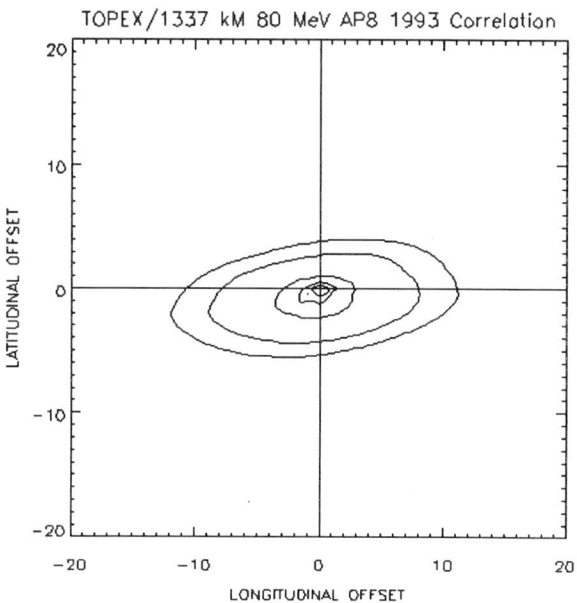

Figure 5. TOPEX SEEs-AP-8 MAX > 80 MeV p^+ cross-correlation contours.

six times per day. In the present study, a total of 2172 Earth sensor anomalies experienced by TOPEX between launch and mid-1994 have been correlated geographically with the SAA.

Two Earth Sensor Assemblies are located in a module on the Earth-facing side of the TOPEX spacecraft. Each sensor consists of a sensor subassembly and an electronics subassembly. Two types of anomalies are seen in these assemblies. One is a spike in the pitch and roll error telemetry, denoted PR. The other is a switch to the Wide Angle Mode, denoted WA. Both of these errors occur in each of the sensors [Rose, 1993]. Since these anomalies are not specifically identified as upsets in digital circuitry, we use the term "SEEs" for "Single Event Effects" in this paper rather than Single Event Upsets, or SEUs. SEEs occur in other instruments on the spacecraft but their rate is too low to be useful for a statistical analysis. Figure 3 is a longitude-latitude plot of the location of the TOPEX SEEs. This is a sparse data set–on average only one out of twenty-two $0.5° \times 0.5°$ latitude-longitude bins in the anomaly region will have a SEE.

3. CORRELATION ANALYSIS

The cross-correlations were produced by constructing X-Y grids with $0.5° \times 0.5°$ cells which contained values of the appropriate parameter (magnetic flux, particle flux, number of single events in that cell, etc.). One grid extended an additional 10° in all directions so that the smaller grid could be shifted ±10° in longitude and latitude with respect to it and still have complete cell overlap. A 20° by 20° X-Y correlation grid, at 0.5° increments, was then constructed by offsetting the smaller parameter grid with respect to the larger parameter grid, multiplying each pair of overlapped cells, and summing. The sum was placed in the correlation grid cell corresponding to the offset X-Y values. When the correlation grid was complete, it was normalized and contours of constant correlation strength were plotted using the CONTOUR call in IDL. Note that the resolution in determining a centroid (as is done in the contouring process) is not limited by the granularity of the input array. The contouring results indicate an effective resolution of better than 0.1° in most cases and better than 0.25° in all cases, even though the granularity was 0.5°.

3.1. General Considerations

For our comparisons, we first produced a cross-correlation between the AP-8 MAX/IGRF 65*/Epoch 1970 (which is the magnetic field model that was used in the construction of AP-8 MAX, the asterisk denoting the fact that the original model actually used the pre-1960 value of 0.311653 for the dipole moment) and the AP-8 MAX/DGRF 65/Epoch 1970, both with > 80 MeV protons. This was necessary because the epoch 1993.5 model we were going to use was internally consistent with DGRF 65 but not with the original IGRF 65*. The cross-correlation showed an offset of 0.25° in latitude and agreement in longitude between the proton maps using the different magnetic field models. This latitudinal offset was used to correct the TOPEX SEE/AP-8 MAX cross-correlation results.

The SAA particle intensity patterns which were used in these correlations were derived from the AP-8 MAX model. The contours were derived using an epoch of 1993.5 corresponding to the mid-point of the SEE data base acquisition period. Note that these AP-8 MAX model intensities will not be correct because of the fact that AP-8 MAX was generated

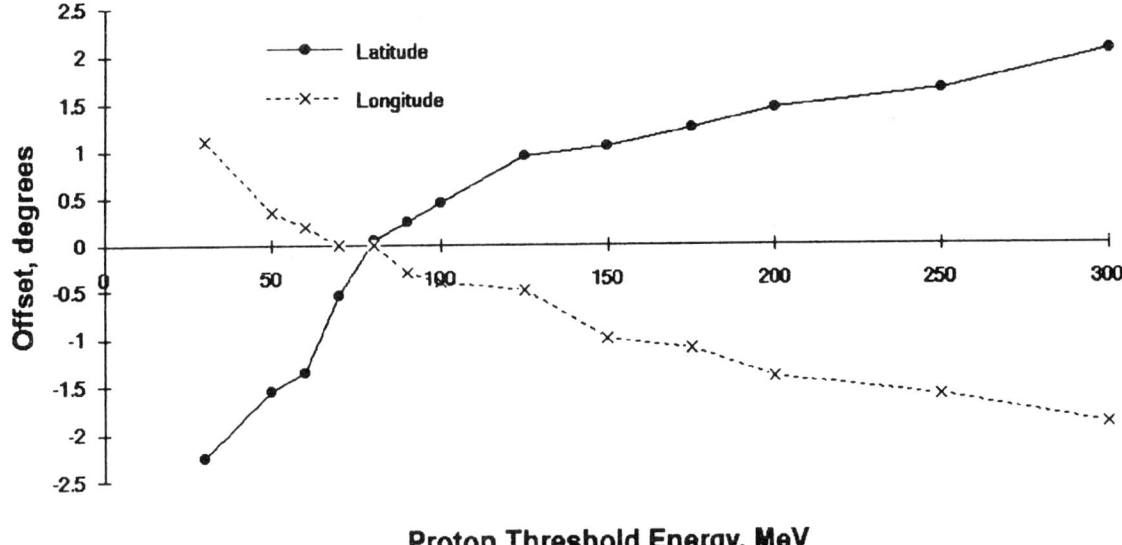

Figure 6. TOPEX SEEs-AP-8 MAX cross-correlation latitude-longitude offsets vs. threshold energy.

using a 1970 epoch and is specified in terms of flux vs. B/B_0 [*Konradi and Hardy*, 1987]. But unless the locations of the flux maxima in L have changed substantially, the longitude-latitude patterns should be approximately correct. A change in L will affect the latitude agreement.

Since the longitudinal agreement between IGRF 65* and DGRF 65 was satisfactory, the predicted "westward drift" between 1970 and 1993.5 was obtained by comparing proton contours from AP-8 MAX using DGRF 65 Epoch 1970 and IGRF 90, Epoch 1993.5. With the > 100 MeV maps, the "drift" was 5.1° (Figure 4), which is in close agreement with the "drift" of the magnetic field minimum at 1336 km (5.0°). The difference in latitude has been noticed previously [*Heynderickx*, 1996]. At > 30 MeV and 1336 km, the "drift" was 4.9°, which is different but is still within quite acceptable agreement with the "drift" of the magnetic field. This difference shows that in testing the "drift" it is essential that the comparison be made between particle maps, or between a particle map and a sensor, with the same threshold energy. The experimental verification mentioned previously [*Konradi et al.*, 1994] resulted in agreement with a 7° "drift". That result was obtained for both a 450 km orbit and a 617 km orbit. Our AP-8 MAX correlation technique with protons > 30 MeV predicts 4.3° and 4.5° for those altitudes, respectively.

3.2. TOPEX SEE - AP-8 MAX Correlation

In order to determine the accuracy with which the AP-8 MAX can predict the present location of energetic particles at low altitude, such as for extra-vehicular activities on Shuttle, the sparse SEE data set was correlated with a normalized AP8MAX flux map using a 2-dimensional function. AP-8 MAX fluxes were determined for each 0.5° × 0.5° bin from $-130°$ to $+60°$ longitude and $-60°$ to $+25°$ latitude.

IRGF 90 Epoch 1993.5 was used as the magnetic field model. The AP-8 MAX and SEE longitude-latitude maps were cross-correlated with offsets from $-10°$ to $+10°$ in both longitude and latitude at 0.5° increments. The resulting variational array was then normalized and contours of constant correlation strength plotted. Figure 5 shows that the centroid of the TOPEX SEEs coincides with the AP-8 MAX Epoch 1993.5 predicted maximum of > 80 MeV protons within the accuracy of the procedure, $\sim 0.2°$.

The contours in Figure 5 were obtained using the integral flux of > 80 MeV protons predicted by AP-8 MAX. If other threshold energies are used, a slightly different correlation pattern is obtained for the two reasons mentioned previously. First, the location of the overall flux intensity pattern in L is a function of proton energy. For higher energy protons, the centroid of the flux intensity pattern is found at lower L. This translates to a center of the SAA flux intensity pattern at 1336 km altitude that moves northward toward the geographic equator as energy increases. The second effect is also energy related. Due to the sharper decrease in flux intensity as a function of altitude with lower energy protons, the centroid of the proton pattern moves eastward with decreasing energy. Thus, when evaluating the "motion" of the SAA, the energy of the particle used in the analysis must be considered.

Since the location of the proton SAA has an energy-latitude-longitude dependence, we made use of this dependence to estimate the energy of the particles that are causing the TOPEX SEEs. Figure 6 shows the latitudinal and longitudinal offsets between the SEEs and AP-8 MAX Epoch 1993.5 as a function of energy. Figure 6 indicates that the threshold energy for producing these SEEs is about 80 MeV. The AP-8 MAX energy-L profile is not quite correct for the 1990 time period (per comparison with CRRES data, [*S. Gussenhoven*, PL/GD, private communication, 1995], so the latitudinal curve may require minor correction.

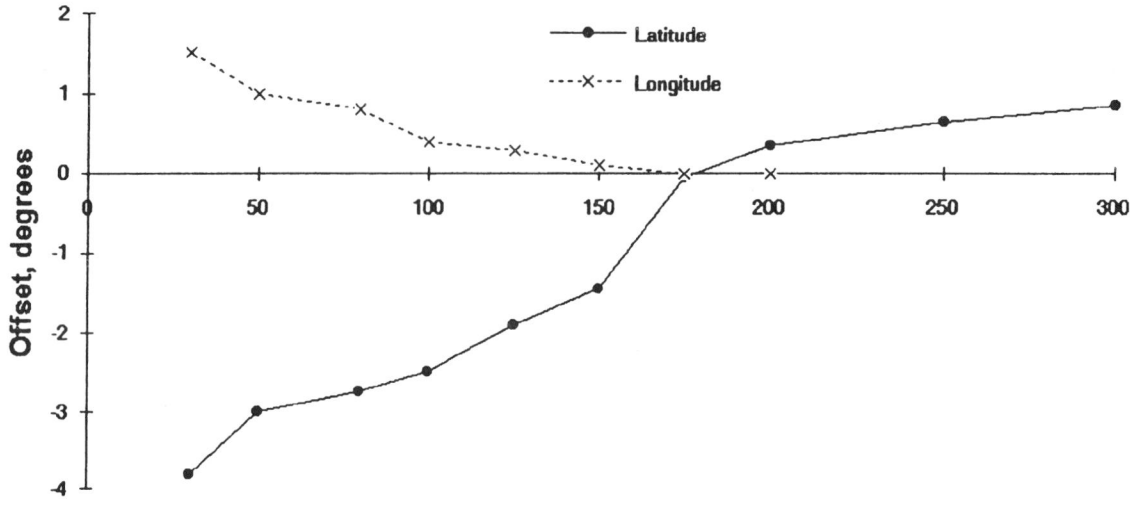

Figure 7. TOPEX PR SEEs-AP-8 MAX cross-correlation latitude-longitude offsets vs. threshold energy.

To some extent, we are engaged in a circuitous argument. We say that the "drift" in the magnetic field predicts a proton environment that agrees with the TOPEX SEE distribution. But this is postulated on a proton threshold for producing the TOPEX SEEs that is 80 MeV, a value that is derived from the SEE data. We get this 80 MeV threshold from the "agreement" with the AP-8 MAX model. The only defense against this criticism is that the latitudinal and longitudinal offsets both agree with the predicted energy-latitudinal-longitudinal dependency. Internally, we get consistent results. Obviously, an updated proton environment derived from in-situ measurements would be a much preferred and a much more reliable quantitative check against the "drift" value.

Figure 6 included all TOPEX SEEs. We separated the SEEs into subgroups, represented by "1", "2", "WA", and "PR", each of which included either all of the SEEs from Assembly 1 or 2, or all of the WA or PR SEEs from both assemblies. The "1", "2", and "WA" subgroups provided results that were indistinguishable from the entire set. But the "PR" subset, Figure 7, were definitely produced by a different particle population. There were only 337 SEEs in the "PR" subset. Figure 7 indicates that the threshold for production of these anomalies is around 175 MeV. The 175 MeV threshold has interesting implications. This probably indicates that the circuit elements that are involved in this type of anomaly are heavily shielded.

3.3. *COBE Sensor Background*

We also applied this procedure to an entirely different type of data set, "glitches" from the COBE/DIRBE sensor. COBE (COsmic Background Explorer) is in a high inclination 890 km circular orbit. The DIRBE sensor consists of a set of infrared detectors at several wavelengths. During passage through the SAA or the low altitude extensions of

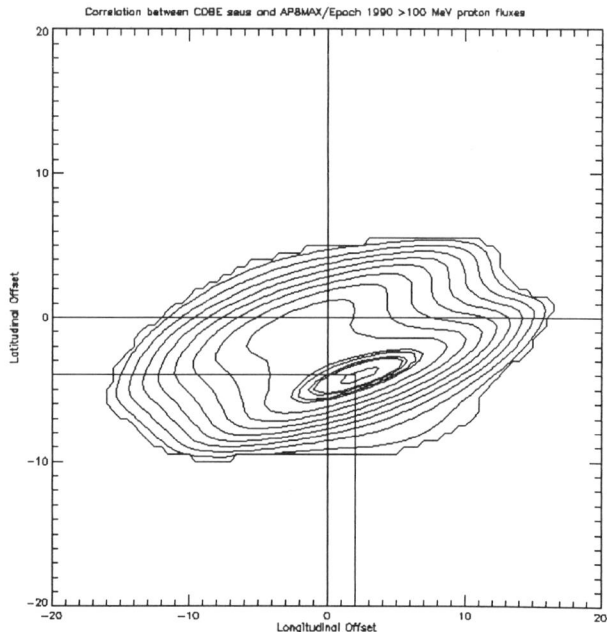

Figure 8. Correlation contours of background effects in the COBE/DIRBE IR sensor with AP-8 MAX $P > 100$ MeV.

the outer zone electron belts, a shutter of several gm/cm^2 is used to protect the detectors. Excess signal in a detector for a single count cycle is considered to be due to particle penetration. The data are stored in 128-second segments, so there is an uncertainty in the location of a "glitch" of about $\pm 3°$ along the orbit path. This uncertainty "smears" the data set for correlation purposes, but one of the advantages of the

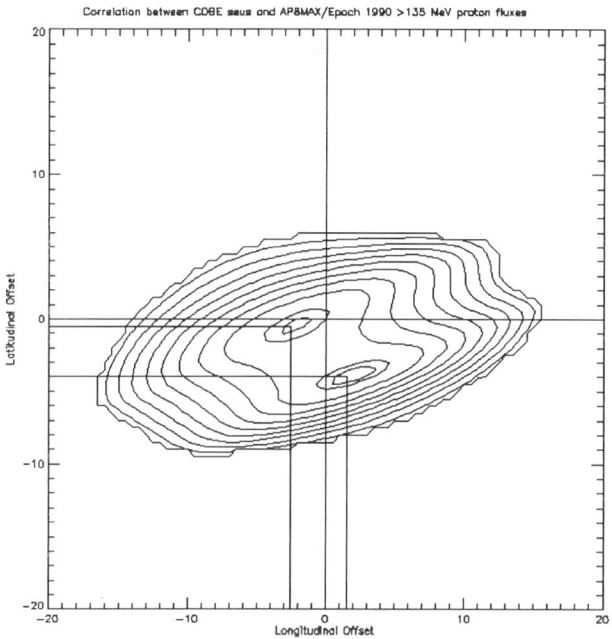

Figure 9. Correlation contours of background effects in the COBE/DIRBE IR sensor with AP-8 MAX $P > 135$ MeV.

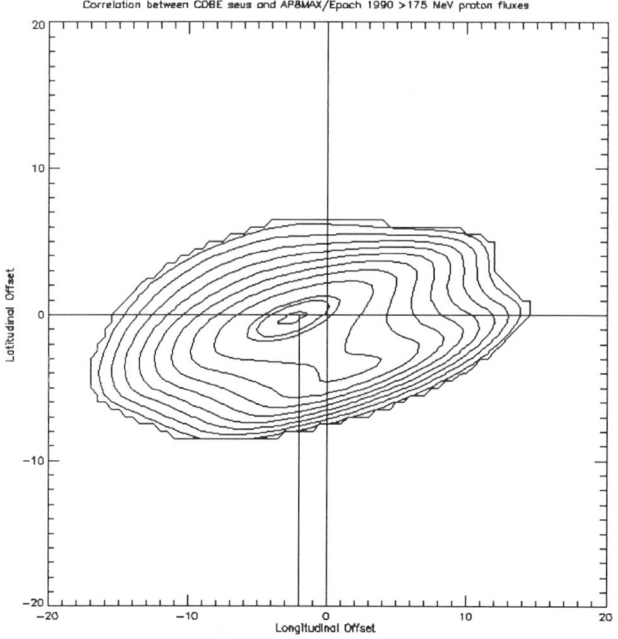

Figure 10. Correlation contours of background effects in the COBE/DIRBE IR sensor with AP-8 MAX $P > 175$ MeV.

2-D correlation is that it locates the centroid of the correlation strength. A large data set will provide accurate results if there are enough samples that the uncertainties average out.

When we apply the correlation procedure to this data set, we get ambiguous results. The correlation procedure produces normal results up to about 100 MeV (Figure 8), at which point a single strong correlation peak is still seen, but a broad plateau begins to appear to the north. By 135 MeV (Figure 9), there are two equal correlation peaks, the original one at lower latitude and a new one closer to the equator. Finally, by 175 MeV, the major correlation is with the northern peak with only a shoulder being seen to the south (Figure 10).

If one produces a plot similar to Figure 6 but for the COBE/DIRBE data, one gets Figure 11. This appears to be a breakdown in the cross-correlation technique. Actually, it illuminates a problem with AP-8. At the COBE altitude in the energy range 120 MeV $< E_p < 150$ MeV, the AP-8 model itself contains two peaks. The two correlation peaks occur because one data set, COBE, has a single peak and the other, AP-8, has two.

Figure 12 shows the 890 km AP-8 flux contours at 135 MeV, which is the energy at which the two peaks are equal in intensity. Below this energy, one peak is dominant, above it the other dominates. This dual-peak characteristic is not seen at much lower or higher altitudes. It is clearly an artifact of the model, since the COBE correlation follows the peaks. If there actually were a dual peak structure in the fluxes, the correlation would have a single peak. One would expect the COBE data to show a dual intensity structure corresponding to the particle structure, and the correlation would be maximum when the two sets of peaks matched. While not as clear cut as Figure 6, Figure 11 does indicate that the background in the DIRBE detectors is due to protons with energies above about 200 MeV. The shutter is being penetrated.

4. DISCUSSION

The agreement between the location of the TOPEX SEEs and the 1993.5 epoch contour confirms the validity of using the change in the magnetic field for predicting the change in the location at which SEEs can be expected to occur. It also shows that the drift in the location of the particle SAA does not coincide precisely with the drift in the magnetic field SAA (a commonly-held misconception). However, due to the fact that the AP-8 map is organized in terms of B/B_0, the weakening of the geomagnetic field during this interval results in an artificial lowering of mirror points of the model fluxes, predicting a larger flux of protons at low altitude than is actually present. A renormalization of AP-8 flux intensities is required to correct for this effect and is in the process of being attempted with NOAA proton spectrometer data (860 km circular polar orbit, $16 < E < 215$ MeV). The new study should also correct deficiencies at low altitude as seen in the comparison with COBE data (Figure 12).

The agreement between the geographic location of the TOPEX SEEs and the location of energetic protons in the SAA as predicted by AP-8 MAX using the current magnetic field epoch shows that a viable interim solution is available for the problem of the secular magnetic field variation on the accuracy of the AP-8 model. We would propose that the INTENSITY of proton fluxes for a given orbit be cal-

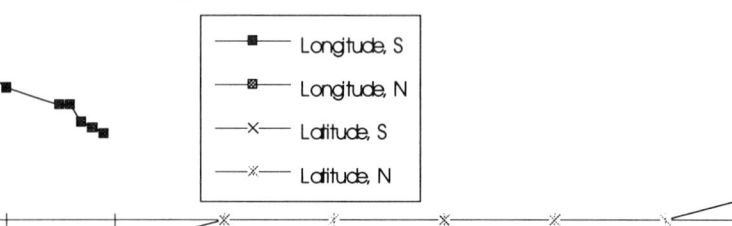

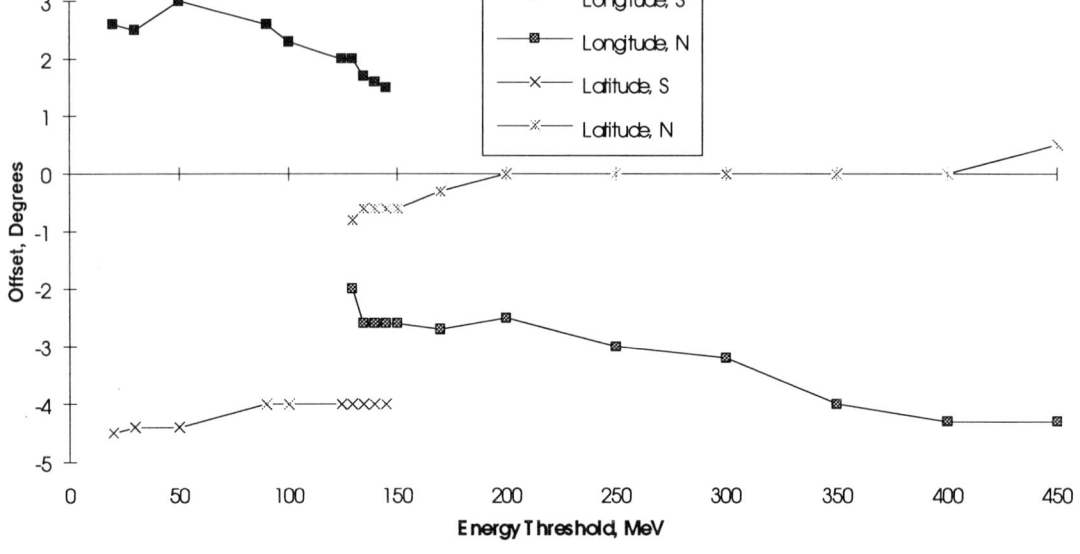

Figure 11. COBE-AP-8 MAX cross-correlation latitude-longitude offsets vs. threshold energy.

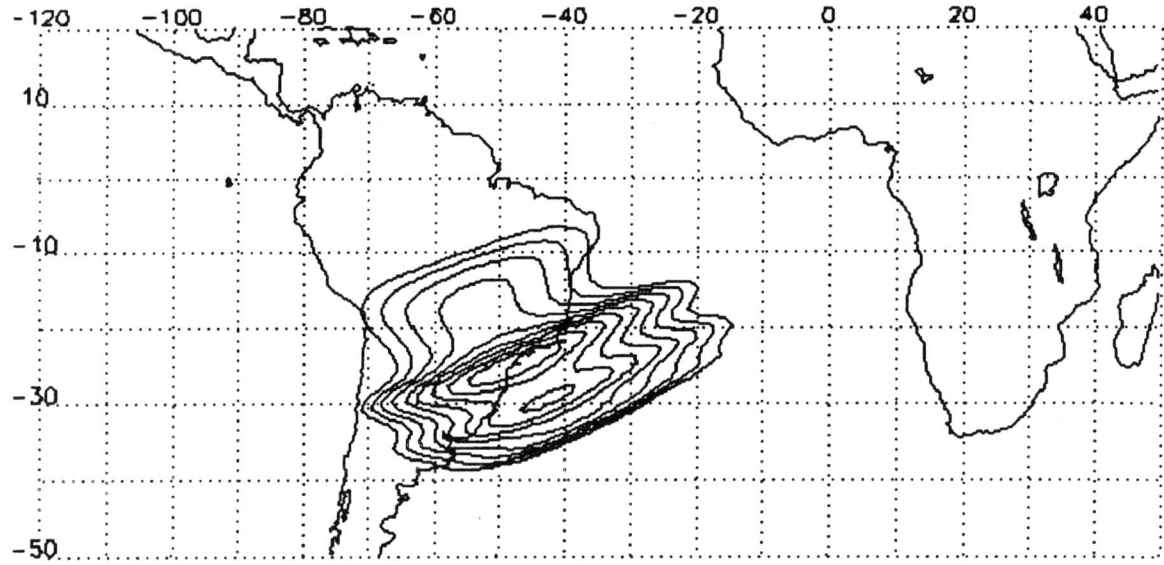

Figure 12. AP-8 MAX Epoch 1990 $P > 135$ MeV flux contours at COBE altitude (890 km).

culated using the AP-8 model with a 1964 (MIN) or 1970 (MAX) epoch, but the LOCATION at which these fluxes is encountered be calculated using current epochs. This avoids the unrealistically large fluxes at low altitudes which are predicted by AP-8 if a current epoch is used for intensity determinations, but retains the accuracy in location obtained by using the current epoch. This approach is already in use in some establishments.

Furthermore, the cross-correlation technique appears to offer a simple means of determining the minimum energy of the particle necessary for producing SEUs and other SEEs on spacecraft in low altitude orbits. This information can then be used to get a better understanding of the SEEs. The availability of this analysis technique adds additional motivation for updating the AP-8 model.

5. SUMMARY

Comparison of the location of SEEs experienced by the TOPEX/POSEIDON spacecraft in low earth orbit shows excellent agreement with the location of the SAA using current-epoch calculations with the AP8MAX proton model. Use of the 1970 epoch with AP-8 MAX, while providing approximate flux intensities at low altitude, does not provide the proper geographic location. A combination of the two approaches, which is already in use at some facilities, is recommended as an interim solution until a new proton model is generated which can take into account the secular variation in the geomagnetic field. The pattern of occurrences of SEEs can be used to estimate the energy of the particle producing them. Correlation with background effects in the COBE infrared detectors disclosed a significant problem in AP-8 at low altitude and high energy (the location of the intensity peak near 900 km shifting bimodally from one location to another as a function of energy).

Acknowledgements. We would like to thank J. Rose and C. Elliott for the data on TOPEX. The COBE data were obtained through private communication with T. Kelsall and B. Franz. This work was supported by NASA's Office of Safety and Mission Assurance.

REFERENCES

Heynderickx, D., Comparison Between Methods to Compensate for the Secular Motion of the South Atlantic Anomaly, *Nucl. Tracks Radiat. Meas.*, in press, 1996.

International Geomagnetic Reference Field, Revision 1987, *J. Geomag. Geoelectr., 39*, 773–779, 1987.

Konradi, A. and A.C. Hardy, Radiation Environment Models and the Atmospheric Cutoff, *J. Spacecraft and Rockets, 24*, 284, 1987.

Konradi, A., G.D. Badhwar and L.A. Braby, Recent Space Shuttle Observations of the South Atlantic Anomaly and the Radiation Belt Models, *Adv. Space Res., 14*, No. 10, 911–921, 1994.

Lemaire, J., E.J. Daly, J.I. Vette, C.E. McIlwain and S. McKenna-Lawlor, Secular Variations in the Geomagnetic Field and Calculations of Future Low Altitude Radiation Environments, in *Proceedings of the ESA Workshop on Space Environment Analysis ESA WPP-23*, 9–12 October 1990, ESTEC, Noordwijk, The Netherlands, Sect. 5.17.

Rose, J., Experiences with Single Event Upsets, TOPEX/POSEIDON Project Memo JPL D-10643, Jet Propulsion Laboratory, April 1993.

Sawyer, D.M. and J.I. Vette, AP8 Trapped Proton for Solar Maximum and Solar Minimum, NSSDC 76-06, National Space Science Center, Greenbelt, Maryland, December 1976.

M. Lauriente, NASA Goddard Space Flight Center, Greenbelt, MD 20771.

A.L. Vampola, K. Gosier, University Research Foundation, Greenbelt, MD 20770.

DISCUSSION

Q: J.B. Blake. By SEE threshold—was it meant proton incident energy plus transport through the S/C to the sensitive part?
A: M. Lauriente. Yes.
Q: M.K. Hudson. Have you calculated the mass density through which particles penetrate on TOPEX to correct for energy loss before reaching silicon?
A: M. Lauriente. Black box, non-available info.
Q: R.A. Mewaldt. Aside from the comparison of the latitude and longitude centroids of SEE's and greater than 80 MeV protons, do the detailed contour maps agree? Are they proportional?
A: A.L. Vampola. They are approximate, but the purpose of using the cross correlation technique is to compare sets which don't correspond exactly in small detail. The cross correlation coefficients are normalized and then contours are plotted.

Low Altitude Trapped Radiation Model Using TIROS/NOAA Data

S.L. Huston, G.A. Kuck and K.A. Pfitzer

McDonnell Douglas Aerospace, Huntington Beach, California

The current NASA trapped radiation models have limited accuracy in the low altitude (250–1,000 km, $L < 1.5$) region because of the large gradients which exist and because of the limited data on which they were based. Under NASA's Space Environment and Effects (SEE) Program, we are developing new models for this region based primarily on data from the TIROS/NOAA polar orbiting spacecraft. We have performed an initial analysis of the data set to determine the variation of the proton flux at the geomagnetic equator as a function of solar activity and altitude (L) for the period 1978–1992. The data show a large variation over the solar cycle, as much as a factor of 10 at $L = 1.12$. There is also an L-dependent phase lag between the solar $F_{10.7}$ flux and the proton flux. These variations should give insight into the source and loss mechanisms at work, as well as into more appropriate coordinate systems for low-altitude models.

1. INTRODUCTION

The trapped radiation environment can have many harmful effects on humans and hardware, and is of great interest to both the scientific and engineering communities. The low-altitude (250–1000 km) region is a particularly important region. Particle interactions with the upper atmosphere constitute one of the major loss mechanisms for the radiation belts, and are a significant source of energy input to the upper atmosphere. It is a particularly important region to engineers because of the large number of spacecraft in orbit there. It is also a particularly difficult region to model accurately. Because the atmospheric interactions and source mechanisms are strongly influenced by solar activity, a good model requires long-term data from at least one complete solar cycle. The current NASA models AP-8 and AE-8 [*Vette*, 1991; *Gaffey and Bilitza*, 1994] have limited accuracy in this region because of the large gradients which exist and because of the nature of the data on which they were based.

Initially under McDonnell Douglas funding, and now under contract to NASA Marshall Space Flight Center, we have been developing an improved model of the low altitude trapped radiation belts. This model will provide improved accuracy at low altitudes ($L < 1.5$) and will provide a true solar cycle dependence. The model will provide omnidirectional integral flux of protons with energy greater than 16 MeV and electrons with energy greater than 30 keV. An objective of the program is to develop and use coordinate systems which are more appropriate to the low altitude regime than the traditional (B, L) coordinates. Above all, the model is intended to be user friendly and useful to both the engineering and scientific communities. The model will be based primarily on data from the US TIROS/NOAA polar orbiting weather satellites, which provide a nearly continuous data set from 1978 to the present. Other data sets are being correlated with the primary set as appropriate.

We have performed an initial assessment of the data set by processing selected files over the period 1978–1992 and have determined the count rates at the magnetic equator as a function of L for about 1.5 solar cycles.

2. INSTRUMENTATION

We are using data from the Medium Energy Proton and Electron Detector (MEPED) aboard the TIROS/NOAA spacecraft [*Seale and Bushnell*, 1987]. Table 1 summarizes the characteristics of this detector. The detector contains two pairs of charged particle telescopes, one pair for protons and one for electrons. Each pair of telescopes has one detector pointed towards the zenith and one perpendicular to both the zenith and the spacecraft velocity vector. There is also an omni-

120 LOW-ALTITUDE TRAPPED RADIATION MODEL

Table 1. The TIROS/NOAA Space Environment Monitor (SEM) package provides particle fluxes in energy ranges important for engineering and scientific studies.

Particle Type	Data Channel	Energy Range (MeV)	Aperture Axis	Cone Half Angle
Protons	P1	.03–.08		
	P2	.08–.25	0° (zenith)	14°
	P3	.25–.80	90°	
	P4	.80–2.5		
	P5	> 2.5		
	P6	>16		
	P7	>36		omni
	P8	>80		
Electrons	E1	> .03		
	E2	> .10	0° (zenith)	14°
	E3	> .30	90°	

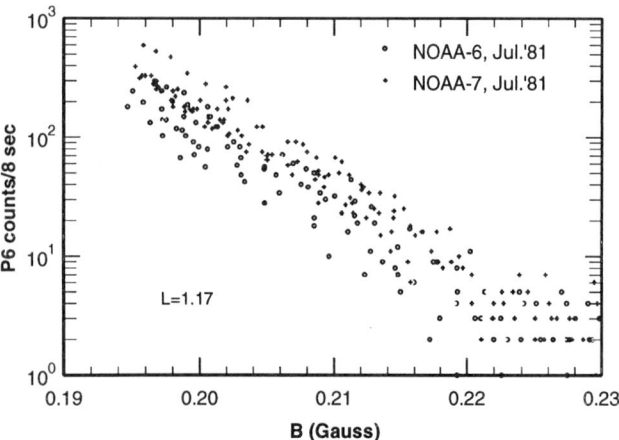

Figure 1. Raw count rate as a function of magnetic field strength at $L = 1.17$, using (B, L) coordinates from the TIROS/NOAA data files. Scatter is about a factor of four for each spacecraft, with an additional factor of three difference between the two spacecraft shown.

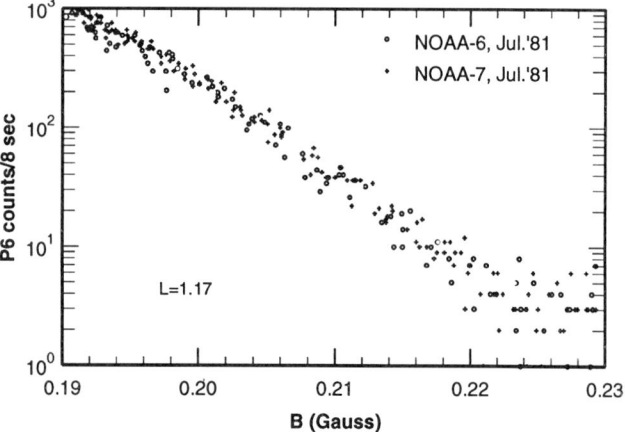

Figure 2. Same data as shown in Figure 1, but with magnetic coordinates re-computed using actual spacecraft altitude. Scatter is reduced to less than a factor of two, and data from the two spacecraft correlate well.

directional proton detector, which measures proton fluxes in three energy channels ranging from 16 to 215 MeV. In this paper we will discuss results from the highest energy channel, 80–215 MeV.

The data are archived at NGDC [*Raben et al.*, 1995] and are available as binary files. The data are accumulated at two-second intervals, and spacecraft ephemeris data are saved at eight-second intervals. Each data file contains approximately ten days' worth of data and is about 30 megabytes in size. We have processed seventeen files spanning the time period from 1978 to 1992. Data have been selected at approximately yearly intervals over this time period in order to determine the variation of the proton flux over the solar cycle.

3. DATA ANALYSIS PROCEDURE

In order to process the data, we first perform a pre-processing procedure in which we sum the counts over the 8 second ephemeris interval and convert the byte order for processing on a PC. We next process the data by computing the vector magnetic field and L parameter using the IGRF internal field model [*Barraclough*, 1987] and the L calculation procedure developed by *Pfitzer* [1991] for the CRRES data analysis. The B and L provided in the data files are calculated for the approximate position of the spacecraft assuming a constant altitude of 870 km. Actual spacecraft altitudes range from about 800 to 850 km, and given the large gradients in this region, the data on the files are not accurate enough for our purposes. Figures 1 and 2 show an example of the dra-

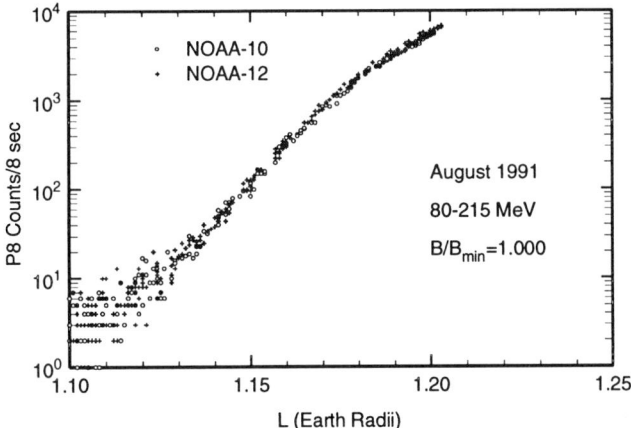

Figure 3. Count rates as a function of L at the magnetic equator $(B/B_{min} = 1)$. A time shift has been introduced as discussed in the text. Scatter is less than 15%.

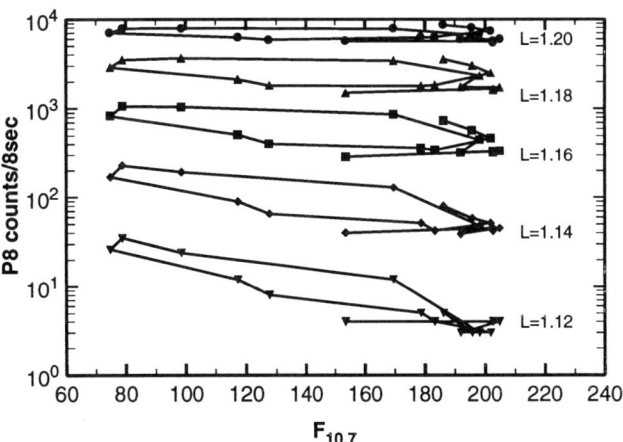

Figure 5. Count rates as a function of $F_{10.7}$ for the same L values as shown in Figure 4. Note the large hysteresis effect.

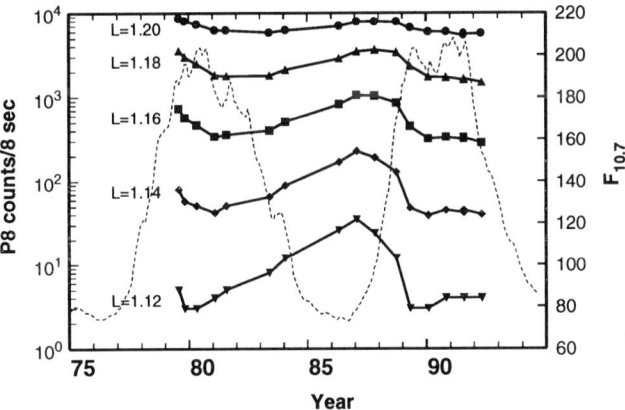

Figure 4. Variation of count rates in the 80–215 MeV channel over the solar cycle as a function of L. The dashed line shows the 13-month smoothed solar $F_{10.7}$ flux.

matic improvement in accuracy obtained by re-computing the magnetic coordinates. Finally, we introduce a time shift to compensate for apparent errors in the spacecraft ephemeris. Figure 3 shows an example of the data following our processing procedure. Scatter in the data is less than 15%, and agreement between the two spacecraft is also within the same range.

4. RESULTS

For this study we have restricted our analysis to the geomagnetic equator $(B/B_{min} = 1)$ for the 80-215 MeV protons. The primary objective of this effort was to establish the variation of the proton flux over the solar cycle. We also are investigating the use of the atmospheric density as an organizing parameter.

Figure 4 shows the variation in the 80–215 MeV proton flux at the geomagnetic equator as a function of time over the last two solar cycles. The dashed line shows the solar $F_{10.7}$ flux. The other lines show the proton flux at five different L values. Several effects are noted. First, the variation over the solar cycle is a strong function of L. Variation between solar minimum and solar maximum is a factor of 10 at $L = 1.12$, and decreases to a factor of about 2 or less at $L = 1.20$. In all cases, the rate of decrease in the proton flux during the rising part of the solar cycle is much faster than the rate of increase during the decreasing phase of the solar cycle. There is also a definite phase lag between the minima and maxima of the $F_{10.7}$ flux and the corresponding maxima and minima in the proton flux. For example, if solar maximum for Cycle 21 is taken to be mid-1980, the minimum proton flux at $L = 1.20$ is not reached until approximately mid-1982. This phase lag decreases to approximately zero at $L = 1.12$. Similarly, if solar minimum is taken to be late 1986, the peak in proton flux at $L = 1.20$ is not reached until late 1987. Again, the phase lag decreases to near zero at $L = 1.12$. Finally, there is a definite difference in the character of the minima in proton flux for the two solar maxima. In Cycle 21, the fluxes decreased, reached a minimum, and started to increase again. In Cycle 22, the proton flux at all L values reached a minimum and remained there for the period 1990–1992.

Figure 5 shows the proton flux at the same L shells plotted as a function of solar $F_{10.7}$ flux. This figure clearly shows the hysteresis effect between the rising and falling phases of the solar cycle. For a given value of $F_{10.7}$, the proton flux can vary by a factor of 3 to 5 between the rising and falling portions. It also appears that the proton flux decreases relatively slowly until $F_{10.7}$ reaches a value of about 170, beyond which the proton flux decreases rapidly. Conversely, during the falling portion of the solar cycle, the proton flux increases slowly until $F_{10.7}$ reaches a value of about 130, after which the proton flux increases much more rapidly.

5. SUMMARY AND CONCLUSIONS

The TIROS/NOAA data comprise a very high quality data set, which is ideal for examining the long-term behavior of the low-altitude trapped radiation environment. Although the instrumentation is limited compared to dedicated scientific spacecraft such as CRRES or UARS, flying nearly identical detectors for more than one solar cycle allows us to unambiguously identify the variations in the trapped proton flux. The sheer quantity of the data provides good statistics and allows the use of small data bins. Scatter in the data is typically less than 15%, and, to the extent that we have determined it, the variations among the different spacecraft are also within about 15%. Given the high quality of the data, we expect that pitch angle distributions can be unfolded for $L < 1.2$.

Care must be taken in using the data, however. Because of the large spatial gradients in the low-altitude region, small errors in spacecraft ephemeris or calculation of magnetic coordinates increase the scatter in the data. Based on our analysis of the data, we estimate that the L calculation is accurate to within about 0.001 R_E, and this accuracy is required to reveal subtleties in the data. These subtleties reveal the need for a new model to supplement or replace the NASA AP-8 models at low altitudes. AP-8 predicts a variation between solar maximum and solar minimum of about a factor of 3, whereas the TIROS/NOAA data reveal a factor of 10 variation at $L = 1.12$. Note that AP-8 MAX was developed for an $F_{10.7}$ flux of 150, whereas the last two solar maxima have been over 200, so it is really not surprising that the model does not predict the actual variation. The data also show L-dependent phase lag and hysteresis effects which are of great importance not only to this modeling effort, but also to improving our physical understanding of the source and loss mechanisms for the radiation belts.

We will continue our processing and analysis of the data set, including the lower energy omnidirectional proton data and the data from the proton and electron telescopes. We will convert the raw counts to flux using the geometric factors for the instruments [*Seale and Bushnell*, 1987]. Data from other spacecraft will be used as appropriate to compare with the NOAA detectors and to determine absolute calibration factors. Of particular importance is the development of coordinate systems which are more appropriate to the low-altitude region, where atmospheric density controls the particle population. We will make use of the ideas developed by *Pfitzer* [1990] and extended by ESA [*Lemaire et al.*, 1995].

Acknowledgements. This work is supported by NASA contract number NAS8 40295; John Watts at MSFC is the project monitor. The authors would like to thank Herb Sauer at NOAA and Dan Wilkinson at NGDC for their help in understanding and accessing the data base.

REFERENCES

Barraclough, D.R., International Geomagnetic Reference Field: The Fourth Generation, *Phys. Earth Planet. Inter., 48*, 279, 1987.

Gaffey, J.D., Jr. and D. Bilitza, NASA/National Space Science Data Center Trapped Radiation Models, *J. Spacecraft and Rockets, 31*, 1994.

Lemaire, J., A.D. Johnstone, D. Heynderickx, D.J. Rodgers, S. Sitza, and V. Pierrard, Trapped Radiation Environment Model Development TREND-2 Final Report, *Aeronomica Acta, A 393*, 1995.

Pfitzer, K.A., Radiation Dose to Man and Hardware as a Function of Atmospheric Density on the 28.5° Space Station Orbit, McDonnell Douglas Space Systems Co. Report No. H5387A, Huntington Beach, CA, March 1990.

Pfitzer, K.A., Improved Models of the Inner and Outer Radiation Belts, Scientific Report No. 1, PL-TR-91-2187, July 1991.

Raben, V.J., D.S. Evans, H.H. Sauer, S.R. Sahm, and M Huynh, TIROS/NOAA Satellite Space Environment Monitor Data Archive Documentation: 1995 Update, NOAA Technical Memorandum ERL-SEL-86, February 1995.

Seale, R.A. and R.H. Bushnell, The TIROS-N/NOAA A-J Space Environment Monitor Subsystem, NOAA Technical Memorandum ERL SEL-75, April 1987.

Vette, J.I. The NASA/National Space Science Data Center Trapped Radiation Model Program (1964–1991), NSSDC World Data Center A for Rockets and Satellites, Report number 91-29, November 1991.

S. Huston, G. Kuck and K. Pfitzer, McDonnell Douglas Aerospace, 5301 Bolsa Ave., Huntington Beach, Calif. 92647, USA. E-mail: stu@huston.mdc.com

DISCUSSION

Q: M. Walt. Since there is an appreciable East-West asymmetry in trapped flux at low altitude, could this explain the different count rates seen on northern and southern passes?
A: S.L. Huston. The detector was omnidirectional and should not be affected by an East-West asymmetry.
Q: J.B. Blake. Can there be a systematic effect due to somewhat different instruments aboard the several spacecraft?
A: S.L. Huston. So far we have performed intercomparisons for three spacecraft. We have found that the differences between spacecraft are of the order of 20–50%, which is about the same as the scatter due to the ephemeris errors noted in the talk. We plan to resolve these inter-spacecraft differences before continuing the data processing.
Q: M. Kruglanski. The time variation of the NOAA data illustrates clearly the effects of the atmospheric loss processes. During solar minimum, the atmosphere deflates, the loss processes decrease but the source remains. At solar maximum, the atmosphere inflates and cuts rapidly the proton fluxes.
A: S.L. Huston. One of the major reasons for doing this study was to investigate this process. We hope that this data will allow us to determine more accurately the mechanisms and magnitudes of the losses.

Modelling He and H Isotopes in the Radiation Belts

R.S. Selesnick and R.A. Mewaldt

California Institute of Technology, Pasadena, California

Nuclear interactions between inner zone protons and atoms in the upper atmosphere produce energetic H and He nuclei that are an additional radiation belt source. We calculate production rates of these isotopes from models of the inner zone proton intensity, the upper atmosphere drift averaged composition and densities, and cross-sections for the various interaction processes. For comparison with observations of radiation belt H and He isotopes, the production rates are combined with a model of the energy loss rate in the residual atmosphere to calculate particle intensities. Although the calculations are in principle straightforward, they depend on a detailed knowledge of the various model inputs, including models for radiation belt protons, and may also depend on the phase of the solar cycle. On the other hand, the results of the calculations, when compared with the observational data, can provide useful tests of the model inputs. Preliminary results show that the atmosphere is a significant source for inner zone ^{4}He, ^{3}He, and d.

1. INTRODUCTION

The inner radiation belt is composed primarily of protons that were produced locally by the cosmic ray albedo neutron decay (CRAND) process, due to collisions between cosmic rays and the neutral atmosphere. The trapped protons similarly create a secondary source of trapped particles by their own nuclear interactions with the atmosphere. This process is less efficient than CRAND for producing protons because the particles must be injected directly whereas the neutrons can propagate to high altitudes, where the atmospheric density is low, before decaying. However, because CRAND produces only protons and electrons, the secondary process can be a significant source of other trapped particles such as isotopes of H and He.

The CRAND source has been the subject of several theoretical calculations for comparison with the trapped proton data (e.g. *Dragt* [1971]; *Farley and Walt* [1971]; *Jentsch and Wibberenz* [1980]; *Jentsch* [1981]). However, probably due to a lack of data on the composition of the high energy trapped ions, the secondary source has not been studied in detail. The new data from SAMPEX [*Cummings et al.*, 1995; *Looper et al.*, 1995] prompted us to begin such an investigation. Although the calculation is analogous to the CRAND case, it is complicated by the need to calculate the source function due to each of the many possible nuclear interactions. In the CRAND calculations the neutron source was generally an empirical model based on neutron flux measurements.

2. CALCULATIONS

If the atmospheric production of trapped particles is balanced only by ionization energy loss in the atmosphere, then their intensity j satisfies a continuity equation [*Jentsch and Wibberenz*, 1980]

$$\frac{1}{v}\frac{\partial j}{\partial t} = S + \frac{\partial}{\partial E}\left(j\left|\frac{dE}{dx}\right|\right), \quad (1)$$

where $v = dx/dt$ is the (non-relativistic) speed of the trapped particle at time t and kinetic energy E. The production rate in $(\text{cm}^3\text{s sr MeV})^{-1}$ is S. The intensity j can be time dependent through S which generally varies due to the solar cycle. In the steady-state case, or if the particle lifetimes are short compared to the 11 year solar cycle, the solution is

$$j = \frac{1}{\left|\frac{dE}{dx}\right|} \int_E^\infty S\, dE. \quad (2)$$

The integral should be cut-off at the maximum energy of adiabatically trapped particles, but this is not significant for a sufficiently soft spectrum. The solution (2) also applies to particles with lifetimes that are long compared to the solar cycle time if S is interpreted as the solar cycle average value. For cases where the lifetimes are comparable to the solar cycle time, a time dependent solution is possible [*Jentsch and Wibberenz*, 1980]. For protons, the continuity equation (1) is valid for low L shells ($L \lesssim 1.3$) beyond which radial diffusion from an external source becomes significant at low energies ($\lesssim 30$ MeV) [*Jentsch*, 1981].

The production rate of secondaries is

$$S = \sum_i \int dE_p \int d\Omega_p\, n_i j_p \frac{d^2\sigma_i}{d\Omega dE}. \quad (3)$$

The summation extends over all interactions that lead to a given type of secondary particle and the integrals cover the range of proton energies E_p and solid angle Ω_p that kinematically can produce secondaries with energy E, pitch-angle α in a solid angle Ω, and at a given L shell. The atmospheric density, n_i, of target atoms for the interaction i is averaged over the drift path of protons for each L and proton pitch angle α_p. The proton intensity is j_p and the cross section for interaction i is σ_i. Both n_i and j_p may be functions of time due to solar cycle variations. Calculating S from Eq. (3) requires knowledge of the atmospheric densities, the trapped proton intensities and drift paths, and the interaction cross-sections. In addition, calculating j from (3) requires knowledge of the energy loss rate in the atmosphere.

For the atmosphere we use the Mass-Spectrometer-Incoherent-Scatter-1986 (MSIS-86) neutral atmosphere model [*Hedin*, 1987] as encoded by the National Space Science Data Center (NSSDC), which provides number densities of He, O, N_2, O_2, Ar, H, and N as functions of day-of-year, local time, altitude, geodetic latitude and longitude, solar 10.7 cm flux ($F_{10.7}$) for the previous day and a 3 month average, and the magnetic A_p index. For the Earth's magnetic field we use the International Geomagnetic Reference Field (IGRF) model extrapolated to 1992 [*Langel*, 1991] from the NSSDC. To calculate the drift averaged densities for each element, trajectories of 100 MeV protons were calculated numerically for a given L shell starting from the minimum magnetic field B on that L shell at a given altitude and continuing for one complete drift in longitude around the Earth. The minimum B was converted to an equatorial pitch angle and the starting altitude was varied to provide the drift-averaged densities as a function of α_p. A typical value of $F_{10.7} = 140$ was used to simulate solar average conditions.

The drift averaged atmospheric densities were also used to calculate energy loss rates in the atmosphere using the formulas compiled by *Salamon* [1980]. We assume that the secondary particles have no bound electrons. Note that for

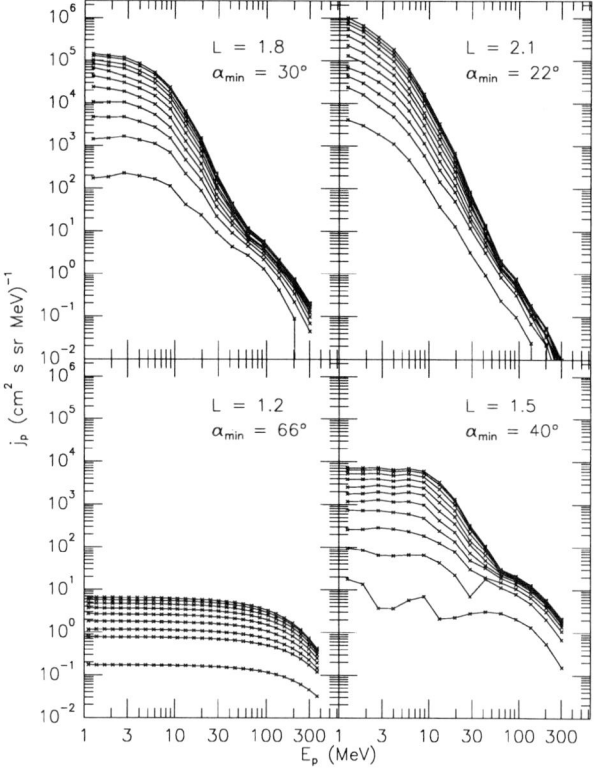

Figure 1. Proton energy spectra for selected L shells from the AP-8 model for solar average conditions. The curves for each L are at equally spaced pitch angles varying from the labeled minimum α to 90°. Some smoothing was done for the $L = 1.2$ curves.

$L \gtrsim 1.3$ energy loss to free electrons in the ionosphere and plasmasphere can be significant [*Jentsch*, 1981], but this is not included in the model. Trapped particle lifetimes against energy loss in the atmosphere can be calculated from dE/dx. They are generally short at low altitudes (small equatorial pitch-angles) but can be long at high altitudes and L shells. Note that the trapped particles observed by a satellite orbiting at low altitude must all have relatively short lifetimes regardless of L shell because all of these particles must reach that altitude.

The proton intensities were derived from the empirical NASA AP-8 models obtained from NSSDC. These provide proton omnidirectional integral intensities, J_p, as a function of L and B/B_0, the ratio of local to equatorial magnetic fields, for solar minimum and solar maximum conditions, although significant solar cycle variations were not found for our region of interest. The equatorial directional differential intensity was calculated by numerically evaluating

$$j_p = \frac{1}{2\pi^2} \frac{\partial^2}{\partial E\, \partial x} \int_0^x \frac{J_p(E, x')}{(x-x')^{1/2}}\, dx', \quad (4)$$

where $x = B_0/B$. Sample proton energy spectra are shown in Figure 1.

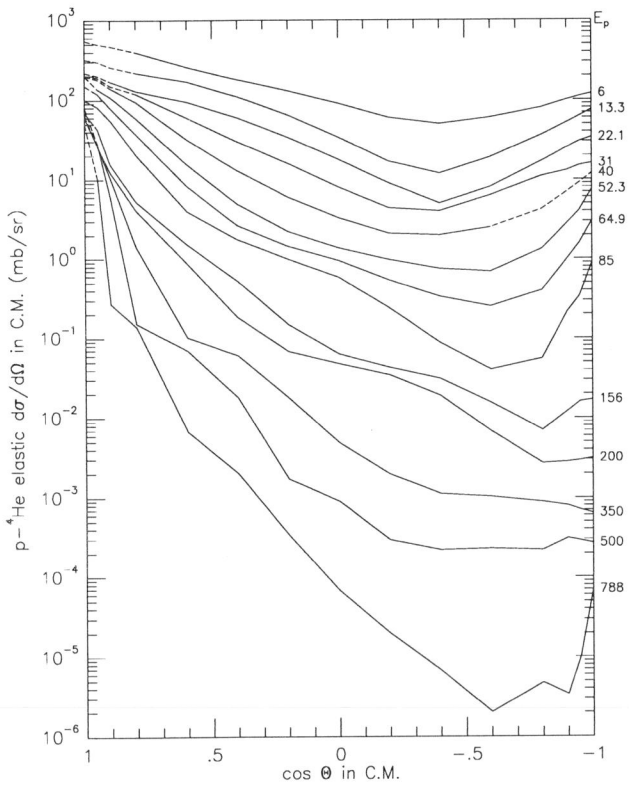

Figure 2. Cross-sections versus scattering angle in the center-of-mass for p-^{4}He elastic scattering [*Meyer*, 1972; *Votta et al.*, 1974; *Comparat et al.*, 1975; *Fong et al.*, 1978; *McCamis et al.*, 1978; *Imai et al.*, 1979; *Moss et al.*, 1980]. Dashed curve segments are extrapolations from the data. Laboratory proton energies are labeled to the right of each curve in MeV.

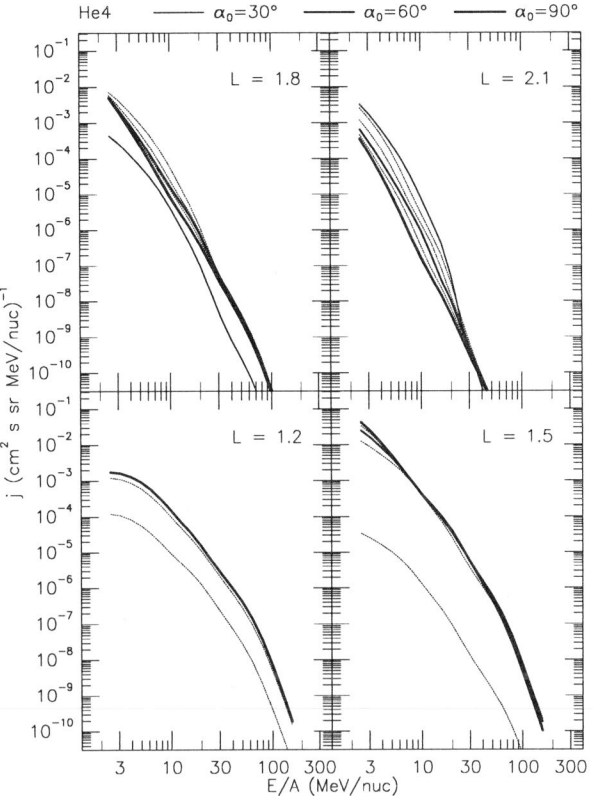

Figure 3. Intensity of elastically scattered ^{4}He as a function of energy-per-nucleon at average solar conditions for the labeled L shells. The equatorial pitch angles α_0 vary from 30° (off scale for low L) to 90° in steps of 10°.

The final input to the source function (3) is the cross-section data for a given nuclear interaction. This should be differential in energy and solid-angle. If the interaction has only two in-going and two out-going particles then the kinematics can be used to simplify the calculation. For the reaction designated 1(2,3)4 involving particles of rest mass m_1, m_2, m_3, and m_4, where m_1 collides with m_2 which is at rest in the laboratory system, the initial 4-momenta are $(E_1, \mathbf{p}_1)$, $(m_2, 0)$ in the lab and $(\varepsilon_1, \mathbf{k})$, $(\varepsilon_2, -\mathbf{k})$ in the center-of-mass, while the final 4-momenta are $(E_3, \mathbf{p}_3)$, $(E_4, \mathbf{p}_4)$ and $(\varepsilon_3, \mathbf{k}')$, $(\varepsilon_4, -\mathbf{k}')$ respectively. If m_4 is the secondary particle of interest, then the scattering angles in the center-of-mass, Θ, and the lab, θ_4 are related to the energies by the Lorentz transformation

$$E_4 = \gamma \varepsilon_4 - \gamma \beta k' \cos\Theta \tag{5}$$

and to each other by

$$\tan\theta_4 = \frac{\sin\Theta}{\gamma(-\cos\Theta + \frac{\beta}{\beta_4})}, \tag{6}$$

where $\beta = p_1/(\gamma E)$ and $\gamma = (E_1 + m_2)/E$ are the center-of-mass speed and Lorentz factor, $E = (2E_1 m_2 + m_1^2 + m_2^2)^{1/2}$ and $\beta_4 = k'/\varepsilon_4$ are the total energy and m_4 speed in the center-of-mass. From Eq. (5)

$$\frac{d\sigma}{dE_4} = -\frac{2\pi}{\gamma \beta k'} \frac{d\sigma}{d\Omega_c}, \tag{7}$$

where Ω_c is the center-of-mass solid angle. In this case the double differential cross-section can be expressed in terms of either of the single differential cross-sections. For example

$$\frac{d^2\sigma}{d\Omega dE_4} = \frac{d\sigma}{dE_4} \frac{\delta(\theta_4 - \theta_4(E))}{2\pi \sin\theta_4} \tag{8}$$

where $\theta_4(E)$ is given by Eq. (6).

To do the source integral (3) over the δ-function in Eq. (8) we change variables from α_p and the proton gyrophase angle to θ and the initial proton phase angle ϕ_p around the secondary particle direction, by a rotation of the axes:

$$S = \sum_i \int dE_p \frac{d\sigma_i}{dE} \frac{1}{\pi} \int_0^\pi d\phi_p \, n_i j_p, \tag{9}$$

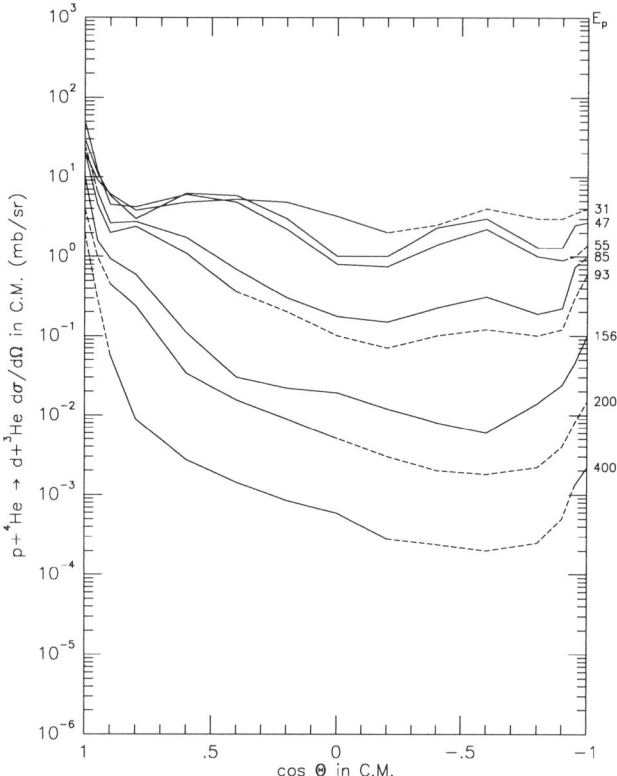

Figure 4. Cross-sections versus scattering angle in the center-of-mass for the reaction $p + {}^4\text{He} \to {}^3\text{He} + d$ [*Meyer*, 1972; *Votta et al.*, 1974; *Alons et al.*, 1986]. Dashed curve segments are extrapolations from the data. Laboratory proton energies are labeled to the right of each curve in MeV.

where $\cos\phi_p = (\cos\alpha_p - \cos\alpha\cos\theta)/(\sin\alpha\sin\theta)$ and θ is the scattering angle θ_4 from Eq. (6). Note that if n_i and j_p are independent of α_p then (9) leads to the expected result

$$S = \sum_i \int dE_p \, n_i j_p \frac{d\sigma_i}{dE} \qquad (10)$$

for the source function due to an isotropic proton flux in a homogeneous atmosphere.

The first interaction that we consider is $p({}^4\text{He},p){}^4\text{He}$, elastic scattering of atmospheric ${}^4\text{He}$. The maximum kinetic energy of the scattered ${}^4\text{He}$ is approximately 16/25 times the proton kinetic energy, so that a 100 MeV proton can produce a ${}^4\text{He}$ of up to 16 MeV/nucleon. Cross-section data are shown in Figure 2 from various sources listed in the figure caption. They are generally peaked in the direction where protons are forward scattered, but the most efficient direction for trapping the scattered ${}^4\text{He}$ is where the protons are scattered backward ($\cos\Theta = -1$), and the scattered ${}^4\text{He}$ follow the original proton trajectories. However, small and intermediate angle scattering can also lead to significant particle trapping if the pitch angle is approximately conserved, especially at the high proton energies where the cross-sections are

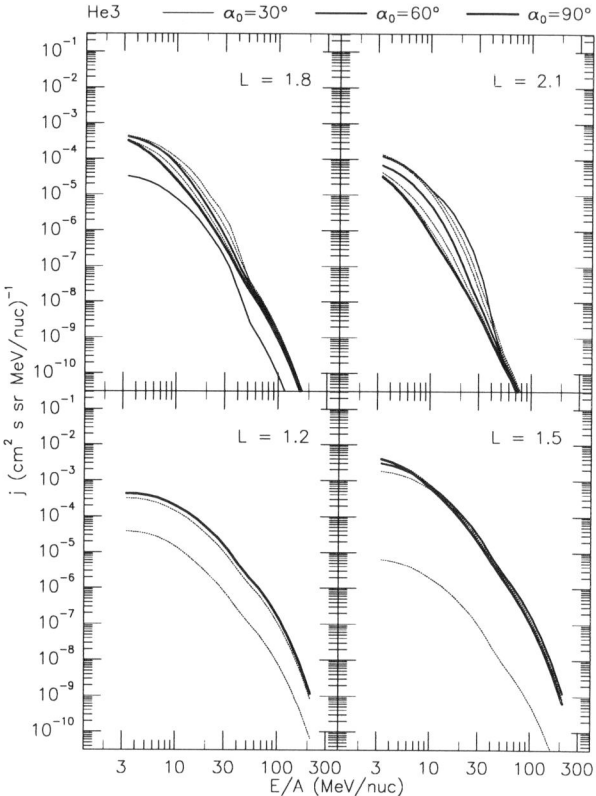

Figure 5. Similar to Figure 3 but for ${}^3\text{He}$ from the reaction $p + {}^4\text{He} \to {}^3\text{He} + d$.

strongly forward-peaked.

Results of the calculation using Eqs. (9) and (2) for elastic scattering of ${}^4\text{He}$ are shown in Figures 3. The ${}^4\text{He}$ intensity at a given energy-per-nucleon varies with equatorial pitch angle due to the corresponding variation in the relative concentration of ${}^4\text{He}$ in the drift-averaged atmosphere. It varies with L primarily due to the variation in the proton intensity. The pitch angle distribution changes from being strongly peaked at 90° for $L = 1.2$ to being peaked near the edge of the loss cone for $L = 2.1$.

We next consider the pickup reaction $p({}^4\text{He},d){}^3\text{He}$. Cross-section data are shown in Figure 4. The backward direction ($\cos\Theta = -1$) is again most efficient for trapping ${}^3\text{He}$ while the forward direction ($\cos\Theta = 1$) is most efficient for trapping d. The ${}^3\text{He}$ and d intensities are shown in Figures 5 and 6 respectively.

The ${}^3\text{He}$ has a harder spectrum than the ${}^4\text{He}$ from elastic scattering due to the differing energy dependencies of the cross sections at backward angles. The deuterium intensity is much higher than those of ${}^4\text{He}$ and ${}^3\text{He}$ because the cross sections are forward peaked.

There are other reactions that can also produce the isotopes considered above. For example, protons can collide with atmospheric O, which is relatively dense at low altitudes, producing evaporation and direct knock-out products including

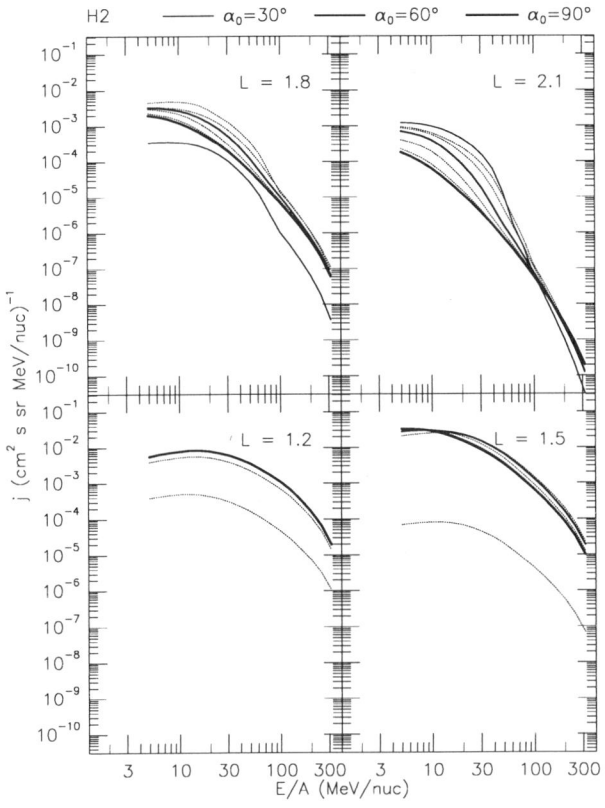

Figure 6. Similar to Figure 3 but for d from the reaction p + ^{4}He → ^{3}He + d.

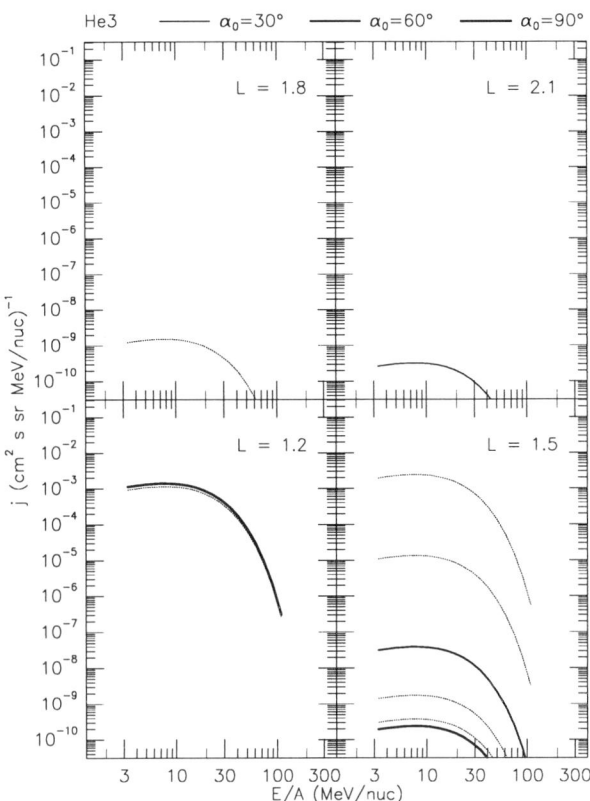

Figure 7. Similar to Figure 3 but for ^{3}He production by p knockout from O.

^{4}He, ^{3}He, and d. The kinematics are not determined as in Eq. (5). Instead, cross-sections and the energy distribution of the products are measured. However, the data are sparse and several approximations must be made. Detailed measurements at $E_p = 90$ MeV have been made by *Wu et al.* [1979]. They found that the evaporation products are generally isotropic and low-energy, so they are not significant here, while the knock-out products are forward-peaked and higher energy. To simplify the calculation we assume that they are produced in the forward direction only:

$$\frac{d^2\sigma}{d\Omega dE} = \sigma(E_p)\frac{\delta(\theta)}{2\pi \sin\theta}F(E_p, E). \quad (11)$$

For the energy distribution F we assume an exponential independent of E_p except for a cutoff at the maximum energy E_{max} which differs from E_p by the binding energy of the knock-out product in the original nucleus

$$F(E_p, E) = \frac{1}{E_0}\frac{e^{-E/E_0}}{1 - e^{-E_{max}/E_0}}H(E_{max} - E). \quad (12)$$

The e-folding energy is E_0 and H is a unit step function. The values of E_0 are taken from the angle-integrated spectra of Wu et al. and are 15.9 MeV for ^{4}He, 29.5 MeV for ^{3}He, and 77.8 MeV for d. The total cross sections $\sigma(E_p)$ are taken from *Cucinotta et al.* [1996].

The source function based on the approximation Eq. (11) is

$$S = \int dE_p n j_p \sigma F, \quad (13)$$

where n, the drift averaged O density, and j_p are evaluated at $\alpha_p = \alpha$. The resulting ^{3}He intensities are shown in Figure 7. Atmospheric O appears to be a significant source at $L = 1.2$ only. At the higher L shells the O products are significant only in a narrow range of equatorial pitch angles near the edges of the loss cones, corresponding to the altitude range where O is the dominant component of the atmosphere. Similar results are obtained for d and ^{4}He. However, because of the differing energy spectra (different values of E_0) the atmospheric O source is relatively more significant for d and less significant for ^{4}He, compared with ^{3}He. Compared with the atmospheric He source at $L = 1.2$, atmospheric O appears to be a comparable source of ^{4}He and d, and a dominant source of ^{3}He. However, the O source is uncertain due to the lack of cross-section data and should be considered only a rough estimate.

3. DISCUSSION

The ^{4}He, ^{3}He, and d intensities described above are comparable in magnitude to the results from SAMPEX reported by *Cummings et al.* [1995] and *Looper et al.* [1995], and to the higher energy CRRES data reported by *Wefel et al.* [1995]. Detailed comparisons between the data and model results will be reported elsewhere. However, it is clear that the atmosphere is a significant and possibly dominant source of these isotopes for the inner zone.

While the elastic scattering and pick-up reactions involving atmospheric He are reasonably well understood, the primary uncertainty in the calculations is due to the lack of experimental cross-section data for the reactions with atmospheric O. Other reactions may also be significant, such as $p(p,d)\pi$ for producing high energy d, while elastic scattering of atmospheric constituents other than He, such as H and O, may also be significant sources for their corresponding radiation belt components.

Other possible improvements in the calculation would be to include solar-cycle variations, to evaluate the time dependence in cases where the lifetimes are comparable to the solar cycle time [*Jentsch and Wibberenz*, 1980], and to evaluate the role of radial diffusion, which is probably significant at the higher L shells [*Jentsch*, 1981].

Acknowledgements. This work was supported by NASA under contract NAS5-30704 and grant NAGW-1919.

REFERENCES

Alons, P.W.F., J.J. Kraushaar, J.R. Shepard, J.M. Cameron, D.A. Hutch-eon, R.L. Liljestrand, W.J. McDonald, C.A. Miller, W.C. Olsen, J.R. Tinsley and C.E. Stronach, ^{4}He(p,d)^{3}He reaction at 200 and 400 MeV, *Phys. Rev.*, *C33*, 406–411, 1986.

Comparat, V., R. Frascaria, N. Fujiwara, N. Marty, M. Morlet, P.G. Roos and A. Willis, Elastic proton scattering on ^{4}He at 156 MeV, *Phys. Rev.*, *C12*, 251–255, 1975.

Cucinotta, F.A., L.W. Townsend, J.W. Wilson, J.L. Shinn, G.D. Badhwar and R.R. Dubey, Light ion components of the galactic cosmic rays: nuclear interactions and transport theory, *Adv. Space Res.*, *17*, (2)77–(2)86, 1996.

Cummings, J.R, R.A. Mewaldt, R.S. Selesnick, E.C. Stone, J.B. Blake and M.D. Looper, MAST observations of high enery trapped helium nuclei (abstract), *EOS Trans. AGU*, Fall Meeting, F 501, 1995.

Dragt, A.J, Solar cycle modulation of the radiation belt proton flux, *J. Geophys. Res.*, *76*, 2313–2244, 1971.

Farley, T.A. and M. Walt, Source and loss processes of protons in the inner radiation belt, *J. Geophys. Res.*, *76*, 8223–8241, 1971.

Fong, J., T.S. Baeur, G.J. Igo, G. Pauletta, R. Ridge, R. Rolfe, J. Soukup, C.A. Whitten, Jr., G.W. Hoffmann, N. Hintz, M. Oothoudt, G. Blanpied, R.L. Liljestrand and T. Kozlowski, p-^{4}He Elastic scattering at 788 MeV, *Phys. Lett.*, *78B*, 205–208, 1978.

Hedin, A.E., MSIS-86 thermospheric model, *J. Geophys. Res.*, *92*, 4649–4662, 1987.

K. Imai, K. Hatanaka, H. Shimizu, N. Tamura, K. Egawa, K. Nisimura, T. Saito, H. Sato and Y. Wakuta, Polarization and cross section measurements for p-^{4}He elastic scattering at 45, 52, 60, and 65 MeV, *Nucl. Phys., A 325*, 397–407, 1979.

Jentsch, V. and G. Wibberenz, An analytic study of the energy and pitch angle distribution of inner-zone protons, *J. Geophys. Res.*, *85*, 1–8, 1980.

Jentsch, V., On the role of external and internal source in generating energy and pitch angle distributions of inner-zone protons, *J. Geophys. Res.*, *86*, 701–710, 1981.

Langel, R.A., International geomagnetic reference field, 1991 revision, *J. Geomag. Geoelectr.*, *43*, 1007–1012, 1991.

M.D. Looper, J.B. Blake, Cummings, J.R, R.A. Mewaldt and R.S. Selesnick, Maps of hydrogen isotopes at low altitudes in the inner zone of the earth's magnetosphere (abstract), *EOS Trans. AGU*, Fall Meeting, F 501, 1995.

McCamis, R.H., J.M. Cameron, L.G. Greeniaus, D.A. Hutch-eon, C.A. Miller, M.S. de Jong, B.T. Murdoch, W.T.H. van Oers, J.G. Rogers and A.W. Stetz, Large angle cross sections and analyzing power for proton-^{4}He elastic scattering between 185 and 500 MeV, *Nucl. Phys., A 302*, 388–400, 1978.

Meyer, J.P., Deuterons and He^3 formation and destruction in proton induced spallation of light nuclei ($Z < 8$), *Astron. Astrophys. Suppl.*, *7*, 417–467, 1972.

Moss, G.A., L.G. Greeniaus, J.M. Cameron, D.A. Hutch-eon, R.L. Liljestrand, C.A. Miller, G. Roy, B.K.S. Koene, W.T.H. van Oers, A.W. Stetz, A. Willis and N. Willis, Proton-^{4}He elastic scattering at intermediate energies, *Phys. Rev., C21*, 1932–1943, 1980.

Salamon, M.H., A range-energy program for relativistic heavy ions in the region $1 < E < 3,000$ MeV/amu, Lawrence Berkeley Laboratory, University of California, 1980.

Votta, L.G., P.G. Roos, N.S. Chant and R. Woody, III, Elastic protons scattering from ^{3}He and ^{4}He and the ^{4}He(p,d)^{3}He reaction at 85 MeV, *Phys. Rev., C 10*, 520–528, 1974.

Wefel, J.P., J. Chen, J.F. Cooper, T.G. Guzik and K.R. Pyle, The isotopic composition of geomagnetically trapped helium, *Proc. Int. Cosmic Ray Conf.*, 24th (4), 1021–1024, 1995.

Wu, J.R., C.C. Chang and H.D. Holmgren, Charged-particle spectra: 90 MeV protons on ^{27}Al, ^{58}Ni, ^{90}Zr, and ^{209}Bi, *Phys. Rev., C 19*, 698–713, 1979.

R.S. Selesnick, R.A. Mewaldt, California Institute of Technology, Pasadena, CA 91125, USA

DISCUSSION

Q: J.B. Blake. Has tritium abundance been calculated?
A: R.S. Selesnick. No.
Q: J.B. Blake. Do you understand why we don't see tritium in the PET data?
A: R.A. Mewaldt. The available cross sections suggest that ^{3}H should be at least an order of magnitude less abundant than ^{2}H. It may be there in the data but not resolved.

Electrons with Energy Exceeding 10 MeV in the Earth's Radiation Belt

A.M. Galper, V.V. Dmitrenko, V.M. Gratchev, Yu.V. Efremova, V.G. Kirillov-Ugryumov, S.V. Koldashov, L.V. Maslennikov, V.V. Mikhailov, Yu.V. Ozerov, A.V. Popov, N.I. Shvets, S.E. Ulin and S.A. Voronov

Kashirskoe shosse 31, MEPhI, Moscow 115409, Russia

Results of experimental research of ratio between electron and positron fluxes, spatial and energy distributions of a stationary flux of high energy ($E_e > 10$ MeV) electron-positron component of particles trapped in radiation belt are discussed; the probable processes of particle generation and trapping of high energy electrons and positrons are considered. The program of study of a belt of high energy electrons for the near future is proposed.

1. INTRODUCTION

In the end of the sixties and beginning of seventies Cosmophysics Laboratory (MEPhI) has carried out balloon born research of high energy electron-photon component of cosmic rays on high altitude. One of new and interesting results was the observation of an increase of a vertical downward flux of electron-positron component of cosmic ray with energy about 100 MeV on a depth of residual atmosphere 5–10 g/cm². The increase depended on the magnetosphere disturbance level and was observed repeatedly [*Galper, 1970*]. In some measurements prevalence of electrons over positrons was shown also [*Voronov, 1975*]. This phenomenon was interpreted as a precipitation of high energy particles from the Earth's Radiation Belt. In the same time the various processes, which demonstrated an opportunity of existence of high energy electrons in radiation belt were considered (here and further, if it is not mentioned specially, electrons mean total flux of electrons and positrons).

From the end of seventies MEPhI began to fulfill the second stage of the search and study of high energy electrons on board satellites directly in the Earth's Radiation Belt. For the first time a flux of trapped electrons with energy more than 40 MeV in the Earth's Radiation Belt ($L = 1.12$–1.8; $B = 0.21$–0.23 Gs) was discovered during experiments on board orbital station SALYUT-6 using the ELENA-F telescope, which could be directed, during the measurements, under various angles to the direction of a vector of magnetic field of the Earth. On Figure 1 one of the measurements in the Brazilian Anomaly Region (BAR) is shown. The observed flux of high energy electrons in the BAR was 10–20 times higher than outside the BAR. The flux was 1000–7000$(m^2 s\, sr)^{-1}$, when the trajectories of particles with pitch-angles about $90°$ were in the aperture of the instrument. The mean value of the flux depends on a place of BAR crossing [*Galper, 1981a*]. Soon these results were confirmed by other scientific groups [*Basilova, 1982; Nikolskiy, 1983*]. Up to today several experiments of MEPhI on board various satellites and orbital stations, directed on the study of new components of radiation belt high energy electrons were carried out. The instruments, conditions of measurements and principal characteristics of a stable belt of high energy electrons, obtained in these experiments are described below.

2. INSTRUMENTS

1. The first experiment on observation of high energy electrons was carried out in 1979 using the ELENA-F instrument on board orbital station SALYUT-6. The instrument consisted of a gas Cerenkov counter, a system of scintillator counters and lead absorbers. The instrument was installed to internal part of the spacecraft closely to a wall. During the flight, the orientation of the instrument varied repeatedly. It allowed to carry out the observation of the electron fluxes at different pitch angle [*Galper, 1981b*]. The second experiment on board orbital station SALYUT-7 using the instrument ELENA-K, analogue of the instrument ELENA-F, was carried out in 1982 [*Alexandrov, 1985*].

2. After discovery of the trapped electrons with energy > 40 MeV (high energy electrons) in the Earth's Radiation Belt on board orbital station SALYUT-6, a new

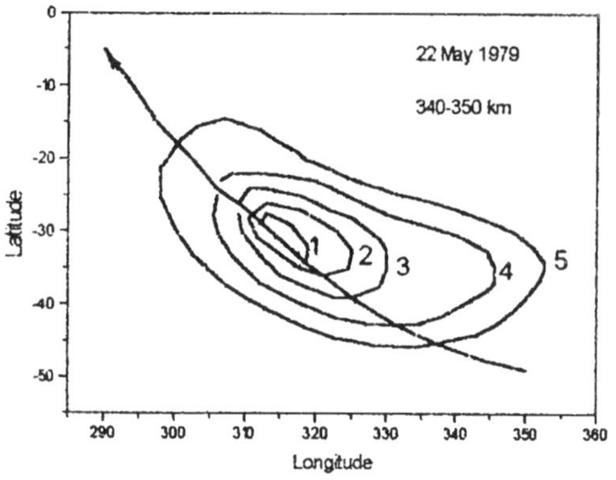

130 ELECTRONS WITH ENERGY EXCEEDING 10 MEV

Figure 1. The distribution of radiation in the Brasilian Magnetic Anomaly and one of the trajectories of the SALYUT-6 orbital station

series of the instruments, intended for the study of electron component at high altitudes was developed and produced. The first experiment from this series, ELECTRON, on board satellite INTERCOSMOS-BULGARIA 1300 (1981) and the second one, ELECTRON-2, on board satellite METEOR-3 (1985) were carried out. The physical schemes of the ELECTRON and ELENA are similar [*Galper*, 1983].

3. The third stage of the study of electron component was principle new one. A new type of the instrument was developed in 1985: the time-of-flight scintillator magnetic spectrometer MARIA. This instrument was delivered by the spacecraft PROGRESS on board orbital station SALYUT-7. The MARIA-2 instrument, improved variant of the MARIA spectrometer, operates on board orbital station MIR from 1988 till present time. At first time magnetic spectrometer is used outside atmosphere for separate observations of high energy (20–200 MeV) electrons and positrons [*Voronov*, 1986; *Alexandrov*, 1988].

4. The GAMMA telescope, included in the GAMMA astrophysics observatory operated in the near-Earth space in 1990–1991, intended, at first, for the study of primary cosmic gamma-radiation with energy 50–5000 MeV, was used in particular for observation of electrons in the same energy range [*Akimov*, 1987].

It should be mentioned that the main characteristics of all above listed instruments were determined at calibration on accelerators in electron, positron and proton beams and by computer simulations. The main characteristics of the instruments and conditions of measurements are presented in Table 1.

3. RESULTS

The experimental data on flux of electrons and positrons are obtained in a wide enough range of latitude, longitudes and altitudes (wide range of B, L). Some results, indicating the existence of a stable belt of high energy trapped electrons, are discussed below.

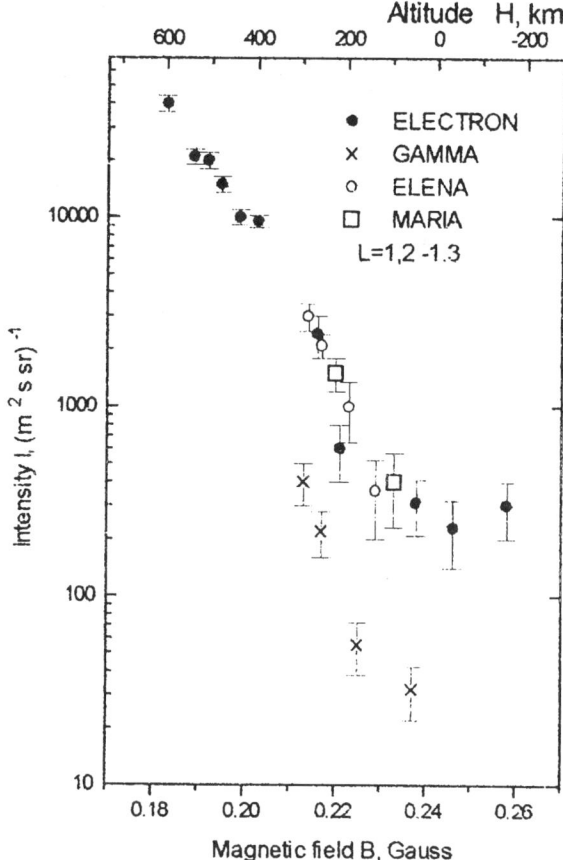

Figure 2. The intensity of trapped electrons (about 90°) versus magnetic field B for $l = 1.2$–1.3. H is the minimum altitude of the mirror points for given B.

3.1. Spatial distribution of a flux of trapped electrons

On Figure 2 the intensity of trapped electron flux vs. B for L-shell range 1.2–1.3 is shown. It is clear from the figure that the experimental data, obtained with various instruments, at different altitudes in different temporal intervals agree among themselves. The availability of similar dependencies for different L permits to obtain the distribution of trapped electron flux with pitch angle 90° for different altitudes from L-shell in BAR. On Figure 3 this dependence for electrons with energy 20–350 MeV on experimental data obtained by the ELECTRON instrument is shown. It is clear from the figure that the trapped electrons concentrate on the L-shells 1.1–1.8 and that the flux intensity exceeds $10^4 (m^2 s\, sr)^{-1}$ [*Galper*, 1983; *Kirillov-Ugryumov*, 1986; *Akimov*, 1993].

3.2. Charge ratio in a trapped flux

Using the time-of-flight scintillator magnetic spectrometer MARIA on board orbital stations SALYUT-7 and MIR there was the opportunity to measure the charge ratio Ie-/Ie+, which largely determines the choice between various origins of

Table 1. List of MEPhI experiments

Experiment	ELENA-F	ELENA-K	ELECTRON	ELECTRON-2	MARIA	MARIA-2	GAMMA
Years	03.19.79–04.17.81	09.18.82–11.14.83	08.08.81–04.15.83	10.24.85–12.20.87	08.09.85–11.20.85	01.01.88, in progress	07.01.88–07.01.91
Satellite	SALYUT 6	SALYUT 7	Intercosmos Bulgaria 1300	METEOR-3	SALYUT 7	MIR	GAMMA module
Altitude (km)	300–350	300–350	825–906	1250	300–350	400	400
Orbit inclination (deg)	51.6	51.6	81.25	82.5	51.6	51.6	51.6
Energy (MeV)	30–350	30–350	20–350	20–350	20–200	20–200	50–40,000
Aperture (deg)	12	12	12	12	20 × 4	20 × 4	12.5
Geometric factor ($cm^2 sr$)	3.5	3.5	4	4	2	2	240
Angular resolution (deg)	12	12	12	12	2	2	12.5
Rejection of protons	10^3	10^3	10^3	10^4	10^5	10^5	10^3
Exposure time (hr)	300	350	390	200	110	1800	4200
Electron selection	Velocity (gas Cerenkov counter), range (scintillator with lead absorber)	Velocity (gas Cerenkov counter), range (scintillator with lead absorber)	Velocity (gas Cerenkov counter), range (scintillator with lead absorber)	Velocity (gas Cerenkov counter), range (scintillator with lead absorber)	Momentum (deflection in magnetic field), velocity (time of flight)	Momentum (deflection in magnetic field), velocity (time of flight)	Velocity (gas Cerenkov counter), range (scintillator with lead absorber), dE/dX (amplitude in scintillator counters)

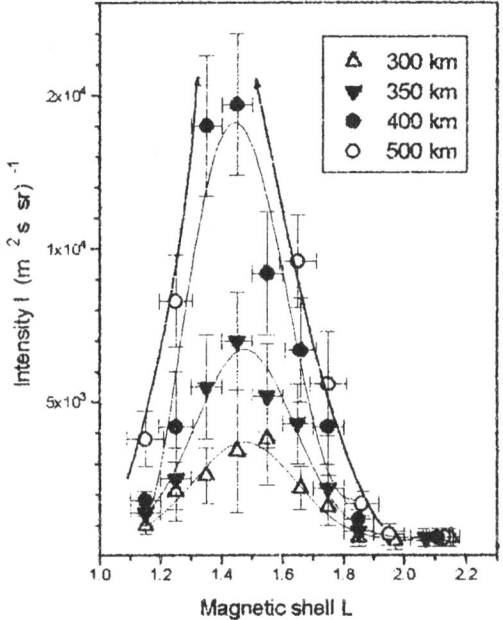

Figure 3. The maximum intensity of trapped electrons versus L shell (ELECTRON)

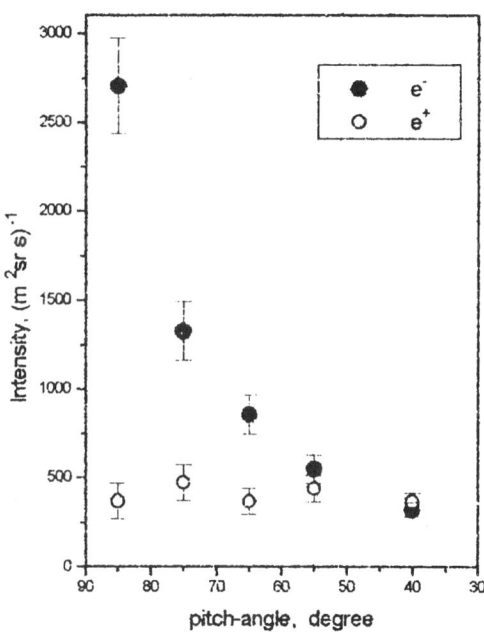

Figure 4. The pitch angle distribution for electrons and positrons (MARIA)

trapped flux. On Figure 4 the pitch angle distribution of electrons and positrons in BAR is shown. Figure 4 demonstrates that the flux of trapped particles is electrons, though the experimental data don't exclude some quantity of trapped positrons [*Voronov*, 1987].

3.3. *Energy distribution of trapped electrons and positrons*

Total energy spectra of the trapped (pitch angle more than 70°) electrons and positrons in the energy range 15–300 MeV are shown on Figure 5. The spectra are obtained using MARIA and GAMMA experimental data. On Figure 6 the separate spectrum of trapped (pitch angle more than 70°) electrons and positrons in BAR and spectra of the same secondary particles on the equator outside of BAR are shown (MARIA experiment). The spectra of electrons and positrons outside of BAR are similar and coincide with the spectrum of positrons in BAR [*Galper*, 1993]. This once more confirms the practical absence of trapped positrons and the small proton contribution in the measured fluxes of positrons and electrons. The estimated proton contamination of the spectra in the BAR is not more than 10% and much less outside of it.

4. DISCUSSION

The obtained results of experimental research indicate that:

- there is a stationary flux of electrons trapped by the magnetic field of the Earth with energy range from 15 up to 250 MeV;
- the flux of trapped electrons is located on magnetic shells $L = 1.1$–1.8;

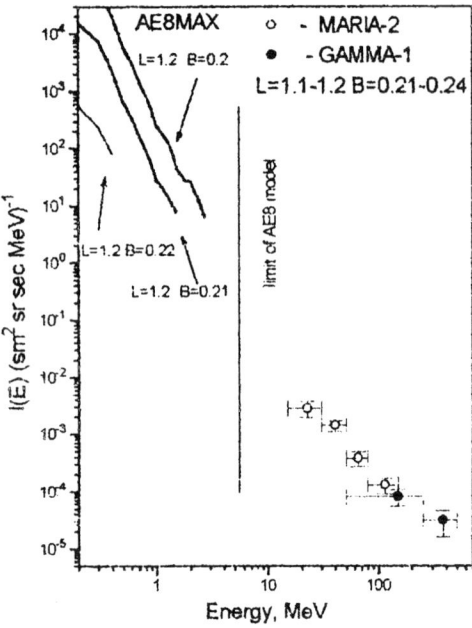

Figure 5. Differential electron spectra. The solid line represents the AE-8 model, the symbols the various experiments.

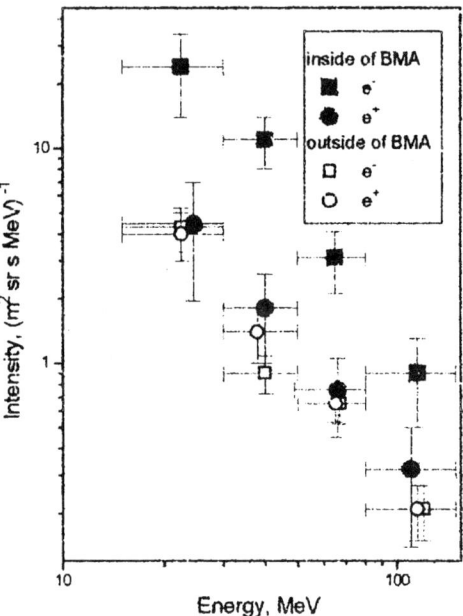

Figure 6. The energy spectrum of electrons and positrons obtained in the MARIA experiment

- its absolute intensity can exceed 10^4 $(m^2 s\, sr)^{-1}$;

- hence, the flux of trapped electrons is significant compared to the flux of trapped protons with the same penetration ability (several percent in the center of BAR and much more on the boundary of BAR).

The choice between models, applying on the explanation of this phenomenon should be based on the experimental data listed in Table 1. Today it is possible to give preference to acceleration processes on a various type of plasma fluctuations [*Tsytovich*, 1963] and specially to radial diffusion model [*Tverskoy*, 1968; *Dmitrenko*, 1993; *Voronov*, 1995]. Probably, during a magnetic disturbance the factor of radial diffusion will increase and it allows to obtain the theoretical value coincides with an observable stationary flux of trapped electrons. It is confirmed by the direct observations of electrons with energy 15–30 MeV on magnetic shells $L = 2$–3 in the time of strong magnetic disturbances [*Gorchakov*, 1984; *Blake*, 1992].

There is a more complicated situation with electrons with energy more than some hundred MeV [*Basilova*, 1982; *Nikolskiy*, 1983], where experimental data contradict one another and need to be checked.

Of course, for complete understanding of the physical nature of a stationary belt of high energy trapped electrons it is necessary to continue experimental as well as theoretical researches. But now it is obvious that available models of the Earth's Radiation Belt (AE-8) must be completed with new experimental data on electrons with energy more than 10 MeV.

5. CONCLUSION

At present the MARIA-2 experiment on board orbital station MIR is in progress. Processing of the experimental data of GAMMA project proceeds and it is possible to hope for expansion of the spectrum of trapped electrons in the large energies. And finally, the Magnetic Spectrometer PAMELLA (the basis of the project Russian-Italian Mission—RIM) [*Adriani*, 1995] now is in the phase of realization. The scientific programme includes the measurements of electrons and positrons with energy exceeding 50 MeV in the Earth's Radiation Belt. This apparatus will be installed onboard the METEOR-3A satellite, which has a circular near-Earth pole orbit with altitude about 700 km. The beginning of the experiment is planned in 1999–2000.

Acknowledgements. The research on the program of "High Energy Trapped Electrons" was carried out with the support, in different years, of GKVSh, MINNAUKI, RFFR, NPO ENERGIA and VNIIEM.

REFERENCES

Adriani, O., B.A. Alpat, G., Barbiellini et al., The magnetic spectrometer PAMELA for the study of cosmic antimatter in space, *24 ICRC*, *3*, 591, 1995.

Akimov, V.V., A.R. Baser-Bashi, V.B Balebanov, et al., The main parameters of the gamma-ray telescope GAMMA-1, *20 ICRC*, *2*, 320–323, 1987.

Akimov, V.V., S.A. Voronov, A.M. Galper, V.M. Zemskov, V.B. Zverev, L.F. Kalinkin, G.E. Kocharov, L.V. Kurnosova, V.E. Nesterov, Yu.V. Ozerov, A.V. Popov, L.A. Razorenov, M.A. Rusakovich, M.N. Soboleva, N.I. Topchiev, M.I. Fradkin, V.Yu. Chesnokov, E.I. Chujkin, A.R. Bazer-Bashi, M. Gro, Zh.M. Lavin', Zh.F. Oliv, and Yu. Yukhnevich, Electrons with energy 50–200 MeV into radiation belt of the Earth, *Izvestia ASSU, Physics Series, 57*, N7, 129, 1993.

Alexandrov, A.P., A.M. Galper, V.M. Grachev, V.V. Dmitrenko, V.G. Kirillov-Ugryumov, V.A. Lyakhov, V.V. Ryumin, S.E. Ulin, and N.I. Shvec, Active orientation of apparatus in the space physics experiment, *Kosmicheskie issledovania, 23*, N6, 941, 1985.

Aleksandrov, A.P., S.A. Voronov, A.M. Galper, V.G. Kirillov-Ugryumov, S.V. Koldashov, M.Kh. Manarov, L.B. Maslennikov, V.V. Mikhailov, P.Yu. Naumov, Yu.V. Romanenko, A.V. Popov, V.V. Titov, V.Yu. Chesnokov, and N.I. Shvec, Investigation of charge particle fluxes in the range of momenta from 20 to 200 MeV/c on orbital station MIR, *Izvestia ASSU, Physics Series, 52*, N12, 2429, 1988.

Basilova, R.M., A.A. Gusev, G.I. Pugacheva, and A.F. Titenkov, High energy electrons in inner radiation belt, *Geomagnetizm i Aeronomia, 22*, 671, 1982.

Blake, J.B., W.A. Kolasinski, R.W. Fillius, and E.G. Mullen, Injection of electrons and protons with energies of tens MeV into $L < 3$ on March 1991, *Geophys. Res. Letts., 19*, N8, 821, 1992.

Dmitrenko, V.V., V.B. Komarov, and B.A. Tverskoy, The radial diffusion as mechanism of creation stable fluxes of high energy electrons in the magnetosphere of the Earth, *Kosmicheskie issledovania, 31*, 83, 1993.

Galper, A.M., V.V. Dmitrenko, V.G. Kirillov-Ugryumov, and B.I. Lutchkov E.M. Shermanzon, Dependence of electron flux in the top atmosphere from a magnetosphere condition of the Earth, *Izvestia ASSU, Physics Series, 34*, 2275, 1970.

Galper, A.M., V.M. Grachev, V.V. Dmitrenko, and S.E. Ulin, Observations of the high energy electron flux near Brasilian anomaly region, *Kosmicheskie issledovania, 19*, 645, 1981a.

Galper, A.M., V.M Grachev, V.V. Dmitrenko, V.G. Kirillov-Ugryumov, V.A. Lyakhov, V.V. Ryumin, S.E. Ulin, and N.I. Shvec, Investigation electron and gamma ray fluxes in the nearest space *Izvestia ASSU, Physics Series, 45*, 637, 1981b.

Galper, A.M., V.M. Grachev, V.V. Dmitrenko, V.G. Kirillov-Ugryumov, and S.E. Ulin, New component of inner radiation belt—high energy electron, *Pisma v Journal Teoreticheskoy i Experimentalnoy Fiziki, 38*, 409, 1983.

Galper, A.M., S.V. Koldashov, V.V. Mikhailov, and S.A. Voronov, Energy distributions of high energy electrons and positrons in the Earth's magnetosphere, *23 ICRC, 3*, 825, 1993.

Gorchakov, E.V., V.A. Iozenas, M.B. Ternovskaja, P.P. Ignatyev, V.G. Afanasyev, and K.G. Afanasyev, Outer belt high energy electrons during geomagnetic storm, *Izvestia ASSU, Physics Series, 48*, N11, 2231, 1984.

Kirillov-Ugryumov, V.G., A.M. Galper, and V.V. Dmitrenko, Discovery of high energy electrons in the radiation belt by devices with gas Cherenkov counters, *NIM, A 248*, 238, 1986.

Nikolskiy, S.I., and V.G. Sinitsyna, High energy electrons in radiation belt, *Short Comm. on Physics, FIAN, 11*, 21 (in Russian), 1983.

Tsytovich, V.N., Acceleration of electrons in radiation belts of the Earth, *Geomagnetizm i Aeronomia, 3*, 616, 1963.

Tverskoy, B.A., Dynamics of the Earth's Radiation Belts, *Science, Moscow* (in Russian), 1968.

Voronov, S.A., B.I. Lutchkov, and V.A. Fedorov, Measurements of charge composition of secondary atmospheric electron flux, *14 ICRC, 4*, 1405, 1975.

Voronov, S.A., A.M. Galper, M.V. Guzenko, V.G. Kirillov-Ugryumov, S.V. Koldashov, A.V. Popov, and V.Yu. Chesnokov, Magnetic scintillate spectrometer of electrons, *Pribory i Technika Experimenta, 2*, 35, 1986.

Voronov, S.A., A.M. Galper, V.G. Kirillov-Ugryumov, S.V. Koldashov, and A.V. Popov, Relation between the intensities of high energy electrons and positrons trapped by geomagnetic field, *20 ICRC, 4*, 449, 1987.

Voronov, S.A., S.V. Koldashov, and V.V. Mikhailov, Nature of high energy electrons in the inner radiation belt, *24 ICRC, 4*, 989, 1995.

A.M. Galper, V.V. Dmitrenko, V.M. Gratchev, Yu.V. Efremova, V.G. Kirillov-Ugryumov, S.V. Koldashov, L.V. Maslennikov, V.V. Mikhailov, Yu.V. Ozerov, A.V. Popov, N.I. Shvets, S.E. Ulin, S.A. Voronov, Kashirskoe shosse 31, MEPhI, Moscow 115409, Russia

DISCUSSION

Q: D.N. Baker. If the positrons observed in the magnetosphere are produced by secondary processes (i.e. pion-muon decay), should there not be a large preponderance of positrons over electrons? Yet, you see about the same flux of electrons and positrons. This seems to be a problem with the assumed source.

A: A.M. Galper. In the radiation belts, as I have demonstrated, the trapped particles are electrons and it is necessary to introduce acceleration mechanisms to explain them. At altitudes below the radiation belt a slight excess of positrons over the electrons is observed, and this in good agreement with the secondary nature (decay of pions, produced as a result of primary cosmic ray interaction with the residual atmosphere) of both electrons and positrons.

Q: M. Walt. Have you calculated the lifetime of the high energy electrons to synchrotron radiation?

A: A.M. Galper. Indeed, the synchrotron losses determine the existence of a stable high-energy electron belt. The calculations performed by V. Dmitrenko, V. Komarov and B. Tverskoy [*Kosmich. issledovanija, 31*, 83, 1993] show that this factor becomes crucial for particle energies exceeding 100–200 MeV.

Low Altitude Models of Radiation Belts Based on Data from Russian Satellites

Yu.V. Mineev and E.D. Tolstaya

Skobeltsyn Institute of Nuclear Physics, Moscow State University

Recent low altitude radiation belt models for the electron component, based on individual satellite experiments, are analysed and compared. A model of the electron component of trapped radiation (energy range 0.04–2.0 MeV, altitude range 350–1,000 km, epochs of solar maximum and minimum), based on two Russian low altitude satellite experiments on COSMOS-1686 and INTERCOSMOS-19, employing identical electron spectrometers, is presented. Detailed comparison of the developed model with AE-8 is made, revealing significant discrepancies.

1. INTRODUCTION

At present it appears to be well recognised that the, for scientific and application purposes extensively used, NASA NSSDC radiation environment models AE-8 [*Vette*, 1991] and AP-8 [*Sawyer and Vette*, 1979] and their Russian analogues [*State Standard of FSU*, 1986; *Getselev et al.*, 1991] require verification and updating. Among the serious deficiencies of the above listed models, most frequently stated in the literature are the following:

1. The data sets used for developing the models are rather old (mostly obtained during the 60-ies and 70-ies).

2. Experimental data sets were contaminated by particles injected in nuclear explosions.

3. Existing models are stationary, i.e. cannot be used to predict flux levels during short-term geomagnetic disturbances, which are known to increase particle flux intensity by several orders of magnitude, creating serious hazards for manned space missions.

4. The models do not describe the temporal variations associated with solar activity, providing only the averaged flux levels for solar maximum and minimum epochs.

The deficiencies listed below are most significant for altitudes below 1000 km:

5. Experimental data sets, used to derive the model flux values for solar maximum were obtained during an anomalously weak solar cycle (the 20th). The intensity of the solar cycle is directly connected with atmospheric density, and, consequently, with flux intensity at low altitudes.

6. Secular variations of the Earth's magnetic field have significantly changed the pattern of geomagnetic field lines at low altitudes [*Lemaire et al.*, 1991], inducing such effects as the longitude shift of the Brazil anomaly region and lowering of the field lines, populated by particles into denser atmospheric layers.

In Table 1 we have made a summary of the comparison of the NASA/NSSDC models with the results of recent satellite experiments and conclusions, drawn from this comparison by the authors.

As it can be seen from Table 1 the common conclusion made from comparison of different experiments with AE-8 (with the exception of the OHZORA results) is, that AE-8 overestimates low energy electron flux values. The agreement between AE-8 and results of prolonged experiments (such as LDEF) is much better than for shorter missions.

2. INSTRUMENTS AND DATA SETS USED IN THE LOW ALTITUDE MODEL

The data sets used for developing the low altitude model were obtained by similar instruments flown onboard two russian satellites—INTERCOSMOS-19 (500–1000 km elliptic orbit, inclination 74°, three-axis stabilised) and COSMOS-1686 (350 and 500 km circular orbit, inclination 51.6°). INTERCOSMOS-19 operated for 6 months in 1979 around solar maximum and the data set for COSMOS-1686 contains 12 months of data for 1986 around solar minimum.

Table 1. Comparison of NASA/NSSDC (AE-8) models with recent satellite data

SC	Coverage, epoch	Instrument	Comparison	Ref
CRRES	1990–1991 (solar max)	spectrometer (0.12–0.199 MeV)	AE-8 fluxes too high	5
CRRES	1990–1991 (solar max)	dosimeter	AE-8 fluxes too high	3
LDEF	1984–1990 (solar max and min)	thermo-luminescent dosimeters, activation samples	generally good agreement with both AE-8 MAX and AE-8 MIN except for low energies	1,2 7
DMSP	1983–1987 (solar min)	dosimeter	AE-8 fluxes too high for energies < 5 MeV	4
OHZORA	1984–1987 (solar min)	spectrometer (0.19–3.2 MeV)	good agreement with AE-8	6

References: 1, *Armstrong, et al.*, [1992]; 2, *Armstrong, et al.*, [1993]; 3, *Gussenhoven et al.*, [1991a]; 4, *Gussenhoven et al.*, [1991b]; 5, *Heck* [1992]; 6, *Kohno et al.*, [1990]; 7, *Watts et al.*, [1993]

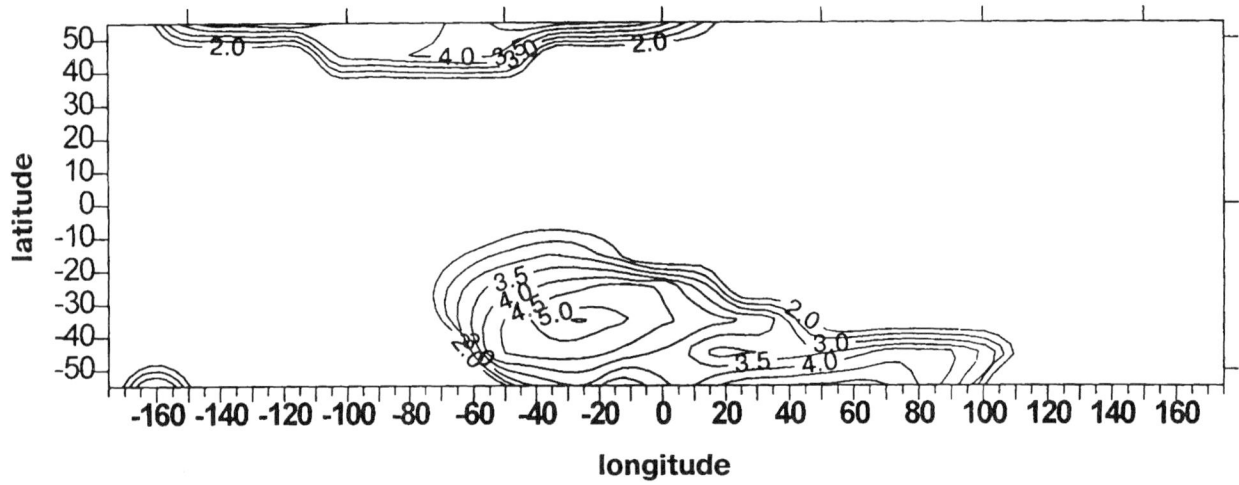

Figure 1. Electron flux distribution at 500 km solar minimum epoch, according to the AE-8 MIN model.

The instruments PERO-3I (flown on INTERCOSMOS-19) and Electron-4 (flown on COSMOS-1686) were practically identical in design and calibration, detailed descriptions are given in [*Gordeev et al.*, 1980] and [*Mineev et al.*, 1986]. The instruments consisted of a telescope with 4 semiconductor detectors mounted in an aluminium collimator with acceptance angle 20°, measuring electrons in the following energy ranges: 0.3–0.6, 0.6–0.9, 0.9–1.2, 1.2–2.0 MeV and two Geiger counters for measuring $E_e > 0.04$ MeV and $E_e > 0.1$ MeV.

In order to develop the model corresponding to the epoch of solar minimum we used the experimental data set obtained on the COSMOS-1686 spacecraft during the time period February–December 1986 (63 data files, 20 hours data coverage each). The orbit altitude for this spacecraft was at first 350 km and later (after August 1986) 500 km. The spacecraft operated mostly in the spinning mode around the horizontal axis at a rate of 0.5 rot/sec, the time interval for integration of counts in the experiment was 2.56 sec. The instrument had two identical detector units (mounted at an angle of 90°), one of which was pointing in the zenith direction, while the other one was located in the horizontal plane. We assumed that the flux, calculated as the average for both detector units, and a time-average for a given location in space, would be averaged over pitch angles. For the model we selected time periods corresponding to magnetically quiet conditions (the selection criterion was the following: the D_{st} index value was not smaller than −40 nT for at least 8 days prior to the

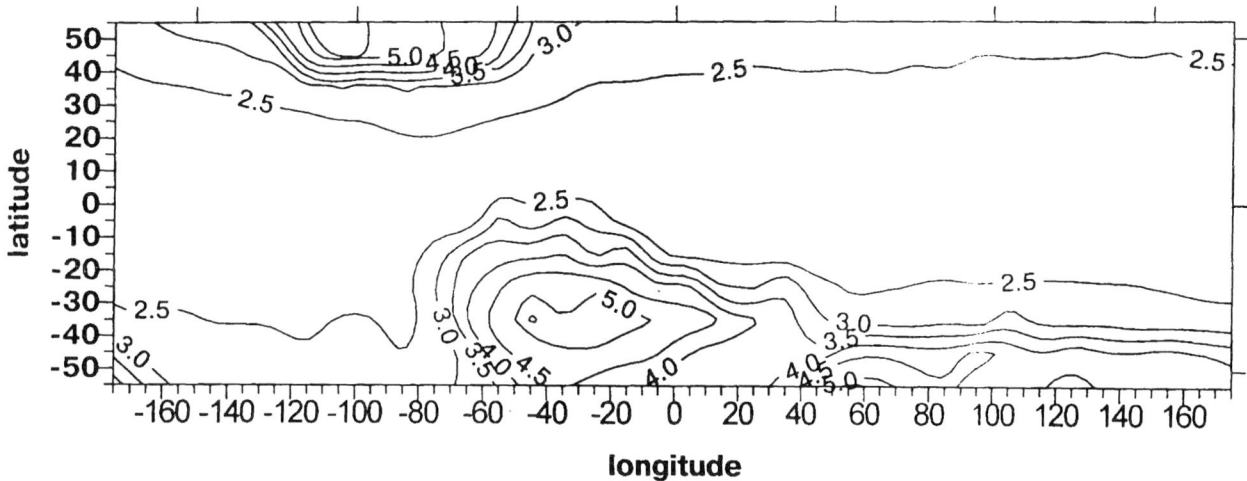

Figure 2. Electron flux distribution at 500 km solar minimum epoch, according to the developed LOWALT model.

chosen interval).

For developing the model of electron fluxes during solar maximum we used the experimental data set obtained on INTERCOSMOS-19. The data set consisted of 145 data files covering the time period of March–August 1979. Unlike COSMOS-1686 this spacecraft was well stabilised, and the location of the instrument was such, that it registered the flux of particles with pitch angles close to 90°.

The selection conditions for the data incorporated into the model were similar to those applied in the model for the solar minimum epoch.

3. MODEL STRUCTURE

At low altitudes electron fluxes can not be adequately described in (L, B) coordinates due to steep gradients—small changes in the magnetic field B can cause electron flux variations of several orders of magnitude. The use of B/B_0 instead of B does not really improve the situation. The (L, B) coordinate system cannot be used to describe the fluxes of quasi-trapped and precipitating electrons (particles with minimum mirror point altitudes below 100 km), which are significant at low altitudes. In order to describe these fluxes it is necessary to introduce additional parameters such as longitude or local time, as it was done in AE-4 [*Singley et al.*, 1976] or the russian State Standard on precipitating particles [*Standard of FSU*, 1990].

We tried several approaches for arranging the model data including (L, B) coordinates + longitude as an additional parameter and using (L, B) coordinates separately for the northern and southern hemispheres. However, in both these cases we had large standard deviations in each model bin. In fact the number of bins where the standard deviation exceeded by a factor of 2 the average flux value in the model bin amounted to 10%. After analysing the various approaches we averaged the experimental data in $10° \times 10°$ meshes in latitude and longitude at three basic altitudes 350, 500 and 800 km. Such synoptic maps at a given altitude are very convenient for estimating radiation conditions for low altitude circular orbiting spacecraft.

The final version of the model consists of 42 tables of omnidirectional electron flux values (21 for solar maximum and 21 for solar minimum) for 7 energy ranges $E_e > 0.04$ MeV, > 0.1 MeV, > 0.3 MeV, > 0.6 MeV, > 0.9 MeV, > 1.2 MeV and > 2 MeV at three basic altitudes of 350, 500 and 800 km.

The model includes the LOWALT software package, which generates integral omnidirectional electron flux values for user-defined geographical coordinates (altitude, latitude and longitude) and a set of energies, using cubic spline interpolation procedures.

4. COMPARISON OF THE DEVELOPED MODEL WITH AE-8

In order to make a general comparison of the models we plotted the distributions of the logarithms of electron fluxes in the energy range 0.3–2.0 MeV at the basic model altitudes of 350 and 500 km for both solar maximum and minimum, obtained from the AE-8 model (using IGRF 1965) and the developed low altitude model. Figure 1 (AE-8 MIN) and Figure 2 (LOWALT model) show how the models compare at 500 km altitude for the epoch of solar minimum. The numbers next to each of the contour lines denote the value of the decimal logarithm of omnidirectional flux intensity. As can be seen from Figures 1 and 2, the areas with intense electron fluxes, according to our low altitude model are much broader than according to AE-8. This is probably due to the fact that the analytic low-altitude cut-off function employed in AE-8 is not very precise.

It should be mentioned that the same situation is observed in all cases: for solar maximum at 500 and for 350 km altitude during both maximum and minimum. Comparison of Figure 1 and Figure 2 also shows that the centres of regions, where most intensive fluxes are observed (in the figures this corresponds to the centre of converging contours) are shifted in longitude with respect to each other. It is obvious that

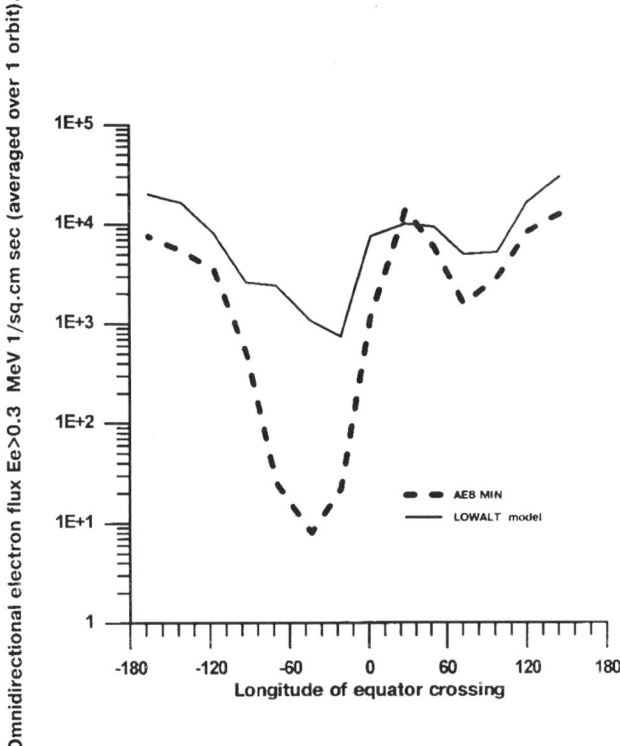

Figure 3. Electron flux values (orbital averages) according to AE-8 and the developed LOWALT model for a sequence of orbits (500 km, 51.6° inclination.)

according to the low altitude model the area with the most intense flux values (Brazil anomaly region) has shifted westwards in longitude, in comparison to AE-8. This effect is connected with the secular variations of the Earth's magnetic field. A similar effect for protons was reported in [*Konradi et al.,* 1992].

As was already mentioned in the introduction, the AE-8 model is based on experimental data sets acquired shortly after nuclear explosions. Theoretical models of the decay of injected electrons were employed when developing the latest versions of the AE models. However, the validity of these theoretical models requires experimental verification. Therefore, it was of interest to compare the electron fluxes according to AE-8 and the developed low altitude model in the inner radiation belt ($L < 2$). In order to conduct such a comparison, we calculated the total (summarised over all model meshes) flux for both models. We also made separate calculations for quasi-trapped and trapped fluxes in the inner belt, according to both models. The regions of quasi-trapped fluxes were taken as those regions where the minimum mirror point altitude was less than 100 km. Comparison shows that electron fluxes in the inner radiation belt according to the developed low altitude model somewhat exceed the AE-8 fluxes both at 500 and 350 km. The reason for this discrepancy could lie in the inaccuracy of theoretical models for decay of artificially injected electrons and the absence of reliable experimental data for the inner belt in AE-8.

A rather interesting result is the obvious presence of relatively large quasi-trapped fluxes in the inner belt. According to AE-8, quasi-trapped fluxes at these altitudes are totally absent.

Detailed comparison was also made for trapped and quasi-trapped fluxes, according to AE-8 and the developed low altitude model in the outer radiation belt ($L > 3$) and the slot region ($2 < L < 3$). The results of this comparison are summarised in the Conclusions section. We also analysed the relative flux values during solar maximum and minimum according to both models.

To compare the models on a smaller time scale we plotted averages of omnidirectional electron flux values with energies $E_e > 0.3$ MeV per one orbital period as a dependence on the longitude of equator crossing for each orbit. Figure 3 shows this comparison for 500 km, 51.6° inclination. As can be seen from Figure 3 the discrepancies can be quite large and attempts to use AE8 for selecting a sequence of orbits with minimum radiation exposure may lead to serious hazards.

5. CONCLUSIONS

Comparison reveals the following significant discrepancies between the developed low altitude model and AE-8:

1. The regions with relatively intensive energetic electron flux values according to the developed model are much larger, giving evidence that the analytical low altitude atmospheric cut-off function employed in AE-8 is not very accurate.

2. In the inner radiation belt, the developed low altitude model predicts a stationary quasi-trapped flux, which is totally absent according to AE-8.

3. In the inner radiation belt at altitudes of 350 and 500 km, during solar minimum higher electron fluxes are observed in the inner radiation belt, according to the low altitude model, than according to AE-8, presumably due to theoretical models of Starfish electron decay employed in AE-8.

4. The values of quasi-trapped electron fluxes, according to the developed model, exceed quasi-trapped fluxes predicted by AE-8 in the slot region and the outer belt region.

5. In all cases a westward longitudinal shift of the Brazil anomaly region (geographical region corresponding to the largest flux intensities) is observed, in agreement with theoretical estimates derived from secular variations of the Earth's magnetic field.

6. Comparison of orbit averaged electron flux values shows significant discrepancies in the models, whereas the differences become less pronounced when averaging is done on a larger time scale.

REFERENCES

Armstrong, T.W., B.L. Colborn, Radiation Model Predictions and Validation Using LDEF Satellite Data, in Proc. of Second LDEF Post-Retrieval Symposium, San Diego, CA, 1–5 June, 1992.

Armstrong, T.W., B.L. Colborn, Predictions of LDEF Radiaoactivity and Comparison with Measurements in Proc. of Third LDEF Post-Retrieval Symposium, Williamsburg, Virginia, November 8–12, 1993.

Getselev, I.V., Gusev A.A., D.A. Darchieva, et al., Model of Space-Energy Distribution of Trapped Particle Fluxes (protons and electrons) in the Earth's Radiation Belts, Preprint INP MSU-91/-37/241, Moscow, 1991.

Gordeev, Yu., Yu.V. Mineev, E.S.Spirkova, Differential Electron Spectrometer PERO-3I in the 0.04–2.0 MeV Energy Range on the Intercosmos-19 Spacecraft, in Intercosmos-19 Spacecraft. Instruments for Studies of the Outer Ionosphere, Published by IZMIRAN, pp. 178–182, 1980.

Gussenhoven, M.S., E. Mullen, D. Brautigam, et al., Preliminary Comparison of Dose Measurements on CRRES to NASA Model Predictions, *IEEE Trans. Nucl. Science, 38*, 1991a.

Gussenhoven, M.S., E. Mullen, D. Brautigam, E. Holeman, Dose Variation During Solar Minimum, *IEEE Trans. Nucl. Science, 38*, 1991b.

Heck, F., Observations of Radiation Environment with CRRES Data, ESA/ESTEC Stage Report, 1992.

Kohno, T., K. Munakata, K. Nagata, H. Murakami, A. Nakamoto, N. Hasebe, J. Kikuchi and T. Doke, Intensity maps of MeV Electrons and Protons Below the Radiation Belts, *Planetary and Space Science, 38*, 483–490, 1990.

Konradi, A., G. Badhwar and L. Braby, Recent Space Shuttle Observations of the South Atlantic Anomaly and the Radiation Belt Models, Preprint, 1992.

Lemaire, J., E. Daly, J. Vette, C. McIlwain and S. McKenna-Lawlor, Secular Variations in the Geomagnetic Field and Calculations of Future Low Altitude Radiation Environments, in Proceedings of the ESA Workshop On Space Environment Analysis, ESA WPP-23, 5.17.

Mineev, Yu. and Spirkova E., Electron Spectrometer for measurements in the Earth's magnetosphere, in Ser. 3. Physics and Astronomy. Vestnik MGU, 22, 1, 91–95, 1981.

Sawyer, D., and J.I. Vette, AP-8 Trapped Proton Environment for Solar Maximum and Solar Minimum, NSSDC/WDC-A-R-S 76-06, 1976.

Signley, G.W. and J.I. Vette, The AE-4 Model of the Outer Radiation Zone Electron Environment, C72-06, August, 1972.

State Standard of FSU. The Earth's Natural Radiation Belts. Space-Energy Characteristics of Proton Densities. (25645.138-86). Space-Energy Characteristics of Electron Fluxes. 25645.139-86. Standard Publ. Moscow, 1986.

State Standard of FSU. Charged quasi-trapped and precipitating particles. Time-Energy dependencies. Standard Publ. Moscow, 1991.

Vette, J.I., The AE-8 Trapped Electron Environment, NSSDC/WDC-A-R-S 1-24, 1991.

Watts, J., Status of LDEF Radiation Modelling, in Proc. Third LDEF Post-Retrieval Symposium, Williamsburg, Virginia, November 8–12, 1993.

Yu. Mineev and E. Tolstaya, Skobeltsyn Institute of Nuclear Physics, Moscow State Universty, 119899, Moscow, Russia.

Comparison Between NASA and INP/MSU Radiation Belt Models

A.A. Beliaev

Institute Nuclear Physics / Moscow State University, 119899 Moscow, Russia

J.F. Lemaire

Belgian Institute for Space Aeronomy, Ave Circulaire 3, B-1180 Brussels, Belgium

Since the 1970s, the Institute of Nuclear Physics Moscow State University (INP/MSU) is developing and updating empirical models for the space radiation environment. The paper describes the models of the Earth's radiation belts developed at the INP/MSU, and compares them with the AP-8 and AE-8 models developed by NASA. Comparison of the formats and grids used for the model storage and the interpolation methods in both models is presented. The influence of these methods on the precision of the model outputs is discussed. Particle fluxes in (B, L) space as well as particle energy spectra in (E, L) space are displayed and compared in a "colour-graph" format. This comparison shows the need for "standardization" of future empirical radiation belt models, i.e. for adopting the methods of storing, gridding and accessing the model entries which are general enough and transportable from one empirical model to the other.

1. INTRODUCTION

The NASA models have remained most useful and popular tools for aeronautical engineers and instrument developers. Evaluations of the omnidirectional flux models labelled AP-1–8 for the protons and AE-1–8 for the energetic electrons can be found in *Lemaire et al.* [1990], *Gaffey and Bilitza* [1994] and *Fung* [1996].

The NASA models provide values of the omnidirectional flux of trapped particles (protons or electrons) as a function of energy (E in MeV), a pseudo-equatorial distance (L measured in Earth radii, and defined by *McIlwain* [1961]) and the B/B_0 coordinate used to measure the latitudinal distance from the equatorial plane; B is the magnetic field intensity at the point of observation (assumed to be a mirror point) and B_0 is the magnetic field intensity at an equatorial distance L: $B_0 = 0.311653/L^3$. B is determined from a magnetic field model which is *Jensen and Cain's* [1960] model for all NASA trapped particle models, except for the AP-8 MAX which was mapped in (B, L) coordinates by using the GSFC 12/66 magnetic field model, updated to 1970 [*Heynderickx et al.*, 1996; *Lemaire et al.*, 1995].

Trapped radiation models similar to the NASA models AP-8 and AE-8 for minimum and maximum solar activity conditions have also been developed since 1970 at the Institute of Nuclear Physics (INP) of Moscow State University (MSU).

The additional particle flux measurements which have been used in the construction of the INP models are from ISEE-1 [*Williams and Frank*, 1984], SCATHA [*Davidson et al.*, 1988], GORISONT [*Grafodatsky et al.*, 1989], COSMOS-900 [*Goriainov et al.*, 1983; *Vlasova et al.*, 1984], INTERCOSMOS-19 [*Volkov et al.*, 1985]. Theoretical considerations based on low altitude satellite data have been used also to build the empirical models INP-PROT-MIN, INP-PROT-MAX, INP-ELEC-MIN and INP-ELEC-MAX which are the counterparts of AP-8 MIN, AP-8 MAX, AE-8 MIN and AE-8 MAX, respectively.

In this article we evaluate the empirical environment models developed at the Institute of Nuclear Physics by comparing them to the NASA models. A detailed study has been repor-

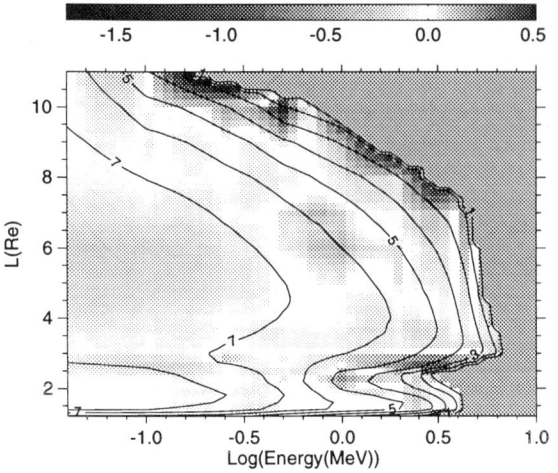

Figure 1. Comparison of equatorial electron omnidirectional fluxes for solar maximum conditions

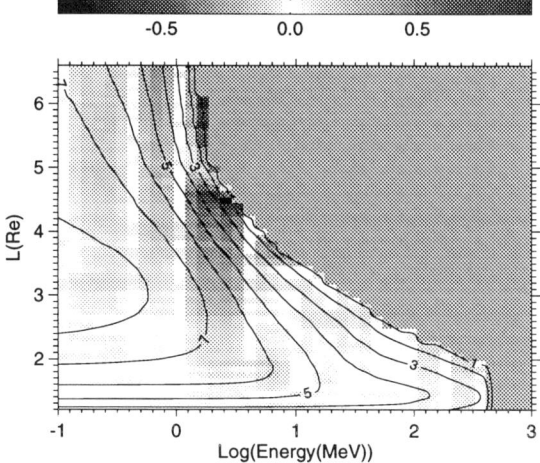

Figure 2. Comparison of equatorial proton omnidirectional fluxes for solar minimum conditions.

ted by *Beliaev and Lemaire* [1994] in a TREND Technical Note prepared under ESTEC contract No. 9828/92/NL/FM. The main differences between both sets of models are summarized in the present paper.

2. COMPARISON OF THE STRUCTURE AND FILE ORGANISATION OF THE NASA AND INP RADIATION BELT MODELS

Since the INP radiation belt models have emerged from the earlier NASA models which have been modified in limited regions of the (E, B, L) space, it is assumed here that the magnetic field models applicable for the Russian models are the same as to those used earlier to build NASA's counter parts. This may be a first source of error and points toward the necessity for future radiation belt modellers to clearly announce the epoch of the IGRF and which external magnetic field model may have been employed to calculate the B and L coordinates. This procedure would avoid future confusion and misuse of new radiation belt models [see *Konradi et al.*, 1987; *Lemaire et al.*, 1990; *Heynderickx et al.*, 1996].

The format of the data files containing the matrix elements for the INP models is similar to that used for the NASA models except that B is used in the Russian models instead of B/B_0. The spacing and numbers of grid points in L and E, are different in the INP and NASA models (see Tables 3 and 4 or Figures 1 and 2 in *Beliaev and Lemaire* [1994]). The ranges in L and E covered by the NASA and INP models are given in Table 1. The grid points in INP models is not as dense as that of the AE-8/AP-8 models, but it is rather comparable to that used in the compressed versions AP-8 MAC and AP-8 MIC.

To interpolate between grid points of the NASA models the FORTRAN programs TRARA1 and TRARA2 are used. These programs perform linear interpolations of the logarithm of the particle fluxes, respectively between energy steps (TRARA1) and between $(L, B/B_0)$ grid points (TRARA2) (see *Vette* [1991], for a literal description of these interpola-

tion algorithms and codes; we found nowhere a quantitative study of the precision of these interpolation algorithms). A smoother non-linear method of interpolation has been implemented by *Daly and Evans* [1993] in a code called TRARAP. This newer and smoother interpolation program is incorporated in the ESA/UNIRAD package [*Heynderickx et al.*, these proceedings]; it is now currently used by us with AP-8/AE-8 models.

A different (simpler) interpolation subroutine (FINTL1) is used with all INP models: it performs logarithmic interpolation in E and linear interpolation in B and L.

3. COMPARISON OF OMNIDIRECTIONAL FLUXES

To compare the omnidirectional fluxes provided by the INP and NASA models, the quantity $z = \log(J_{\text{INP}}/J_{\text{NASA}})$ was calculated for pairs of models, e.g. INP-ELEC-MIN and AE-8 MIN or INP-PROT-MAX and AP-8 MAX, etc.

While calculating $z(E, B, L)$, the minimum flux values for both models were taken into consideration i.e. $J_{\text{INP}} \geq 1\,\text{cm}^{-2}\text{s}^{-1}$ and $J_{\text{NASA}} \geq 10\,\text{cm}^{-2}\text{s}^{-1}$. When the fluxes for given (L, B) and E are smaller than these lower limits, the value of z is disregarded and the pixel corresponding to this point in Figures 1 to 3 is then colored in gray. When $z(E, B/B_0, L) > 0$ the Russian model (INP) gives a larger flux than the American model (NASA) and the corresponding pixel is then colored in red with an intensity proportional to the value of $|z|$. On the other hand when $z < 0$ the flux of the American model is larger than that of the Russian one; in this case the pixel $(E, B/B_0, L)$ is painted in blue. Where the pixel is white, $z = 0$, and $J_{\text{INP}} = J_{\text{NASA}}$, i.e. both models give the same flux of particles. Using complementary colors (blue and red), Figure 5 in *Beliaev and Lemaire* [1994] shows where the INP-ELEC-MAX equatorial electron fluxes are larger/smaller than the corresponding AE-8 MAX fluxes. Figure 1 is a gray shaded version of the color figure taken from the original publication. The dark areas around $E = 1$ MeV,

Table 1. Limits of NASA and INP trapped radiation models

Type of limit	AE-8/AP-8 Models	INP Models
Energy Range	protons: 0.1–400 MeV electrons: 0.04–7 MeV	protons: 0.1–400 MeV electrons: 0.04–7 MeV
L Range	protons: 1.15–6.6 electrons: 1.2–11	protons: 1.2–6.6 electrons: 1.2–7.0
B/B_0 Range	implicit limitation imposed by implemented atmospheric cut-off	—
B Range	—	implicit limitation imposed by implemented atmospheric cut-off
Flux Range	$\geq 10\,\mathrm{cm}^{-2}\mathrm{ster}^{-1}\mathrm{s}^{-1}$	$\geq 1\,\mathrm{cm}^{-2}\mathrm{ster}^{-1}\mathrm{s}^{-1}$

$L = 6$ and at $E = 0.2\,\mathrm{MeV}$, $L = 9$ should be colored in red, while all other darker areas in this figure should be painted in blue.

The solid lines in this and in the following figures give equiflux contours corresponding to the NASA model: they always correspond to the flux values of the second listed model. The range of color intensities is choosen to accomodate for the dynamic range of all values of z within the limits of the map. The color scale shown at the top of the figures enables to determine how large the differences are: on the blue side where $z = -0.5$ one has $J_\mathrm{INP}/J_\mathrm{NASA} = 10^{-0.5}$ on the red side where $z = 0.5$ one has $J_\mathrm{INP}/J_\mathrm{NASA} = 10^{0.5}$. The maximum fluxes corresponding to the inner and outer electrons zones are located at $L = 1.5$ and $L = 3$–4 in the AE-8 MAX model. It can be seen in Figure 2 that the location of the peak proton flux in the AP-8 MIN model is energy dependent: it varies from $L = 1.4$ to 2.5 when E changes from 300 MeV to 1.5 MeV.

These figures show that for wide ranges in the (E, L) space, the fluxes of the INP and NASA agree within a factor of 2: i.e. $|z| < 0.3$. However, differences of more than one order of magnitude are observed in the regions of step gradient: i.e. where $|\partial J/\partial E|$ and $|\partial J/\partial L|$ are large. Note however, the good agreement of both models at small L values despite the steep gradient of J versus L at low altitudes.

The white vertical strips, observed at $E = 0.4$, 1, 4 and 10 MeV in Figure 2 correspond to grid lines of constant energy which are common in the INP and NASA models. More significant discrepancies are found in regions of the (E, L) space where the grid points are sparser in the INP model than in the NASA models. The largest discrepancies shown in the electron models (Figure 1) occur at the proximity of the boundary of applicability of the AE-8 model, i.e. where the electron flux drops below the threshold, i.e. at the limits of the gray shading areas where ($J < 10\,\mathrm{elec/cm^2 s}$). Note that the minimum flux threshold of the INP is a factor 10 lower (i.e. $J < 1\,\mathrm{elec/cm^2 s}$). Therefore at the rims of the the empirical radiation belt models the INP is probably more reliable than the older NASA models. It should be pointed out however that these regions are those where the fluxes are smallest and do not contribute significantly to the integral flux

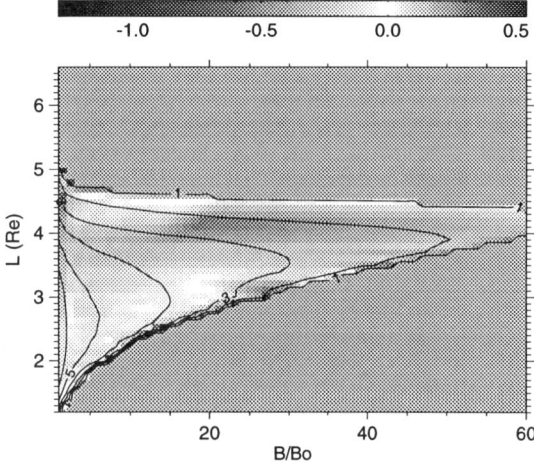

Figure 3. Comparison of 2 MeV proton fluxes for solar maximum conditions

and radiation doses.

4. MAPS OF OFF-EQUATOR OMNIDIRECTIONAL FLUXES

The solid lines in Figure 3 show the iso-contours of the AP-8 MIN omnidirectional proton flux in the $(B/B_0, L)$ plane for a constant proton energy $E = 2\,\mathrm{MeV}$. The distribution along magnetic field lines correspond approximately to constant L cuts (horizontal lines). The flux is maximum at the equator where $B/B_0 = 1$. Its intensity decreases when the latitude or B/B_0 increases. The low-altitude atmospheric cut-off B_c/B_0 increases with L as a power law: $B_c/B_0 = 0.66\,L^{3.452}$: this corresponds to Vette's atmospheric cut-off definition [*Daly and Evans*, 1993]. The dark/red spot at $L = 3$ and near the atmospheric cut-off is where the Russian model gives larger proton fluxes than the American model. Similar plots can be generated for all protons energies between 0.1 and 400 MeV, as well as for all electron energies between 0.04 and 7 MeV (see *Beliaev and Lemaire* [1994]).

The differences between the Russian proton models and the American ones are much larger for lower energies that for higher energies. The comparison of the electron fluxes in both models indicates that a much better agreement is found in the inner electron belt ($L < 2.5$) than in the outer electron belt. A poor agreement is again observed at the edges of the radiation belts, where the gradients of flux are large.

The dark/blue vertical strip seen in Figure 3 near $B/B_0 = 1$ is due to the differences in the interpolation algorithms employed for both models when $B/B_0 = 1$ and when $B/B_0 \neq 1$. Indeed, the interpolation subroutine FINTL1 used for the Russian models is based on a different algorithm, than TRARA2.

Similar "color" maps in the $(E, B/B_0)$ plane (see *Beliaev and Lemaire* [1994]) indicate that the differences between both models are smaller where the energy grid points of the INP models coincide with those of the NASA ones. This confirms that both interpolations algorithms work differently and do not offer equivalent accuracies.

5. COMPARISON OF FLUENCES

Similar differences are found in the fluences of the electrons and protons. Figures 26 and 27 in *Beliaev and Lemaire* [1994] show orbit integrated fluxes, respectively, for protons and electrons of energies.

The fluences predicted with the INP proton model are less than a factor of 2 smaller than those obtained by the corresponding NASA model, for a circular low earth orbit at 500 km altitude and $51.6°$ inclination. The largest discrepancy is found for the smallest proton energies.

It is difficult to say whether these differences are due to a secular change between the time the NASA models were developed and the time the INP models updating occured, or whether these different fluences are consequences of the limited accuracies of the interpolations routines employed with both models. To clarify this issue a detailed mathematical and numerical study of the interpolation algorithms would be needed. In the present study we intended to compare the INP and NASA models when they are used with their own software environment i.e. with the interpolation tools currently linked to those models.

6. CONCLUSIONS

Although orders of magnitude differences are reported in the values of particles fluxes at the edges of domains of application of these models e.g. near the atmospheric cut-off, where the omnidirectional flux is relatively small and where the spatial gradient (versus L or versus B) of these fluxes are largest (see Figures 1 and 2), it can be concluded from this study that the discrepancies in flux and fluences between both models are less than a factor 2 in all regions of the (E, B, L) space where the particle fluxes are highest. Owing to the large statistical fluctuations these differences between the average values is within the limits stated by the developers of these empirical models [*Vette*, 1991].

This study also points out that the different algorithms and subroutines used to interpolate between the grid points of the INP and NASA empirical trapped radiation models do not have comparable accuracies: the best agreement between both models being obtained at those grid points which are common in the INP and NASA models; elsewhere, the interpolated flux values differ sometimes significantly. Related problems with the NASA interpolation subroutine TRARA2 had already been pointed out by *Daly and Evans* [1993].

Acknowledgements. We thank Prof. M.I. Panasyuk, Director of the Institute for Nuclear Physics of the University of Moscow for giving us the opportunity to use the INP models. This paper was prepared while A.A. Beliaev visited BISA, Brussels in 1994. This study was supported by ESA/ESTEC/WMA contract No. 9828/92/NL/FM. We wish to thank Dr. E.J. Daly from the Space Systems Environment Analysis Section of ESA for his suggestions and support during this study.

REFERENCES

Beliaev, A. and J. Lemaire, Evaluation of the INP radiation belts models, TREND, Technical Note A, ESTEC contract no 9828/92/NL/FM, June 1994.

Daly, E.J., Evans, H.D.R., Problems in Radiation Environment Models at Low Altitudes, Memorandum ESA/ESTEC/WMA/93, 1993.

Davidson, G.F., Filbert, P.C., Nightingale, R.W., Imhof, W.L., Reagan, J.B. and Whipple, E.C., Observations of intense trapped electrons fluxes at synchronous altitudes, *J. Geophys. Res., 93*, 77–95, 1988.

Fung S.F., Recent Development in the NASA Trapped Radiation Models, (this issue), 1996.

Gaffey, J.D. Jr. and Bilitza, D., *NASA/National Space Science Data Center Trapped Radiation Models, Journal of Spacecraft and Rockets, 31*, 2, 1994.

Goriainov, M.F., Dronov, A.V., Kovtykh, A.S., Sosnovets, E.N., Spatial, Spectral and Angular Structure of low-Altitude 30–210 keV Electron Fluxes during Magnetically Quiet Time, *Kosmicheskie Islledovaniya*, **21**, 609–618 (in Russian), translated in *Cosmic Research, 21*, 494–500, 1983.

Grafodatsky, O.S., Islayev, Sh.N., Panasyuk, M.I., et al., Magnetospheric Plasma Fluxes Registration on Board of GORISONT Satellite, *Issledovaanya po geomagnetismu, aeronomii, i physike Solntsa (Geomagnetism, Aeronomy and Solar Physics Studies), Issue 86*, 99–130 (in Russian), 1989.

Heynderickx, D., J. Lemaire and E.J. Daly, Historical review of the different procedures used to compute the L-parameter, *J. Nucl. Tracks Radiat. Meas.*, 1996, in press.

Jensen, D.C. and J.C. Cain, An interim geomagnetic field, *J. Geophys. Res., 67*, 3568, 1962.

Konradi, A., A.C. Hardy and A. Atrell, Radiation environment models and the atmospheric cut-off, *Journal of Spacecraft and Rockets, 24*, 284–285, 1987.

Lemaire, J.F., A.D. Johnstone, D. Heynderickx, D.J. Rodgers, S. Szita and V. Pierrard, Trapped radiation environment model development, *Final report of TREND-2 study (ESTEC contract No. 9828/92 NL/FM)*; also *Aeronomica Acta, A 386*, 1995.

Lemaire, J., M. Roth, J. Wisemberg, P. Domange, D. Fonteyn, J.M. Lesceux, G. Loh, G. Ferrante, C. Garres, J. Bordes, S. McKenna-Lawlor and J.I. Vette, Development of improved models of the Earth's radiation environment, *Final report of TREND study (ESTEC contract No. 8011/88/NL/MAC)*, September 1990.

McIlwain, C.E., Coordinates for mapping the distribution of geomagnetically trapped particles, *J. Geophys. Res., 66*, 3681–3691, 1961.

Vette, J.I., The AE-8 trapped electron model environment, NSSDC/WDC-A-R&S 91-24, November 1991.

Vlasova, N.A., Knyasev, B.N., Kovtykh, A.S., Kozlov, A.G., Panasyuk, M.I., Reyzman, S.Ya., and Sosnovets, E.N., Protons with E larger than 30 keV at Low Heights near Geomagnetic Equator during a Magnetically Quiet Time, *Kosmicheskie Islledovaniya, 22*, 53–66 (in Russian), translated in *Cosmic Research, 22*, 46–58, 1984.

Volkov, I.B., Dronov, A.B., Kratenko, Yu.P. et al., Outer Electron Belt Dynamics according Simultaneous Measurements on Board of INTERCOSMOS-19 and COSMOS-900 Satellites, *Kosmicheskiye Islledovaniya (Cosmic Researches), 23*, 642–646 (in Russian), 1985.

Williams, D.J., Frank, C.A., Intense low energy ion populations at low equatorial altitudes, *J. Geophys. Res., 89*, 3903–3911, 1984.

A.A. Beliaev, Inst. Nuclear Physics / Moscow State University, 119899 Moscow, Russia.

J.F. Lemaire, Belgian Institute for Space Aeronomy, Ave Circulaire, 3, B-1180 Brussels, Belgium.

Empirical Radiation Belt Models
Report of Discussion Group B

Reporter: D.J. Rodgers

Mullard Space Science Laboratory, University College London

Participants: S.M. Gussenhoven (chair), E.D. Tolstaya (co-chair), D.J. Rodgers (reporter), A.L. Vampola

The discussions on empirical radiation belt modelling centred on the need for a new standard model for engineering use. The current NASA models are deficient in a number of ways and a new global standard model for ions and electrons is required. The properties of this new model were outlined and possible ways of organising its creation were discussed.

1. AIMS AND LIMITATIONS

The group aimed to find what kind of models were needed by the users, to assess whether the existing models fulfilled these needs and, if not, what kind of new models were needed. In addition, we sought to produce guidelines for how these new models could be created.

The choice of coordinate systems and the availability of suitable data are clearly essential elements in any effective radiation belt model. These aspects were not addressed by this discussion group, however, because they were discussed in detail in other discussion groups taking place simultaneously. The existence of suitable coordinate systems and data are thus implicit in the conclusions reached here.

It is not appropriate for a discussion group to propose to the community at large a course of action that does not have wide support within the group or which could not be fully discussed in the short time available. Hence we have left many details unspecified and have not issued a complete prescription for the development of new models. Nevertheless there was wide agreement on the state of current empirical radiation belt models and the approach to improving them.

2. CONCLUSIONS

It was considered that the pressing need was for an easy-to-use global standard model that could be accessed by engineers. Models for scientific simulation do not need to be the same everywhere and will probably be created in accordance with their authors' particular interest. Users of this type of model expect to set up a complicated description of the magnetospheric state and to exercise caution as to where the model is valid. Spacecraft engineers do not have the time to investigate the detailed properties of a model and need results compatible with colleagues and competitors throughout the world. For engineering purposes the state of the magnetosphere is generally unknown, except in a statistical sense.

An engineering empirical model need not involve a sophisticated physical description of magnetospheric processes. Adding more physics to the model increases complexity and should be brought in only where it clearly improves the accuracy of the model.

2.1. *The need to replace the NASA models*

The NASA AE-8 and AP-8 models have become the standard electron and proton radiation belt models in use in the west. However, these models have a number of shortcomings:

- The proton model has inadequate resolution at low altitudes. This is because, near the atmospheric cut-off, the (B, L) coordinates become a poor coordinate system because particle fluxes are far more strongly influenced by atmospheric density than by the magnetic field.
- These is no variability on timescales less than a solar cycle. This is particularly important for electrons and for ions above $L = 2.2$.
- Protons up to 5 MeV below $L = 2.5$ are very poorly represented.
- AE-8 electron fluxes are significantly higher in the outer belt than fluxes observed in recent years.

These deficiencies are sufficient that a new model for both electrons and ions is required. Maximum acceptance of the replacement model and minimum disruption for users would be accomplished if the structure of the existing model could be retained. However, this is not feasible given the wide scope of the changes required. The new model will, in time, require updating and the ability to accommodate updated elements is an important characteristic that the model should possess.

2.2. *The structure of the new model*

The group identified a number of desired properties of the new model that were incompatible with a single structure like the $(B/B_0, L)$ grid used by the NASA models. Hence the new model needs to be a composite of sub-models covering different areas of space. For instance, the low altitude region could be described in terms of a function related to atmospheric density while geostationary orbit could include detailed flux probabilities which would be hard to calculate throughout the magnetosphere. This scenario lends itself to a modular design of software which is an integral part of the model itself. The model is thus easily updated by the replacement or addition of subroutines. Updates should occur as often as significant discrepancies occur. A top-level module would determine which subroutine to call and would present the user with a very simple interface.

The model should include:

- An increased proton energy range, both to lower and higher energies (> 100 MeV).

- A higher electron energy range (to 10 MeV).

- Directionality, i.e. east-west asymmetry at low altitude and pitch angle dependence elsewhere. Pitch angle dependence requires a magnetic field model and a suitable external field. The purely internal magnetic field of the present NASA models is not adequate at geostationary orbit and beyond. However, a dynamic magnetic field model including, say, K_p is not justified.

- Probability of flux levels in addition to mean fluxes. This may not be available throughout the magnetosphere since it requires a high level of data coverage.

- Variability described in terms of a state vector, such as K_p or $A_{p_{15}}$, plus information on the probability of occurrence of the state vector.

- Heavy ions. Knowledge of these ions is likely to be infrequently necessary. However, new data and models already exist and would be straightforward to include for completeness.

2.3. *Production of the new model*

The discussion group heard about some new approaches to modelling that have resulted from networking and the increased power of computers.

There is no longer the need for a database to be held at one site. A globally distributed database could be formed simply by allowing internet access on the data owners' local computers. It is possible also for models to be produced from the raw database at the time of use. Whilst this does not fulfil the aim of producing a single standard model, it does allow the evolution of the model as better data become available and lets the user customize the model according to the immediate requirements.

The group concluded that a new standard model is best produced under the supervision of a respected competent international body, such as the International Standards Organization or COSPAR. As input to the models, empirical data from around the world would be needed and agreement on coordinate systems in which to organize the data. Production of the model could be performed by one supervised group using a single common database. Alternatively, individual groups could be allowed to develop their own models based on their own data and these could be combined after comparisons and discussions.

The approach favoured by the group was for the production of a common globally distributed database which individual groups would be able to access to produce their own version of all or part of the model. The final model would be the best or a combination of these. This approach allows the maximum supervision and involvement by modelling groups within the community. However, a real doubt exists as to whether data would be released freely for modelling by other groups, particularly where cross-border transfers are involved.

D.J. Rodgers, Mullard Space Science Laboratory, University College London

DISCUSSION

Q: M. Walt. Did your group also consider whether the current models are adequate for scientific purposes, as opposed to engineering requirements?
A: M.S. Gussenhoven. Greg Ginet and Jay Albert in our group are looking at the theoretical implications in the CRRES models and find evidences of dynamical processes that are not explained by steady state sources and diffusion. Some of their results are presented here. I think it is clear that the existing models fall well short of representing dynamical processes in the radiation belts for time scales shorter than a solar cycle.
Q: J.F. Lemaire. The state vector should contain independent solar wind and geomagnetic parameters or indices, not interdependent ones.
A: S. Fung. Yes, the example of a magnetospheric state vector given in my presentation does contain the solar wind "driver" parameters and the magnetospheric "response" parameters. A set of time delays, represented by T, is also included to anticipate the "histeresis" of the magnetosphere. The specific state parameters used are only examples of parameters which may reflect magnetospheric processes operating at various time scales.

Introduction to Trapped Particle Flux Mapping

J.G. Roederer

Geophysical Institute, University of Alaska-Fairbanks

The radiation belts consist of energetic charged particles which in absence of any perturbations would remain permanently trapped in the earth's magnetic field. The motion of these particles is described adequately by the so-called adiabatic theory, in which the three main periodic components of their motion—cyclotron, bounce and drift—are "averaged out", i.e., the concepts of guiding center, guiding field line and (closed) drift shell are introduced as geometric entities describing the average location of the particles in question. By working with the instantaneous position of the guiding center of a particle, information about the cyclotron phase is lost but the average electromagnetic effect of the gyration is retained by attaching a magnetic moment M to the guiding center "particle"; M happens to be the first adiabatic invariant. Likewise, by specifying the instantaneous guiding field line, we lose information on the bounce phase but the dynamics of the bouncing process is represented in the form of the second adiabatic invariant J; in the representation of a drift shell the information on the drift phase is lost, but an average dynamic property is retained in the form of the third adiabatic invariant Φ (flux invariant) [*Northrop*, 1963].

The three adiabatic invariants are the "natural" coordinates to represent a distribution of radiation belt particles that is isotropic in all phases (only under this condition can the original distribution in 6-dimensional phase space be converted into one in the 3-dimensional M, J, Φ space). However, there are situations, particularly involving static configurations of the magnetic field, in which some appropriate functions of M, J, Φ can be used as more "intuitive" coordinates for the representation of trapped particle fluxes; e.g., McIlwain's I-value and B, L coordinate system [*McIlwain*, 1961]; Kaufmann's $K = I\sqrt{B}$ invariant [*Kaufmann*, 1965]; Roederer's generalized L^* value [*Roederer*, 1972].

Spatial and temporal variations of the trapping field and other external forces acting on the particles can violate the conservation of the adiabatic invariants and thus introduce irreversible changes in the trapped particle population. There is a hierarchy of decreasing time scales of change determined by the drift, bounce and cyclotron periods of a trapped particle, by which the third, second and first invariants are violated, respectively. Equivalently, for the spatial extents of perturbations, the hierarchy is determined in decreasing order by the size of the drift shell, the length of the guiding field line and the Larmor radius. As long as all invariants are conserved (the external variations are slow with respect to the longest drift period) the distribution function in invariant space is constant and the behavior of the trapped population in "real" space (e.g., the pitch angle, energy and spatial distributions at different times in a slowly varying B-field) is governed by a field-dependent (and therefore time-dependent but reversible) transformation. A subtle but important point should be made here: particle detectors are usually much smaller in size than a typical radiation belt particle Larmor radius. Therefore, what one measures experimentally is a "*true*" particle flux, whereas adiabatic theory really works with magnetized *guiding center particles*. It is important to distinguish among the two descriptions, particularly when the particle density gradients are large (the density varies appreciably over a Larmor radius): a perfectly gyrotropic spatially inhomogeneous ensemble leads to a particle distribution that is azimuthally asymmetric perpendicular to **B** (just as a static but inhomogeneous single-particle fluid in a uniform field carries diamagnetic currents).

Adiabatic theory can still be used to determine the effect of changes that violate the third invariant but conserve the first two (i.e., that are still slow with respect to the bounce and cyclotron periods). In that case, it is necessary to determine the behavior of the particles (their guiding field lines) *during* the change by integrating the bounce-average drift velocity; however, this must be done for all possible initial drift phases (longitudinal positions on the shell—since these are random the end result may be chaotic). When the second adiabatic invariant is violated (but M still conserved) the instantaneous drift and parallel velocities of the guiding center can be integrated to determine the behavior during the time or in the region where the perturbation is being felt (e.g., Fermi acceleration of cosmic rays). When the first invariant breaks down (e.g., when a trapped particle enters a region in which the field lines are sharply bent), the "real" particle velocity must be integrated to determine the actual orbit. Again, this must be done for the different possible cyclotron phases at the beginning of the perturbation and the result will usually be chaotic.

There are different kinds of non-adiabatic perturbations acting on a trapped particle population. Even if the external perturbation is continuous rather than stochastic, its end effect will usually be stochastic because, as stated above, the outcome will depend on the instantaneous phases in which the particles were caught at the start of the perturbation. Quite

Radiation Belts: Models and Standards
Geophysical Monograph 97
Copyright 1996 by the American Geophysical Union

generally, the integral effect of perturbations on a trapped particle population is equivalent to that of a diffusion process, whether these perturbations consist of many random "microscopic" actions on individual particles (as in Coulomb scattering in the atmosphere, or wave-particle interactions in the plasmasphere), or whether they consist of a discrete set of large-scale disturbances simultaneously affecting the entire population (as in a succession of magnetospheric field compressions). Therefore, perturbations can often be parameterized in the form of diffusion coefficients, and the problem treated with diffusion theory (e.g., Fokker Planck equations) (e.g., *Schulz and Lanzerotti* [1974]). A special case is the treatment of atmospheric scattering and absorption at low altitudes, which makes the particle distribution longitude-dependent.

One of the most important practical problems in radiation belt physics is that of *trapped particle flux mapping*, a paradigm in which, ideally, information on a trapped population is represented in such a way that it can be used to predict the trapped particle flux incident from a given direction within a given solid angle, at any given point and time in the trapping region. Since what we measure is not the "natural" distribution function in M, J, Φ space but the particle flux in "real" space, we need to have knowledge of the complete trapping magnetic field at the time of the measurement, to be able to map in invariant space, and we also need to know this field at the time for which information in "real" space is to be retrieved. In addition, ideally, we must have knowledge of all the effects of non-adiabatic, irreversible changes and perturbations that have occurred between the time the flux was mapped and the time for which the prediction (or "hindcast") is wanted. Of course, these requirements cannot be fulfilled because of the eminently time-varying nature of the magnetosphere and the relative scarceness of information about it.

All one can do is work with approximations and averages. For instance, one can restrict the study to equatorial (90° pitch angle) particles as it is frequently done in quantitative studies, or use analytical approximations for near-equatorial particles and simplified magnetic field models. In addition, many studies so far have assumed a magnetic field constant in time. Such limitations and approximations are indeed very useful to gain physical insight, but they may not yield much information of a practical value, because the real radiation belt is three-dimensional and the real magnetospheric field is quite asymmetric and temporally variable. Even *during* the primary experimental information acquisition this must be taken into account: particle measurements are usually carried out only along discrete orbital segments at different times—during which the magnetospheric field is bound to change!

A key ingredient for *practical* radiation belt studies is a magnetic field model with time-adjustable parameters that describes as accurately as possible the field in the trapping region at the times of interest. This of course also requires information on the "state of the magnetosphere" at the critical times so that the model parameters can be adjusted accordingly. With such a model any (differential and directional) trapped particle flux measurement at a point $\mathbf{r}$ and time t can be converted into a value of the distribution function in M, J, Φ (or some other invariant) space. This requires a computer code that calculates with pre-fixed accuracy the invariants M, J, Φ (or, in practice, some more convenient functions thereof) given the point $\mathbf{r}$ and the energy and direction (or velocity $\mathbf{v}$) of the particles (with the field model parameters corresponding to time t). While the determination of J (or L, K) requires only one integration along a field line, an accurate determination of Φ requires the determination of the full particle shell, which could demand thousands of field line integrations (depending on the accuracy wanted). Localizing a shell of prefixed M, J, Φ values in a given field configuration could require an iterative procedure involving up to hundreds of calculations of Φ, or hundred-thousand integrations! This may be the reason why Φ, or $L^* \propto \Phi^{-1}$, have not been widely used so far (except in simple cases where analytical approximations are possible); indeed, a Cray may take a minute or more for just one reasonably accurate Φ calculation.

Assuming that we have mapped the radiation belt accurately in invariant space, we can retrieve the flux values for any other field configuration, provided the field transition has been *adiabatic*, i.e., slow compared to the drift periods. This in itself can yield much useful information, such as the prediction of the secular variation and 11-year cycle effects, seasonal and diurnal (magnetic axis tilt) effects, and other slow changes of the magnetospheric configuration. Yet one may argue: What for? Would it not be hopeless to determine all the irreversible changes that were bound to occur in the interim? This is not an argument if we recognize that the main scientific objective of trapped flux mapping is *to provide a means to obtain a consistent, comprehensive set of standard reference values*.

Having a credible flux map for a reference state of the magnetosphere, a measurement made at a different time can be transformed back to the reference state *for comparative studies*—for instance, to determine the effects of prototype time variations on the flux, to distinguish adiabatic from non-adiabatic variations and identify irreversible changes that could have occurred, or to test external or internal field models by using the trapped particles as "remote sensing" probes. As stated earlier, while general characteristics of the dynamic behavior of trapped particles may be determined using only crude approximations and simplifications, the determination of dynamic effects, such as particle precipitation in the South Atlantic anomaly, trapping boundary changes, the effects of an electric field on the lower energy particles, the effects of internal magnetic field distortions, and, most importantly, the test of the new theories of source and loss mechanisms, all require a more accurate approach.

Last but not least, accurate flux mapping and accurate adiabatic transformation algorithms are of critical practical importance to develop user-friendly, trustworthy radiation flux models for the quantitative prediction of radiation effects on technological systems and humans in space.

Adiabatic theory is not difficult conceptually—but it can be tricky! I have noticed in the recent literature some confusion about "old" physical concepts, and some misinterpretations of the meaning and the limits of the different orders of approximation in this theory. At meetings I sometimes hear papers that are nothing but a "re-discovery" of (formerly) well-known (and published) facts. This may be the result of a tendency nowadays to take a "cook book recipe" approach of doing physics, that is, to perform research with primary attention to achieving some quick practical results, and less concern for clear physical understanding. I firmly believe that these two goals are *not* mutually exclusive. The present Workshop is a clear demonstration that both can, indeed, be achieved at the same time!

REFERENCES

Kaufmann, R.L., Conservation of the first and second adiabatic invariants, *J. Geophys. Res., 70*, 2181, 1965.

McIlwain, C.E., Coordinates for mapping the distribution of magnetically trapped particles, *J. of Geophys. Res., 66*, 3681–3691, 1961.

Northrop, T.G., *The Adiabatic Motion of Particles*, Interscience, New York, 1963.

Roederer, J.G., Geomagnetic field distortions and their effects on radiation belt particles, *Rev. Geophys. and Space Phys., 10*, 599–630, 1972.

Schulz, M., and L.J. Lanzerotti, *Particle Diffusion in the Radiation Belts*, Springer-Verlag, New York, 1974.

J.G. Roederer, Geophysical Institute, University of Alaska-Fairbanks (e-mail : JGR@giuaf.gi.alaska.edu)

Canonical Coordinates for Radiation-Belt Modeling

M. Schulz

Space Sciences Laboratory, Advanced Technology Center, Lockheed Martin Missiles & Space, Palo Alto, California

Expressed as a time-dependent function of the three adiabatic invariants (M, J, Φ), the phase-space density f constitutes a "canonical" description of the radiation environment. Even more convenient than J, however, is the energy-independent but adiabatically invariant quantity $K \equiv (8m_0 M)^{-1/2} J$. The $(M, K, \Phi; t)$ system provides an optimally convenient kinematical framework for specifying the new generation of radiation-belt models and for studying the dynamics of geomagnetically trapped radiation. These coordinates accommodate the changes in energy and mirror-field intensity experienced by trapped particles during temporal variations in magnetospheric configuration (e.g., between quiet and disturbed times). Their use makes it possible to separate stormtime transport and loss from storm-associated adiabatic modulations of particle intensities. The (M, K, Φ) coordinates even accommodate secular variations in the main geomagnetic field (whose dipole moment has decreased already by 3% during the Space Age). It is important in this context to distinguish between magnetic shells and drift shells. A magnetic shell is a surface of constant α when the field model $\mathbf{B} = \nabla\alpha \times \nabla\beta$ is expressed in terms of the usual Euler potentials (α, β). A drift shell is a surface of specified (M, K, Φ) in general, which reduces to a surface of constant (K, Φ) at high particle energies (such that adiabatic $\mathbf{E} \times \mathbf{B}$ drifts are unimportant). The connection $\alpha = \Phi/2\pi$ applies in axisymmetric geometry, but not in general.

1. BACKGROUND

The goal of radiation-belt modeling is to provide a reliable means of specifying the expected particle flux as a function of energy, direction of incidence, location in space, and geomagnetic disturbance level (past and present) for the various particle species and charge states of interest. Radiation-belt models can be empirical (based on data), dynamical (based on theory), or preferably some combination of the two.

This exposition addresses the question of how best to organize the growing body of *in situ* data on radiation-belt particle intensities. It puts forward an argument for the operational use of canonical coordinates derived from the three adiabatic invariants as independent kinematic variables and the phase-space density as the dependent variable. The three adiabatic invariants (M, J, Φ) of charged-particle motion are directly proportional (respectively) to the canonical (Hamilton-Jacobi) action integrals [*Haerendel*, 1968]

$$J_i \equiv \oint_i [\mathbf{p} + (q/c)\mathbf{A}] \cdot d\mathbf{s}_i, \qquad (1)$$

($i = 1, 2, 3$) corresponding to the underlying quasi-periodic motions (gyration, bounce, and drift). As usual, the integrand in Eq. (1) is the canonical momentum $\mathbf{p} + (q/c)\mathbf{A}$, where $\mathbf{p}$ is the particle momentum, $\mathbf{A}$ is the vector potential, q is the particle charge, and c is the speed of light.

The first invariant M (= $p_\perp^2/2m_0 B$, with B evaluated at the guiding center) is equal to $\gamma\mu$, where γ is the ratio of relativistic mass to rest mass and μ is the magnetic moment of the particle. The corresponding action integral ($J_1 = \pi p_\perp^2 c/2qB$) entails partially cancelling contributions from the two terms in Eq. (1), the second term being equal to q/c times the amount of magnetic flux enclosed by the orbit

Radiation Belts: Models and Standards
Geophysical Monograph 97
Copyright 1996 by the American Geophysical Union

of gyration. The second adiabatic invariant J ($= J_2$) is evaluated along the guiding-center trajectory from one mirror point to the other (and back). Being field-aligned, this path encloses no magnetic flux. The third adiabatic invariant Φ ($= cJ_3/q$) is equal to the magnetic flux enclosed by the guiding-center drift path, as the kinetic contribution is negligible for radiation-belt particles (i.e., for particles of much less than cosmic-ray energy).

The validity of M, J, and Φ as kinematic variables requires essentially that the corresponding action integrals defined by Eq. (1) make sense (i.e., that the motion be separable into gyration, bounce, and drift). This requires in turn that $\Omega_3 \ll \Omega_2 \ll \Omega_1$ over the entire drift shell, as in the adiabatic theory of charged-particle motion. However, the required inequalities ($\ll$) need not be especially strong. Inequalities among the drift, bounce, and (equatorial) gyration frequencies ($\Omega_i/2\pi$) are controlled by the "smallness" parameter [cf. *Northrop*, 1963, pp. 2–3]

$$\varepsilon \equiv (pc/qB_0) \max(|\partial \hat{\mathbf{B}}/\partial s|, |\nabla \ln B|)_0, \qquad (2)$$

where the subscript 0 denotes evaluation at the magnetic equator, and the condition for requisite separability of the constituent motions can be expressed as $\varepsilon \lesssim \varepsilon^*$ (Otherwise the particle motion becomes chaotic, at least the first two adiabatic invariants are strongly violated, and the particle lifetime is reduced to a few bounce periods times the ratio B_c/B_0 of the foot-point B to the equatorial B on the field line of interest.). The estimate $\varepsilon^* \approx 1/3$ made by *Schulz* [1991, p. 202] was based largely on guesswork, but *Il'in et al.* [1986, 1993] have shown in a systematic study that ε^* should actually vary from 3/4 for particles mirroring at the equator ($B_m = B_0$) to $\varepsilon^* \approx 1/9$ in the limit $B_m \gg B_0$.

A dipolar field line satisfies the equation $r = La \sin^2 \theta$, where a is the planetary radius and θ is the magnetic colatitude. A dipolar magnetic shell of dimensionless label L thus encloses a magnetic flux $\Phi = 2\pi\mu_E/La$, where μ_E is the planetary magnetic moment (the inverse proportionality between L and Φ being most easily verified by calculating the equal and opposite amount of magnetic flux that crosses the equatorial plane *outside* the same shell, so as to avoid the offsetting infinities encountered at $r = 0$, the idealized location of the point dipole). This relationship prompted *Roederer* [1970, pp. 107–115] to propose that a dimensionless parameter $L_R = 2\pi\mu_E/a\Phi$ actually be defined (in lieu of McIlwain's more familiar L parameter, discussed below) for compiling and organizing data on radiation-belt particle intensities in the more complicated field geometry that prevails in a planetary magnetosphere. This approach has much to recommend it, especially since L_R is easy to visualize from a geometrical standpoint.

However, the use of Φ itself instead of L_R would maintain adiabatic invariance of the coordinate system even over the time scale of geomagnetic secular variation (the geomagnetic moment μ_E having decreased by about 3% since 1957). The influence of geomagnetic secular variation on the radiation environment is an important topic. Particle energization by the electric field it induces is significant only for inner-zone protons of energy 10–100 MeV [*Farley et al.*, 1972]. This could well be treated separately from the mainstream of radiation-belt model development. Another consequence

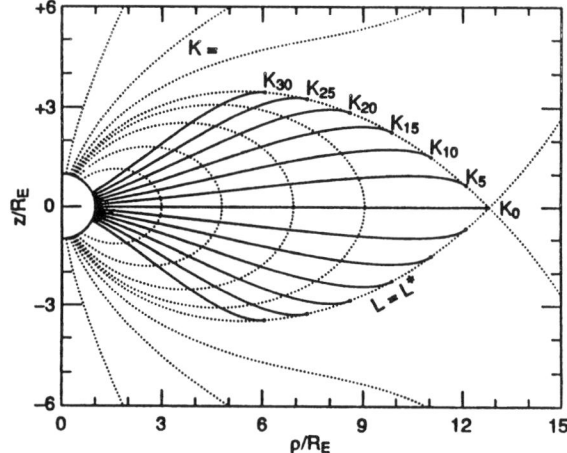

Figure 1. Contours of constant $K \equiv (8m_0 M)^{-1/2} J$, corresponding to mirror-point surfaces if first two adiabatic invariants (M and J) are conserved [*Chen and Schulz*, 1995]. Subscript on K specifies mirror-point latitude (in degrees) at $L = 3$. Numerical values: $K_0^2 = 0$, $K_5^2 = 0.00022\mu_E/R_E = 0.27$ MWb, $K_{10}^2 = 0.0038\mu_E/R_E = 4.7$ MWb, $K_{15}^2 = 0.021\mu_E/R_E = 26$ MWb, $K_{20}^2 = 0.079\mu_E/R_E = 98$ MWb, $K_{25}^2 = 0.23\mu_E/R_E = 0.28$ GWb, $K_{30}^2 = 0.61\mu_E/R_E = 0.76$ GWb. The field model for this illustration is inspired by *Dungey* [1961] and obtained by adding uniform southward ΔB to dipolar magnetic field (e.g., *Hill and Rassbach* [1975]). Field lines are closed for $L < L^* = 8.547$ and open for $L > L^*$.

of the secular variation, however, is an increasing offset $\sim$ 2 km/yr between the dipole center and the geocenter. This amounts to several atmospheric scale heights since 1957 and has the effect of progressively enlarging the loss cone for inner-zone electrons as well as protons at any value of Φ or L_R. The inner boundary of trapped radiation is thus displaced progressively outward in L_R and hence toward larger values of Φ. Apart from the secular variation, moreover, Earth-induction currents can cause the effective geomagnetic dipole moment to vary by an amount $\sim 0.3 D_{st} R_E^3$ during the course of a geomagnetic storm (i.e., by up to 0.3% over just a few days). This is another reason to consider using Φ instead of L_R for radiation-environment modeling. Numerical values of Φ (≈ 7.78 GWb/L_R) conveniently range from 1 GWb (10^9 webers) to 7 GWb over the region occupied by geomagnetically trapped radiation.

Since $p^2 = 2m_0 M B_m$, the kinetic energy of a trapped particle varies directly with Φ (i.e., inversely with L) if M and J are conserved (the mirror-point field B_m at fixed M and J increases with increasing $|\Phi|$). The second invariant J vanishes for equatorially mirroring particles but (because it depends also on p) is not a good indicator of mirror-point latitude. A better invariant for this purpose [*Kaufmann*, 1965; *McIlwain*, 1965] is $K \equiv (8m_0 M)^{-1/2} J$. Mirror-point trajectories corresponding to selected values of K are shown in Figure 1 [*Schulz and Chen*, 1995] for a field model constructed by adding a uniform southward $\Delta \mathbf{B} = \hat{\mathbf{z}} \Delta B_z$ to the magnetic field of a point dipole [*Dungey*, 1961; *Hill and*

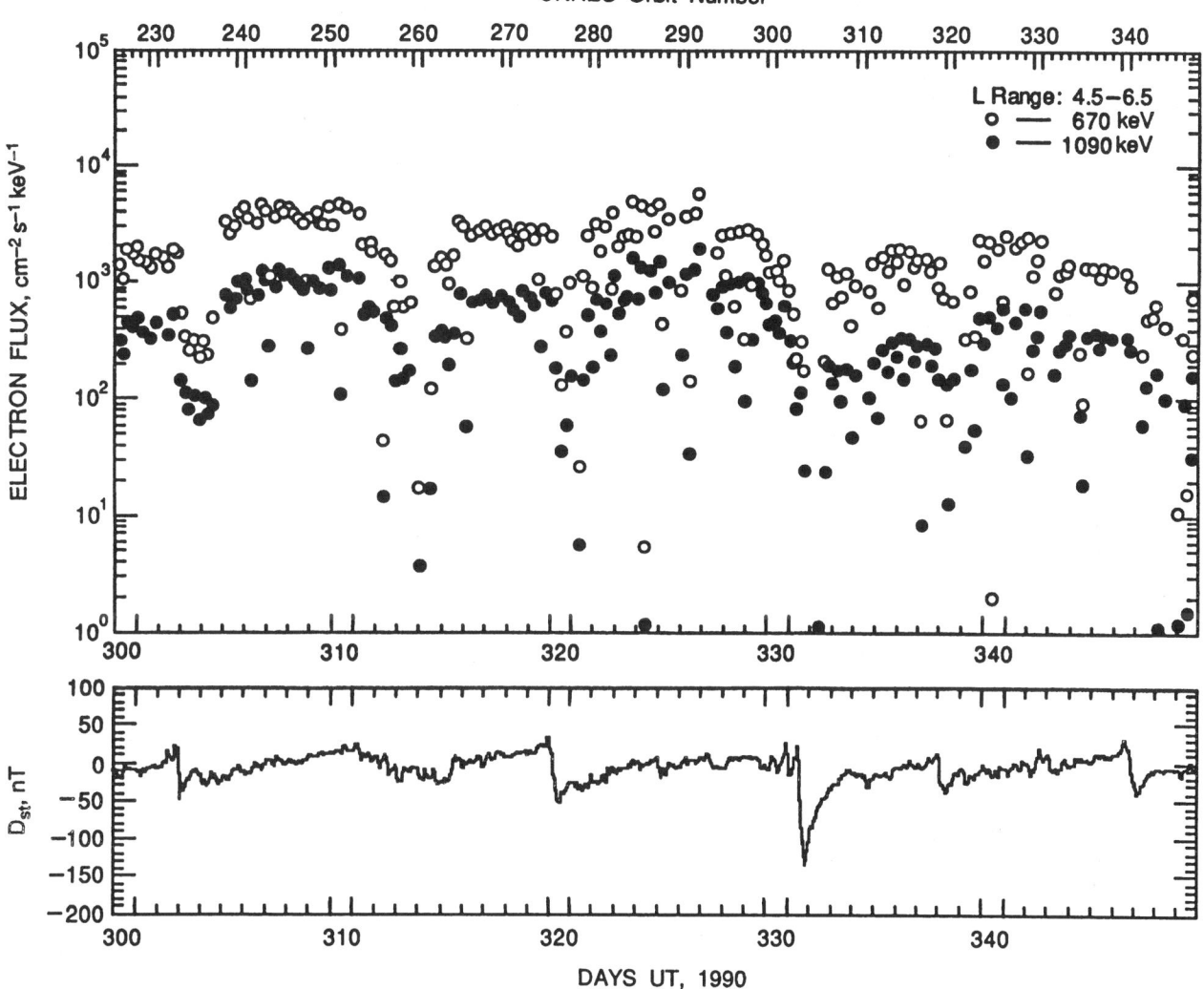

Figure 2. Omnidirectional electron fluxes in two energy channels, averaged over the time spent by CRRES between $L = 4.5$ and $L = 6.5$ on each orbit (full period is 9.85 hr). For comparison: histogram showing hourly values of D_{st} for the same 50 days, 27 October 1990 through 6 December 1990 (this figure was plotted August 1993 by M.A. Rinaldi, using CRRES data provided by R.W. Nightingale and D_{st} data obtained from World Data Center A for Solar-Terrestrial Physics).

Rassbach, 1975]. Possible values of K^2 range from zero to about 16.4 GWb, but a representative value is 0.76 GWb, corresponding to a particle having a 30° mirror latitude at $L = 3$ (The limit $K^2 \to \infty$ is often instructive from a mathematical standpoint. However, particles with $K^2 > 16.4$ GWb would mirror inside the Earth even at $L = 8.547$, the magnetic shell that intersects the Earth at $\theta = 20°$ and corresponds to the last closed field line in Figure 1.).

The phase-space density f (= $j_\perp/p^2$, where $j_\perp$ is the differential unidirectional particle flux) is likewise adiabatically invariant, in accordance with Liouville's theorem. This is a properly relativistic specification for f (e.g. *Schulz and Lan-*

zerotti [1974, p. 39–40]), whereas the commonly encountered approximation $f \approx j_\perp/2mE$ is nonrelativistic. As conventional dimensions (g^{-3}sec^{3}cm^{-6} or kg^{-3}sec^{3}m^{-6}) typically lead to inconveniently large numerical values for f, it is common to plot "phase-space density" profiles representing $m_0^3 f$ instead. A popular alternative is to plot $2m_0 f = j_\perp/MB_m$, but with units of cm^{-2}sec^{-1}ster^{-1}MeV^{-2}.

Canonical phase space is six-dimensional in principle. The variables respectively conjugate to the three canonical action integrals (J_1, J_2, J_3) and associated adiabatic invariants are the three canonical phases ($\varphi_1, \varphi_2, \varphi_3$). The gyrophase φ_1 advances at a rate Ω_1, equal to 2π times the "local" gyrofrequency (evaluated at the guiding center). The bounce phase φ_2 represents 2π times the fraction of a bounce period $(2\pi/\Omega_2)$ completed. The drift phase φ_3 (not precisely the same concept as magnetic local time) represents 2π times the fraction of a drift period $(2\pi/\Omega_3)$ completed. The function $f(M, J, \Phi; \varphi_1, \varphi_2, \varphi_3; t)$ constitutes (in principle) an optimal specification for any population of geomagnetically trapped particles. Drift-periodic echoes [*Lanzerotti et al.*, 1967; *Brewer et al.*, 1969; *Chanteur et al.*, 1978] are a familiar consequence of φ_3-dependent phase-space densities. Auroral x-ray and electron microbursts [*Anderson and Milton*, 1964; *Lampton*, 1967] contain what is possibly a signature of bounce-phase (φ_2) organization. Directly observable organization of f with respect to φ_1 is rare (The east-west effect seen in cosmic-ray fluxes is not a signature of gyrophase organization. It arises instead from the fact that different directions of incidence on a given particle detector can correspond to significantly different guiding-center locations.). For transport studies, as well as for the construction of radiation-environment models, the phase-space density should be averaged over any of the conjugate phases $(\varphi_1, \varphi_2, \varphi_3)$ on which f might depend. The result, called $\bar{f}(M, J, \Phi; t)$ satisfies a Fokker-Planck equation but does not obey Liouville's theorem in most cases, as it represents an average over constituent particles that have followed various dynamical trajectories.

2. ADVANTAGES OF THE CANONICAL FORMULATION

This work addresses the question of how best to organize the growing body of *in situ* data on radiation-belt particle intensities. It constitutes an argument for the operational use of canonical coordinates derived from the three adiabatic invariants (along with universal time t) as independent kinematic variables and the phase-space density as the dependent variable.

A self-evident advantage of the canonical formulation is that M, K, and Φ are adiabatically invariant. A further advantage is that $df/dt = 0$ (according to Liouville's theorem) under the influence of Hamiltonian dynamical processes. Even if M, K, and Φ are not conserved, however, they serve as the best possible coordinate system for describing non-adiabatic transport. Moreover, the $\{M, K, \Phi\}$ system leads (see below) to simple Jacobian factors in the corresponding transport equation for $\bar{f}$. It even accommodates the effects of geomagnetic secular variation $(d\mu_E/dt \neq 0)$ on trapped particle populations, such as long-lived inner-zone protons produced by the beta decay of cosmic-ray-albedo neutrons.

Even if the ε parameter specified by Eq. (2) is sufficiently small, an adiabatic invariant J_i is typically conserved only in the absence of "frictional" and absorptive processes, and only by Hamiltonian processes whose forces vary on time scales long compared to the periodicity $2\pi/\Omega_i$ associated with that J_i ($i = 1, 2, 3$). The conservation of K and Φ by "frictional" processes constitutes an exception to this rule because K and Φ are energy-independent. Moreover, the third invariant Φ is conserved by axisymmetric disturbances of the magnetospheric **B** field, even if such disturbances vary on time scales short compared to the drift period $2\pi/\Omega_3$.

Even so, the third invariant is the most easily violated by storm-associated magnetospheric dynamical processes, since the azimuthal drift period is by far the longest of the three characteristic periodicities for radiation-belt particles.

Processes that violate one or more of the adiabatic invariants inherently lead to transport in the sense contemplated by *Schulz and Lanzerotti* [1974, pp. 7–9, 46–48], and particle transport in radiation belts is canonically described [*Haerendel*, 1968] by an equation of the form

$$\frac{\partial \bar{f}}{\partial t} + \sum_i \frac{\partial}{\partial J_i}\left[\left\langle\frac{dJ_i}{dt}\right\rangle_\nu \bar{f}\right] = \sum_{i,j} \frac{\partial}{\partial J_i}\left[D_{ij}\frac{\partial \bar{f}}{\partial J_j}\right] - \frac{\bar{f}}{\tau_q} + \bar{S}, \quad (3)$$

where $\langle dJ_i/dt\rangle_\nu$ is a transport coefficient describing "frictional" processes such as Coulomb drag (the subscript ν connoting friction) and D_{ij} is the diffusion tensor (encompassing pitch-angle diffusion as well as radial diffusion). The lifetime τ_q characterizes any sudden loss process such as charge exchange, and the distributed source $\bar{S}$ represents the drift-averaged contribution to $\partial \bar{f}/\partial t$ from (for example) the beta decay of cosmic-ray-albedo neutrons [*Dragt et al.*, 1966] or of nuclear-fission products (e.g., *Christofilos* [1959]). The angle brackets surrounding dJ_i/dt and the bar above S denote averages over gyration, bounce, and drift, just as $\bar{f}$ denotes the phase-average of f over $\{\varphi_1, \varphi_2, \varphi_3\}$.

The absence of intervening Jacobian factors in Eq. (3) stems precisely from the fact that the $\{J_i\}$ are canonical coordinates (action integrals) in Hamiltonian mechanics. However, it may be convenient [*Haerendel*, 1968] to transform from the $\{J_i\}$ to a set of more easily contemplated coordinates $\{Q_i\}$, in which case the transport terms in Eq. (3) become

$$\sum_i \frac{\partial}{\partial J_i}\left[\left\langle\frac{dJ_i}{dt}\right\rangle_\nu \bar{f}\right] = \frac{1}{\Im}\sum_i \frac{\partial}{\partial Q_i}\left[\Im\left\langle\frac{dQ_i}{dt}\right\rangle_\nu \bar{f}\right] \quad (4)$$

and

$$\sum_{i,j} \frac{\partial}{\partial J_i}\left[D_{ij}\frac{\partial \bar{f}}{\partial J_j}\right] = \frac{1}{\Im}\sum_{i,j} \frac{\partial}{\partial Q_i}\left[\Im\tilde{D}_{ij}\frac{\partial \bar{f}}{\partial Q_j}\right], \quad (5)$$

respectively. The factor $\Im \equiv \det\{\partial J_i/\partial Q_j\}$ in Eqs. (4)–(5) is the Jacobian of the transformation from the canonical variables $\{J_i\}$ specified by Eq. (1) to the "new" variables $\{Q_j\}$, and the tensor $\tilde{D}_{ij}$ is the "transformed" diffusion coefficient appropriate to the "new" variables, which are chosen (if possible) so as to make $\tilde{D}_{ij}$ diagonal (Since the "new" variables thus simplify the transport problem, the "transformed" diffusion coefficient is usually calculated directly, without actually transforming D_{ij} to obtain it.). For "new" variables $\{Q_j\} = \{M, K, \Phi\}$ it is easy to show that $\Im = 4\pi(2m_0^3 M)^{1/2} \propto M^{1/2}$, the constant factors in $\Im$ being unimportant to Eqs. (4)–(5).

Energetic particle fluxes in the magnetosphere show considerable time-variability, often in association with geomagnetic storms. Figure 2 shows an example drawn from CRRES data on relativistic electrons. The usual pattern is that the energetic electron flux in any energy channel decreases sharply as the ring current (inferred from the D_{st} index) develops, and then recovers (often to a level above the pre-storm value) as the ring current decays. The phenomenon thus described is partially adiabatic (reversible) and partially non-adiabatic (quasi-diffusive in this case).

The canonical formulation, based on $\bar{f}(M, K, \Phi; t)$, offers an ideal means of separating adiabatic modulation from diffusive transport, at least within the context of a global model for the magnetospheric **B** field, expressed as a function of **r** (spatial coordinates) and t (universal time). Purely adiabatic variations would leave $\partial \bar{f}/\partial t = 0$ at fixed (M, K, Φ), thereby producing a calculable modulation of the observed $j_\perp$ ($= 2m_0 M B_m \bar{f}$) at any energy. As the energy of an individual particle typically varies in concert (monotonically, in positive correlation) with D_{st} at fixed (M, K, Φ), a steeply decreasing spectrum ($\partial \ln \bar{f}/\partial \ln M \ll -1$) would tend to produce the modulation signature seen in Figure 2 (Account must also be taken of the radial gradient and pitch-angle distribution, as indicated by $(\partial \ln \bar{f}/\partial \ln \Phi)_{M,K}$ and $(\partial \bar{f}/\partial K)_{M,\Phi}$, respectively.). Deviations from $\partial \bar{f}/\partial t = 0$ at fixed (M, K, Φ) must be attributed either to statistics, to observational error, or to genuine transport (conceived by *Schulz and Lanzerotti* [1974, pp. 7–9, 46–48] as entailing violation of one or more adiabatic invariants).

3. MCILWAIN'S L PARAMETER

Thirty years ago it would have been computationally burdensome to determine third adiabatic invariants on a routine basis. For this reason *McIlwain* [1965] introduced a widely popular alternative parameter called L or sometimes (as in the present work) L_m. The subscript "m" can stand for either "McIlwain" or "mirror-point" (just as B_m here and often elsewhere represents the magnetic field strength at a particle's mirror point). *McIlwain* [1965] was mainly concerned with organizing measured fluxes of very energetic inner-zone radiation-belt particles, whose adiabatic motions are not greatly affected by electric fields (either induced or electrostatic). The main source of difficulty was the presence of anomalies (deviations from dipolarity) in the Earth's main field **B**(**r**). Accordingly, *McIlwain* [1965] reasoned that the drift shell of any particle could be identified uniquely by specifying the mirror-point field B_m and the ratio $I \equiv J/2p$, where J ($= J_2$) is the second adiabatic invariant and p (the square root of $2m_0 M B_m$) is the scalar momentum (both p and B_m are constants of the motion if effects of electric fields are neglected.)

The main impediment to using I as a drift shell label under the above assumptions is the difficulty of visualizing it (The use of J presents a similar difficulty, as has been noted above.). Although energy-independent, I still seems to depend (roughly speaking) on both the mirror latitude and the drift-shell radius. *McIlwain* [1965] accordingly proposed to label the drift shell with the parameters B_m and L_m, where L_m would have been the radius (measured in R_E) of a dipolar drift shell corresponding to the same values of B_m and I (The equation of a dipolar field line is $r = La\sin^2\theta$, with $a = 1\,R_E$, where the radial distance r is measured from the point dipole and the colatitude θ is measured from the dipole axis.). *Hilton* [1971] later showed that this L_m can be approximated within 0.01% by using the interpolation formula

$$\frac{B_m L_m^3 R_E^3}{\mu_E} \approx 1 + \frac{3X}{\pi}\sqrt{2} + 0.465380 X^2 + \left(\frac{X}{2.760346}\right)^3, \quad (6)$$

where $X^3 \equiv I^3 B_m/\mu_E$ (the full arc length of a dipolar field line being $2.760346 La$).

McIlwain's L parameter is widely used in space research, even well beyond the limited domain for which it was developed (viz., for particle energies $E \gtrsim 300$ keV at $L \lesssim 3$). The main pitfall in using McIlwain's L so widely is its lack of adiabatic invariance for $\partial \mathbf{B}/\partial t \neq 0$. Indeed, neither B_m nor L_m is adiabatically invariant in this sense, nor is I for that matter. This means that I, B_m, and L_m are not useful for separating adiabatic particle-intensity modulations from diffusive transport. They are also not adiabatically invariant in principle under the geomagnetic secular variation, nor even under the transformation that defines L_m by associating particles of specified I and B_m in the model **B** field with particles of the same I and B_m in a dipolar **B** field (Hypothetical attenuation of the higher harmonics would induce hypothetical electric fields so as to change the particle energies somewhat.). Moreover, the value of L_m (as defined via I and B_m) can vary along a field line even in an axisymmetric (but non-dipolar) field geometry (such that drift shells coincide with magnetic shells). The canonical coordinates (M, K, Φ) excel with respect to all these tests. Such objections to the use of B_m and L_m are commonly dismissed by noting that the errors thus introduced are small compared to observational uncertainties in the region originally of interest [*McIlwain*, 1965]. However, the lack of adiabatic invariance is disadvantageous in principle, whenever and wherever B_m and L_m are used for organizing particle data, even if the resulting errors seem small. Also, there remains the need to organize CRRES and other particle data on the outer radiation belt, for which purpose the "coordinates" B_m and L_m are decidedly unsuitable for the reasons noted. Their supposed advantage (ease of computation, with no need to follow drift shells) is illusory, being a direct consequence of neglecting the third adiabatic invariant.

Quantities such as I, B_m, and L_m are also not adiabatically invariant around drift shells made asymmetric (as for ring-current particles) by the convection electric field. Figure 3 [*Chen et al.*, 1994] illustrates some of the drift-shell configurations that can result from a 50 kV cross-magnetospheric potential drop, modeled as by *Stern* [1973] and *Volland* [1973] (The particles depicted in Figure 3 are nonrelativistic ions with $K = 0$, hence mirroring at the magnetic equator. A first invariant $M = \mu = 10$ MeV/G thus corresponds to a kinetic energy $E \approx 110$ keV at $L = 3$ and to a kinetic energy $E \approx 10$ keV at $L = 6$.). *Chen et al.* [1994, Appendix A] showed how to organize phase-space densities on such trajectories, labeled according to the amount of magnetic flux enclosed.

Of course, the differential flux $j_\perp$ at specified E is not adiabatically invariant either. The quantity actually preserved in an adiabatic transformation is the drift-averaged phase-space density $\bar{f}$ at specified (M, K, Φ). This suggests that the quantity appropriate for organization by canonical coordinates is $\bar{f}(M, K, \Phi; t)$ and not $j_\perp(E, B_m, L_m; t)$. It is even

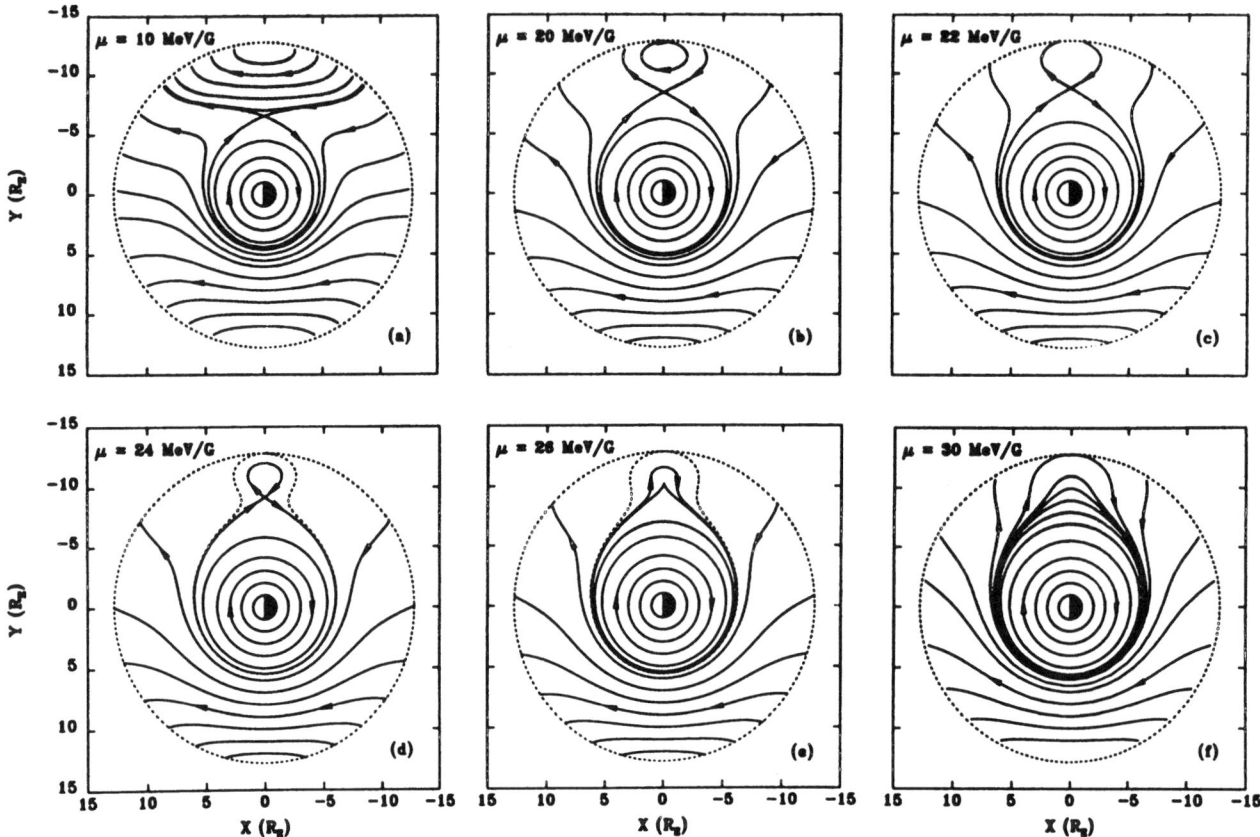

Figure 3. Quiet-time trajectories of equatorially mirroring ions (singly charged), as obtained by *Chen et al.* [1994] for selected values of first adiabatic invariant M (equal to magnetic moment μ in the nonrelativistic approximation). The Sun is at the left here.

less reasonable, of course, to organize omnidirectional fluxes (integrated over the cosine of the local pitch angle) in terms of local B and L_m, with I computed as if the mirror points of all particles comprising the local omnidirectional flux had coincided at the point of interest. McIlwain's prescription for computing L_m requires as inputs the value of B_m at the mirror points of the particle of interest and the value of I ($= J/2p$) obtained by integrating between the mirror points of the particle of interest (Adiabatic drift shells come to depend on K and Φ alone in the limit of large M, such that electric-field effects are negligible. McIlwain's prescription applies only in this limit, whereas the canonical coordinates are valid in general.).

4. EULER POTENTIALS AND STRETCH TRANSFORMATIONS

The third adiabatic invariant Φ, advocated here as a coordinate for purposes of radiation-belt modeling, is usefully related to the Euler potentials α and β commonly invoked in magnetospheric modeling (e.g., *Stern* [1968]). The Euler potentials are (in principle) functions of $\mathbf{r}$ and t, constructed so that $\mathbf{B} = \nabla \alpha \times \nabla \beta$ (It follows that $\hat{\mathbf{B}} \cdot \nabla \alpha = \hat{\mathbf{B}} \cdot \nabla \beta = 0$, which means that α and β constitute field-line labels.). It is usual to regard α as a (magnetic) flux-like coordinate and β as a quasi-azimuthal coordinate. Thus, for example, the field model illustrated in Figure 1 can be regarded as derivable from the Euler potentials

$$\alpha = (\mu_E/r)\sin^2\theta + \frac{1}{2}r^2\sin^2\theta\Delta B_z \qquad (7)$$

and $\beta = \varphi$, where μ_E ($\approx -0.302\,\mathrm{GR_E^3}$) is the geomagnetic dipole moment, ΔB_z ($\approx -14.332\,\mathrm{nT}$) is the strength of the added uniform southward field (corresponding topologically to the geomagnetic tail field), and φ is the magnetic local time (Most investigators would divide the above expression for α by the planetary radius a and specify $\beta = a\varphi$.).

It can be said of the foregoing construction that $\mathbf{B}$ ($= \nabla \times \mathbf{A}$) has been derived from the (magnetic) vector potential $\mathbf{A} = \alpha\nabla\beta$. Since the magnetic flux Φ enclosed by an arbitrary closed path is thus equal (by Stokes' theorem) to the line integral of $\mathbf{A}$, it follows [*Northrop and Teller*, 1960, p.219] that

$$\Phi = \oint_3 \alpha\,d\beta \qquad (8)$$

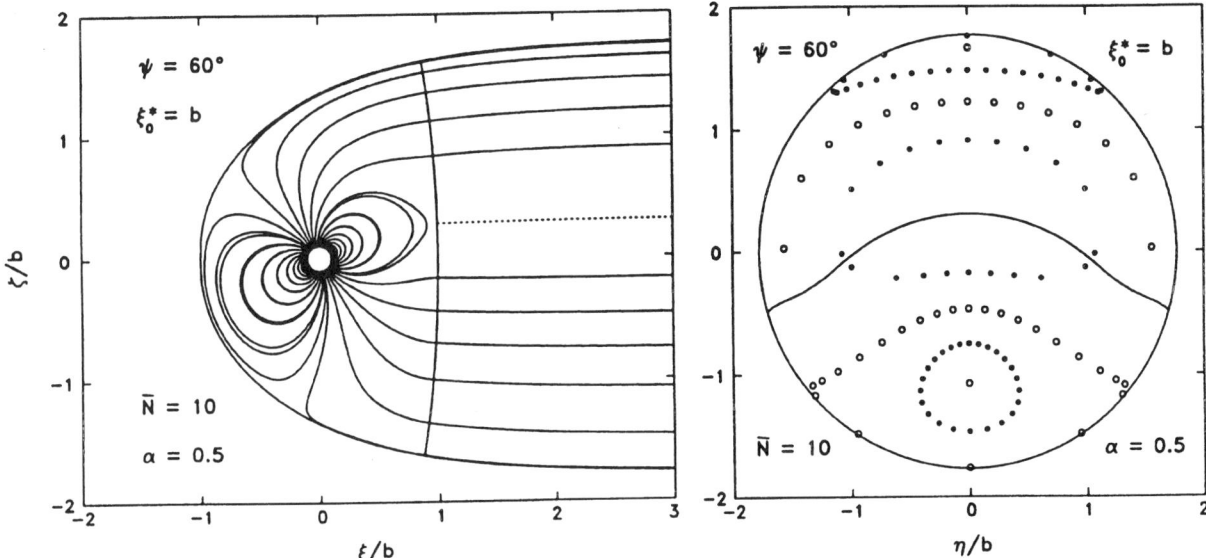

Figure 4. The left-hand panel shows representative field lines in the noon-midnight meridional plane for near-solstice conditions (Ψ is the angle between dipole moment and solar-wind velocity) in the field model of *Schulz and McNab* [1996]. The right-hand panel shows intersections of representative field lines with asymptotic cross section of distant magnetotail. Selected field lines emanate from planetary surface at 5° intervals of magnetic latitude Λ and 15° intervals of magnetic local time.

if α is specified (either analytically or numerically) as a function of β around any path such as those seen in Figure 3. Thus, it follows that $\Phi = 2\pi\alpha$ for an axisymmetric path, such as would be attained at very high M or at very low L values in this model (The qualifier "3" on the above integral sign refers to azimuthal drift, which corresponds to $i = 3$ in the adiabatic hierarchy.).

As *Stern* [1968] has shown, the utility of α and β in this context is not limited to axisymmetric **B**-field geometries. A magnetospheric geometry such as that found in Figure 4 [*Schulz and McNab*, 1996] has been derived in the near-Earth region from a scalar potential expanded in 65 spherical harmonics (up to degree and order $\overline{N} = 10$) and in the tail region from a "nested" geometrical construction of the field lines. However, the resulting field model can be regarded as having been derived from Eq. (7) by "stretching" the various field lines into the desired configuration (cf. *Stern* [1987]). As long as the field model is dipole-dominated in the limit $r \to 0$, the optimal prescriptions for α and β are the limiting values of $(\mu_E/r)\sin^2\theta$ and φ, respectively, as $r \to 0$. This means that formulas such as Eq. (8) remain valid (and useful) in realistic models of the magnetosphere. Complications ensue if account must be taken of higher-order harmonics in the Earth's main magnetic field, but perturbation methods for treating such effects have been described by *Pennington* [1961].

5. SUMMARY

The main objective of this work has been to promote use of adiabatically invariant quantities (M, K, Φ) as optimal co-ordinates for organizing the growing body of *in situ* data on drift-averaged phase-space densities $(\bar{f})$ of radiation-belt particles, and thus for providing statistically valid empirical models of the Earth's radiation environment. This was essentially the program advocated by *Roederer* [1970, pp. 106–109]. It remains valid as a geophysical ideal. Meanwhile, the necessary modeling and computational capabilities have vastly improved, to the extent that implementation of Roederer's program has become practical from an operational standpoint.

Even so, any program for modeling the Earth's radiation environment must rely for its success on a quantitatively realistic model for the magnetospheric **B** field, especially under disturbed conditions. The development of such a data-based field model is the continuing project of *Tsyganenko* [1987,1989,1995] and colleagues. While some uncertainties still remain, their project has led to continuous improvement in field modeling over the past 15 years and offers the best prospect of future progress in this essential research area.

Acknowledgements. The author thanks Joseph Lemaire for the kind invitation to present these ideas at the Workshop on Radiation Belt Models and Standards, held 17–20 October 1995 in Brussels. He thanks the (Belgian) Fonds National de la Recherche Scientifique for partial coverage of travel expenses. The work itself was supported largely by the Independent Research and Development (IR&D) program of the Lockheed Martin Missiles & Space Company.

REFERENCES

Anderson, K.A., and D.W. Milton, Balloon observations of X rays in the auroral zone, 3, High time resolution studies, *J. Geophys.*

Res., 69, 4457–4479, 1964.
Brewer, H.R., M. Schulz, and A. Eviatar, Origin of drift-periodic echoes in outer-zone electron flux, *J. Geophys. Res., 74*, 159–167, 1969.
Chanteur, G., R. Gendrin, and S. Perraut, High energy electron drift echoes at the geostationary orbit, *J. Atmos. Terr. Phys., 40*, 367–371, 1978.
Chen, M.W., L.R. Lyons, and M. Schulz, Simulations of phase space distributions of storm time proton ring current, *J. Geophys. Res., 99*, 5745–5759, 1994.
Christophilos, N.C., The Argus experiment, *J. Geophys. Res., 64*, 869–875, 1959.
Dragt, A.J., M.M. Austin, and R.S. White, Cosmic ray and solar proton albedo neutron decay injection, *J. Geophys. Res., 71*, 1293–1304, 1966.
Dungey, J.W., Interplanetary magnetic field and the auroral zones, *Phys. Rev. Lett., 6*, 47–48, 1961.
Farley, T.A., M.G. Kivelson, and M. Walt, Effects of the secular magnetic variation on the distribution function of inner-zone protons, *J.Geophys.Res.,77*, 6087–6092, 1972.
Haerendel, G., Diffusion theory of trapped particles and the observed proton distribution, in *Earth's Particles and Fields*, edited by B. M. McCormac, pp. 171–191 (especially pp. 172–176, based on work done in collaboration with L. Davis, Jr.), Reinhold, New York and London, 1968.
Heckman, H.H., and P.J. Lindstrom, Response of trapped particles to a collapsing dipole moment, *J. Geophys. Res., 77*, 740–743, 1972.
Hill, T.W., and M.E. Rassbach, Interplanetary magnetic field direction and the configuration of the day side magnetosphere, *J. Geophys. Res., 80*, 1–6, 1975.
Hilton, H.H., L parameter, a new approximation, *J. Geophys. Res., 76*, 6952–6954, 1971.
Il'in, V.D., I.V. Il'in, and S.N. Kuznetsov, Stochastic instability of charged particles in a magnetic trap, *Kosm. Issled., 24(1)*, 88–96, 1986 (in Russian); English translation: *Cosmic Res., 24*, 69–76, 1986.
Il'in, V.D., I.V. Il'in, and S.N. Kuznetsov, Accumulation of oxygen ions in a geomagnetic trap, *Kosm. Issled., 31(6)*, 115–117, 1993 (in Russian); English translation: *Cosmic Res., 31*, 687–689, 1993.
Kaufmann, R.L., Conservation of the first and second adiabatic invariants, *J. Geophys. Res., 70*, 2181–2186, 1965.
Lampton, M., Daytime observations of energetic auroral-zone electrons, *J. Geophys. Res., 72*, 5817–5823, 1967.
Lanzerotti, L.J., C.S. Roberts, and W.L. Brown, Temporal variations in the electron flux at synchronous altitudes, *J. Geophys. Res., 72*, 5893–5902, 1967.
McIlwain, C.E., Magnetic coordinates, *Space Sci. Rev., 5*, 585–598, 1966; also in *Radiation Trapped in the Earth's Magnetic Field*, edited by B.M. McCormac, pp. 45–61, Reidel, Dordrecht, 1966.
Northrop, T.G., *The Adiabatic Motion of Charged Particles*, Interscience, New York, 1963.
Northrop, T.G., and E. Teller, Stability of the adiabatic motion of charged particles in the Earth's field, *Phys.rev.,117*, 215–225, 1960.
Pennington, R.H., Equation of a charged particle shell in a perturbed dipole field, *J. Geophys. Res., 66*, 709–712, 1961.
Roederer, J.G., *Dynamics of Geomagnetically Trapped Radiation*, Springer, Heidelberg, 1970.
Schulz, M., The magnetosphere, in *Geomagnetism*, vol. 4, edited by J.A. Jacobs, pp. 87–293, Academic Press, London, 1991.
Schulz, M., and M.W. Chen, Bounce-averaged Hamiltonian for charged particles in an axisymmetric but non-dipolar model magnetosphere, *J. Geophys. Res., 100*, 5627–5653, 1995.
Schulz, M., and L.J. Lanzerotti, *Particle Diffusion in the Radiation Belts*, Springer, Heidelberg, 1974.
Schulz, M., and M.C. McNab, Source-surface modeling of planetary magnetospheres, *J. Geophys. Res., 101*, 5095–5118, 1996.
Schulz, M., and G.A. Paulikas, Secular magnetic variation and the inner proton belt, *J. Geophys. Res., 77*, 744–747, 1972.
Stern, D., Euler potentials and geomagnetic drift shells, *J. Geophys. Res., 73*, 4373–4378, 1968.
Stern, D.P., A study of the electric field in an open magnetospheric model, *J. Geophys. Res., 78*, 7292–7305, 1973.
Stern, D.P., Tail modeling in a stretched magnetosphere, 1, Methods and transformations, *J. Geophys. Res., 92*, 4437–4448, 1987.
Tsyganenko, N.A., Global quantitative models of the geomagnetic field in the cislunar magnetosphere for different disturbance levels, *Planet. Space Sci., 35*, 1347–1358, 1987.
Tsyganenko, N.A., A magnetospheric magnetic field model with a warped tail current sheet, *Planet. Space Sci., 37*, 5–20, 1989.
Tsyganenko, N.A., Modeling the Earth's magnetospheric magnetic field confined within a realistic magnetopause, *J.Geophys.Res., 100*, 5599–5612, 1995.
Volland, H., A semiempirical model of large-scale magnetospheric electric fields, *J. Geophys. Res., 78*, 171–180, 1973.

M. Schulz, Space Sciences Laboratory, Advanced Technology Center, Lockheed Martin Missiles & Space, 3251 Hanover Street, Palo Alto, California 94304, U.S.A. (e-mail: schulz@agena.space.lockheed.com or LOCKHD::SCHULZ)

DISCUSSION

Q: J. Albert. Could you discuss the description of pitch angle scattering in terms of the canonical coordinate formulation?
A: M. Schulz. A transformation of variables should be made for pitch angle scattering; in the new variables, a 2×2 diffusion tensor is needed to describe the physics of the scattering.
Q: R.S. Selesnick. With modern computers we can calculate Φ fairly easily, so should one give up using the McIlwain L parameter?
A: M. Schulz. Yes, I think so. And we should try to recast previously acquired data in terms of canonical coordinates also.

Magnetic Field Models in the inner Magnetosphere

T.I. Pulkkinen

Finnish Meteorological Institute, Helsinki, Finland

The connectivity between various plasma regions in the coupled magnetosphere-ionosphere system is discussed in light of the new empirical magnetic field models for the terrestrial magnetosphere that have recently become available. Variations of the magnetic field configuration from the dipolar shape and their effects on the motion of charged particles are shortly reviewed. It is demonstrated that during magnetically active periods, the large and rapid magnetic field changes can extend to quite close to the Earth.

1. INTRODUCTION

The magnetospheric magnetic field has two different sources: The quasi-dipolar internal field dominates in the near-Earth region, whereas further out the contribution from the extra-terrestrial current systems becomes dominant. The transition from dipole-dominated to current system dominated fields takes place typically in the region between the geostationary orbit and $\sim 10\,R_E$. However, the location of the transition region is strongly dependent on the intensity of the external currents, which enhance and decay as a response to the driving solar wind. Furthermore, the external currents can influence the innermost magnetosphere during periods when the currents are strong.

The internal field can be represented as the gradient of a scalar potential ($\mathbf{B} = -\nabla V$), and can thus be represented as an expansion in spherical harmonics. When the coefficients of the expansion are determined from observations, the resulting IGRF models [*Langel*, 1991] show that at the Earth's surface, the higher-order multipole terms that decay rapidly as a function of altitude can contribute up to 20% of the field intensity. On the other hand, the external field is created by the magnetospheric currents, and can be expressed by a vector potential ($\mathbf{B} = \nabla \times \mathbf{A}$). Because the currents are created by the interaction between the internal field and the solar wind and the interplanetary magnetic field (IMF), these currents flow fixed (although time varying) in magnetic local time.

Thus, there are two different natural coordinate systems in the magnetosphere: At very low altitudes, where the higher order terms in the IGRF model are significant and where the external currents produce negligible variations in the field intensity, a geographic coordinate system rotating together with the Earth is appropriate. However, when the external current contribution increases and the higher multipoles become small, a coordinate system fixed by the Sun-Earth line and the dipole axis is appropriate for describing the symmetries in the field. In the transition region, the field intensity is a complex function of both geographic position and local time. In this paper we concentrate on effects of the external current systems, and consider only regions where the higher multipole contribution can be assumed to be small.

The geostationary altitude (6.6 R_E) is an interesting region where the relative contributions from the internal and external field sources is highly variable (e.g., *Hones et al.*, [1996]). Being close to the outer edge of the ring current and the inner edge of the tail plasma/current sheet satellites at geostationary orbit probe temporal variations in both of these regions. In the dayside, variability in the magnetopause currents, which are influenced by the IMF direction and solar wind pressure, can also be measured at geostationary orbit.

The shape of the magnetospheric magnetic field has received considerable attention recently, when both new analytical, data-based models (see reviews by *Tsyganenko* [1990] and *Jordan* [1994]) and global MHD simulations [*Walker et al.*, 1993; *Raeder*, 1994; *Fedder et al.*, 1995] have become available. The MHD models describe the field evolution through the interaction process between a dipole and the solar wind and IMF. The empirical models define the field by modeling the current sources, whose intensity and location are then determined from observations.

The answer for "what is the best field model" depends on the purpose for the model use. Analytical magnetic field models are often used in organizing and analyzing data. Automated analysis of large amounts of data calls for simple field

Radiation Belts: Models and Standards
Geophysical Monograph 97
Copyright 1996 by the American Geophysical Union

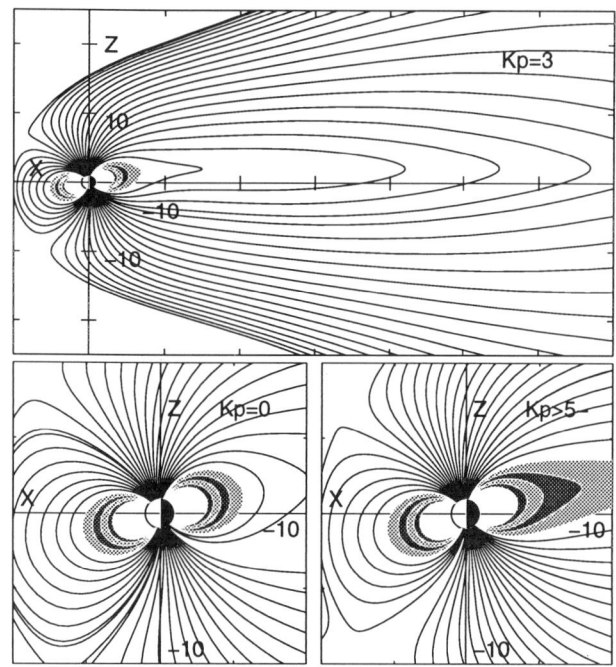

Figure 1. *Tsyganenko* [1989] models: Noon-midnight meridian view for $K_p = 0$, 3, and $\geq 5^-$. The shading highlights flux tubes between 56° and 64°.

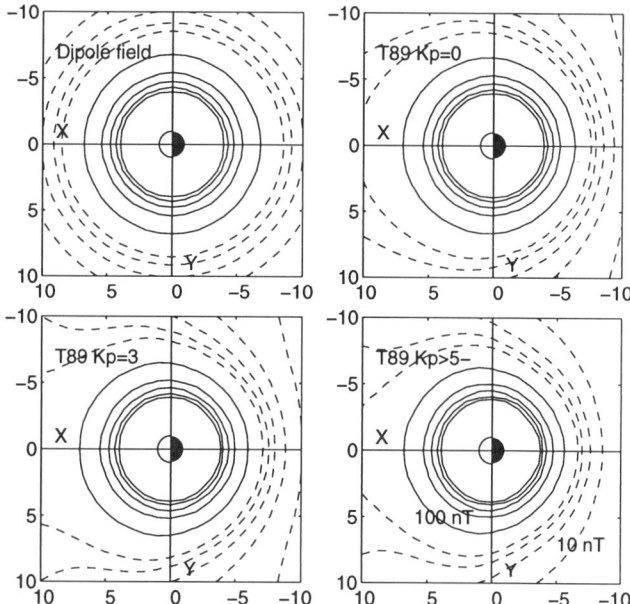

Figure 2. Contours of constant equatorial magnetic field for the dipole field (top left) and *Tsyganenko* [1989] models for $K_p = 0$, 3, and $\geq 5^-$. Contours for 10–50 nT at 10 nT intervals are shown with dashed lines and contours for 100–500 nT are shown with the solid lines.

models, which depend only on few easily available parameters. Similar requirements can be set for models that are used for prediction of configurations and conjunctions in multi-instrument operations, because the detailed magnetospheric state cannot be predicted in advance (see, e.g., *Lockwood and Opgenoorth* [1995]). In such cases, models like the *Tsyganenko* [1987, 1989] model are at their best. Issues related to space weather forecasting typcially involve large variations from the average configuration, as most of the operational problems are caused by storm-type variations. The *Hilmer-Voigt* [1995] model includes parameters to describe the magnetopause location and current wedge currents, which allows one to study such more disturbed periods. However, for detailed event analysis, even more complex models are required: The current sheet thicknesses and current intensities vary a great deal during magnetically active periods, and these changes have important effects on the particle motion within the inner magnetosphere. Field models tailored to describe such variations and to provide a best fit to available magnetic field measurements for each event separately have been developed by *Pulkkinen et al.* [1992, 1994].

This paper addresses the inner magnetosphere effects of the external current systems under different activity conditions. Various mappings are presented, which compare the results of the dipole field, the *Tsyganenko* [1989] models, and substorm-time fields [*Pulkkinen et al.*, 1992, 1994].

2. STATISTICAL, DATA-BASED FIELD MODELS

The *Tsyganenko* [1987, 1989] (T87, T89) models describe the external field in terms of four current systems, the ring current, the cross-tail current, the Chapman-Ferraro current, and the nightside magnetopause current. The ring current and the cross-tail current flow close to the equatorial plane, whereas the other two systems produce currents close to the nominal magnetopause. The field values are functions of two external parameters: the dipole tilt angle defining the orientation of the dipole with respect to the Sun-Earth line and the K_p index giving the average level of magnetospheric activity.

Figure 1 shows the T89 magnetic field lines in the noon-midnight meridian view for different activity levels. In particular, the nightside field undergoes substantial changes when the magnetic activity level is increased. The field lines are stretched much further out in the tail, and the plasma sheet is compressed also well inside $10 R_E$. The effects of the external current systems are even more clearly visible in Figure 2, which shows contours of constant magnetic field intensity in the equatorial plane for different field models. Note how the day-night asymmetry becomes more pronounced as the magnetic activity increases.

When the external currents increase during magnetically active periods, the equatorial field in the tail becomes small even in the near-geostationary region. The transition region over which the change from dipolar to tail-like field lines becomes much narrower, resulting in closely spaced constant-B contours and large magnetic field gradients in the bottom right panel of Figure 2.

Changes in the magnetic field also change the particle drift paths. Whereas particles with all pitch-angles drift along L

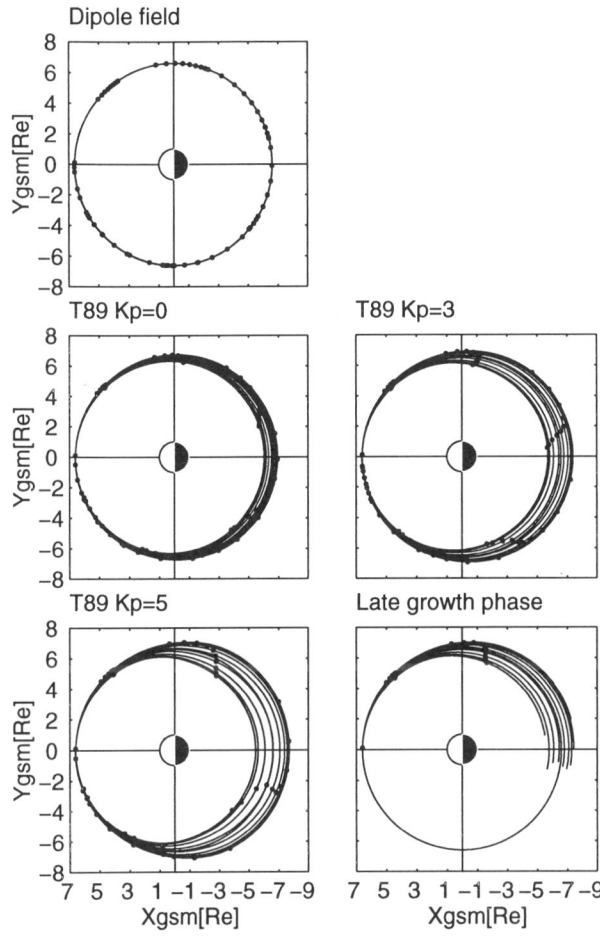

Figure 3. Effects of the field models to energetic ion drift paths. Drift trajectories of protons with pitch angles 5–90° are shown in the dipole field, in the *Tsyganenko* [1989] models, and in a model for the late growth phase.

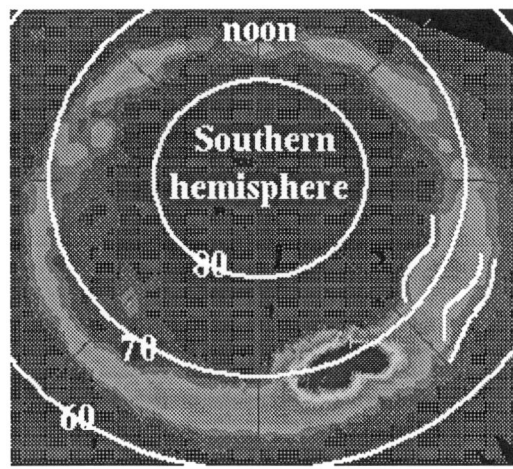

Figure 4. DE-1 oval image on May 8, 1986, at 12.10 UT (CDAW-9 Event E).

shells in the dipole field, already the weak external currents of the T89 ($K_p = 0$) model are sufficient to produce significant displacements in the trajectories of different pitch-angle particles. Figure 3 illustrates the drift paths of 100 keV protons launched at the geostationary orbit at local noon. The external current systems cause the small pitch-angle particles to drift further out than the particles that have pitch-angles closer to 90°. The bottom left panel shows similar calculations in a time-evolving magnetic field during the last half hour of the substorm growth phase. Note that because the field stretching increases during the growth phase, the effects actually seem smaller than those computed using the (time-invariant) maximally stretched Tsyganenko model ($K_p > 5^-$) field.

One of the key questions for magnetospheric dynamics is to resolve where in the magnetosphere the particles precipitating in the auroral regions come from. Figure 4 shows a UV image of the auroral oval at the time of a substorm onset as recorded by the Dynamics Explorer 1 satellite above the southern hemisphere. To study the mappings of the precipitation, three boundaries were identified based on constant intensity levels in the image. These boundaries are illustrated by the white lines in part of the oval, similar constant-intensity boundaries were traced around the oval. The top panel of Figure 5 shows the equatorial mapping of this pattern using the dipole field. The dayside oval is at higher latitudes, and thus maps further away from the Earth than the nightside oval. The bottom panels show the mapping using two versions of the Tsyganenko model. It is evident that much of the dayside precipitation comes from regions close to the boundary. Furthermore, as the field lines are bent tailwards, there is a large difference in the longitudes of the field line footpoint in the ionosphere and its crossing point in the equatorial plane. In the nightside, the ring current shifts the nightside mapping further outward than in the pure dipole model.

3. EVENT-ORIENTED MODELS

The *Tsyganenko* [1989] models are constructed by fitting analytical functions with several free parameters to a large number of observations, which are binned according to the average level of magnetospheric activity given by the K_p index. Since K_p is a 3-hour index, these fits include observations from the full range of magnetospheric states that can occur during a single 3-hour period. Especially, the fittings include observations from different substorm phases. Thus, it is clear that any specific substorm phase is not well represented by these average models. The observations show that during the substorm growth phase, the magnetotail is in very stretched configuration, and that strong currents flow within a thin current sheet in the relatively near-Earth region [*Kaufmann*, 1987; *Pulkkinen et al.*, 1992]. On the other hand, the recovery phase is characterized by strongly dipolarized field and expanded plasma sheet. During the substorm expansion phase the current system is very complex, and a large portion of the field disturbances is caused by field-aligned currents.

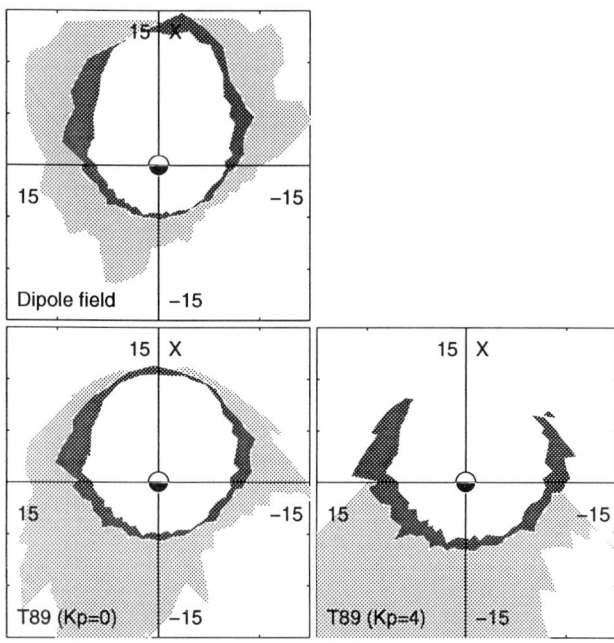

Figure 5. Mapping of the auroral distribution shown in Figure 4 using the dipole field (top) and the *Tsyganenko* [1989] models (bottom).

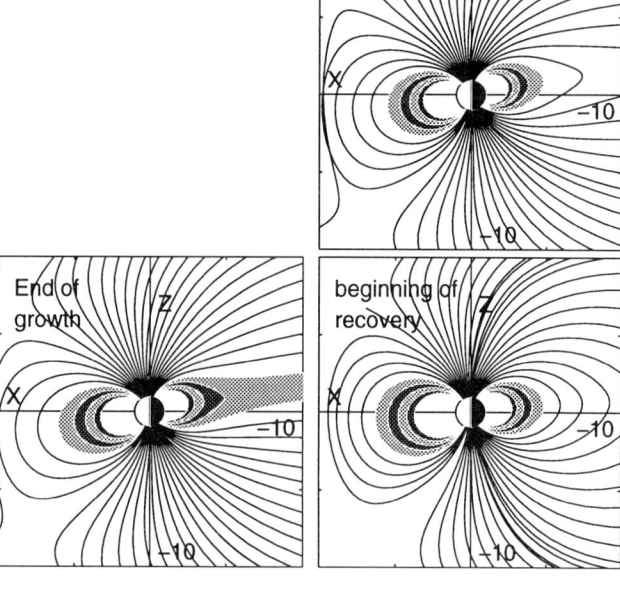

Figure 6. Event-oriented models [*Pulkkinen et al.*, 1995]: Substorm growth and recovery phases. The shading highlights flux tubes between 56° and 64°.

Pulkkinen et al. [1992, 1994] have modified the Tsyganenko models to include variations during the substorm growth and recovery phases. For the growth phase model, the current sheet was thinned, the currents were intensified, and new current systems were included to represent the strong enhancement of the near-Earth currents (for details of the model see the original publications). The recovery phase was modeled by weak current systems and expanded plasma sheet thickness over large portion of the inner magnetosphere. These changes were quantified by several free parameters. The parameter values were set by a least-squares fitting technique using in-situ magnetic field observations from preferably several spacecraft in the inner magnetotail for each event separately. Time dependency was built into the models by assuming that the configuration during the growth phase evolved linearly from the original *Tsyganenko* [1989] model to the maximally stretched model at substorm onset. Correspondingly, during the recovery phase the configuration evolved linearly from the maximally dipolarized model after the expansion phase to the original Tsyganenko model at the end of the recovery phase.

Figure 6 shows magnetic field lines during three steps of a well-observed substorm (see *Pulkkinen et al.* [1995], *Baker et al.* [1993]). The top configuration shows the midnight meridian field lines at the beginning of the growth phase (represented by the T89 model for $K_p \geq 5^-$), bottom left represents the situation at the end of the growth phase (maximally stretched configuration), and bottom right at the beginning of the recovery phase (maximally dipolarized configuration). Note that the substorm-associated variations affect field lines crossing the equatorial plane well inside the geostationary orbit.

Figure 7 shows the constant-B contours for these three cases and the dipole field. Especially at the end of the growth phase, the strong gradients near and even inside 5 R_E have strong effects on the particle drifts. Note also that during the recovery phase, the field is dipolarized such that the constant-B contours are shifted further tailward than in the pure dipole model.

4. DISCUSSION

With few examples, this paper illustrates the strong time variability of the magnetospheric field configuration. Situations discussed here cover a variety of average configurations for different activity levels (the *Tsyganenko* [1989] models) as well as substorm growth and recovery phases (the modified T89 models [*Pulkkinen et al.*, 1992; 1994]). However, during large geomagnetic storms, the inner magnetosphere is considerably disturbed by the inward shift of the dayside magnetopause and the strongly enhancing ring current in addition to the simultaneous substorm activity. Such periods will introduce effects stronger and closer to the Earth than illustrated here.

The magnetic field models are important in organizing data, as the magnetic field in some form (L value, magnetic latitude, etc.) is often used as a coordinate in sorting the data. In many cases, the actual and model field configurations are close enough for sufficiently accurate mapping. However, in cases when the measurements are made close to boundaries, even small errors can lead to physically incorrect interpretations.

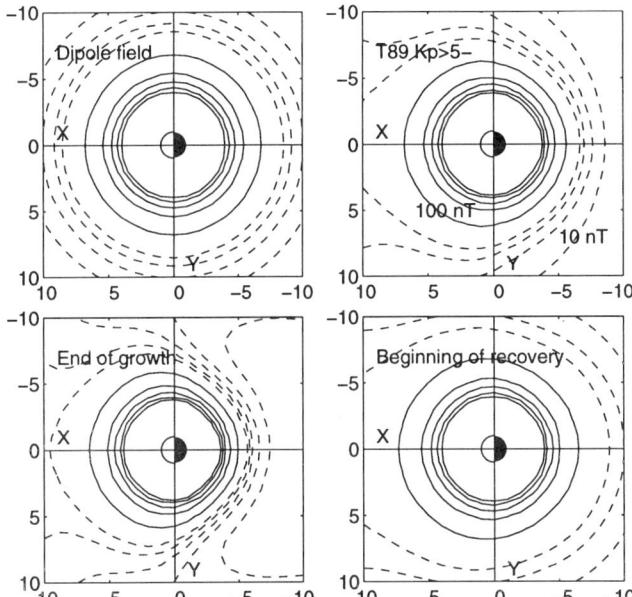

Figure 7. Dipole field and event-oriented models for substorm growth and recovery phases [*Pulkkinen et al.*, 1995]. Contours of constant magnetic field intensity, contours for 10–50 nT are shown with dashed lines and contours for 100–500 nT are shown with solid lines.

Furthermore, as illustrated here, during disturbed times the mappings can vary significantly from the average predicted by the statistical models (see also *Fairfield* [1991]; *Peredo and Stern* [1991]; *Peredo et al.* [1993]). Thus, effort should always be devoted to evaluating the obtained mapping results.

The time-evolving magnetic field models developed by *Pulkkinen et al.* [1992] do not include description of the expansion phase. *Hilmer and Voigt* [1995] have developed a magnetic field model, which includes several magnetospheric activity parameters as input parameters. The field configuration is defined as a function of the dipole tilt angle, the magnetopause standoff distance, the midnight equatorward boundary of the diffuse auroras, and the D_{st} index. Furthermore, the model is capable of including the effects of the current sheet disruption in the midnight sector during the substorm expansive phase. However, the contribution from the field-aligned currents are not included. These models can be used to estimate the mappings also during periods of strong activity and during the substorm expansion phase away from the regions where the field-aligned currents are assumed to be strongest.

A new version of the Tsyganenko models [*Tsyganenko*, 1995] is under development. This model includes several important parameters such as the solar wind dynamic pressure, D_{st}, AE, and IMF B_Y and B_Z components as input parameters. The new model also includes an explicit magnetopause, Region 1, and Region 2 field-aligned currents. This model will provide more flexibility into the mapping under various dynamic situations for events where the input parameters are all available and where the data set used for model specification is sufficiently representative.

The ISTP era with several spacecraft in the high-altitude magnetosphere (e.g., Geotail, Interball, Polar, Cluster) complemented by several missions at low altitudes and an extensive ground-based program will provide an unprecedented opportunity for magnetospheric science. However, much of our ability to analyze this vast amount of data depends on our ability to relate the observations from vastly different regions to each other. The empirical magnetic field models provide an excellent tool for such mappings, and will be of key importance in the these analysis efforts.

Acknowledgements. TP thanks the staff of NSSDC for organizing the CDAW 9 effort and R.L. McPherron for leading the analysis of Event E. The DE-1 image was obtained from the CDAW-9 CDROM, L.A. Frank is acknowledged for providing the imager data to the CDAW-9 database. Petri Toivanen kindly provided the particle drift computations for Figure 3.

REFERENCES

Baker, D.N., T.I. Pulkkinen, R.L. McPherron, J.D. Craven, L.A. Frank, R.D. Elphinstone, J.S. Murphree, J.F. Fennell, R.E. Lopez, T. Nagai, and G. Rostoker, CDAW-9 analysis of magnetospheric events on 3 May 1986: Event C, *J. Geophys. Res.*, 98, 3815, 1993.

Fairfield, D.H., An evaluation of the Tsyganenko magnetic field model, *J. Geophys. Res.*, 96, 1481, 1991.

Fedder, J.A., and J.G. Lyon, The Earth's magnetosphere is 165 R_E long: or self-consistent currents, convection, magnetospheric structure and processes for northward IMF, *J. Geophys. Res.*, 100, 3623, 1995.

Hilmer, R.V., and G.-H. Voigt, A magnetospheric magnetic field model with flexible internal current systems driven by independent physical parameters, *J. Geophys. Res.*, 100, 5613, 1995.

Hones, E.W., M.F. Thomsen, G.D. Reeves, L.A. Weiss, D.J. McComas, and P.T. Newell, Observational determination of magnetic connectivity of the geosynchronous region of the magnetosphere to the auroral oval, *J. Geophys. Res.*, 101, 2629, 1996.

Jordan, C.E., Empirical models of the magnetospheric magnetic field, *Rev. Geophys.*, 32, 139–157, 1994.

Kaufmann, R.L., Substorm currents: growth phase and onset, *J. Geophys. Res.*, 92, 7471, 1987.

Langel, R.A., International Geomagnetic Reference Field, 1991 Revision, *J. Geomag. Geoelectr.*, 43, 1007, 1991.

Lockwood, M., and H.J. Opgenoorth, Opportunities for magnetospheric research using EISCAT/ESR and CLUSTER, *J. Geoelect. Geomag.*, submitted, 1995.

Peredo, M.D., and D.P. Stern, On the position of the near-Earth neutral sheet: A comparison of magnetic model predictions with empiric formulas, *J. Geophys. Res.*, 96, 19,521, 1991.

Peredo, M.D., D.P. Stern, and N.A. Tsyganenko, Are existing magnetospheric models excessively stretched?, *J. Geophys. Res.*, 98, 15,343, 1993.

Pulkkinen, T.I., D.N. Baker, R.J. Pellinen, J. Büchner, H.E.J. Koskinen, R.E. Lopez, R.L. Dyson, and L.A. Frank, Particle scattering and current sheet stability in the geomagnetic tail during the substorm growth phase, *J. Geophys. Res.*, 97, 19,283, 1992.

Pulkkinen, T.I., D.N. Baker, P.K. Toivanen, R.J. Pellinen, R.H.W. Friedel, and A. Korth, Magnetospheric field and current distributions during the substorm recovery phase, *J. Geophys. Res.*, 99, 10955, 1994.

Pulkkinen, T.I., D.N. Baker, P.K. Toivanen, J.S. Murphree, and L.A. Frank, Mapping of the auroral oval and individual arcs during substorms, *J. Geophys. Res., 100*, 21987, 1995.

Raeder, J., Global MHD simulations of the dynamics of the magnetosphere: Weak and strong solar wind forcing, in *Substorms II*, p. 561, Geophysical Institute, University of Alaska Fairbanks, Fairbanks, Alaska, 1994.

Tsyganenko, N.A., Global quantitative models of the geomagnetic field in the cislunar magnetosphere for different disturbance levels, *Planet. Space Sci., 30*, 1007, 1987.

Tsyganenko, N.A., Magnetospheric magnetic field model with a warped tail current sheet, *Planet. Space Sci., 37*, 5, 1989.

Tsyganenko, N.A., Quantitative models of the magnetospheric magnetic field: Methods and results, *Space Sci. Rev., 54*, 75, 1990.

Tsyganenko, N.A., Modeling the Earth's magnetospheric magnetic field confined within a realistic magnetopause, *J. Geophys. Res., 100*, 5599, 1995.

Walker, R.J., T. Ogino, J. Raeder, and M. Ashour-Abdalla, A global magnetohydrodynamic simulation of the magnetosphere when the interplanetary magnetic field is southward: The onset of magnetotail reconnection, *J. Geophys. Res., 98*, 17235, 1993.

T.I. Pulkkinen, Finnish Meteorological Institute, P.O. Box 503, FIN 00101 Helsinki, Finland (e-mail: tuija.pulkkinen@fmi.fi)

DISCUSSION

Q: X. Li. How was the error determined in the two colour-code plots for the X-component and the Y-component from the Tsyganenko model?

A: T.I. Pulkkinen. A "data-based" mapping was constructed using data from the Fairfield et al. [1994] database. The errors given represent the average statistical, systematic errors that were present when comparing mappings using the T89 model and mappings using the database only.

Q: D.N. Baker. Users may prefer rather simple models and tools, whereas your presentation emphasized rather sophisticated scientific issues. How does our community reconcile the needs of users with the complex scientific issues that we know underlie the magnetic configurations in the outer magnetosphere?

A: T.I. Pulkkinen. I think the problem is twofold. We scientists should put our best understanding of the physics and best available models into the system. Then, from there, we should develop an interface that gives unique and relatively simple answers to the questions posed by the model users.

A Quantitative Test of Different Magnetic Field Models Using Conjunctions Between DMSP and Geosynchronous Orbit

G.D. Reeves, L.A. Weiss, M.F. Thomsen and D.J. McComas

Los Alamos National Laboratory, Mail Stop D 436, Los Alamos, New Mexico

We report here on a study which tests the magnetic field line mapping between geosynchronous orbit and the ionosphere. The mapping is determined both observationally and from five magnetospheric magnetic field models. The mapping is tested observationally by comparing electron energy spectra obtained by the Magnetospheric Plasma Analyzer (MPA) at geosynchronous orbit and by the DMSP spacecraft. Because the orbits are nearly perpendicular, in general, the spectra match well for only a few seconds providing a good determination of when DMSP crosses the geosynchronous drift shell. In this way the mapping between geosynchronous orbit and the ionosphere can be determined to better than one degree. We then compare the measured magnetic footpoints of geosynchronous orbit with the footpoints predicted by five magnetospheric field models: Tsyganenko-89, Tsyganenko-87, Tsyganenko-82, Olsen-Pfitzer, and Hilmer-Voigt. Based on a set of over 100 measured magnetic conjunctions we find that, in general, there are significant differences between the mappings predicted by various magnetic field models but that there is no clear "winner" in predicting the observed mapping. We find that the range of magnetic latitudes at which we measure conjunctions is much broader than the range of latitudes which the models can accommodate. This lack of range is common to all magnetic field models tested. Although there are certainly cases where the models are not sufficiently stretched, we find that on average all magnetic field models tested are too stretched. This technique provides an excellent opportunity for testing future magnetic field models and for determining the appropriate parameterizations for those models.

1. INTRODUCTION

A crucial aspect of the modeling of the Earth's radiation belts is modeling the Earth's magnetic field. This study uses an observational criteria for determining the mapping between three low-altitude DMSP spacecraft and high-altitude geosynchronous spacecraft. It also compares the results of the measured mapping with the mapping predicted by several commonly used magnetic field models.

The current generation of empirical magnetic field models (including those of Tsyganenko) are statistical fits to single point measurements of the magnetic field measured by spacecraft in the magnetosphere. The models have commonly been tested by comparing model magnetic field vectors to the magnetic field measured by one or more spacecraft in the magnetosphere (e.g. *Tsyganenko* [1989]; *Fairfield* [1991]; *Peredo and Stern* [1991]; *Peredo et al.* [1993]; *Pulkkinen et al.* [1994]; *Thomsen et al.* [1996]). However, one of the most common uses for magnetic field models is to trace the magnetic field lines to determine the magnetic connectivity between two points in space or between a spacecraft and the ground and the *mapping* predicted by these empirical models has not been systematically tested. Tests of the magnetic field mapping are, in some ways, more sensitive because they

168 MAGNETIC FIELD MAPPING

integrate along the entire field line.

We apply this test of the field models to geosynchronous orbit for several reasons. Firstly, geosynchronous orbit is an inherently interesting region of space. It lies in the transition region between dipole-like and tail-like magnetic field lines and near the edge of the trapping region for energetic particles. It is sensitive to the effects of substorms and is a source region for the injection of particles into the ring current and radiation belts. Geosynchronous orbit is also heavily populated by satellites. The effects of the radiation environment on the operation of those satellites is a fundamentally important application of magnetospheric physics. Finally, there is a large and continuous database of plasma and energetic particle measurements from a series of Los Alamos instruments on geosynchronous satellites which make a large, systematic, and statistical study possible.

2. FINDING MAGNETIC CONJUNCTIONS

The technique we use to establish magnetic conjugacy between the low-altitude DMSP spacecraft and the geosynchronous satellites is to compare electron energy spectra on the two satellites as a function of time and to look for times when the spectra are very nearly identical. DMSP orbits at an altitude of approximately 850 km in a nearly polar, circular orbit with a period of approximately 90 minutes. Therefore DMSP crosses the geosynchronous L-shell approximately once every 23 minutes. The DMSP orbits are also sun-synchronous and therefore each DMSP satellites samples a nearly fixed region of local time.

The geosynchronous satellites orbit at a geocentric distance of 6.6 R_E with a period of 24 hours and therefore must pass through the local times sampled by the DMSP satellites. In this sense the orbits are perpendicular to each other and for each geosynchronous-DMSP satellite pair there are numerous possible conjunctions each day. To further increase the statistics we use data from two geosynchronous satellites (1989-046 and 1990-095) and from three DMSP satellites (DMSP F8, F9, and F10).

We define a "nominal conjunction" as a time when one DMSP and one geosynchronous satellite are within $\pm 10°$ in magnetic longitude and when DMSP is between 50° and 80° magnetic latitude. During a nominal conjunction we examine the electron energy spectra from the SSJ/4 instrument on DMSP which measures precipitating electrons in 20 energy bins from 30 eV to 30 keV [*Hardy et al.*, 1984]. One complete energy spectrum is obtained every second which provides excellent resolution in latitude. At the same time we examine electron energy spectra from the Magnetospheric Plasma Analyzer (MPA) instrument at geosynchronous orbit. The MPA is a spherical-sector electrostatic analyzer which measures electrons from about 1 eV to 40 keV [*Bame et al.*, 1993]. The MPA measures in a fan of 6 look directions and takes 24 azimuthal sectors in each 10-second spin of the spacecraft. The spacecraft spin axis points toward the center of the earth so excellent pitch angle coverage is obtained.

In comparing spectra it is important that several conditions be met. First, since DMSP measures only that portion of the distribution that is in the loss cone we use only the geosynchronous MPA spectrum that is most nearly field aligned (as described by *Weiss et al.* [1996] and *Thomsen et al.* [1996]).

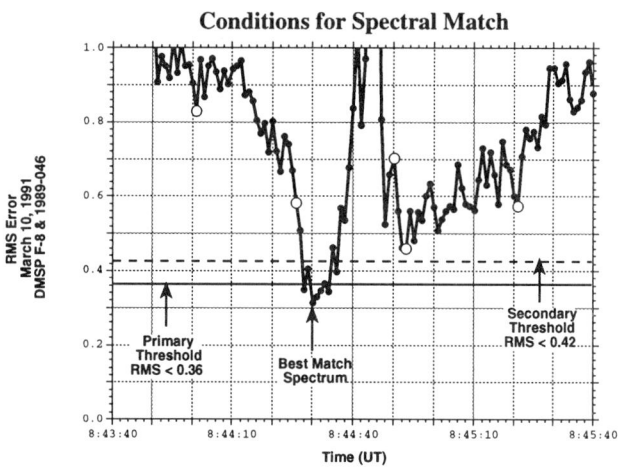

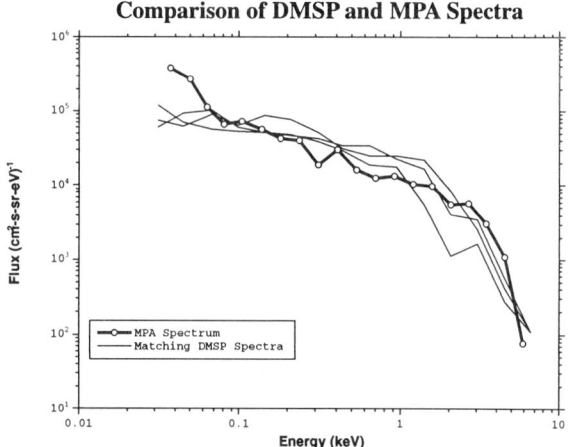

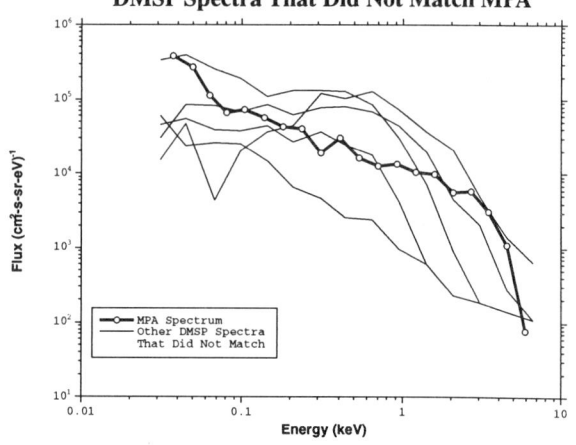

Figure 1. An illustration of how good spectral matches are chosen. The top panel shows the RMS difference between the MPA and DMSP spectra as a function of time. The second panel shows MPA and DMSP spectra that met our matching criteria for this event. The bottom panel shows several DMSP spectra in this interval that did not match the MPA spectrum. The times of those spectra are shown with open circles in the top panel.

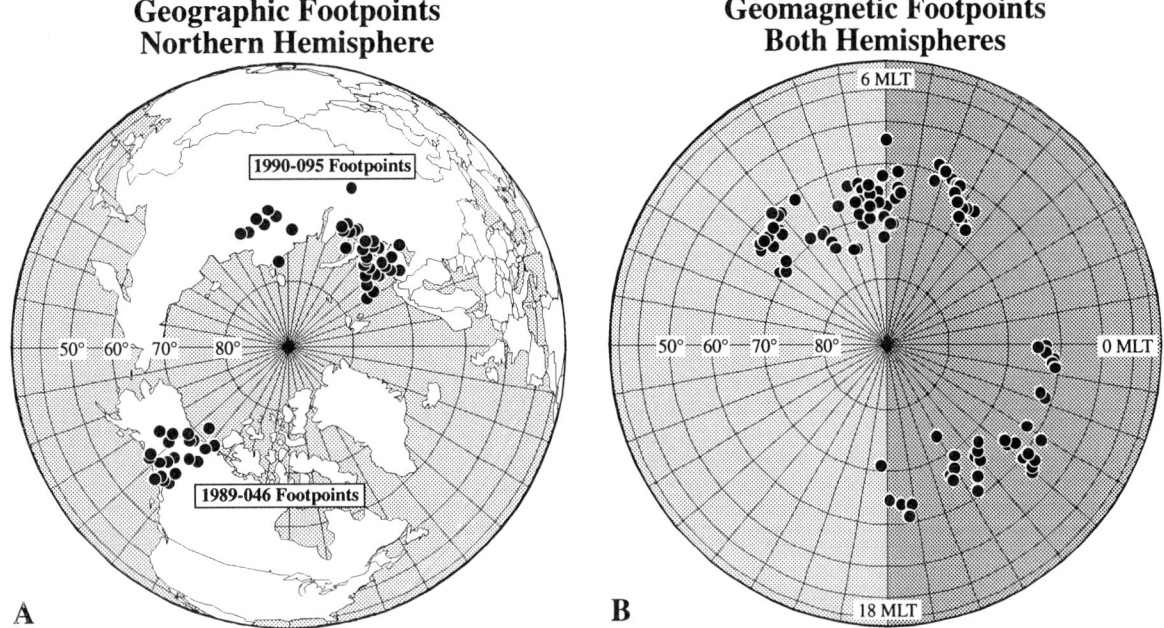

Figure 2. The location of DMSP when it measured a magnetic conjunctions with a geosynchronous satellite. (A) The position of DMSP in geographic coordinates. Only northern hemisphere conjunctions are shown. (B) The position of DMSP in geomagnetic coordinates (magnetic latitude and magnetic local time). Conjunctions from both hemispheres are shown.

Second, our technique assumes that the magnetic field is unchanging in the time it takes DMSP to cross the geosynchronous field line so we limit analysis to times when the MPA spectrum is constant for several minutes. This criterion also assures that there are not strong variations in local time in the vicinity of the geosynchronous spacecraft. Finally, we also eliminate times when there is a field-aligned potential drop. This occurs naturally as a result of our spectral comparison because if DMSP is measuring an accelerated population and MPA is not then the spectra will not match. We assure this by adopting an extremely strict condition for spectral matching.

Figure 1 shows a typical spectral match and the criteria used to define it. In the top panel we plot the RMS difference between the spectrum measured by DMSP and by MPA for two minutes of a nominal conjunction. We define a "measured conjunction" as the times at which the RMS difference in the spectra falls below a threshold of 0.36. By examining a large number of spectra we determined that this threshold is actually more selective than a visual examination of the spectra. Furthermore we require that a measured conjunction last for less than 30 seconds. This assures that the longitude range over which the spectra match is sufficiently small that we can distinguish between one magnetic field model and another. Finally, we also require that the RMS error does not fall below a second threshold of 0.42 for more than 60 seconds which assures that the minimum in the RMS is sharp and well-defined.

In this case our "measured conjunction" (as defined by the criteria above) lasted for 6 seconds. The middle panel in Figure 1 shows the MPA spectrum and DMSP spectra that satisfied our matching criteria (we note that the energy spectra use the nominal calibrations of the instruments and are not normalized in any way). In the bottom panel of Figure 1 we plot the same MPA spectrum along with 5 other DMSP spectra. The times of those spectra are marked with open circles on the RMS plot in the top panel. Clearly the spectral match for those times is significantly worse. We also point out that this is only a 2-minute portion of the DMSP crossing which occupies a very small portion of a typical DMSP spectrogram plot. Outside this 2-minute interval the DMSP spectrum did not even resemble the MPA spectrum which is what one would expect since during those times DMSP is mapping to very different parts of the magnetosphere.

This type of two-point spectral comparison has been used before to establish magnetic conjugacy [*Sharp et al.*, 1971; *Mende and Shelly*, 1976; *Meng et al.*, 1979; *Lundin and Evans*, 1985; *Schumaker et al.*, 1989; *Mauk and Meng*, 1991; *Hones et al.*, 1996]. What distinguishes this study from those earlier studies is that we have established an automated algorithm for identifying magnetic conjunctions and we have applied it to a set of satellites that have frequent conjunctions. This allows us to study the magnetic mapping from geosynchronous orbit to the ionosphere in a statistical manner.

3. MEASURED CONJUNCTION STATISTICS

Using the technique described in the previous section we examined three months of data (March, September, and December, 1991) for nominal conjunctions between one of three

DMSP satellites (F8, F9, and F10) and one of the two geosynchronous satellites (1989-045 and 1990-095). Out of over 3,500 nominal conjunctions we found 102 that satisfied our selection criteria. For each of the 102 conjunctions we identified the times of all spectra that met our criteria as well the single 1-second DMSP spectrum that best matched the MPA spectrum (as defined by the minimum RMS error).

Figure 2 shows the location of DMSP at the times when the best matched spectra were observed. Figure 2a shows DMSP's geographic location and Figure 2b shows its location in magnetic local time and magnetic latitude. Because we are using geosynchronous satellites we are limited to the geographic longitudes those satellites sample. 1989-046 had footpoints near Alaska. 1990-095 has two clusters of footpoints—over western and central Russia—because it was moved in the middle of 1991. Likewise, because the DMSP satellites are sun-synchronous they sample only a limited range of local time. But, thanks to the rotation of the earth's dipole DMSP is able to sample about one half of the possible magnetic local times (as shown in Figure 2b).

It is apparent from Figure 2 that the footpoint of geosynchronous orbit generally lies in the auroral ionosphere. Most often it is in the region of diffuse aurora but it frequently lies in the region of discrete aurora. It is also apparent from the figure that the footpoint of geosynchronous orbit can be quite variable, spreading over more than 10° in magnetic latitude. In a related study *Weiss et al.* [1996] investigate how well the measured location of the geosynchronous footpoint correlates with various magnetospheric indices such as K_p, AE, D_{st}, the local tilt of the field at geosynchronous orbit, and the equatorward edge of the auroral boundary. In this paper our emphasis is on evaluating how well various magnetic field models predict the location of the measured footpoint. An important point about Figure 2 is that no magnetic field models have been used to determine the magnetic footpoints of geosynchronous orbit. Therefore we have a completely model-independent data set of field line mappings which we can use to evaluate magnetospheric magnetic field models.

4. COMPARISON WITH MAGNETIC FIELD MODELS

We now compare the magnetic field mapping determined from the DMSP and MPA spectra with the magnetic field mapping predicted by various magnetic field models. The models we use are the Tsyganenko-89 [*Tsyganenko*, 1989], Tsyganenko-87 [*Tsyganenko*, 1987], Tsyganenko-82 [*Tsyganenko and Usmanov*, 1982], Olsen-Pfitzer [*Olsen and Pfitzer*, 1974], and Hilmer-Voigt [*Hil-mer and Voigt*, 1995] magnetic field models. For each of these models we use the IGRF representation of the earth's internal field. This is very important for mapping studies because, as shown by *Reeves and Weiss* [1996], the deviation of the earth's field from a dipole can have a significant effect on the magnetic mapping from geosynchronous orbit to low altitudes.

Figure 3 illustrates the method we used to compare the footpoints predicted by the field models with the measured footpoint determined from DMSP and MPA spectra. This event is the same event shown in Figure 1. It was a conjunction between the DMSP F9 satellite and the geosynchronous satellite 1989-046. The best spectral match was recorded by DMSP at 08:44:30 UT when DMSP was at $-59.78°$ magnetic latitude. DMSP was in the southern hemisphere moving poleward and during the 6-seconds of spectral match it moved only 0.4° in magnetic latitude.

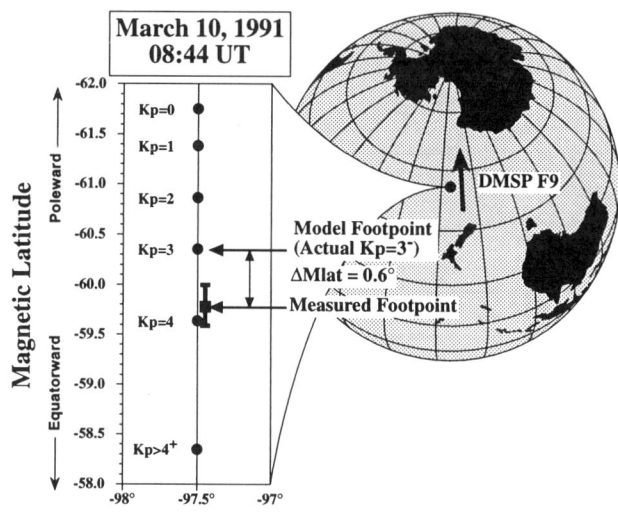

Figure 3. Comparing a measured footpoint with the footpoint predicted by the Tsyganenko-89 model. The globe shows the location of DMSP when it measured a good spectral match with MPA. The inset shows the magnetic latitude and longitude of the measured footpoint with a bar for the location of DMSP during the entire spectral match and a square for the location of the best spectral match. The footpoints predicted by various K_p levels of the Tsyganenko-89 model are shown with circles. The actual K_p was 3^- which gives a difference for this event of only 0.6°.

An expanded plot of region of the conjunction is shown in the inset. Here we have plotted the location of the footpoint measured by DMSP with a square and the location of the footpoints predicted by the Tsyganenko-89 magnetic field model with circles. In both cases the footpoint is defined at an altitude of 100 km. The Tsyganenko-89 field model comes in five versions for integral values of the magnetic activity parameter K_p. For this event $K_p = 3^-$ and the Tsyganenko-89 model predicted a footpoint which was 0.6° further poleward than the actual measured footpoint which is excellent agreement.

We calculated the difference in magnetic latitude of the footpoint for each of the five models used in this study and for each of the 102 conjunctions in our data set. We used each model "as advertised". In other words, for the Tsyganenko family of field models we used the actual K_p parameter for each conjunction to specify what version of the model to use. The Hilmer-Voigt model is specified by three parameters: D_{st}, the stand-off distance of the magnetopause (given by solar wind pressure), and the equatorward boundary of the auroral oval (given by DMSP electron precipitation signatures). For the Hilmer-Voigt model we again used the parameters that were appropriate for each event. The Olsen-Pfitzer model has no free parameters so the same model applies to all of our cases.

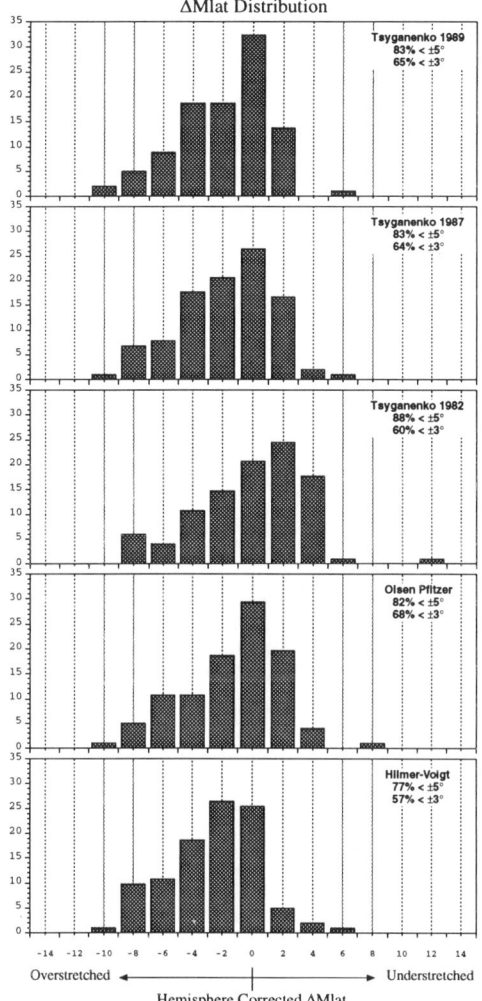

Figure 4. Histograms of the difference between the measured and model footpoints. All models share the same strengths and weaknesses as described in the text.

A histogram of the difference between the measured and model footpoints for each of the five models is plotted in Figure 4. The top panel shows the statistics for the Tsyganenko-89 model. Here, 32% of the model footpoints agreed with the measured footpoints to within $\pm 1°$, 65% were within $\pm 3°$, and 83% were within $\pm 5°$. Put another way, if one needs to know the location of the footpoint of geosynchronous orbit to within $1°$, the Tsyganenko-89 model has a 32% probability of being correct. However, it also has a 17% chance of being off by more than $5°$ and the statistical uncertainty in the mapping is approximately $3°$.

A fairly surprising result of this study is that, on average, all the field models tested perform about equally well—or equally poorly. The three generations of Tsyganenko magnetic field models contain various refinements and improvements but the changes in model did not improve the accuracy of the mapping from geosynchronous orbit to the ionosphere.

This is in part because the measured footpoints have a much larger range of latitudes than is accommodated in the models (compare Figures 2 and 3). However, the Tsyganenko models have a larger range than the Olsen-Pfitzer model (which has none) yet the Olsen-Pfitzer model does as good a job of predicting the location of the geosynchronous footpoint as any of the Tsyganenko models. This is no doubt in part because K_p is a poor parameter to use for determining the amount of stretching in the field. In fact, *Hones et al.* [1996] and *Weiss et al.* [1996] have shown that the K_p value needed for the Tsyganenko-89 model to reproduce the observations is completely uncorrelated with the actual K_p measured for an event. It may, then, be somewhat surprising that the Hilmer-Voigt model, which uses a set of parameters which might be expected to be better correlated with the measured footpoint of geosynchronous orbit, still does not perform significantly better than its rivals.

Finally we note that the distributions are also not symmetric. The difference in magnetic latitude is calculated such that, regardless of what hemisphere DMSP was in, negative values represent cases where the measured footpoint was poleward of the model footpoint and positive values represent cases where the measured footpoint was equatorward of the model footpoint (for example the conjunction shown in Figure 3). The more stretched the field model is the further equatorward the model footpoint moves (see Figure 3). Therefore negative values mean that the model is too stretched compared to the observations and positive values mean that the model is not sufficiently stretched.

It is apparent from Figure 4 that all the models are, on average, too stretched. Our cases include a variety of types of magnetospheric conditions which include quiet times, growth phases, expansion phases, and recovery phases and a variety of activity levels from $K_p = 0$ to $K_p = 9^-$. We note, however, that we have very few conjunctions in the midnight local time sector due to the limited range of local times sampled by the DMSP orbits. Therefore these results do not imply that the field models are too stretched compared to a growth phase field at midnight. Rather we suspect that in order to better represent the conditions at midnight the modelers may have made the models too stretched at other local times.

5. CONCLUSIONS

We have compiled a set of 102 conjunctions between one of three low-altitude DMSP satellites and one of two geosynchronous satellite using an algorithm that compares the field-aligned electron energy spectra measured at each location. Excellent spectral matches can be found for a subset of nominal conjunctions. Those conjunctions that meet our spectral matching criteria provide a sensitive and field model-independent determination of the magnetic footpoint of geosynchronous orbit.

We compared the measured footpoint obtained by the spectral matching technique with the footpoint predicted by the Tsyganenko-89, Tsyganenko-87, Tsyganenko-82, Olsen-Pfitzer, and Hilmer-Voigt field models. Surprisingly, we found that no one field model performed significantly better than any of the other models. The statistical uncertainty in the footpoint predicted by all of the field models tested was approximately $\pm 3°$. Only about 25–30% of the time did the field model predict the conjunction to within $\pm 1°$ and as much as 20% of the time the field model could be off by more than $\pm 5°$.

We found that the footpoint of geosynchronous orbit varies over more than $10°$ of magnetic latitude. This is a larger range of latitudes than any of the field models tested can accommodate. This suggests that the next generation of magnetic field models should allow a greater range in the amount of stretching that they allow. However, we also found that all the field models were, on average, too stretched compared to the measured footpoints therefore adding more stretching to the models will only aggravate the problem.

This database of conjunctions is useful not only for evaluating the models that we have used here but can also be used to evaluate any magnetospheric magnetic field model. The database can be extended to include a larger number of cases and a larger number of seasons or magnetospheric conditions as needed. By using other low-altitude satellites it may be possible to get broader coverage in local time and by using other high-altitude satellites, such as CRRES or POLAR or CLUSTER it may be possible to systematically test other ranges of L. In addition to testing existing field models we are also using this database to help determine what magnetospheric parameters actually control the mapping between geosynchronous orbit and the ionosphere and therefore what measurements might be most useful for parameterizing future magnetospheric magnetic field models.

Acknowledgements. We are grateful to P.T. Newell for providing the DMSP data and to the US Department of Energy Office of Basic Energy Science for financial support.

REFERENCES

Bame, S.J., D.J. McComas, M.F. Thomsen, B.L. Barraclough, R.C. Elphic, J.P. Glore, J.T. Gosling, J.C. Chavez, E.P. Evans and F.J. Wymer, Magnetospheric plasma analyzer for spacecraft with constrained resources, *Rev. Sci. Instrum., 64*, 1026, 1993.

Fairfield, D.H., An evaluation of the Tsyganenko magnetic field model, *J. Geophys. Res., 96*, 1481, 1991.

Hardy, D.A., L.K. Schmitt, M.S. Gussenhoven, F.J. Marshall, H.C. Yeh, T. L. Schumaker, A. Huber and J. Pantazis, Precipitating electron and ion detectors SSJ/4 for the block 5D/flights 6-10 DMSP satellites: Calibration and data presentation, Rep. AFGL-TR-84-0317, Air Force Geophys. Lab., Hanscom AFB, Bedford Mass., 1984.

Hilmer, R.V. and G.-H. Voigt, A magnetospheric magnetic field model with flexible current systems driven by independent physical parameters, *J. Geophys. Res., 100*, 5613, 1995.

Hones, E.W., M.F. Thomsen, G.D. Reeves, L.A. Weiss, D.J. McComas and P.T. Newell, Observational determination of magnetic connectivity of the geosynchronous region of the magnetosphere to the auroral oval, *J. Geophys. Res.*, in press, 1996.

Lundin R. and D.S. Evans, Boundary layer plasmas as a source for high latitude, early afternoon auroral arcs, *Planet. Space Sci., 32*, 1389, 1985.

Mauk, B.H. and C.-I. Meng, The aurora and middle atmospheric process, in Auroral Physics, edited by C.-I. Meng, M.J. Rycroft and L.A. Frank, Cambridge Univ. Press, NY, 223, 1991.

Mende, S.B. and E.G. Shelly, Coordinated ATS 5 electron flux and simultaneous auroral observations, *J. Geophys. Res., 81*, 97, 1976.

Meng, C.-I., B. Mauk and C.E. McIlwain, Electron precipitation of evening diffuse aurora and its conjugate electron fluxes near the magnetic equator, *J. Geophys. Res., 84*, 2545, 1979.

Olsen, W.P. and K. A. Pfitzer, A quantitative model of the magnetospheric magnetic field, *J. Geophys. Res., 79*, 3739, 1974.

Peredo, M.D., D.P. Stern and N.A. Tsyganenko, Are existing magnetospheric models excessively stretched?, *J. Geophys. Res., 98*, 15,343, 1993.

Peredo, M.D. and D.P. Stern, On the Position of the Near-Earth Neutral Sheet: A Comparison of Magnetic Model Predictions With Empirical Formulas, *J. Geophys. Res., 96*, 19,521, 1991.

Pulkkinen, T.I., D.N. Baker, D.G. Mitchell, R.L. McPherron, C.Y. Huang and L.A. Frank, Thin current sheets in the magnetotail during substorms: CDAW-6 revisited, *J. Geophys. Res., 99*, 5793, 1994.

Reeves, G.D. and L.A. Weiss, Seven magnetic field models compared: Mapping from geosynchronous orbit to the auroral zone, *J. Geophys. Res.*, submitted, 1996.

Schumaker, T.L., M.S. Gussenhoven, D.A. Hardy and R.L. Carovillano, The relationship between diffuse auroral and plasma sheet electron distributions near local midnight, *J. Geophys. Res., 94*, 10,061, 1989.

Sharp, R.D., D.L. Carr, R.G. Johnson and E.G. Shelly, Coordinated auroral-electron observations from a synchronous and a polar satellite, *J. Geophys. Res., 76*, 7669, 1971.

Thomsen, M.F., D.J. McComas, G.D. Reeves and L.A. Weiss, An observational test of the Tsyganenko (T89a) model of the magnetospheric field, *J. Geophys. Res.*, submitted, 1996.

Tsyganenko, N.A., A magnetospheric magnetic field model with a warped tail current sheet, *Planet. Space Sci., 37*, 5, 1989.

Tsyganenko, N.A., Global quantitative models of the geomagnetic field in the cislunar magnetosphere for different disturbance levels, *Planet. Space Sci., 35*, 1347, 1987.

Tsyganenko, N.A. and A.V. Usmanov, Determination of the magnetospheric current system parameters and development of experimental geomagnetic field models based on data from IMP and HEOS satellites, *Planet. Space Sci., 30*, 985, 1982.

Weiss, L.A., M.F. Thomsen, G.D. Reeves and D.J. McComas, An examination of the Tsyganenko (T89a) field model using a database of two-satellite magnetic conjunctions, *J. Geophys. Res.*, submitted, 1996.

D.J. McComas, G.D. Reeves, M.F. Thomsen and L.A. Weiss, Los Alamos National Laboratory, Mail Stop D 436, Los Alamos, NM 87545, USA (E-mail: reeves@lanl.gov)

DISCUSSION

Q: X. Li. When comparing the flux spectra from two spacecraft, one in geosynchronous orbit can measure all particles, the other spacecraft, DMSP, measures particles only with small pitch angles at the equator. Do you assume that particles at different pitch angle have the same spectra?

A: G.D. Reeves. We compare the DMSP spectrum with only the field aligned portion of the geosynchronous distribution.

C: R. Friedel. One needs to use total field data when building a radiation model to decide if the magnetic field model used is valid. Only select valid points to go into the model.

A: G.D. Reeves. Yes, it's paramount that this is done at the time of measurement when the model is built.

A New Tool for Calculating Drift Shell Averaged Atmospheric Density

D. Heynderickx, M. Kruglanski and J.F. Lemaire

Belgisch Instituut voor Ruimte-Aëronomie/Institut d'Aéronomie Spatiale de Belgique, Brussels, Belgium

E.J. Daly

ESA/ESTEC, Postbus 299, NL-2200 AG Noordwijk, The Netherlands.

The coordinate system (B, L) has proved suitable for mapping trapped particle fluxes in most of the region covered by the Van Allen belts, but is not very adequate for the low-altitude regions where the Earth's atmosphere interacts with the trapped particle population. The steep flux gradients in the region of the upper atmosphere are better represented by using as a coordinate the weighted average of the atmospheric density over the drift shells of trapped particles. This coordinate appears rather efficient in mapping fluxes for low L values. A new computer programme has been written that traces drift shells with a very high precision and integrates an arbitrary function over a shell. The programme is applied to the calculation of drift shell averaged atmospheric densities for trapped protons. The results are compared to the results obtained with a similar programme developed in the sixties. The comparison shows the importance of accurate drift shell tracing, especially at low altitudes.

1. INTRODUCTION

The widely used coordinate systems (B, L) and $(B/B_0, L)$ have proved very suitable for most of the region covered by the Van Allen belts, but are not well suited for the low-altitude regions where the Earth's atmosphere interacts with the trapped particle population.

Pfitzer [1990] has found that the atmospheric density at space station altitudes is a better variable than B/B_0 to organise the AP-8 MIN and AP-8 MAX fluxes. He found that the AP-8 MIN and AP-8 MAX proton fluxes for Space Station altitudes (350–500 km) fall on almost the same curve when plotted as a function of the atmospheric densities for minimum and maximum solar activity conditions, respectively. *Pfitzer's* [1990] study confirms that at low altitudes the atmospheric density distribution governs the flux distribution of trapped protons.

The atmospheric density at a given altitude does not determine, however, the total mass of material traversed by a particle detected at that altitude. Instead, one should consider an average of the atmospheric density over an azimuthal drift path of particles of a given species and with a given energy.

Hassitt [1964, 1965] has developed a computer code at UCSD Physics Department which calculates the number density of atoms, ions, and molecules given by appropriate atmospheric and ionospheric models over a drift shell (B, L). A weighted average density is then determined by multiplying the resulting number densities with the collision cross section σ_i of the trapped particles with the constituents i, summing over i and integrating the sum over the drift shell (B_m, L), where B_m is the magnetic field intensity at the particles' mirror points. In the following sections, we discuss the definition and calculation of the drift shell average in more detail.

Even after extensive modifications, we found that Hassitt's code has difficulties with the integration procedure near the geomagnetic equator and near the atmospheric cutoff. We

Radiation Belts: Models and Standards
Geophysical Monograph 97
Copyright 1996 by the American Geophysical Union

therefore decided to write a new code where the drift shell tracing and the integration over the drift shell are well separated. This renders the code more transparent, more flexible and more efficient. We describe the workings of our new code and compare results with the Hassitt code. It appears that unsufficiently fine field line tracing in the Hassitt code leads to a strong overestimate of the averaged density for low L values.

2. DEFINITION OF DRIFT SHELL AVERAGES

Protons in the energy range 1–100 MeV lose energy in the neutral atmosphere mainly by interactions which ionise or excite target molecules. Charge exchange and nuclear collisions are negligible in this energy range. The main quantities to be considered are the combined number N of ionisation and excitation interactions over one drift period and the corresponding total energy loss ΔE. N is defined as

$$N = \oint \sum_i n_i \sigma_i(E) \, v \, dt, \tag{1}$$

where i represents the components of the neutral atmosphere, n_i is the number density and $\sigma_i(E)$ the sum of the ionisation and excitation cross section for constituent i and proton energy E, v is the total velocity of the proton, and $dt = dr/v$ where dr is an element of length along the proton's helicoidal trajectory. We make the approximation that E remains constant during one full drift motion.

The energy loss over one drift period due to the N interactions can be written as

$$\Delta E = \oint \frac{dE}{dt} \, dt, \tag{2}$$

where dE/dt is the energy loss per unit time and is given by the Bethe formula for non-relativistic protons:

$$\frac{1}{v}\frac{dE}{dt} = \sum_i n_i A_i \left(\frac{Z_i}{A_i} \ln \frac{2m_e v^2}{\mathcal{I}_i}\right), \tag{3}$$

where $\mathcal{I}_i$ is the mean ionisation and excitation potential of constituent i [*Fano*, 1963].

The motion of a trapped particle can be decomposed into three components: gyration around the field line, bounce motion along field lines between conjugate mirror points, and azimuthal drift. With the assumption that the three motions are independent, the time-average of a function $f(\mathbf{r})$ over a drift shell (B_m, L) can be written as

$$\langle f \rangle \equiv \frac{\oint f(\mathbf{r}) \, dt}{T} = \frac{\iiint f(\mathbf{r}) \, dt_1 dt_2 dt_3}{\iiint dt_1 dt_2 dt_3}, \tag{4}$$

where the indices 1, 2 and 3 refer to the gyration, bounce and azimuthal drift motion, respectively, and $T = \int dt_3$ is the time needed to cover a drift shell. In the guiding-centre approximation, Eq. (4) is reduced to

$$\langle f \rangle = \frac{\iint f(\mathbf{r}) \, dt_2 dt_3}{\iint dt_2 dt_3} \tag{5}$$

$$= \frac{\iint f(\mathbf{r}) \frac{ds}{v_\parallel} \frac{dl}{v_d}}{\iint \frac{ds}{v_\parallel} \frac{dl}{v_d}},$$

where ds and dl are elements of length along the field line and along the azimuthal drift direction, respectively, and $v_\parallel$ and v_d the corresponding velocities.

The parallel velocity follows from the conservation of the first adiabatic invariant:

$$\frac{v_\parallel}{v} = \sqrt{1 - \frac{B}{B_m}}. \tag{6}$$

Lew [1961] gives the following expression for the drift velocity:

$$\frac{v_d}{m\gamma v^2/2q} = \left(2 - \frac{B}{B_m}\right) \left|\frac{\mathbf{B} \times \nabla B}{B^3}\right|. \tag{7}$$

With these definitions, the quantity

$$S(f, B_m, L) \equiv v \frac{m\gamma v^2}{2q} \iint f(\mathbf{r}) \frac{ds}{v_\parallel} \frac{dl}{v_d} \tag{8}$$

mainly depends on the magnetic field configuration, and on energy only through the possible energy dependence of the function f. Since the velocity v is conserved, we have that

$$\langle f \rangle = \frac{S(f, B_m, L)}{S(1, B_m, L)}. \tag{9}$$

The quantity $S(f, B_m, L)$ can be calculated by numerically tracing the drift shell (B_m, L).

The evaluation of the double integral in Eq. (8) is very time-consuming. However, a significant simplification has been suggested by *Hassitt* [1965]: the drift time between two neighbouring field lines depends only weakly on the position on the field line, so that for a given field line dl/v_d can be considered a constant. Consequently, dl/v_d has to be evaluated only once for each field line in the integration in Eq. (8).

3. THE DRIFT SHELL TRACING PROGRAMME

The code developed by *Hassitt* [1964] calculates the drift shell average defined in Eq. (9) for the atmosphere model of *Anderson and Francis* [1964], using the *Jensen and Cain* [1962] magnetic field model for the drift shell tracing. In Hassitt's code, the drift shell tracing is not completely separated from the integration of the atmospheric density. Hence, we found it very difficult to extend this code to other magnetic

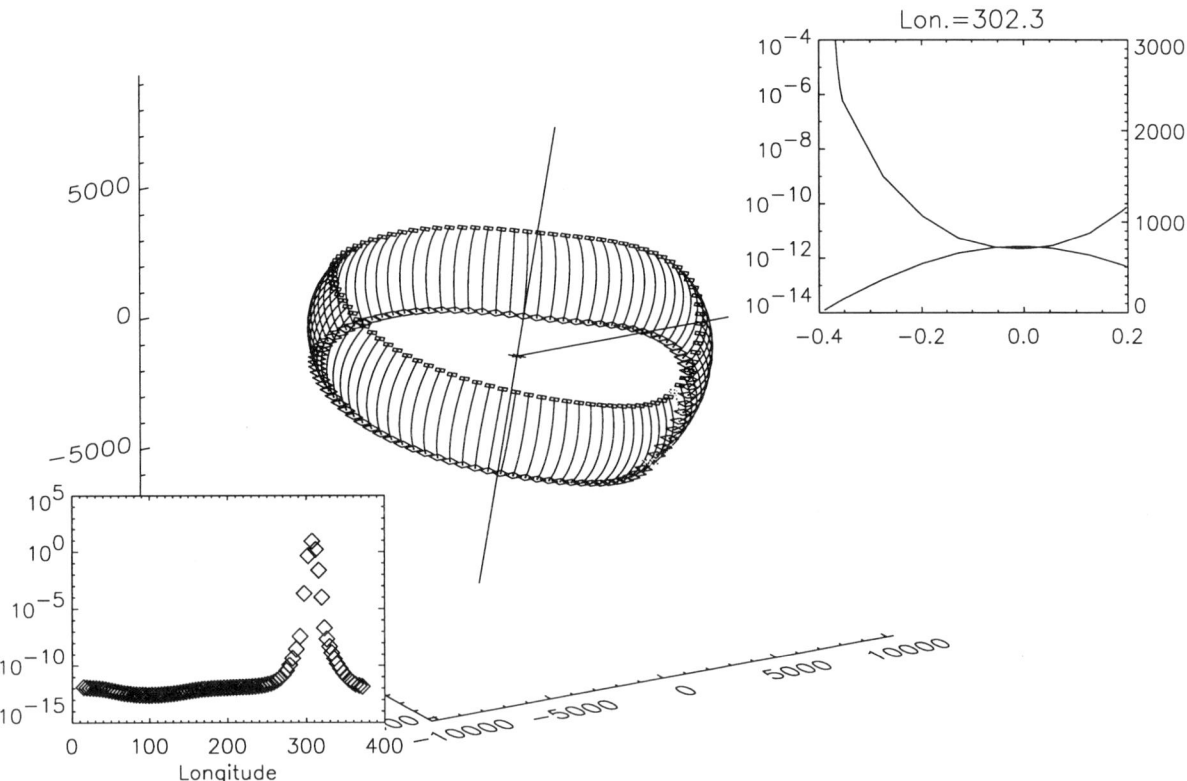

Figure 1. Representation of a drift shell traced with the new software package described in the text. The main figure shows the field line segments making up the drift shell in geographic coordinates, and the magnetic dipole axis. The horizontal line points towards the Greenwich meridian. The right hand insert plot shows, as a function of distance along the field line (in R_E, 0 corresponds to the magnetic field minimum), the altitude above the Earth's surface (in km) and the locally averaged atmospheric density, for a field line through the heart of the SAA. The left hand insert shows the density averaged over each field line segment in the main plot as a function of longitude.

field representations and atmospheric models. In addition, we found that the integration procedure in Hassitt's code is not stable near the geomagnetic equator nor near the atmospheric cutoff.

In the end, we decided to write a new code, partly based on the methods formulated by *Hassitt* [1964], in which the tracing of a drift shell is done first, after which an arbitrary function can be integrated over the shell. This procedure has the additional advantage that one can run the software just to produce a grid of points on a drift shell, which has many applications besides drift shell averaging. The user can choose between the magnetic field models implemented in the UNIRAD package [*Heynderickx*, these proceedings], i.e. all the DGRF/IGRF models, the *Jensen and Cain* [1962] model, and the GSFC 12/66 model [*Cain et al.*, 1967].

3.1. Atmosphere and Ionosphere Models

The new code includes contemporary models for the neutral atmosphere, the ionosphere and the plasmasphere. We have added the ionosphere and plasmasphere densities to make our code as general as possible. The three models can be selected and combined at will.

The Mass-Spectrometer-Incoherent-Scatter (MSIS) neutral atmosphere model describes the neutral temperature and the densities of He, O, N_2, O_2, Ar, H, and N. The model version implemented in our software is MSISE-90 [*Hedin*, 1991]. We have also implemented a model developed by McDonnell Douglas Astronautics Co. which *Pfitzer* [1990] has used in his study of the radiation dose at Space Station altitudes.

The International Reference Ionosphere (IRI) describes monthly averages of electron density, electron temperature, ion temperature, ion composition and ion drift in the altitude range from 50 km to 1000 km for magnetically quiet conditions in the non-auroral ionosphere [*Bilitza*, 1990].

Since the IRI-90 model is limited in altitude, an extension of the ion density into the magnetosphere is needed. Several three dimensional models have been proposed to describe the equatorial and field-aligned ionization density

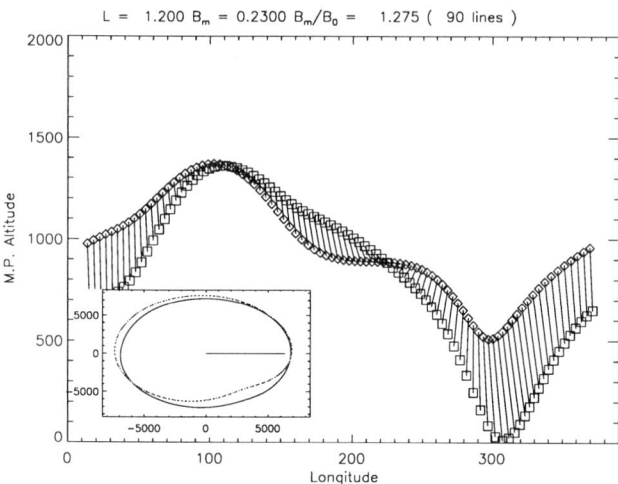

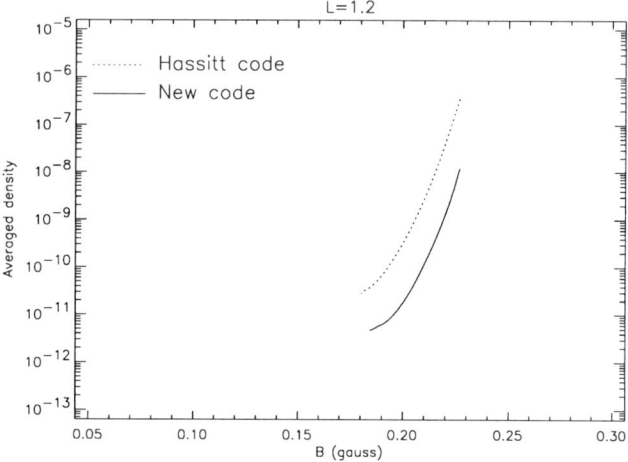

Figure 2. Representation of the mirror point altitudes of the field line segments shown in Figure 1. The main plot shows the altitude of the northern (diamonds) and southern (squares) mirror points as a function of longitude. Conjugated mirror points are connected by a solid line. The insert plot is the projection of the locus of the northern (solid line) and southern (dotted line) mirror points in the plane of the geographic equator. The horizontal line points towards the Greenwich meridian.

Figure 3. Comparison of the drift shell averaged density obtained with the Hassitt code and the new code, respectively, for $L = 1.2$ and a number of B values from the magnetic equator to the atmospheric cut-off region

in the plasmasphere and plasmatrough. We have used the model of *Carpenter and Anderson* [1992] for the equatorial electron density. This model describes, in piecewise fashion, the "saturated plasmasphere", i.e. the region of steep plasmapause gradients, and the plasmatrough. The electron densities at non-equatorial latitudes above 1000 km were extrapolated along dipole field lines to fit the equatorial density distribution of *Carpenter and Anderson* [1992].

From a literature study, *Pierrard* [1994] has compiled for each interaction of charged particles in the neutral atmosphere and ionosphere analytic expressions for the collision cross section as a function of the kinetic energy of the incident particle, and has implemented these expressions in a computer programme called CROSS. CROSS takes as input the kinetic energy of the incident particle, the type of the incident and the target particles, and a list of interactions, and returns the total cross section.

3.2. *The new shell tracing algorithm*

For a given pair of (B_m, L) values, the new algorithm makes an iterative search (in distance) in the plane of the geomagnetic equator at a fixed geographic longitude for a point in space and the field line passing through it which yield the same (B_m, L) values. When this point has been found, the coordinates of a set of points on the magnetic field line are stored. This procedure is repeated for a set of fixed longitudes. The result of this tracing algorithm is a grid of points covering the whole drift shell.

The next step is the evaluation of Eq. (9). For each longitude, the programme evaluates the function f and $v_\parallel/v$ at the points of the stored field line and at a number of inter-

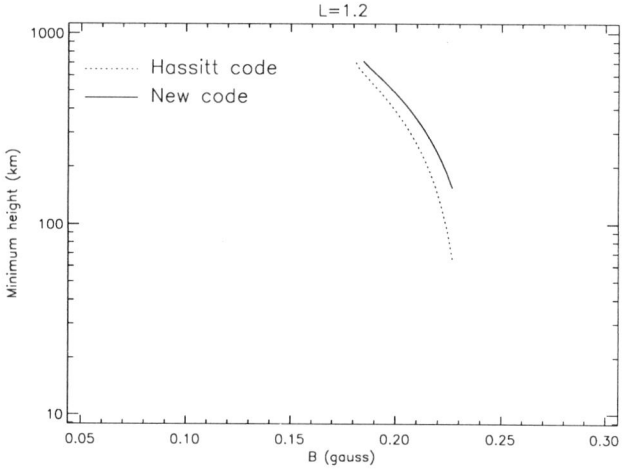

Figure 4. Comparison of the minimum altitude on the drift shell obtained with the Hassitt code and the new code, respectively, for $L = 1.2$ and the B values used for Figure 3

mediate, interpolated, points. The integration along the field line is then carried out, the value of the integral is stored, and Eq. (7) is evaluated. Finally, the integration in longitude is carried out.

4. ILLUSTRATION OF THE PROGRAMME

In order to test our new software, we have calculated the drift shell averaged density of the *Anderson and Francis* [1964] model, summed over all constituents and weighted with energy dependent cross sections, for a number of (B_m, L) pairs and compared the results to the results obtained with the Hassitt code.

Figure 1 represents the drift shell ($B_m/B_0 = 1.275$, $L = $

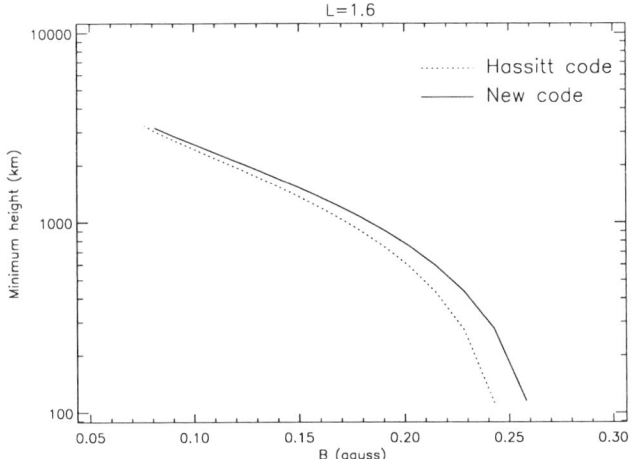

Figure 5. Same as Figure 4, but for $L = 1.6$

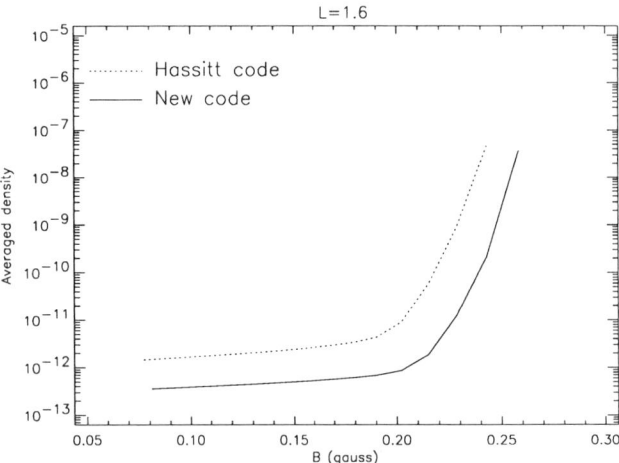

Figure 6. Same as Figure 3, but for $L = 1.6$

1.2) calculated by the new programme. The field line segments making up the drift shell are separated in longitude by 4°, which is a reasonable compromise between accuracy and computing time. The deformation of the shell in the region of the South Atlantic Anomaly (SAA) can be clearly seen. The insert plots in Figure 1 represent the mean atmospheric density along a field line passing through the heart of the SAA (right hand insert) and the mean density integrated along the field lines as a function of longitude (left hand insert). The insert plots confirm that the grid of points covering the drift shell is sufficiently fine to obtain smooth curves of the density averages, both along field lines and as a function of longitude.

Figure 2 shows the altitudes of the conjugate mirror points on the field lines making up the drift shell, as a function of longitude. The "dipping down" of the southern hemisphere mirror points in the region of the SAA is striking. Figure 2 is in perfect agreement with a similar figure presented by *Roederer* [1970, Figure 76].

In Figure 3 the drift shell averages—as defined in Eq. (9), i.e. the number of interactions per second, in arbitrary units— of the *Anderson and Francis* [1964] model, obtained with the Hassitt code and with the new code, respectively, are shown for a number of B_m values (ranging from the equator to the atmospheric cutoff) with a fixed value $L = 1.2$. The averages obtained with the Hassitt code are consistently much higher than those obtained with the new code, increasingly so with increasing B_m (i.e. closer to the cutoff region, where there is about a factor fifty difference). It turns out that this discrepancy between the two codes is almost entirely due to the performances of the respective shell tracing algorithms. To illustrate this point, Figure 4 shows the minimum altitude reached on the drift shells in Figure 3, for both tracing algorithms. The drift shells generated by the Hassitt code reach consistently lower than those obtained with our new code. As the atmospheric density increases exponentially with decreasing altitude (in fact, as the main contribution to the density average comes from the SAA region, we are studying the feasibility of replacing the average over the drift shell with the density at the lowest point on the shell, which would result in a significant gain in computing time), the corresponding density averages are overestimated by the Hassitt code. Hassitt's code is built around McIlwain's field line tracing algorithm developed in the early 60's, which we found has a rather coarse mechanism for stepping on a field line. More particularly, when McIlwain's routine reaches the magnetic equator, the step size, which increases away from the first mirror point, continues to increase, instead of which it should gradually diminish to guarantee a good coverage of the field line near the conjugate mirror point. McIlwain's tracing algorithm is in general adequate for calculating L values since the main contribution to the second adiabatic invariant comes from the equatorial region. However, the algorithm is not capable to locate the mirror points with sufficient accuracy. We have therefore implemented a new field line tracing algorithm in our new code: this algorithm is based on a procedure developed by *Pfitzer* [1991], which we have further refined and improved (the resulting algorithm has also been implemented in our UNIRAD package [*Heynderickx*, these proceedings]). We feel confident that our code traces drift shells to a high degree of accuracy, and have therefore abandoned Hassitt's code.

As Figure 5 shows, the discrepancy between Hassitt's shell tracing routine and our own persists for higher values of L (i.e. $L = 1.6$), although it decreases near the equator. The corresponding drift shell averaged densities, shown in Figure 6, diverge strongly in the cut-off region and display a significant difference even near the equator.

5. CONCLUSIONS

The new code presented in this paper has proven to be a very powerful tool to generate drift shells and to integrate arbitrary functions over shells. We are continuing our study of the drift shell averaged atmospheric density and total energy loss as low altitude coordinates for trapped radiation models. To this end, the new software will be applied to the AP-8 models and to radiation maps built from measurements made by the AZUR, SAMPEX and UARS satellites.

As we have made the guiding centre approximation, the drift shell tracing ignores the finite gyroradius of energetic protons. *Heckman and Brady* [1966] have estimated that taking into account the helicoidal trajectory of protons at 125 MeV for $L = 1.38$ can raise the drift shell averaged density by a factor of two. We plan to evaluate and include this effect in future versions of the software.

Acknowledgements. We thank C.E. McIlwain for making available a copy of the program developed by A. Hassitt. This work was funded by ESA/ESTEC/WMA TRP Contract Nos. 9828/92/NL/FM and 10725/94/NL/JG(SC).

REFERENCES

Anderson, A.D. and Francis, W.E., A semitheoretical model for atmospheric properties from 90 to 10,000 km, Lockheed Missiles and Space Company 6-74-64-19, 1964.

Bilitza, D., International Reference Ionosphere 1990, National Space Science Data Center, NSSDC/WDC-A-R&S 90-20, 1990.

Cain, J.C., Hendricks, S.J., Langel, R.A. and Hudson, W.V., A Proposed Model for the International Geomagnetic Reference Field-1965, *J. Geomag. Geoelectr., 19*, 335–355, 1967.

Carpenter, D.L. and Anderson, R.R., An ISEE/Whistler Model of Equatorial Electron Density in the Magnetosphere, *J. Geophys. Res., 97*, 1097–1108, 1992.

Fano, V., Penetration of Protons, Alpha Particles, and Mesons, *Ann. Rev. Nucl. Sci., 13*, 1–66, 1963.

Hassitt, A., *An average atmosphere for particles trapped in the Earth's magnetic field*, University of California at San Diego, 1964.

Hassitt, A., Average Effect of the Atmosphere on Trapped Protons, *J. Geophys. Res., 70*, 5385–5394, 1965.

Heckman, H.H. and Brady, V.O., Effective Atmospheric Losses for 125-MeV Protons in South Atlantic Anomaly, *J. Geophys. Res., 71*, 2791–2798, 1966.

Hedin, A.E., Extension of the MSIS thermosphere model into the middle and lower atmosphere, *J. Geophys. Res., 96*, 1159–1172, 1991.

Jensen, D.C. and Cain, J.C., An Interim Geomagnetic Field, *J. Geophys. Res., 67*, 3568, 1962.

Lew, J.S., Drift rate in a dipole field, *J. Geophys. Res., 66*, 2681–2865, 1961.

Pfitzer, K.A., Radiation Dose to Man and Hardware as a Function of Atmospheric Density in the 28.5 Degree Space Station Orbit, MDSSC Report No. H5387 Rev A, 1990.

Pfitzer, K.A., Improved Models of the Inner and Outer Radiation Belts, Phillips Laboratory, PL-TR-91-2187, 1991.

Pierrard, V., Cross sections for collisions between electrons or protons and the main atmospheric components, *Aeronomica Acta, 54*, 1994.

Roederer, J.G., *Dynamics of Geomagnetically Trapped Radiation*, Springer-Verlag, 1970.

D. Heynderickx, M. Kruglanski and J.F. Lemaire, Belgisch Instituut voor Ruimte-Aëronomie/Institut d'Aéronomie Spatiale de Belgique, Ringlaan 3, B-1180 Brussels, Belgium (E-mail: d.heynderickx@oma.be, m.kruglanski@oma.be, j.lemaire@oma.be)

E.J. Daly, ESTEC/WMA, the Space Systems Environment Analysis Section of ESA, Postbus 299, NL-2200 AG Noordwijk, The Netherlands.

DISCUSSION

Q: G.D. Reeves. What atmosphere model is used? Does it include solar cycle effects? Does it reproduce the NOAA low altitude fluxes?

A: D. Heynderickx. The user can choose between the MSISE-90 and the MDAC model for the neutral atmosphere, and can add IRI-90 and a plasmaspheric extension. All these models contain parameters to incorporate solar cycle effects. We haven't done that yet.

Q: R. Selesnick. Is the Coulomb drag included in the calculation of the drift average density?

A: D. Heynderickx. Yes, through the IRI-90 electron and ion densities.

Q: J.B. Blake. Is the guiding centre used in calculating drift averaged density?

A: D. Heynderickx. Yes.

Radiation Conditions Modelling at the Geostationary Orbit

G.V. Popov, V.I. Degtjarev and S.S. Sheshukov

Inst. Solar-Terr. Phys., Irkutsk, Russia

It's well known that radiation models for the region of the geostationary orbit do not meet contemporary requirements of users. It seems that the progress in the radiation condition modelling is not possible without reconsideration of some basic definitions and principles, namely: What is the radiation model as itself, who are its makers and users? Is it possible to develop a unique radiation model, which will satisfy all users? Are the input parameters used before optimal ones or not? These questions are discussed briefly on the basis of tendencies, observable in Russian researches.

A recent result of the ISTP-group is presented concerning the correlation of electron fluxes observed at the geostationary orbit with solar wind parameters.

1. INTRODUCTION

At present it is widely discussed the necessity to develop a new radiation belt model. Clear understanding of necessity of the radiation models upgrading is not accompanied by clear ideas of which the model is necessary to be developed, for which purposes, which properties and what form it should have—i.e. there is not yet the full and agreed understanding what is a model of radiation; each group of researchers understands this term in its own way.

At the same time, it is very important to have conventional and precisely formulated concept of the radiation model, since it allows to plan the research activity on future, to define the list of key questions and to develop such model, which is necessary to users.

Some reasonings and definitions are listed below, which are rather well-known but are discussed rather rarely.

2. MISUNDERSTANDINGS BETWEEN MODEL MAKERS

We'll start with statement, that one hardly will find some publication, containing such definition of term "the radiation model" which one likes and which will completely coincide with one's understanding of this term. Even the author of any such definition will find very soon its limitation and imperfection.

The reason of it is in the extraordinary diversity of makers and users of radiation models. Each participant of this makers and users community has its own ideas about the structure and the output of model; even the terminology and languages used are different.

Radiation belts are in the focus of several scientific fields. Each of them has its own goals, specific theoretical and experimental methods of research.

Among these scientific disciplines the heaviest activity is displayed by nuclear physics, plasma physics, solar-terrestrial physics and applied geophysics. Each of these fields uses theoretical and experimental methods of research, on-board and ground based facilities.

But as far as the scientific interests of representatives of these scientific fields are different, they do not have full understanding of one another. This mutual misunderstanding has sometimes quite contrary effect—an excessive "trustfulness" to results and conclusions of experts of other field. Even within the same scientific field the experimentalists, interpreting their results, frequently do not know the limits of applicability of theories used and, on the contrary, theorists do not know frequently the "weaknesses" of experimental data and trust them excessively.

3. WHAT IS NEEDED FOR USERS: A SINGLE MODEL OR A SET OF MODELS?

The most important user of radiation belt models is the aerospace industry, and particularly, engineers-designers and operators of spacecrafts. For what purposes do they use the space environment model? The answer on this question determines the requirements to models.

Engineers use the space environment model both in design process and during spacecraft operation. When designing the evaluation of radiation doses is carried out, which determine the thickness and material of screens. In last years the practice of the spacecraft designing included also the evaluation of the surface and bulk charging which requires more detailed information about characteristics of particles, interacting with spacecraft. Obviously, that for first and for second evaluations (which, by the way, are executed the different experts) one needs information about particles of different energy ranges. Different are also requirements of engineers to the resolution of date in time and space.

Old static models are rather suitable for dose calculations but charging calculation codes require new information about dynamics of spectra and fluxes of particles influencing the spacecraft in different point of its orbit.

Designer of spacecraft has main interest to most dangerous regimes of environmental effects. As for the operator, it is necessary for him to know the real situation in the spacecraft vicinity. Best of all for him to know the forecast of the situation, for beforehand to manage such actions, which would exclude the harmful environmental effects. Such information partially can give the environment monitoring with the help of facilities, located on-board, or with the help of special environment model.

It is hardly realistic to try to construct some unique environment model which would satisfy such diverse requirements of users-engineers! It is more reasonable to say about creation of a set of specialized models of new generation.

Our research group concentrates the attention on development of one of such specialized models, which we name "dynamic model". In our understanding the dynamic environment model for geostationary orbits is a set of formulas (or data base), which, after inputting of some set of source parameters, gives on exit (output) the information about flux and spectrum of charged particles, influencing the geostationary apparatus at given time, in given point of the orbit and for given geomagnetic conditions. The output information concerns primarily particles of such energies, which most effectively control the electrostatic charging of the spacecraft.

4. INPUT INDICES

We are convinced, that the dynamic model of the environment on the geostationary orbit should be based on data- and knowledge-bases of the solar-terrestrial physics (STP).

The contemporary STP integrates sophisticate theoretical models of magnetospheric and ionospheric processes, extensive data bases of ground-based and on-board measurements of geophysical quantities; some of these data-bases ara represented in the form of empirical models. The knowledges base of the STP includes quantitative information about long chains of cause-effect phenomena and, in particular, information about relations between geophysical quantitaties, registered on ground, and physical characteristics of environment in the bulk of the magnetosphere (including the region of the geostationary orbit).

That is why a set of input parameters of dynamic model must include indices determined from ground-based measurements and describing both the instant state of the magnetosphere and its subsequent behaviour.

Generally speaking, indices, determined from ground-based measurements, have long been used in radiation belt models. These are well-known indices of the magnetic activity: K_p, A_p, D_{st}, AE. But now they are used mainly "on inertia", without assesment of their suitability for model and for user. Actually, all aforementioned magnetic indices are inaccessible for user in real time scale and already on this reason they can not be used as input parameters in dynamic model. Furthermore, one does not frequently take into account that the K_p index has very low time resolution—three hours—and cannot be used therefore in a model, attempting describe dynamics of particle fluxes in the course of the substorm. The K_p index characterizes mainly processes in the high-latitude magnetosphere, and D_{st} in its low latitude part.

Even these brief comments can illustrate our statement, that the selection of input parameters of dynamic model requires special and rather detailed research.

So, one of key problems of dynamic environment modelling is the problem to choose a set of input parameters (indices). One may state such general requirements to such indices:

- They should be linked with characteristics of particles, attacking the geostatonary spacecraft, i.e. they should describe well physical processes in the region of the geostationary orbit.
- A set of such indices should identify the state of the magnetosphere uniquely.
- Values of indices should be determined on the basis of measurements, accessible for users of the model in real time (ground-based observations are most suitable for that).
- It is desirable that the chosen indices have the well investigated relations with indices traditionally used before (K_p, D_{st}, etc.) that will give the opportunity to continue the use of existing long series of geophysical observations.
- It is desirable also, that indices had the forecasting potential.

5. CORRELATION OF ELECTRON FLUXES AT GEOSTATIONARY ORBIT WITH SOLAR WIND CHARACTERISTICS

In attempts to develop the dynamic environment model for geostationary orbit we are focusing now on the problem of input indices. In particular, we are searching for some substitute for magnetic indices like K_p; as the source of indices we consider ground-based measurements of fields of geophysical quantities, and/or solar wind characteristics. The last "source" is of the prime interest due to its possible forecasting potential. Some example of results received by authors recently in this field is given below and is disussed in detail in [*Degtjarev et al.*, 1995ab].

Table 1 demonstrates results of our searching for relations between electron fluxes on geostationary orbit and some solar

Table 1. Maximum values of crosscorrelation function and time lags (in days) between day-averaged solar wind parameters, indices of magnetospheric activity and electron fluxes on geostationary orbit

Input values	Electron energy, kev					
	30–45	65–95	140–200	200–300	430–630	930–1360
K_p	0.733 (1)	0.779 (1)	0.782 (2)	0.781 (2)	0.691 (3)	0.593 (3)
AE	0.725 (0)	0.746 (1)	0.734 (2)	0.726 (2)	0.654 (3)	0.603 (3)
$-D_{st}$	0.726 (0)	0.723 (1)	0.751 (1)	0.742 (1)	0.720 (2)	0.614 (2)
v	0.733 (0)	0.746 (0)	0.772 (1)	0.780 (1)	0.731 (2)	0.593 (2)
nv^3	0.496 (1)	0.585 (2)	0.592 (2)	0.590 (2)	0.486 (4)	0.405 (5)
vs	0.498 (2)	0.560 (2)	0.632 (2)	0.642 (2)	0.457 (3)	0.340 (4)
e	0.478 (2)	0.523 (2)	0.568 (2)	0.558 (2)	0.531 (3)	0.457 (3)
vB^2	0.451 (2)	0.542 (2)	0.565 (2)	0.575 (2)	0.403 (3)	0.257 (3)
$-vB_z$	0.413 (0)	0.421 (1)	0.371 (2)	0.353 (2)	0.427 (3)	0.425 (3)
nv^2	0.385 (2)	0.476 (2)	0.478 (3)	0.484 (3)	0.396 (5)	0.369 (7)

Notations: n=SW density; v=SW velocity; $B_z = z$-component of IMF; s=normal deviation of IMF; e: Akasofu index

wind parameters. For comparison relations between electron fluxes and magnetic indices are presented.

Three-hour averaged data on electron fluxes registered by geostationary LANL spacecraft 1977-007 [*LANL*, 1981] and King catalogue [*King*, 1979] was used as input data. For investigation two time intervals was chosen: 11.01.78–14.08.78 (includes 8 Sun revolutions) and 15.08.78–27.12.78 (includes 5 Sun revolutions). Missed data were recovered by linear interpolation.

In our study we tested 20 different combinations of SW parameters but in the table was included only that, which demonstrate the best correlation features and/or have definite physical sense.

6. CONCLUSIONS

1. It's hard to imagine, that the unique model of RB may be constructed! A set of models for different users is a more realistic goal.
2. Input parameters (indices), which will control the future dynamical model of radiation must be choosen after additonal detailed investigations.
3. Input indices must be accessible to users in real time!
4. It is desirable that input parameters should have predictive potential.
5. Solar wind parameters may be included in a list of candidates for being input indices. They must be investigated in detail.

Acknowledgements. This work was supported by the Russian Foundation for Fundamental Investigations (RFFI), grant no. 94-05-16005-a.

REFERENCES

Degtjarev, V.I., Zherebtzov, G.A. and Sheshukov, S.S. The dynamics of electron fluxes at geostationary orbit and their relations with SW parameters and geomagnetic activity, *Issl. po geomagn., aeron. and Solar phys., 103*, 62–72, 1995.

Degtjarev, V.I., Platonov, O.I., Popov, G.V., et al., Relation of electron fluxes (E>200 kev) at geostationary orbit with solar and geomagnetic activity. *Issl. po geomagn., aeron. and Solar phys., 103*, 72–76, 1995.

King, J.H., Interplanetary medium date book-supplement 1975–1978. NSSDC-A-RC, 79-08, NASA, GSFC, 1979.

The Los Alamos geostationary orbit synoptic data set. A compilation of energetic particle data. Preprint, LA-8843, 1981.

G.V. Popov, V.I. Degtjarev, S.S. Sheshukov, Inst. Solar-Terr. Phys., P.O. Box 4026, Irkutsk, 664033, Russia; E-mail: root@sitmis.irkutsk.su)

Dynamics of energetic electrons in the radiation belts

L.V. Tverskaya

Scobeltsyn Institute of Nuclear Physics, Moscow State University, Moscow, Russia

The dependence of the $E > 1\,\text{MeV}$ radiation belt electron behavior on geomagnetic activity and on the 11- and 22-year solar cycles is analysed. The empirical dependence which defines the position of the outer radiation belt maximum for relativistic electrons injected into the magnetosphere during a magnetic storm ($L_{\max}$) as a function of the largest D_{st}-variation amplitude for each storm ($|D_{st}|_{\max}$) is shown to be of the form: $|D_{st}|_{\max} = 2.75 \times 10^4 / L_{\max}^4$. A case of rapid diffusion of electrons with $E_e > 1\,\text{MeV}$ into the inner radiation belt is considered. New radiation belt formation according to CRRES and DMSP data are discussed.

1. INTRODUCTION

The storm-time behavior of the $< 100\,\text{keV}$ and $> 1\,\text{MeV}$ radiation belt electrons has been shown by diverse experiments to be very different. The demarcation energy is 200–300 keV [*Tverskoy*, 1968]. Even during moderate magnetic disturbances the electron fluxes with energies of dozens of keV fill the slot between the inner and outer radiation belts. The $\sim 1\,\text{MeV}$ electrons are injected to $L \sim 3$ during strong magnetic storms only when, as a rule, the storms are in their recovery phase, whereupon the radial diffusion makes the electrons move to deeper L-shells.

The estimates obtained in terms of a simple nonsteady model for electric fields of substorms [*Bondareva and Tverskaya*, 1973] making allowance for possible potential differences across the magnetosphere [*Tverskoy*, 1969, 1972] have shown that electrons with energies of dozens of keV can be injected to $L \sim 3$ during an individual moderate-intensity substorm. Frequent repetition of substorms during a storm gives rise to enhanced diffusion of these electrons to the inner belt.

Electrons with $E \sim 1\,\text{MeV}$ probably are injected to the inner magnetosphere as a result of betatron acceleration. This occurs when a portion of the magnetotail plasmasheet field lines are drawn into the trapped radiation region [*Tverskoy*, 1969]. In that case electrons may also be accelerated during the storm recovery phase. The particles are injected within a field which proves to be much weakened by the ring current. After the ring current decays and the field recovers on the respective L-shells, the particle energy must markedly rise adiabatically. This mechanism can probably account for the few-day delay of the relativistic electron diffusion waves with respect to magnetic storm onset [*Tverskoy*, 1968]. The intensity of injected electrons is controlled by the conditions in interplanetary space [*Gorchakov et al.*, 1985; *Baker et al.*, 1986]. The injected-particle intensity peak ($L_{\max}$) is formed near the boundary between the dipole field lines and the field lines extended pronouncedly to the magnetospheric tail. Besides, this boundary defines the position of the discrete auroral form, polar electrojets, field-aligned currents, and inverted-V particle precipitations [*Tverskoy*, 1982]. Approximately in this region must lie the minimum latitude storm-time position of the magnetosphere cut-off boundary of solar protons with $\sim 1\,\text{MeV}$ energy and ring current maximum [*Sosnovets and Tverskaya*, 1986]. The $L_{\max}$ peak for storm-time injected radiation belt protons and ions must be situated there too. Knowing the position of the electron belt maximum, we can thus predict the extreme storm-time positions of various magnetosphere structure elements [*Tverskaya*, 1986, 1993, 1996].

2. THE DEPENDENCE OF L_{MAX} ON GEOMAGNETIC DISTURBANCES AND SOLAR CYCLE

The dependence $L_{\max} = f(|D_{st}|_{\max})$ was inferred from the data of five storms with $|D_{st}|_{\max}$=30–140 nT by [*Williams et al.*, 1968] for the first time and appeared to be linear. In later studies it was shown that the dependence is essentially

Radiation Belts: Models and Standards
Geophysical Monograph 97
Copyright 1996 by the American Geophysical Union

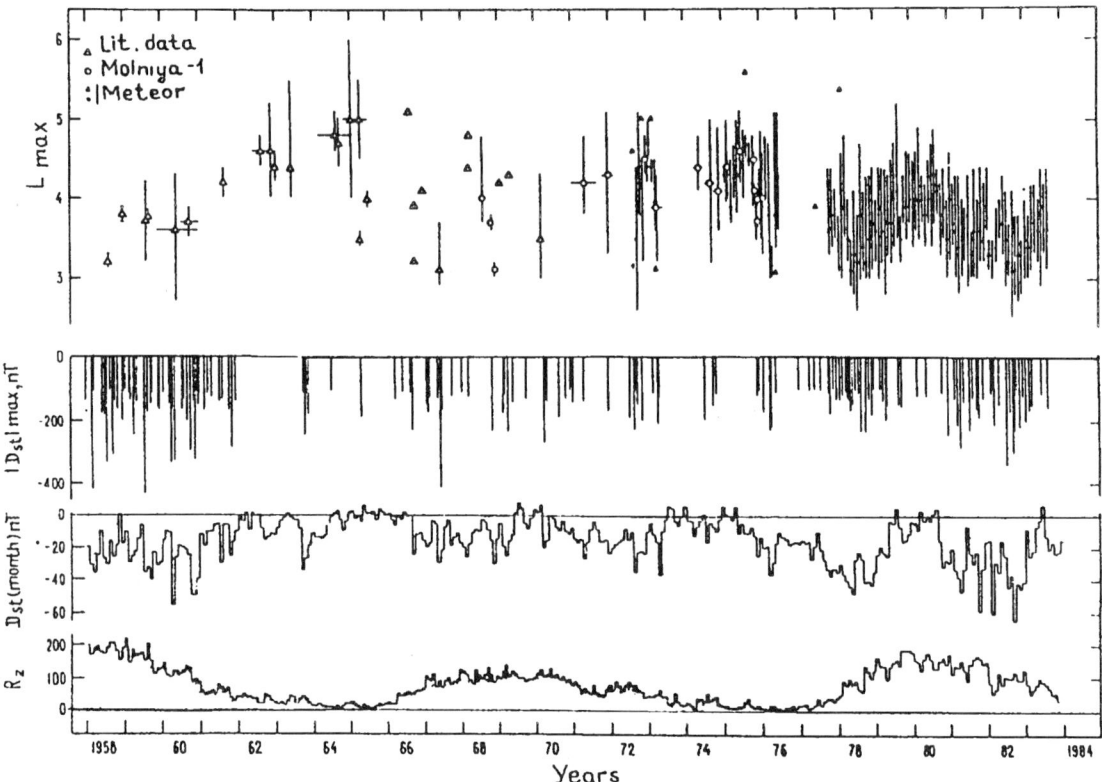

Figure 2. The position of the outer radiation belt maxima ($L_{\max}$) during the period 1958–1984 (top), the storms with $D_{st} > 100$ nT and monthly averaged D_{st} (middle), sunspot number R_z (bottom).

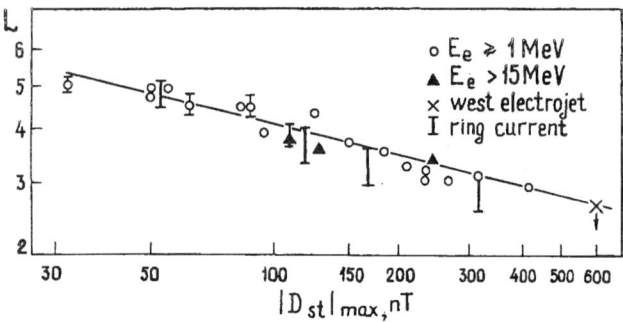

Figure 1. The position of the outer-belt maximum, $L_{\max}$, of storm-time injected > 1 MeV electrons as a function of storm-time D_{st}-variation amplitude. The open circles are $L_{\max}$ values for > 1 MeV electrons, the triangles for 15 MeV electrons. The vertical bars designate the maxima of the ring current ions energy density. The cross indicates the extreme position of the westward polar electrojet during the March 13–14, 1989 storm [*Tverskaya*, 1996].

nonlinear and is of the form given by [*Tverskaya*, 1986]:

$$|D_{st}|_{\max} = \frac{2.75 \times 10^4}{L_{\max}^4}. \quad (1)$$

Figure 1 represents this empirical dependence.

We have no data on the radiation belt position after the giant March 13–14, 1989 storm with $|D_{st}|_{\max} = 600$ nT. The predicted position $L_{\max}$ according to Eq. (1) is 2.6 ± 0.2. The dependence of the outer radiation belt maximum on solar cycle and geomagnetic storm amplitude is represented in Figure 2 [*Kirdina et al.*, 1992]. It is seen that there is no direct correlation of the $L_{\max}$ position with solar activity, but it is influenced by the D_{st}-variation. For a period of some years with permanent Molniya-1 and Meteor data, the correlation coefficient of $L_{\max}$ with R_z is -0.2 (no correlation), while there is a high correlation coefficient (-0.7) for $L_{\max}$ and monthly averaged D_{st}. There is a very interesting feature in Meteor data during the 1978–1984 period, namely, the activation of magnetic storms before and after solar maximum (1979–1980) and corresponding displacement $L_{\max}$ to lower L.

3. THE RAPID DIFFUSION OF RELATIVISTIC ELECTRONS INTO THE INNER BELT

Figure 3 illustrates the variations of energetic electrons with Molniya-1 data in March–May 1973 [*Vakulov et al.*, 1976]. The data show the displacement of electrons with different energies to lower L values after the strong magnetic storm. This effect on differential spectrometer data was observed

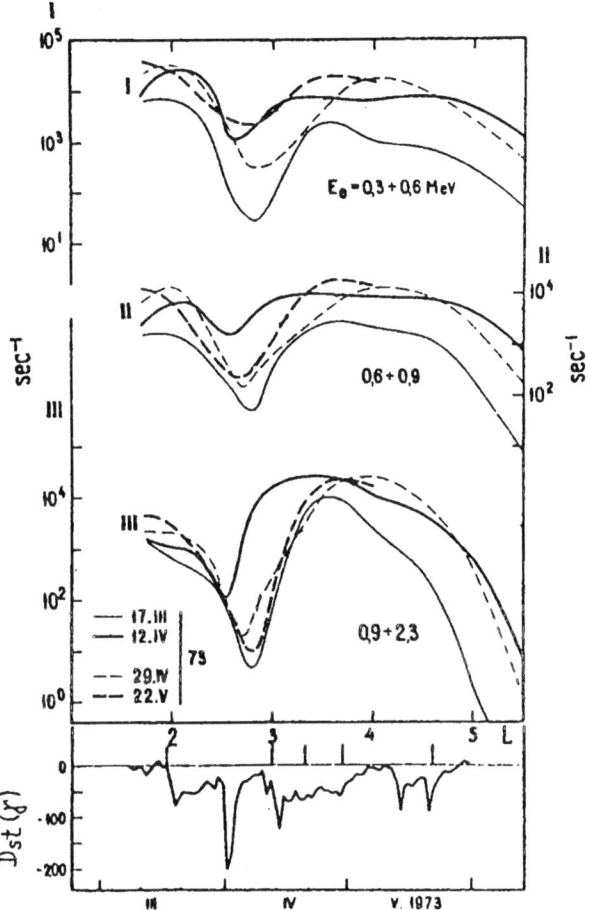

Figure 3. Count rate distribution for three channels of the electron differential spectrometer according to Molniya-1 data during March 17, April 12, 20, 29 and May 22, 1973 (top); D_{st}-variation (bottom).

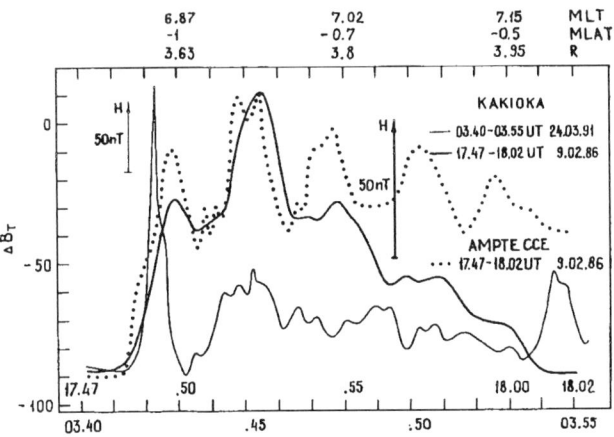

Figure 4. H-component magnetograms at Kakioka. ΔB_T of AMPTE CCE indicates a difference of the total field from the IGRF-80 model field [*Yumoto et al.*, 1989].

earlier by [*Pfitzer et al.*, 1968]. But after the September 1966 storm there were no variations in the 690–1700 keV channel in the inner belt. Our data demonstrate rapid diffusion of electrons with $E > 0.9$ MeV at $L < 2$. The reason of these variations may be quasi-periodic variations of magnetosphere disturbances with corresponding periods (the drift period for $E_e \sim 1$ MeV at $L = 2$ is 40 m). The magnetograms from ground-based stations for April–May 1973 display large amplitude global variations of DP-2 type with a ten minutes period. Simultaneously we observed particle flux variations at the distant part of the Molniya-1 orbit [*Kovrigina et al.*, 1976]. Giant quasi-periodical DP-2 variations may be the cause of the resonant redistribution of radiation belt electrons and protons [*Vernov et al.*, 1972]. During sudden compressions of the magnetosphere (SC and SI) series of micropulsations arise. Such global micropulsations give rise to enhanced diffusion of particles with suitable drift periods [*Tverskoy*, 1968; *Dmitrenko et al.*, 1993].

Untypical behavior of relativistic electrons during the moderate geomagnetic storm was observed at the low altitude by Intercosmos-19 and Cosmos-900: the intensity didn't decrease in the storm main phase. Moreover, it increased and electrons with different energies simultaneously moved to $L \sim 3$ in 2 days [*Volkov et al.*, 1989]. A similar effect was observed by SAMPEX and CRRES [*Baker et al.*, 1994]. See discussion of CRRES data further on.

4. CRRES AND DMSP NEW RADIATION BELTS

The most interesting phenomenon in radiation belts during the last years was the appearance of a "new" radiation belt during the March 24, 1991, SC registered by CRRES [*Blake et al.*, 1992].

More than 2 years there was no explanation of this effect. The idea that in this case there must be pair (positive and negative) sudden impulses was first proposed by *Chuchkov et al.* [1993] and *Pavlov et al.* [1993]. The analytical solution was based on the theory of particle drift in the electric and magnetic field of sudden impulses [*Tverskoy*, 1964]. Some parameters and the duration of the pair impulses, which are necessary to explain the experimental data, were predicted. Subsequent analysis of real ground-based magnetograms from the Kakioka station corroborates the theoretical predictions.

Practically simultaneously the results of computer simulations developing an analogous idea on the basis of real magnetic and electric field data were published [*Li et al.*, 1993].

In Figure 4 two giant sudden impulses, 24 March 1991 and 9 February 1986, are compared. The magnetosphere was strongly compressed several times during the February 1986 storm. These compressions excited SC and SI-associated oscillation with a 1–3 m period [*Yumoto et al.*, 1989]. During this SC and SI the diffusion coefficient for radiation belt electrons must increase and there may be resonance acceleration of electrons with drift periods of a few minutes. According to experimental data at low altitudes [*Volodichev et al.*, 1991] during the February 1986 storm unusually rapid diffusion and appearance of electrons with energies > 15 MeV on $L = 1.2$–1.3 were observed.

The structure of the sudden impulse of February 9, 1986 was very complex. Practically it contains some pair positive and negative impulses. The duration of each of them is longer than that of the March 24, 1991 SC. If this SI has formed a new radiation belt, electrons and protons must have lower energies than during the CRRES case. Higher-energy particles will move adiabatically. According to low altitude DMSP data, during the February 1986 storm a second proton belt with energies up to 50 MeV was found at $L = 2.8$. There were no "new" protons with energies > 50 MeV [*Gussenhoven et al.*, 1989].

It is known that under the influence of sudden impulses there is a strong dependence of the diffusion coefficient D_0 on the impulse amplitude. For example, the mean statistical value of D_0 increases by two orders of magnitude when the impulse amplitude is 100 nT [*Tverskoy*, 1965, 1968]. Some cases when the slot region has been filled with relativistic electrons in a few days were observed near the equatorial plane by CRRES in the summer of 1991 [*Baker et al.*, 1994]. One of the causes of these occurences may be the unusual increase in D_0. As is seen from the low-latitude station data many sudden impulses with an amplitude of tens nT and three giant sudden impulses with the >100 nT amplitude really were registered in June-August 1991. It is obvious that the giant fluctuations of D_0 do take place. Such fluctuations may also be a possible cause of the faster inward displacement of a "new" radiation belt [*Blake et al.*, 1992] than in the period 1963–1965, when a second radiation belt of energetic protons moved from $L = 2.25$ to $L = 2.15$ in two years [*Williams*, 1970]. So, when unusual radiation belt variations are studied, the corresponding global geomagnetic disturbances must be carefully analysed.

Acknowledgements. The author is indebted to L.N. Kainara for the Moscow magnetic records of March 13–14, 1989, to T. Kohno, E.P. Kharin and T. Kamei for their help in obtaining Kakioka data.

REFERENCES

Baker, D.N., J.B. Blake, R.W. Klebesadel and P.R. Higbie, Highly relativistic electrons in the Earth's outer magnetosphere. 1. Lifetimes and temporal history 1974–1984, *J. Geophys. Res., 91,* 4285, 1986.

Baker, D.N., J.P. Blake, L.B. Callis, J.R. Cummings, D. Hovestadt, S. Kanekal, B. Klecker, R.A. Mewaldt and R.D. Zwickl, Relativistic electron acceleration and decay time scale in the inner and outer radiation belts: SAMPEX, *Geophys. Res. Lett., 21,* 409, 1994.

Blake, J.B., M.S. Gussenhoven, E.G. Mullen and R.W. Fillius, Identification of an unexpected space radiation hazard, *IEEE Trans. Nucl. Sci., 39,* 1761, 1992.

Bondareva, T.B. and L.V. Tverskaya, On the radiation belt particle drift during substorms, *Geomagn. i Aeron., 13,* 723, 1973.

Chuchkov, E.A., N.N. Pavlov, L.V. Tverskaya and B.A. Tverskoy, The variations of radiation belts during strong geomagnetic storm in March 24–26, 1991, Workshop on Space Radiation Environment: Empirical and Physical Models, Dubna, 2–7 June 1993, Abstracts, 35, 1993.

Dmitrenko, V.V., V.B. Komarov and B.A. Tverskoy, Radial diffusion as mechanism of the formation of the stationary high-energy electron fluxes in geomagnetosphere, *Cosmicheskie Issled., 31,* 83, 1993.

Gorchakov, E.V., V.A. Iozenas, M.V. Ternovskaya, P.P. Ignatiev, V.G. Aphanasiev and K.G. Aphanasiev, Injection of hard electrons into the outer radiation belt during magnetic storms, *Geomagn. i Aeron., 25,* 738, 1985.

Gussenhoven, M.S., E.G. Mullen and E. Holeman, Radiation belt dynamics during solar minimum, *IEEE Trans. Nucl. Sci., 36,* 2008, 1989.

Kirdina, T.A., L.M. Kovrygina, V.A. Kuzmina, A.B. Malyshev, E.N. Sosnovets and L.V. Tverskaya, The influence of solar activity cycle on the position of outer belt electron maximum, *Cosmic Rays, 88,* 1992.

Kovrygina, L.M., M.I. Panasyuk, E.N. Sosnovets, L.V. Tverskaya and O.V. Khorosheva, Fast quasiperiodic variations of particles in high latitude regions of the magnetosphere according to Molniya 1 data, *Solar Terr. Phys. Symp. Tbilisi, Thesis,* 113, 1976.

Li, X., I. Roth, M. Temerin, J.R. Wygant, M.K. Hudson and J.B. Blake, Simulation of the prompt energization and transport of radiation belt particles during the March 24, 1991 SSC, *Geophys. Res. Lett., 20,* 2423, 1993.

Pavlov, N.N., L.V. Tverskaya, B.A. Tverskoy and E.A. Chuchkov, Variations of radiation belt energetic particles during strong magnetic storm March 24–26, 1991, *Geomagn. i Aeron., 33,* N6, 41, 1993.

Pfitzer, K.A. and J.R. Winckler, Experimental observation of a large addition to the electron inner radiation belt after a solar flare event, *Geophys. Res., 73,* 5793, 1968.

Sosnovets, E.N. and L.V. Tverskaya, Dynamics of the ring current on direct measurements and on solar cosmic rays in the magnetosphere, *Geomagn. i Aeron., 26,* 107, 1986.

Tverskaya, L.V., On the injection boundary of electrons into the magnetosphere, *Geomagn. i Aeron., 26,* 864, 1986.

Tverskaya, L.V., Dynamic model of the relativistic electron injection boundary in the magnetosphere, Workshop on Space Radiation Environment: Empirical and Physical Models, Dubna, 2–4 June 1993, Abstracts, 13, 1993.

Tverskaya, L.V., The latitude position dependence of the relativistic electron maximum as a function of D_{st}, *Adv. Space Res., 18,* 135, 1996.

Tverskoy, B.A., Dynamics of the Earth's radiation belt I, *Geomagn. i Aeron., 4,* 224, 1964.

Tverskoy, B.A., Transfer and acceleration of particles in the Earth's magnetosphere, *Geomagn. i Aeron., 5,* 793, 1965.

Tverskoy, B.A., Dynamics of the Earth Radiation Belt (in Russian), Nauka Publ., Moscow, 1968; NASA Techntranslation F-635, June 1971.

Tverskoy, B.A., Main mechanisms in Formation of Earth's Radiation Belts, *Rev. Geophys., 7,* N1, 2,219, 1969.

Tverskoy, B.A., On the electric fields in the Earth's Magnetosphere, *Doklady Akad. Nauk SSSR, 188,* 575, 1969.

Tverskoy, B.A., Electric fields in the magnetosphere and origin of trapped radiation, Solar-terrestrial Physics/1970, Reidel Publ. Co., 297, 1972.

Tverskoy, B.A., On the field-aligned currents in the magnetosphere, *Geomagn. i Aeron., 22,* 991, 1982.

Vakulov, P.V., L.M. Kovrygina, Ju.V. Mineev and L.V. Tverskaya, Variations in intensity and spectrum of energetic electrons in earth radiation belts during strong magnetic disturbances, *Space Res., 16,* 529, 1976.

Vernov, S.N., I.Ya. Kovalskaya, M.I. Panasyuk, I.A. Rubinshtein, E.N. Sosnovets, L.V. Tverskaya and O.V. Khorosheva, Proton radiation belt variations in July–August, 1970, *Space Res., 12,* 1493, 1972.

Volkov, I.B., A.V. Dronov, L.M. Kovrygina, Ju.P. Kratenko, Ju.V. Mineev, E.N. Sosnovets and L.V. Tverskaya, Dynamics of energetic electrons in the outer radiation belt on simultaneous Intercosmos-19 and Cosmos-900 data, *Cosmicheskie Issled., 23*, 642, 1985.

Volodichev, N.N., A.A. Gusev, Ju.V. Mineev, G.I. Pugacheva, K. Kudela and L. Yust, Dynamics of fluxes and spectra of the energetic electrons during strong magnetic storm, *Izvestia Akad. Nauk SSSR, 55*, 2000, 1991.

Williams, D.J., I.F. Arens and L.T. Lanserotti, Observations of trapped electrons at low and high altitudes, *J. Geophys. Res., 73*, 5673, 1968.

Williams, D.J., Sources, Losses and Transport of Magnetospherically Trapped Particles, ESSA Technical Report ERL 180-SDL 16, Boulder, Co., August, 1970.

Yumoto, K., K. Takahashi, T. Ogawa and T. Watanabe, SC and SI-Associated ULF and HF-Doppler Oscillations during the Great Magnetic Storm on February 1986, *J. Geomagn. Geoelectr., 41*, 871, 1989.

L.V. Tverskaya, Scobeltsyn Institute of Nuclear Physics, Moscow State University, Moscow 119899, Russia

Field Modeling Methods for the Inner Magnetosphere

D.P. Stern

Laboratory for Extraterrestrial Physics, Goddard Space Flight Center, Greenbelt, Maryland

The motion of particles trapped near Earth is governed by the magnetic field and in particular by the structure of its field lines. The field line structure can be described mathematically by Euler potentials: their form in a dipole field is well known, and this form can be deformed to provide a wide variety of asymmetric models of the magnetosphere. Their practical use, unfortunately, is hampered by their nonlinearity and by the difficulty of fitting them to a given pattern of currents. The main models in use are based on different representations, especially those by Tsyganenko which fit defined current patterns to large sets of observed fields. The frequently used L parameter is basically a function of the invariants I and B, and does not describe field lines. However, it is possible to derive drift surfaces of particles with a given (L, B) and show that they depend weakly on B, so that particles of a given L stay on a certain surface expressible in Euler potentials. If these potentials are known, one could in principle derive L from them, and such derivations exist. Since the derivations are only approximate, the alternative procedure, requiring the derivation of I by a line integral, is likely to continue.

1. INTRODUCTION

Magnetic fields form the framework for magnetospheric research, and to obtain accurate interpretations and predictions, one generally needs accurate information about the magnetic field $\mathbf{B}$. Near Earth almost all of $\mathbf{B}$ comes from the "main field" originating in the Earth's core, typically all but 15–35 nT out of 30–60,000 nT [*Fischbach et al.*, 1994, Table 1]. This field can be represented by a scalar potential expanded in spherical harmonics (a is the Earth's radius):

$$\mathbf{B} = -\nabla \gamma \quad (1)$$

$$\gamma = a \sum_{n,m} \left(\frac{a}{r}\right)^{n+1} P_n^m(\theta) \left[g_n^m \cos m\phi + h_n^m \sin m\phi\right]. \quad (2)$$

The largest terms are those representing the dipole component; higher harmonics with (g_n^m, h_n^m) create some irregularities near Earth but decrease with distance faster than the dipole field, so that at 2–4 R_E the field is fairly close to a dipole (plus a small nearly constant field due to external sources). An approximate rule by *Tsyganenko* [1990, Eq. (6)] states that in calculating $\mathbf{B}$ of the main field at a distance r to an accuracy of 1 nT, one only needs to include in Eq. (2) terms up to $n = 1 + 9(a/r)$.

With increasing r the external field becomes appreciable. The four sources of this field are the current systems of the magnetopause, ring current, tail and Birkeland currents; the usual modeling approach is to represent the field of each in a way that best suits its features, derive the parameters of each representation from data and then add up the results [*Tsyganenko*, 1989a, 1990, 1995; *Stern*, 1994].

The magnetopause field, for instance is best represented by a scalar potential as in Eq. (1), since it is current-free in the interior of the magnetosphere; however, the expansions representing that potential differ from Eq. (2) and use spheroidal, cylindrical or parabolic harmonics. The ring current and tail current in recent models [*Tsyganenko*, 1989b, 1995] are represented by a combination of vector potentials $\mathbf{A}^{(i)}$ of suitable standard current distributions; Birkeland currents may be assumed to flow in thin sheets and their fields can then be described by scalar potentials, differing on each side of the sheet, or by other descriptions of curl-free fields, e.g. image dipoles. The expression for each such component contains

Radiation Belts: Models and Standards
Geophysical Monograph 97
This paper is not subject to U.S. copyright.
Published in 1996 by the American Geophysical Union

certain adjustable parameters, and in most cases these are fitted by least squares using a large set of magnetic observations in space.

An exception to this fitting procedure is the magnetopause field, which—in a closed magnetosphere, when all internal magnetospheric sources are given, including the main field—is uniquely determined by the shape of the magnetopause. That shape has been studied by *Sibeck et al.* [1991], *Roelof and Sibeck* [1993] and others, using satellite crossings. The size and shape vary with the dynamic pressure p of the solar wind and with the interplanetary magnetic field $\mathbf{B}_{IMF}$, and these results can be incorporated in the model.

Thus $\mathbf{B}$ is generally expressed as $-\nabla\gamma$ or as $\nabla\times\mathbf{A}$. Another representation, related to vector spherical harmonics [*Backus*, 1986], is

$$\mathbf{B} = \nabla\times\psi_1\mathbf{r} + \nabla\times\nabla\times\psi_2\mathbf{r} \qquad (3)$$

and has been developed by *Kosik* [1984] into a relatively crude model. All these models are linear, and any two fields of these types are easily combined.

2. EULER POTENTIALS

However, it is often desirable in magnetospheric physics to know about the structure of magnetic field lines (e.g., given a point in space, to find where its field line is likely to cross the equator, or reach Earth). It is possible to do so by using a model to trace field lines numerically, but the information is not given explicitly by γ or by $\mathbf{A}$.

That information is however available if one uses Euler potentials (α,β), satisfying

$$\mathbf{B} = \nabla\alpha\times\nabla\beta. \qquad (4)$$

Since $\mathbf{B}$ at any given point P is orthogonal to $\nabla\alpha$, it is tangential there to the local surface of constant α, whose equation may be written as $\alpha(x,y,z)=\alpha_c$. In a similar way $\mathbf{B}$ at P is tangential there to some surface $\beta(x,y,z)=\beta_c$. Hence the intersection line of those two surfaces is everywhere tangential to $\mathbf{B}$, making it a field line, and anywhere along that line the functions (α,β) have the same values (α_c,β_c).

As an example, if the z-directed dipole coefficient g_1^0 were the only non-zero term in Eq. (2), then the dipole field $\mathbf{B}_0$ would satisfy (4) with

$$\alpha = \alpha_0(x,y,z) = ag_1^0\frac{a}{r}\sin^2\theta \qquad \beta = \beta_0(x,y,z) = a\phi \qquad (5)$$

(some users transfer the factor a from β to α). Because of the affinity between Euler potentials and field lines, many equations become simpler when expressed in terms of (α,β). For instance, the variation of a magnetic field "frozen" to a moving plasma satisfies

$$\frac{D\alpha}{Dt}=0 \qquad \frac{D\beta}{Dt}=0, \qquad (6)$$

where D/Dt is the convective derivative

$$\frac{D}{Dt} = \frac{\partial}{\partial t} + \mathbf{v}\cdot\nabla. \qquad (7)$$

The representation is non-unique—for instance, adding any $f(\alpha)$ to β leaves (4) valid as before—but this is ignored for now, as is the possibility of (α,β) to be many-valued; see *Stern* [1994] and references given there.

Suppose a map is given on which each point on the Earth's surface is labeled by its values of (α,β). Suppose also that we have some method of locally calculating (α,β) at any given point P, without having to trace the field line through P. Then, given any point in the magnetosphere, we can quickly find where its field line intersects the Earth (or more precisely, the two conjugate points where this happens), simply by finding where on Earth do (α,β) match those of the given point.

Unfortunately, such a method is hard to come by, because Eq. (4) is nonlinear, involving products of the derivatives of α and β. Given $\mathbf{B}' = \nabla\alpha'\times\nabla\beta'$ and $\mathbf{B}'' = \nabla\alpha''\times\nabla\beta''$, if we require the Euler potentials of $\mathbf{B}'+\mathbf{B}''$, then in the general case knowing (α',β') and (α'',β'') is of no help whatsoever. However, in fields where one Euler potential is shared—e.g. axisymmetric fields with $\beta=\phi$, or 2-dimensional ones with $\beta=y$—the other one may be superposed.

The situation is a little better for the Earth's field, where the axially symmetric dipole field $\mathbf{B}_0=\mathbf{B}'$ is much larger than the added field $\mathbf{B}_1=\mathbf{B}''$ due to higher harmonics. In that case one can approximately obtain (α,β) by perturbation. Let Eqs. (1)–(2) be rewritten

$$\mathbf{B} = -\nabla\gamma_0 - \nabla\gamma_1, \qquad (8)$$

where γ_0 refers to the dipole and γ_1 to the rest of the expansion, small by comparison. The method converges better if the z-axis is first rotated to coincide with the dipole axis, so that g_1^1 and h_1^1 vanish. Let (α,β) be similarly resolved: then

$$-\nabla\gamma_0 - \nabla\gamma_1 = \nabla(\alpha_0+\alpha_1)\times(\beta_0+\beta_1). \qquad (9)$$

Neglecting the 2nd order term $\nabla\alpha_1\times\nabla\beta_1$ gives

$$-\nabla\gamma_0 = \nabla\alpha_0\times\nabla\beta_0, \qquad (10)$$
$$-\nabla\gamma_1 = \nabla\alpha_0\times\nabla\beta_1 + \nabla\alpha_1\times\nabla\beta_0. \qquad (11)$$

If α_1 and β_1, like γ_1, are linear in the harmonic coefficients (g_n^m, h_n^m), then Eq. (11) can be resolved into individual equations, one for each harmonic term. If γ_1 is then expanded as in Eq. (1), solutions for α_1 and β_1 can be obtained analytically [*Stern*, 1967]. The first-order potentials $(\alpha_0+\alpha_1, \beta_0+\beta_1)$ can be tested by mapping them onto the surface of the Earth. If the conjugate points obtained by matching the values of these potentials are compared with the ones obtained from field line tracing, the discrepancy is found to be about $0.5°$.

In summary, Euler potentials provide a field representation with attractive features, useful in describing trapped particle motion, but hard to implement because of its inherent nonlinearity. It may also be noted that (α,β) are useful in theory, especially in characterizing adiabatic motions [e.g. *Northrop and Teller*, 1960] and in convection theory [e.g. *Harel et al.*, 1981; *del Pozo and Blanc*, 1994]. Further below they will be used to provide the theoretical grounding of the L-parameter and to derive its limitations.

3. DEFORMATIONS

One interesting application of (α, β) is the deformation of magnetic fields, the "stretch transformation" [*Stern*, 1987]. The magnetospheric field is obviously a deformed dipole field, compressed at noon and stretched out on the tail side. Does a mathematical procedure exist which produces such a deformation? One can easily visualize a physical process which does it: embed the dipole in a perfectly conducting fluid, to which the field lines become "frozen", and then let the fluid move.

Suppose such a flow drags some point $P = (X, Y, Z)$ in the dipole field, with field $\mathbf{B}(X, Y, Z)$ and Euler potentials $\alpha(X, Y, Z)$ and $\beta(X, Y, Z)$, to a new point $P' = (x, y, z)$. Since by Eqs. (6)–(7) the values of the Euler potentials associated with a moving fluid element stay the same throughout its motion, the new Euler potentials $\alpha'(x, y, z)$ and $\beta'(x, y, z)$ at the mapped point P' are the same as those at P. The entire transformation $\alpha \to \alpha', \beta \to \beta'$ is then characterized by three functions which relate any P' to its P, namely $X(x, y, z), Y(x, y, z)$ and $Z(x, y, z)$, and it satisfies

$$\left. \begin{array}{c} \alpha'(x,y,z) = \alpha(X,Y,Z) \\ \\ \beta'(x,y,z) = \beta(X,Y,Z) \end{array} \right\} \quad (12)$$

Given those three functions, the new Euler potentials (on the left above) are easily expressed, and so is the deformed field $\mathbf{B}'$. The coordinate axes are the same for $\mathbf{B}$ and for $\mathbf{B}'$.

But what about deforming a magnetic field $\mathbf{B}$ when its Euler potentials are not known? Remarkably, it turns out that when one takes Eqs. (12) through the algebra of expressing the new field $\mathbf{B}'$ in terms of (α, β) and the old field $\mathbf{B}$, the final relation between $\mathbf{B}$ and $\mathbf{B}'$ does not involve (α, β) explicitly but only requires the functions (X, Y, Z) and their derivatives. For instance, a simple "stretch transformation" which stretches out the tail (or compresses the dayside) is given by

$$X = f(x, y, z) \qquad Y = y \qquad Z = z \quad (13)$$

and yields

$$\left. \begin{array}{l} B_{x'} = B_x(f,y,z) - \dfrac{\partial f}{\partial y} B_y(f,y,z) - \dfrac{\partial f}{\partial z} B_z(f,y,z) \\ B_{y'} = \dfrac{\partial f}{\partial x} B_y(f,y,z) \\ B_{z'} = \dfrac{\partial f}{\partial x} B_z(f,y,z) \end{array} \right\} \quad (14)$$

The deeper reason for such a direct relation between the original and deformed magnetic fields is a theorem by Cauchy [1816], originally applied to the freezing-in of vorticity in ideal hydrodynamics (see Appendix A of *Stern* [1994]).

Deformation is a powerful tool in the modeling of magnetic fields but has so far been little used, because it is hard to control with it the current density $\mathbf{j}$. An additional problem in deforming the dipole field into a magnetospheric model (assuming for convenience a long but closed tail) is that such a mapping must replace singular dipole lines, which extend from the poles to infinity, with field lines which pass the cusps and which have different positions in the dipole field.

4. THE L PARAMETER

One reason why Euler potentials are of interest is that trapped particles are guided by magnetic field lines. If $\mathbf{E} = 0$, the energy of such particles is conserved, and in the guiding center approximation their orbits are characterized by two quantities—the mirror field B_m, and the "invariant integral" $I = \int (1 - B/B_m)^{1/2} ds$, calculated between mirror points. The "drift surface" followed by the guiding center of a trapped particle is everywhere tangential to field lines, and since its shape depends on the above two parameters, one can formally write its equation as

$$F(\alpha, \beta, I, B_m) = 0. \quad (15)$$

From this $\mathbf{B} \cdot \nabla F = 0$. A more convenient form is

$$\alpha = G(\beta, I, B_m). \quad (16)$$

Let subscript zero again denote the dipole field. In such a field drift surfaces are axially symmetric and hence β drops out, leaving

$$\alpha_0 = G_0(I, B_m) = \frac{a^2}{R_0} g_1^0, \quad (17)$$

where R_0 is the equatorial radius of the surface. The L-parameter is then defined as

$$L(I, B_m) = \frac{a\, g_1^0}{G_0(I, B_m)} = \frac{R_0}{a}, \quad (18)$$

with R_0/a giving R_0 in Earth radii. For the family of particles trapped on the same dipole field line, even though their values of I and B_m cover a wide range, the combination $L(I, B_m)$ always has the same value.

McIlwain (1961) assumed (and demonstrated in numerical experiments) that in a near-dipole field—such as the field of the inner radiation belt—all particles threaded by the same field line still keep nearly the same $L(I, B_m)$ throughout their drift around the Earth. That helped solve a practical problem, namely, deciding when two satellites (or the same satellite at different locations) sampled the same particle population.

In principle one should match both I and B_m at those two locations, a 2-parameter fit. However, if at the two locations $L(I, B_m)$ is the same (say, for locally mirroring particles), then to a good approximation the same family of orbits is sampled in both cases. Note that specific orbits observed at one point could be absent at the other, having mirrored closer to the equator: a near-Earth polar satellite at the same $L(I, B_m)$ as an equatorial one intercepts only a small part of the orbits seen by the equatorial spacecraft, only the part near the loss cone.

This method is still widely used. Given a spacecraft at some (x, y, z), and neglecting external field sources, I must be derived by field line integration, using the entire harmonic expansion of Eqs. (1)–(2). After that I and the local magnetic field B are used in an analytical approximation of $L(I, B_m)$, e.g. one due to McIlwain.

One may wonder if the same result could be obtained more directly, by expressing the shape of the average local drift surface and then seeing when the other satellite crosses it. After all, the equation of the surface is already known to the the zeroth order, namely as $r/\sin^2\theta = r_0$. If first order corrections are incorporated, linear in the higher harmonics (g_n^m, h_n^m) of the Earth's field, a fairly good approximation should result.

Such an approximation was in fact derived by *Pennington* [1961] and was expressed in the Euler potential formalism by *Stern* [1965, 1967, 1968]. Resolving Expression (3) into the zero-order approximation (4) and a first order correction, linear in (g_n^m, h_n^m) and denoted by subscript "1", gives

$$\alpha_0 + \alpha_1 = G_0(I, B_m) + G_1(\beta, I, B_m), \qquad (19)$$

where α_1 is derived from the solution of Eq. (11). In G_1 only higher-order inaccuracies are caused if β is replaced by ϕ (or by $\alpha\phi$), and the dependence on (I, B_m) may formally be replaced by one on (L, B_m). Then by Eq. (18)

$$\frac{1}{L(I, B_m)} = \frac{r}{a}\sin^2\theta + \frac{1}{ag_1^0}[\alpha_1 - G_1(\phi, L, B_m)]. \qquad (20)$$

The first term is the dipole approximation, while the second is a series with one term for each harmonic coefficient, combining contributions from α_1 and G_1 [*Stern*, 1965, 1967, 1968, 1976, 1994]. Because G_1 is small, the value of L appearing in it may be replaced by its zero-order approximation $r/a\sin^2\theta$. Note that in this approximation, the values of L encountered at a given point vary slightly with B_m: a better representation might average G_1 over B_m. I do not think anyone has developed this approach or compared it to conventional derivations.

How about including the effect of external fields? McIlwain's method can in principle be extended to include in the derivation of I not just the main field but also the contributions of other sources, using (for instance) some model by *Tsyganenko* [1989a, 1995]. To similarly extend the perturbation method, the external field must be approximated by additional harmonics with positive powers of r, and this was worked out for the two leading terms by *Stern* [1968]. In practice, the tail and magnetopause introduce appreciable asymmetry and at synchronous orbit, particles with the same $L(I, B_m)$ and different B_m already have significantly different drift surfaces.

One final comment pertains to the so-called strong-scattering limit, sometimes used in convection theory (especially for the geotail). Suppose trapped particles have their pitch angles constantly changed, sampling all directions in such a way that their distribution always remains isotropic (and neglecting the loss cone). Such particles conserve neither I nor B_m, but rather (e.g. *Harel et al.* [1981], Appendix 1) the specific flux tube volume V

$$V(\alpha, \beta) = \int \frac{ds}{B} \qquad (21)$$

evaluated between the ends of the field line. It might be of interest to compare the equatorial intersections of surfaces of constant V with those of surfaces of constant L, but again I am not aware of this having been done.

5. CONCLUSION

Modeling the magnetic field of the radiation belt region raises some interesting issues. The field is relatively stable and the scalar potential defined by Eqs. (1)–(2) yields a good approximation, which may be improved by including the contributions of external sources, from a data-based model such as one by *Tsyganenko* [1989a, 1995].

Euler potentials are a theoretically interesting tool, expressing the field-line structure mathematically, but in the general case are hard to derive because of their nonlinearity. In the inner magnetosphere they may be approximated by a perturbation method. One interesting technique related to Euler potentials is the deformation of magnetic fields, and perhaps one day the inflation and compression of the inner magnetosphere might be modeled that way.

The function $L(I, B_m)$ (also known as "L parameter") is widely used to characterize drift surfaces. Its conceptual foundations involve the (α, β) formalism, which gives a somewhat clearer insight into its properties and limitations. As far as classifying drift surfaces, however, the form of $L(I, B_m)$ now used is probably close to the best single-parameter representation possible, and at most only minor improvements can be expected.

Acknowledgements. The author thanks Drs. Martin Walt and Michael Schulz for useful comments on a draft of this article.

REFERENCES

Backus, G.E., Poloidal and toroidal fields in geomagnetic field modeling, *Rev. Geophys., 24*, 75–109, 1986.

Cauchy, A.L., Théorie de la Propagation des ondes à la Surface d'un Fluide Pesant d'une Profondeur Indéfinie, *Mém. Divers Savants, (2)*1, 3, 1816; œuvres, *(1)*1, 5.

del Pozo, C.F. and M. Blanc, Analytical self-consistent model of the large-scale convection electric field, *J. Geophys. Res., 99*, 4053–4068, 1994.

Fischbach, E., R.A. Langel, A.T.Y. Lui and M. Peredo, New Geomagnetic Limits on the Photon Mass and on Long-Range Forces Co-existing with Electromagnetism, *Phys. Rev. Lett., 73*, 514–517, 1994.

Harel, M., R.A. Wolf, P.H. Reiff, R.W. Spiro, W.J. Burke, F.J. Rich and M. Smiddy, Quantitative simulation of a magnetic substorm, 1. Model logic and overview, *J. Geophys. Res., 86*, 2217–2241, 1981.

Kosik, J.-C., Quantitative magnetospheric magnetic field modeling with toroidal and poloidal vector fields, *Planet. Space Sci., 32*, 965–973, 1984.

McIlwain, C.E., Coordinates for mapping the distribution of magnetically trapped particles, *J. Geophys. Res., 66*, 3681–3691, 1961.

McIlwain, C.E., Magnetic Coordinates, pp. 45–61 in Radiation Trapped in the Earth's Magnetic Field, ed. by Billy McCormac, D. Reidel, Dordrecht, The Netherlands, 1966; also in *Space Sci. Rev., 5*, 585–598, 1966.

Northrop, T.G. and E. Teller, Stability of the adiabatic motion of charged particles in the Earth's field, *Phys. Rev., 117*, 215–225, 1960.

Pennington, R.H., Equation of a charged particle shell in a perturbed dipole field, *J. Geophys. Res., 66*, 709–712, 1961.

Roelof, E.C. and D.G. Sibeck, Magnetopause shape as a bivariate function of the interplanetary magnetic field B_z and solar wind dynamic pressure, *J. Geophys. Res., 98*, 21,421–21,450, 1993.

Sibeck, D.G., R.E. Lopez and E.C. Roelof, Solar wind control of the magnetopause shape, location and motion, *J. Geophys. Res., 96*, 5489–5495, 1991.

Stern, D.P., Classification of magnetic shells, *J. Geophys. Res., 70*, 3629–3634, 1965.

Stern, D.P., Geomagnetic Euler potentials, *J. Geophys. Res., 72*, 3995-4005, 1967.

Stern, D.P., Euler potentials and geomagnetic drift shells, *J. Geophys. Res., 73*, 4373–4378, 1968.

Stern, D.P., Representation of magnetic fields in space, *Rev. Geophys., 14*, 199–214, 1976.

Stern, D.P., Tail modeling in a stretched magnetosphere 1. Methods and transformations, *J. Geophys. Res., 92*, 4437–4448, 1987.

Stern, D.P., The art of mapping the magnetosphere, *J. Geophys. Res., 99*, 17,169–17,198, 1994.

Tsyganenko, N.A., A magnetospheric magnetic field model with a warped tail current sheet, *Planet. Space Sci., 37*, 5–20, 1989a.

Tsyganenko, N.A., Solution of the Chapman-Ferraro problem for an ellipsoidal magnetopause, *Planet. Space Sci., 37*, 1037–1046, 1989b.

Tsyganenko, N.A., Quantitative models of the magnetospheric magnetic field: Methods and results, *Space Sci. Rev., 54*, 75–186, 1990.

Tsyganenko, N.A., Modeling the Earth's magnetospheric magnetic field within a realistic magnetopause, *J. Geophys. Res., 100*, 5599–5612, 1995.

David P. Stern, Laboratory for Extraterrestrial Physics, Code 695, Goddard Space Flight Center, Greenbelt, MD 20771, USA

Use of (B, L) coordinates in radiation dose models

M. Kruglanski

Belgian Institute for Space Aeronomy, Brussels, Belgium

A good part of the data on the proton radiation belts consists of dosimeter measurements from which information on energy spectrum and pitch angle cannot be deduced easily. Usually these measurements are organized in the (B, L) coordinate system. However the procedure to compare measurements of different epochs or to use old measurements for predictions is not clearly established. We investigate such procedures to highlight their degree of accuracy. As an illustration, we apply them to the comparison of the DMSP/F7 dosimeter data of December 1985 to doses obtained with the NASA AP-8 MIN combined with a depth-dose database. To understand the significant differences obtained by these procedures, we look at the relation between the coordinates (B, L) and the adiabatic invariants (M, J, Φ), and suggest on this basis, an alternative comparison procedure.

1. INTRODUCTION

The secular variation in the Earth's magnetic field has significant effects on predictions of trapped proton exposure, especially at low altitude. Generally, the geomagnetically trapped radiation is mapped with the (B, L) coordinates developed by *McIlwain* [1961]. The secular variation has an effect on the (B, L) coordinates as well as on the trapped proton population: the secular decay of the geomagnetic field induces, at low L value, a radially inward advection and an energy gain for the high-energy protons [*Heckman and Lindstrom*, 1972; *Schulz and Paulikas*, 1972; *Farley et al.*, 1972], and a temporal decrease of the altitude of a (B, L) point. Indeed, *Lindstrom and Heckman* [1968] showed that, for instance, the minimum mirror-point altitude at $L = 1.4$ decreases with a rate of about 7 km per year.

In this paper, we study the procedures to intercompare radiation dose maps or measurements of different epochs. If the trapped radiation environment were static, the appropriate procedure would only have to take into account the geographic variation of the (B, L) coordinates depicted by *Lindstrom and Heckman* [1968]. However, the actual situation is more complicated than that. To investigate the different procedures, we compare DMSP/F7 dose measurements of December 1985 to doses obtained with the AP-8 MIN model related to the solar minimum period of 1964.

The DMSP/F7 satellite was launched in November 1983, into a sun-synchronous circular polar orbit with an altitude of 840 km and an inclination of 98.8° [*Gussenhoven et al.*, 1986, 1987]. The mission ended around December 1987 and covers a portion of the solar cycle corresponding mainly to solar minimum. Twice per day, about five successive orbits of the satellite pass through the South Atlantic Anomaly.

The satellite carried four dosimeters measuring radiation dose from both electrons and protons behind different hemispherical aluminium shieldings (0.55, 1.55, 3.05 and 5.91 g cm^{-2} Al). Each detector distinguished two levels of dose deposition. Here, we consider only the highest level of the first dome. In this channel, doses were due mainly to protons of energy comprised between 20 and ∼ 100 MeV. Usually, there was one measurement every four seconds.

The NASA trapped radiation model AP-8 MIN [*Vette*, 1991] is distributed as a table of omnidirectional fluxes in function of particle energy E, B/B_0 and L, where B is the local magnetic field intensity, L is the *McIlwain* [1961] parameter and $B_0 = 0.311653 \, \text{gauss}/L^3$. The *Jensen and Cain* [1962] magnetic field model has been used to compute B and L [*Heynderickx et al.*, 1996]. From the AP-8 MIN fluence spectrum, we have calculated doses behind a spherical aluminium shielding of $0.55 \, \text{g/cm}^2$ using pre-calculated, mono-energetic depth-dose data generated by a Monte Carlo transport code [*Seltzer*, 1979].

2. METHODS

Two methods are investigated to compare the DMSP/F7 measurements to the AP-8 MIN model. In the first method, the DMSP/F7 measurements are compared to predictions obtained with AP-8 MIN. This allows us to compare, point to point, the data recorded by the satellite and the predictions on a short time period (e.g. one day). In the second method, the dose model obtained with AP-8 MIN is compared to radiation maps built with the DMSP/F7 measurements. This method allows us to compare the model to an average of the measurements over a larger period (e.g. one month).

The two methods are described below and summarized in Table 1.

2.1. Prediction method

In this method, the measurements of the DMSP/F7 satellite are compared to three different predictions (P1, P2, P3) obtained with the AP-8 MIN model. For each prediction, the same ephemeris as that of the DMSP/F7 satellite is used, but different geomagnetic field models are applied to compute the (B, L) coordinates. The predictions are calculated for the DMSP/F7 measurements of 4 December 1985 in the morning.

For prediction P1, the geomagnetic field model of *Jensen and Cain* [1962] is used as recommended by *Heynderickx et al.* [1996]. In this case, the geomagnetic field model used is the same model (and for the same epoch) as the one used to build AP-8 MIN.

For prediction P2, the *Jensen and Cain* [1962] model is used as well, but rotated as proposed by *Heynderickx* [1996]. The rotation takes into account the secular westward drift of the South Atlantic Anomaly [*Merrill and McElhinny*, 1983].

Finally, for prediction P3, the geomagnetic field model IGRF 1985 is used. Prediction P3 corresponds to the most widespread, but incorrect, use of the NASA models which implies the evaluation of B and L with a contemporary geomagnetic field model. This procedure artificially enhances the particle fluxes at low altitudes [*Daly*, 1989; *Heynderickx et al.*, 1996].

2.2. Mapping method

In this method, the doses deduced from the AP-8 MIN model are compared to two different radiation maps (M1, M3) built with the DMSP/F7 measurements which differ according to the geomagnetic field model used. The maps are built from an average over all the December 1985 measurements and are organized in (α_0, L) coordinates where α_0 is the equatorial pitch angle.

For map M3, a contemporary geomagnetic field is used to compute the (B, L) coordinates, i.e. IGRF 1985. For map M1 the (B, L) coordinates are evaluated with the same old geomagnetic model as the NASA model [*Jensen and Cain*, 1962]. We have made no map called M2.

Map M1 is the counterpart of prediction P1. Prediction P1 corresponds to the radiation dose observed by a satellite that would have flown in the year 1960 while map M1 corresponds to the radiation dose model produced by the DMSP measurements if the geomagnetic field had been unchanged

Table 1. Summary of the two methods used to compare the DMSP/F7 measurements and the AP-8 MIN model

Id.	Magnetic field model
Predictions obtained with the AP-8 MIN model	
P1	Jensen & Cain [1962]
P2	Jensen & Cain [1962], rotated
P3	IGRF 1985
Radiation maps built with the DMSP/F7 data	
M1	Jensen & Cain [1962]
M3	IGRF 1985

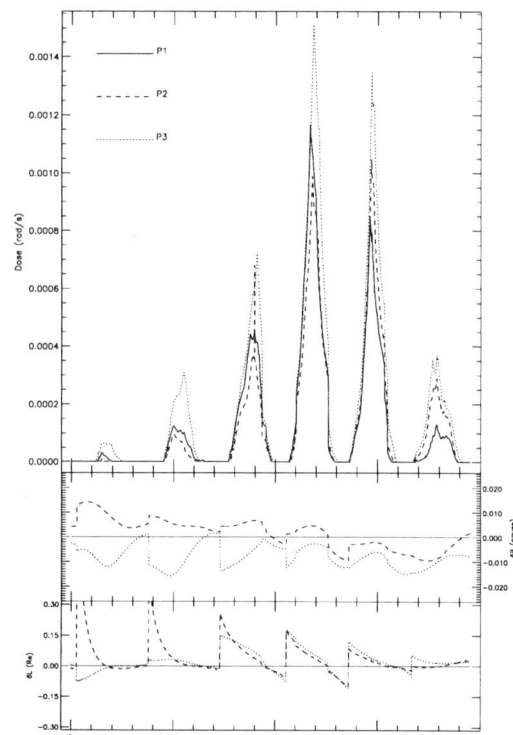

Figure 1. Predictions of radiation dose under an aluminium shielding of $0.55 \, \text{g/cm}^2$ obtained with the AP-8 MIN model for the DMSP/F7 ephemeris of 4 December 1985 in the morning. The predictions differ by the geomagnetic field models used (see Table 1). The two lowest panels show the (B, L) variation between the predictions; P1 is chosen as reference.

since 1960.

Map M3 and Prediction P3 are similary related.

3. RESULTS

Using the methods summarized in Table 1, we first made comparisons among the three predictions. The predicted radiation doses and the variation of the (B, L) coordinates along the trajectory of DMSP/F7 are shown in Figure 1. The

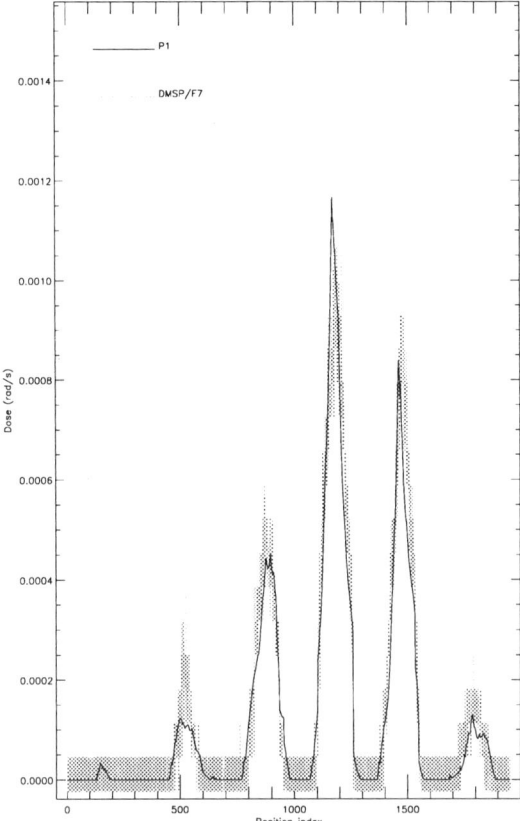

Figure 2. Comparison between the radiation dose encountered by DMSP/F7 and the prediction P1 from Figure 1.

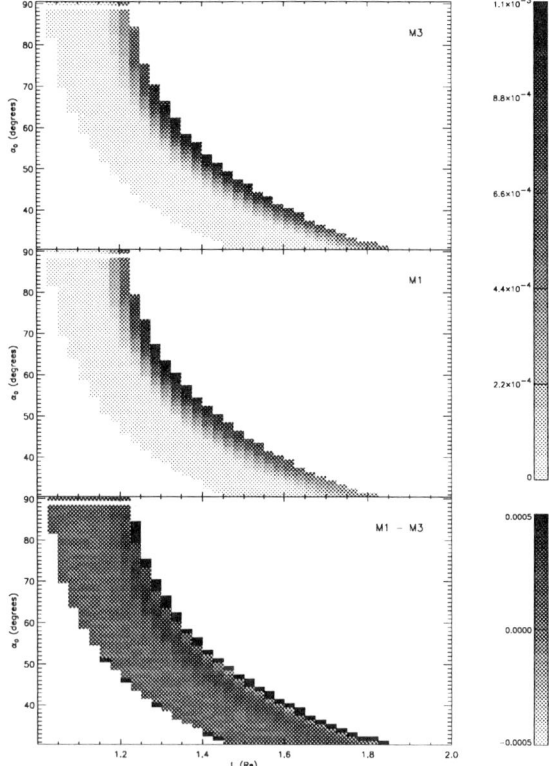

Figure 3. Maps of radiation dose below an aluminium shielding of 0.55 g/cm^2 obtained from the DMSP/F7 measurements of December 1995. The maps differ by the geomagnetic field model used (see Table 1). The lowest panel shows the differences between the two maps. In this panel, the solid black boxes represent the cells of the binning which are covered by only one of the two maps.

figure corresponds to location of the DMSP/F7 satellite in the latitude range from 60° S to 30° N and longitude range from 90° W to 60° E. The IGRF 1985 model produces the lowest geomagnetic field intensity in Figure 1 because of the intervening decrease in dipole strength.

In all the predictions, six peaks appear and all the peaks are located on the same position. Nevertheless, their intensities differ. As expected, the prediction P3 clearly overestimates the radiation dose. The two calculations based on the *Jensen and Cain* [1962] geomagnetic field model (P1 and P2) produce the same long-term average radiation dose, but differences appear for individual orbit because the geomagnetic field has been rotated to obtain P2.

In Figure 2, prediction P1 is compared with the radiation doses encountered by the DMSP/F7 satellite. The prediction agrees surprisingly well with the measurements, given that P1 predicts the radiation dose such as it would have been seen in 1964. The agreement of Figure 2 will clearly not be met when the comparison is done with prediction P3. Therefore predictions P1 and P2 seem to be the more trustworthy ones. Nevertheless one should note that the radiation dose seen by the DMSP/F7 satellite increased systematically during the life of the satellite from 1984 to 1987 [*Gussenhoven et al.*, 1991]. Since the rate of increase was about 7% per year, the agreement between the measurements and prediction P1 will not be so good for a different epoch of the satellite life.

The increase of the radiation dose from 1984 to 1987 is due to the varying influence of the atmospheric density on the radiation belt population. At solar maximum, the upper atmosphere is inflated and particles on the inner edge of the radiation belts are scattered more efficiently. During solar minimum, the Earth's atmosphere contracts and the inner edge of the radiation belts is slowly replenished by cosmic-ray and solar-proton albedo neutron decay. To take into account this atmospheric influence, *Pfitzer* [1990] has proposed to organize the inner-zone proton data according to the atmospheric density encountered locally by the protons. This approach would allow to evaluate the radiation doses during different solar conditions than solar minimum and solar maximum.

As a complement of the comparison between the prediction P1, P2, P3 and the DMSP/F7 measurements, the binning M1 and M3 of the DMSP/F7 data may be compared to the dose map obtained with AP-8 MIN.

In Figure 3, the two maps M3 and M1 are shown and compared. Due to the polar circular orbit of the satellite, a large part of the (α_0, L) space is left empty. In the part of the maps

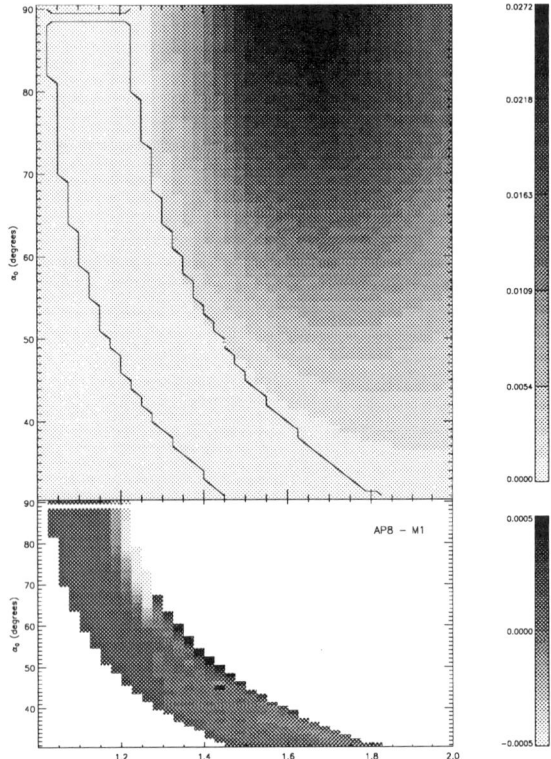

Figure 4. Comparison between the radiation dose model obtained with AP-8 MIN and the map M1 of Figure 3.

where data exist, the highest radiation doses are encountered around $L = 1.4$ for the largest equatorial pitch angle available. In the lowest panel of Figure 3, the difference between maps M3 and M1 is plotted. The solid black boxes represent the cells of the binning which are covered by only one of the two maps. The map M3 apparently covers more cells than the map M1.

Between the two maps, M1 seems to show the higher doses. But the total dose is the same, only its repartition differs. Map M3 covers a larger part of the (α_0, L) space than map M1. For instance, when the binning M3 is artificially shifted by the translation $L' = L - 0.05$ and $\alpha'_0 = \alpha_0 + 1°$, the two maps become more similar.

As a counterpart of the comparison of Figure 2, we compare in Figure 4 the map M1 to the radiation doses obtained with AP-8 MIN. The dose model obtained with AP-8 MIN is represented in the upper panel of Figure 4. From this model, the highest radiation doses—below an aluminium shielding of $0.55\,\mathrm{g/cm^2}$—are encountered at $L \approx 1.7$ along the equator. The solid line encloses the region in the (α_0, L) space where data are available from the DMSP/F7 satellite. This satellite obviously covers only the inner edge of the proton radiation belt.

In the lower panel of Figure 4, the difference of the radiation doses between the AP-8 MIN map and map M1 is shown. Unfortunately, this difference is difficult to interpret because such a large region of (B, L) space is not covered by the DMSP/F7 satellite. Nevertheless, it appears clearly that AP-8 MIN produces lower radiation doses below $L = 1.28$ than those observed by DMSP/F7 and higher doses above $L = 1.3$. Since the two maps are built with the same geomagnetic field model, their difference indicates a modification in the geographical distribution of the proton radiation environment between 1964 and 1986 (both being solar-minimum epochs).

4. DISCUSSION

The different results show that there is not an obvious way to compare measurements and models of different epochs. This drawback is probably caused in part by the coordinate system used to organize the data.

Usually, radiation dose models are organized in the (B, L) coordinate system or in a related one like $(B/B_0, L)$ or (α_0, L). This usage is justified by a simple link [*Hilton*, 1971] between the *McIlwain* [1961] parameter L and the first two adiabatic invariants, given for nonrelativistic kinematics by

$$M = \frac{E}{B_m} \quad (1)$$

$$J = \sqrt{2mE} \oint \sqrt{1 - \frac{B(s)}{B_m}}\,ds, \quad (2)$$

where m, q, E are respectively the mass, charge and energy of the particle, and where ds denotes the element of arc length along the field line of interest. The integral in Equation (2) extends from one mirror point to the other and back.

However, the equivalence between (B, L) and (M, J) is exact only when the magnetic field is static, the particle energy is fixed and B represents the magnetic field intensity B_m at the mirror point. In the case of trapped radiation dose models, these conditions are not fulfilled. The B coordinate is set equal to the magnetic field intensity at the measurement location and the recorded doses are caused by particles of different energies which are not necessarily mirroring at B.

Nevertheless, the (B, L) coordinate system appears to be well-suited for modeling trapped proton radiation doses in the average environment. This discrepancy can be understood by looking more carefully at the relation between (B_m, L) and the adiabatic invariants. For a given realization of the geomagnetic field, the pair (B_m, L) determines a unique drift shell, independently of the energy E. The same pair (B_m, L) thus determines, in this case, a unique value for Φ, the third adiabatic invariant [*Roederer*, 1970, pp. 76–79]. For a fixed realization of the Earth's magnetic field, therefore, (B_m, L) is equivalent to the pair (K, Φ) where

$$K^2 = \frac{J^2}{8mM} \quad (3)$$

depends only on B_m and McIlwain's L. The adiabatic invariant quantity K was proposed by *Kaufmann* [1965]. It has the advantage of being energy-independent and of accommodating secular as well as other slow variations in the geomagnetic field [*Schulz*, 1996], against which B_m and L_m are not adiabatically invariant.

Since K and Φ accommodate the adiabatic variations of the geomagnetic field such as the secular decay, the coordinate system (K, Φ) should be preferred to the (B, L) system to compare different radiation dose models or measurements, with different epoch of the magnetic field. In view of this, we plan to elaborate a subroutine to transform, for a given magnetic field, a (B_m, L) coordinate pair into a (K, Φ) pair. This subroutine would have to be applied to the comparison made in this paper between measurements and models.

Acknowledgements. This work was funded by ESA/ESTEC WMA contract No. 10725/94/NL/JG (SC). We wish to thank Dr. E.J. Daly from the Space Systems Environment Analysis Section of ESA for his suggestions and support during this study.

REFERENCES

Daly, E.J., Effects of Geomagnetic Field Evolution on Predictions of the Radiation Environment at Low Altitudes, ESTEC Working Paper WP-1531, 1989.

Farley, T.A., M.G. Kivelson and M. Walt, Effects of the secular magnetic variation on the distribution function of inner-zone protons, *J. Geophys. Res., 77*, 6087–6092, 1972.

Gussenhoven, M.S., R.C. Filz, K.A. Lynch, E.G. Mullen and F.A. Hanser, Space Radiation Dosimeter SSJ* for the Block 5D/Flight 7 DMSP Satellite: Calibration and Data Presentation, Air Force Geophysics Laboratory, Hanscom AFB, MA, AFGL-TR-86-0065, 1986.

Gussenhoven, M.S., E.G. Mullen, R.C. Filz, D.H. Brautigam and F.A. Hanser, New Low-Altitude Dose Measurmements, *IEEE Trans. Nucl. Sc., NS-34*, 676, 1987.

Gussenhoven, M.S., E.G. Mullen and E. Holeman, Radiation Belt Dynamics during Solar Minimum, *IEEE Trans. Nucl. Sc., NS-36*, 2008, 1989.

Gussenhoven, M.S., E.G. Mullen, D.H. Brautigam and E. Holeman, Dose Variation during Solar Minimum, *IEEE Trans. Nucl. Sc., NS-38*, 1671–1677, 1991.

Heckman, H.H. and P.J. Lindstrom, Response of trapped particles to a collapsing dipole moment, *J. Geophys. Res., 77*, 741–743, 1972.

Heynderickx, D., Comparison between methods to compensate for the secular motion of the south atlantic anomaly, *Nucl. Tracks Radiat. Meas.*, 1996, in press.

Heynderickx, D., J. Lemaire and E.J. Daly, Historical review of the different procedures used to compute the L-Parameter, *Nucl. Tracks Radiat. Meas.*, 1996, in press.

Hilton, H.H., L Parameter, A New Approximation, *J. Geophys. Res., 76*, 6952–6954, 1971.

Jensen, D.C. and J.C. Cain, An Interim Geomagnetic Field, *J. Geophys. Res., 67*, 3568, 1962.

Kaufmann, R.L., Conservation of the First and Second Adiabatic Invariants, *J. Geophys. Res., 70*, 2181–2186, 1965.

Lindstrom, P.J. and H.H. Heckman, B-L space and geomagnetic field models, *J. Geophys. Res., 73*, 3441–3447, 1968.

McIlwain, C.E., Coordinates for Mapping the Distribution of Magnetically Trapped Particles, *J. Geophys. Res., 66*, 3681, 1961.

Merrill, R.T. and M.W. McElhinny, The Earth's Magnetic Field, *International Geophysics Series, 32*, Academic Press, 1983.

Pfitzer, K.A., Radiation Dose to Man and Hardware as a Function of Atmospheric Density in the 28.5 Degree Space Station Orbit, *MDSSC Report No. H5387 Rev A*, McDonnell Douglas Corporation, 1990.

Roederer, J.G., *Dynamics of Geomagnetically Trapped Radiation*, Springer-Verlag, 1970.

Schulz, M. and G.A. Paulikas, Secular magnetic variation and the inner proton belt, *J. Geophys. Res., 77*, 744–747, 1972.

Schulz, M., Canonical Coordinates in Radiation Belt Modeling, these proceedings, 1996.

Seltzer, S.M., Electron, Electron-Bremsstrahlung and Proton Depth-Dose Data for Space Shielding Applications, *IEEE Trans. Nucl. Sci., NS-26*, 4896, 1979.

Vette, J.I., The NASA/NSSDC Trapped Radiation Environment Model Program (1964–1991), National Space Science Data Center, NSSDC/WDC-A-R&S 91-29, 1991.

M. Kruglanski, IASB/BIRA, 3 avenue Circulaire, B-1180 Brussels, Belgium, e-mail: m.kruglanski@oma.be.

Coordinates and Indices
Report of Discussion Group C

Reporter: D. Heynderickx

Belgian Institute for Space Aeronomy, Brussels, Belgium

Participants: E.J. Daly (chair), G. Popov (co-chair), D. Heynderickx (reporter), M. Kruglanski, M. Schulz, A.J. Sims, T. Pulkkinen

1. COORDINATES

The proceedings started with a discussion on the use of canonical coordinates with empirical radiation belt models. Since there is a one-to-one correspondence between the third adiabatic invariant and L^* and this coordinate takes into account the secular variation of the geomagnetic field, L^* should be more appropriate than L. However, the user community is used to L, so "black box" interfaces between particle models organised in canonical coordinates and user input and output in (B, L) coordinates have to be provided.

Even canonical coordinates do not solve all problems. At low altitudes, the Earth's atmosphere acts as a moving boundary problem, depending on solar activity. At high altitudes, the magnetic field models are not able to reconstruct the field lines on which trapped particles travel. One way out of these difficulties is to order empirical particle data in terms of geocentric coordinates (as is done in the recent Russian low altitude electron model). However, to cover the whole region of the trapped particle belts, huge model arrays would be needed.

To accurately map the low altitude region, there is a clear need for coordinates that take into account the influence of the atmosphere. Possible coordinates are the atmospheric density averaged over a satellite orbit, and the density averaged over a drift shell (or the density at the mirror point closest to the Earth). Even these coordinates may not be sufficient, since there is a lag between the solar activity cycle and the emptying and refilling of the radiation belt at low altitudes. To take this effect into account, further study has to be made of the CRAND effect. Finally, the uncertainties of the atmosphere models should be taken into account. More generally, it is important to provide error estimates on newly developed trapped particle models and probabilities of exceeding fluxes (in terms of geomagnetic indices). Also, the original data points that go into the models should be archived in geocentric coordinates, so that reconstruction or re-evaluation of the models remains possible.

At high altitudes, particles that are not on closed field lines may still be mapped to closed field lines with the current external magnetic field models. This raises the question of how to treat these data: they should not be excluded from the models, but should be flagged and treated in a different way, possibly in geocentric coordinates. Also, the there should be more study into the question whether the external magnetic field models really take out the longitude dependence as they claim. An alternative approach would be to use an internal magnetic field only and keep the longitude dependence as an additional parameter in the models. This would also relieve the problem of which external magnetic field model to use (for the internal field, there is general agreement that the IGRF models should be used).

It is of vital importance for model builders and users to specify in detail how they constructed or used trapped particle models, i.e. they should specify which magnetic field models they use (including epochs), which coordinates (and how they calculate them), the binning procedures, etc. It also is important to have consistent and clear definitions of the coordinates and parameters that are generally used (cf. the confusion on the definition and use of L). There should be a consensus on how to treat omnidirectional data: they should not simply be binned in terms of pitch angle independent (B, L) values; deconvolution in terms of pitch angle seems to be unavoidable.

2. INDICES

The D_{st} and R_{so} indices seem to be more appropriate for trapped particle modelling than the K_p index (D_{st} and R_{so} will be incorporated in the next version of the Tsyganenko model). Even D_{st} and K_p combined to not suffice to model the whole magnetosphere. The development of new indices should be considered. However, the existing indices should not be discarded, as they are the only ones available for historical data sets, and they have been accumulated over a very long time.

New indices should satisfy the following requirements:

1. Indices should serve as input of dynamic models; they should control the output.

2. Indices should describe the state of the magnetosphere

and of the processes which control the dynamics of particles. Preferably they should have a clear physical sense.

3. Indices should be accessible to users both in archives and on line.

4. Indices should have properties to forecast the magnetospheric activity and dynamics of particles.

5. New indices should be compatible with old ones (i.e. there should be proven direct or statistical relations between them) for preservation of long series of observations.

It is reasonable to continue and re-activate the research work for choosing new indices, satisfying the requirements above, paying specific attention to the following items:

1. The resolution of indices on time (for example, K_p has a resolution of only 3 hours, the resolution of D_{st} or AE is 1 minute and even better). So, the question arises which resolution is necessary to users and which resolution should be optimal, as excessive resolution is unprofitable from an economical point of view.

2. Integrated and local indices. K_p is an integrated index and is obtained after processing of data from a network of stations. A local K index is based on data of one station and can be made in a regime of real time. The solar wind parameters, measured in one point, have the integrated nature: how effectively can they control flows of particles on given orbits?

3. Which space areas characterize the indices? For example, D_{st} describes the processes in the inner radiation belt but AE describes the outer magnetosphere.

4. For forecasting it is reasonable to use indices constructed on the basis of solar activity observations. Which manifestations of the solar activity are the most geo-effective? How can they be described quantitatively and how should the long series of observations be established?

The research on and choosing of new indices should be conducted in close cooperation with experts on magnetosphere physics and solar physics, and make use the organizational structure (working groups, sections and subsections) available in IAGA, COSPAR, IAU and other international scientific organizations.

The existing data bases of indices are not completely centralised, i.e. one still has to consult different data bases to obtain different indices. Perhaps a new, general data bases combining all existing indices should be envisaged. Also, user interfaces should be developed which allow easy access to the indices in terms of searching as a function of time. Satellite data bases should contain at least the more generally used indices so that recalculation of the indices is avoided.

Finally, the current activities on predicting indices using neural networks look very promising. They could be combined with probabilities of exceeding fluxes in the particle models to provide worst case estimates and estimates of the influence of the real time space weather.

D. Heynderickx, Belgian Institute for Space Aeronomy, Ringlaan 3, B-1180 Brussels, Belgium (E-mail: dh@oma.be)

Availability of Radiation Belt Data and the Need for New Sources

A.D. Johnstone

Mullard Space Science Laboratory, Dept. of Space and Climate Physics, University College London

The data sets on which radiation belt models are based has been obtained from a relatively small number of missions in an even smaller number of orbits. We review the three dimensional coverage which these orbits provide and assess the degree to which extrapolation is required. In looking to the future we review the spacecraft which are likely to need radiation belt data as well as assessing the scientific requirements for understanding the radiation environment. The questions we address are a) where are the gaps in coverage which need to be filled? b) what type of missions are needed and what type of instruments should they carry? c) should all operational spacecraft carry radiation detectors and what types should they be?

1. INTRODUCTION

Our concern here is whether the available data is adequate to construct a model of the energetic electron population in the outer radiation belts. To do that we must establish the purpose of the model. There are many ways a model can be used but we take the view here that it should:

1. provide comprehensive information about the radiation belts in a compact form which allows an engineering assessment of the hazards to spacecraft operation;
2. predict levels of radiation likely to be encountered in specific circumstances, ultimately in advance, so that action can be taken to avoid damage.

The hazards as they are currently understood include

1. damage to electronic components;
2. generation of noise by deep dielectric charging;
3. background noise in sensitive detectors;
4. human well being.

The characteristics of the radiation which are important for assessing the risks vary from one hazard to the next. For example the damage to electronic components may depend on the accumulated dose over the duration of the mission. On the other hand deep dielectric charging depends on the fluence over a relatively short interval of time. Each mechanism probably also has its own specific dependence on the energy spectrum of the electrons. Therefore the information contained in a model must include at least the energy spectrum of the electrons, their spatial distribution and the statistical variability of the fluxes. In the future new modes of failure may be encountered which require further types of information.

The types of model can be summarized as follows:

1. empirical models provide a compact summary of the data which is available and use various methods of interpolation and extrapolation to estimate the radiation in regions where no measurements have been made.
2. physical models depend on an understanding of the physical processes controlling the intensity of the radiation. They provide the basis for extrapolation and smoothing of the empirical models.
3. predictive models take a number of measurable parameters and attempt to predict radiation levels to be encountered as far ahead as possible.

Ultimately the only type of model which is useful for determining the risk from hazards is an empirical model i.e. one which is based on real measurements. Predictive models require a good physical understanding and a good empirical base to be valuable and raise special problems which are some way from being solved. We limit the discussion here to models designed to assess the engineering risks.

Table 1. Traditional Operational Users

Type	Orbit	Incl.	L range
Communications	Geosynchronous	0	6.6
High latitude Molniya	1600 × 26571	63.5	1.25–26
Meteosat	Geosynchronous	0	6.6
Nimbus	800 circular	82	1.12–58
GPS/Glonass	20232 circular	55	4.17–12
Shuttle/Space Station	300 circular	51	1.05–2.6
Astronomical Observat. (ISO, XMM)	1000 × 71000		1.2–12

Table 2. New LEO Communications Constellations

System	Altitude	Inclination	Number
Inmarsat P	10400	~60	12
Teledesic	700	high	840
Globalstar	1389	52,47	24–48
Iridium	780	86.4	70
Odyssey	10600	~60	12
CCI	885	low	10
Ellipso	1250 × 500	some high	12
	2903 × 426	some equatorial	12
Orbcomm	785	polar	2
		45	24
Gonets	1300 × 1500	83	36
CTA(VITA)	670	polar	8
Starsys	1300	60	24

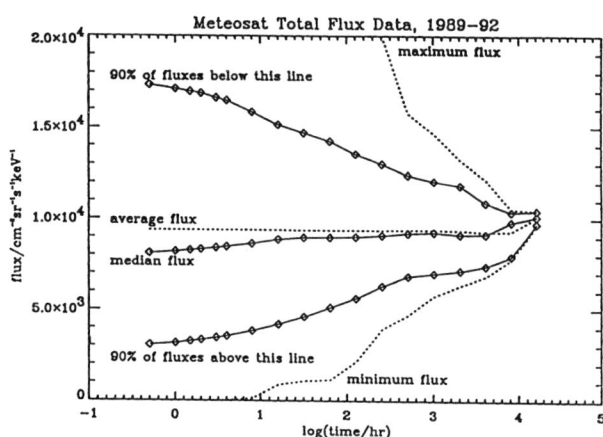

Figure 1. A model of the statistical variation as a function of accumulation period for electron $42.9\,\text{keV} < E < 300\,\text{keV}$ in geosynchronous orbit [*Szita*, 1996].

2. POTENTIAL USERS

We need to evaluate to what extent the observations have been made within the regions where modern users will be deploying their spacecraft and if not, whether the current physical models provide a reliable basis for extrapolation. In Table 1, some of the traditional operational orbits are listed to give an idea of the extent to which these spacecraft have ventured into the radiation belts. The simple fact is that for many years users avoided the worst of the radiation belts.

The most heavily used orbits have been geosynchronous, and LEO (Low Earth Orbit) with the latter generally below 800 km. Recently astronomical satellites have been using highly elliptical orbits since these orbits take the spacecraft beyond the range of penetrating radiation, which generates unwelcome background in the detectors, for many hours at a time. This allows long accumulations with low background noise which effectively improves the sensitivity. The radiation intensity in all these orbits is relatively well known. Recently spacecraft have started using intermediate orbits and "braving" the radiation which they encounter by making use of radiation-hardened electronic devices. An example is the constellation of navigational satellites of the GPS and Glonass type. In the near future a series of satellite constellations in low to medium earth orbit will be deployed for direct communications to the ground to customers with mobile phones etc. A potential list is given in Table 2. These satellites are generally in orbits which are above the usual LEO level and can therefore expect to encounter higher radiation levels. Some, such as Inmarsat P and Odyssey, are in orbits which are deliberately located at the slot between the radiation belts. Again this will take them into regions where the radiation is much higher than in the better known orbits. Thus users are pushing their spacecraft into regions which have been traditionally avoided because of the technical advantages which the new orbits confer. The question is—is information about the radiation belts in those regions adequate to make a proper risk assessment and to set reliable engineering standards?

3. CURRENT SOURCES OF OUTER RADIATION BELT ELECTRON DATA

A list of sources is given in Table 3. The traditional source of engineering information, the AE-8 model [*Vette*, 1991] is based mainly on low altitude data, on some data from

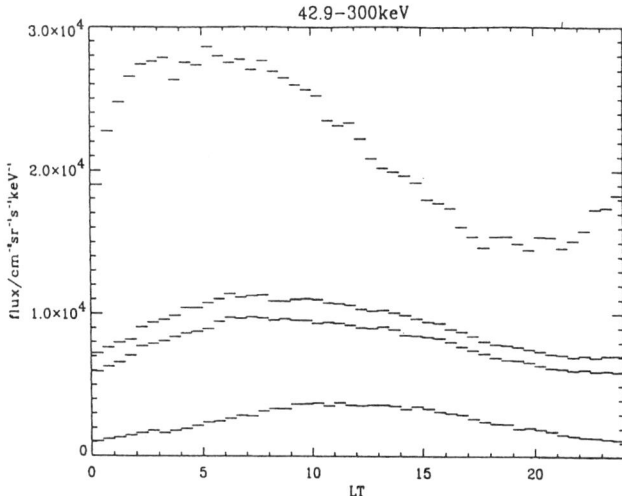

Figure 2. The local time dependence of the distribution of 30 min averages of the electron flux 42.9 keV $< E <$ 300 keV in geosynchronous orbit [*Szita*, 1996].

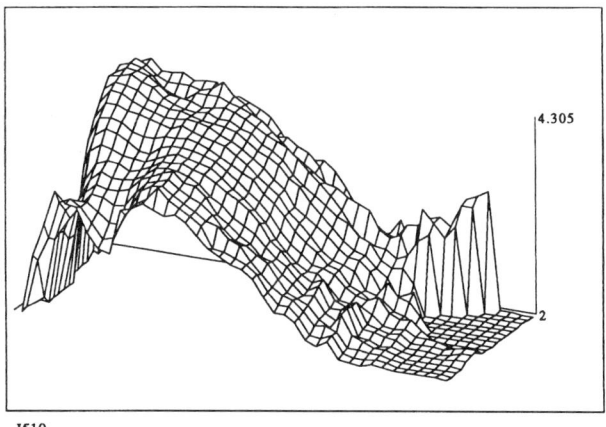

Figure 3. An empirical model of energetic electron fluxes measured by the MEA instrument on CRRES. The model is assembled in a coordinate system based on equatorial pitch angle α_0 and L value. The diagram shows a stacked plot of α_0 versus L for a single energy of 510 keV [*Rodgers*, 1966].

elliptical orbits and some from geosynchronous orbit. Very little direct information from the intermediate regions where the new satellites are going has been collated into the form of models. On the other hand there is abundant data from geosynchronous orbit.

There are three aspects in which the current data sets are inadequate. They are:

1. there is no information about the local time variation which becomes most significant at high altitude;
2. there is very little information about pitch angle distributions which would enable the measurements made at high altitude on a field line to be projected along the rest of the field line;
3. the temporal coverage is very limited which does not allow the information about the amount of variation from magnetic activity etc. to be built up.

For example the data from most of the low altitude polar orbiting satellites in the table was gathered over such a short period of time that they did not scan through all the variation in height, latitude, local time, and magnetic activity that satellites in those orbits would usually encounter.

Data has now been gathered from geosynchronous orbit for such a long time that the local time, and magnetic activity variation is well-established. An example of a model of statistical variability from geosynchronous orbit is shown in Figure 1 [*Szita*, 1996]. It is based on data from the SEM-2 detector on Meteosat 3. It shows that the longer the period over which the fluence is averaged the less the variability. For example, at the 90% probability level the total flux averaged over 3 hours will be less than 1.7×10^5 which is more than twice the average. However averaged over 10^4 hours, the 90% average is less than 1.0×10^5 which is only 1.1 times the average. The implication of this result is that, deep dielectric charging which may, say, require a flux of more than 1.7×10^5 persisting for 3 hours to generate a discharge is likely to occur 10% of the time. On the other hand, the dose which an electronic chip will receive over one year can be predicted with some accuracy.

Another presentation of the same data is shown in Figure 2 which shows the diurnal variation in the range of intensities averaged over 30 minutes. The plot is obtained from 3 years of data and the four curves drawn give the levels below which the total flux in the energy range 40 kev to 300 keV is found 5%, 45%, 55% and 95% of the time. This plot shows the influence of substorm injections on the nightside of the earth which lead to occasional high values but which only have a minor effect on the median value. Such statistical results can be produced readily for geosynchronous orbit because of the amount of data which has been collected and because of the lack of variability in the orbit itself. However even though the models are valuable for that orbit they cannot be extrapolated to other regions.

4. WHAT DATA IS REQUIRED?

In principle, the measurement of the complete pitch angle distribution and energy spectrum of the energetic electron distribution from a spacecraft in an equatorial, elliptical orbit over a long period of time would provide all the data necessary. All trapped particles pass through the magnetic equator and can therefore be measured there. Given these data the flux in any other region of the radiation belts can be deduced by using:

1. a magnetic field model to trace field lines;
2. the first adiabatic invariant to trace pitch angles along field lines.

The CRRES satellite was in a nearly ideal orbit for such observations [*Johnson and Kierein*, 1992]. A model obtained from the pitch angle distribution and energy spectrum

Table 3. Sources of Outer Zone Electron Radiation Measurements

Spacecraft	Epoch	Orbit	Incl.	Pitch angle data
AE-8 Data Sets				
Equatorial–Highly Elliptical				
Explorer 6	8/59–9/59	245 × 42400	47	No
Explorer 12	8/61–12/61	790 × 76620	33.3	No
Explorer 14	10/62–8/63	281 × 98530	33	No
IMP-1	11/63–5/64	192 × 197616	33.3	No
ERS-13	7/64–11/64	220 × 105000	36.7	No
OGO-1	9/64–6/67	281 × 149385	31.2	Yes
Explorer 26	1/65–12/65	316 × 26191	20.1	Yes
ERS-17	7/65–11/65	153 × 112694	34.4	Yes
OGO-3	6/66–12/67	319 × 122173	31.4	Yes
Low altitude, Polar				
Injun 3	12/62–9/63	235 × 2785	70.3	No
1963 38C	9/63–1/68	1066 × 1124	89.8	No
AZUR	11/69–3/70	387 × 3150	103	No
OV1-19	3/69–1/70	466 × 5764	105	Yes
OV3-3	8/66–9/67	360 × 4492	81	No
Geosynchronous				
ATS-5,6		∼ 35300	0	No
Processed Data Sets				
LANL series	since 78	Geosynchronous	0	Yes
Meteosat 3	89/95	Geosynchronous	0	Partial
ISEE	78/84	370 × 146000	28.7	Yes
CRRES	8/90–10/91	300 × 36000	18	Yes
Current Missions				
SAMPEX	since 7/92	520 × 670	82	Yes
APEX	since 6/94	400 × 2300	polar	?
STRV1a,b	since 7/94	300 × 36000	5	No
GPS		20232 circular	55	
Future Missions				
Ørsted		700	80	

of energetic electron fluxes measured by the MEA instrument [Vampola et al., 1992] on that spacecraft is shown in Figure 3 [Rodgers, 1966]. The model is an empirical model which was assembled in a coordinate system based on equatorial pitch angle α_0 and L value. The diagram shows a stacked spectrum plot of α_0 versus L for a single energy of 510 keV. This shows a number of interesting features of which we would like to comment on the following. First there is a strong and distinctive pitch angle variation at all L values. This is shown explicitly in Figure 4. It is not adequate to assume that the distribution is for example isotropic. It is also unreliable to project measurements made at low altitude upwards because there is no direct information about particles which mirror at an altitude above the point of observation. For AE-8 an upward extrapolation was made, based on the slope of the pitch angle distribution at the point of measurement. The data were fitted with a function of the form $sin^m\alpha$. Such an extrapolation would be more useful if we had good physical models of the processes which shape the pitch angle distribution. Secondly, the flux at geosynchronous orbit is much lower than that at the peak which means that communication spacecraft designed for geosynchronous orbit are going to find a much more dangerous environment if they are deployed in the intermediate range.

The main problems with this model from CRRES MEA data are:

1. the pitch angle resolution is not adequate to project to low altitudes at high latitudes;

2. the magnetic field models at high latitudes do not take account of the effects of magnetic substorm activity;

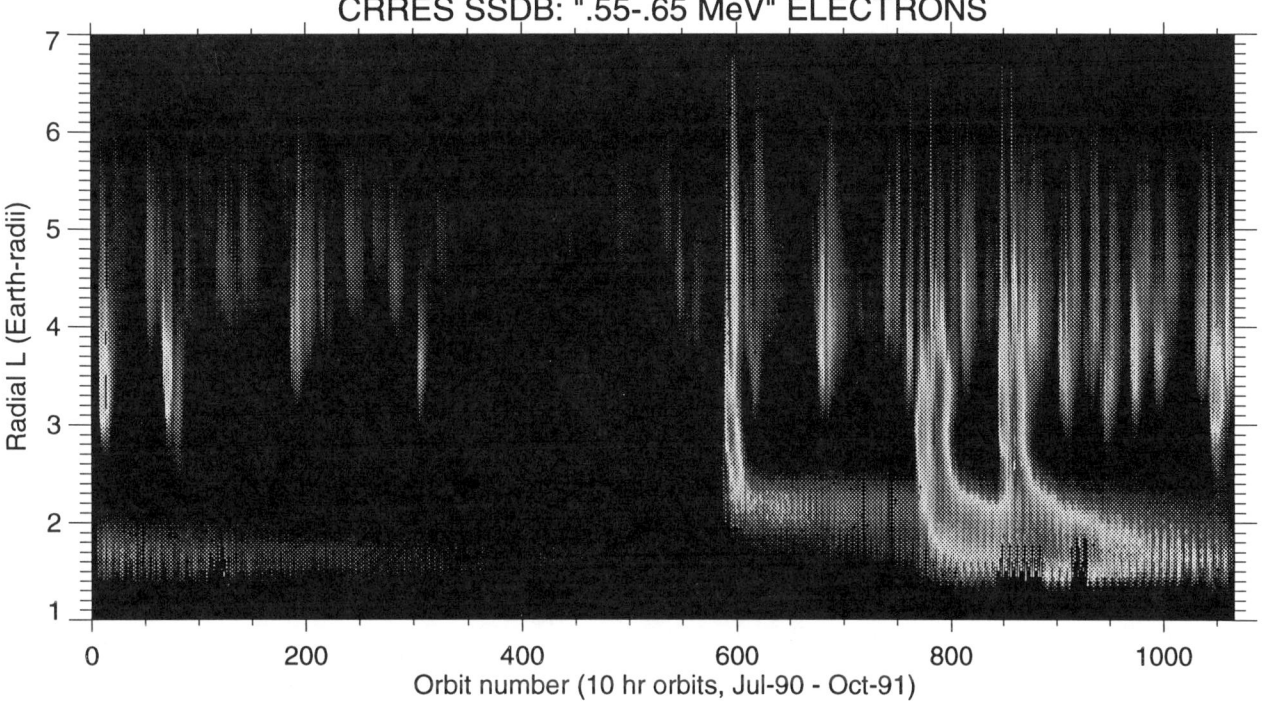

Figure 5. The intensity of 510 keV electrons measured by the MEA instrument versus radial distance for successive orbits of CRRES.

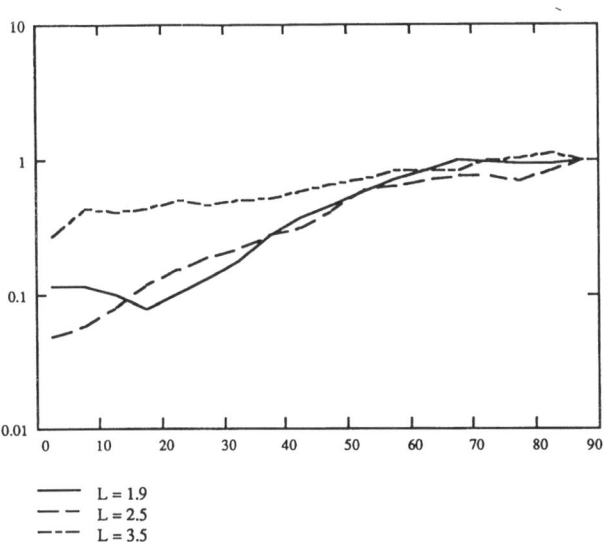

Figure 4. The equatorial pitch angle distribution at three different L values, at the inner peak, in the slot and at the outer peak, for an energy of 510 keV from the model of *Rodgers* [1996]. The intensities are plotted as a ratio to the flux at $87.5°$ in order to emphasize the relative change between $90°$ and $0°$.

3. the duration of the mission was not enough to give a complete scan of local times and so there is no assessment of the local time effects which are so noticeable, for example, in Figure 2;

4. likewise, there is not enough information on the variation with magnetic activity although CRRES showed how significant that variation could be even at relatively low altitudes.

The last point is illustrated by Figure 5 which shows the intensity at 510 keV measured by the MEA instrument as a function of radial distance and orbit number. There are a number of large events during the mission lifetime which increase the fluxes in the slot between the belts by an order of magnitude within a few hours. It had not been thought previously that the flux could increase some much so rapidly at such a low altitude. Similar increases were also observed at LEO by SAMPEX [*Baker et al.*, 1994].

5. CONCLUSIONS

There is not enough direct information about the radiation intensities and variations in the intermediate altitude ranges, from ~ 800 km up to geosynchronous orbit. In scientific terms more information is needed to understand:

1. the variation of intensity with local time at altitudes below geosynchronous. While the average variation is probably small and can be deduced from the trace

of drift shells around the Earth, information on the influence of substorm injections on the variability, as illustrated by Figure 2 for geosynchronous orbit, is not available.

2. the relation of the massive injections into $L = 2$ or less to substorms, their probability, and their dependence on local time;

3. the pitch angle distribution, particularly at small equatorial pitch angles which correspond to mirror points below an altitude of 10000 km so that the altitude variation of the radiation intensity can be obtained.

The need for the data is demonstrated by the increased exploitation of intermediate orbits. It will not be possible to obtain the data before it is needed because the exploitation will not wait. Meanwhile the value of the present information could be enhanced if there was a better scientific understanding, in the form of physical models of the following processes:

1. the loss of particles from the radiation belts, which would provide better models of the pitch angle distributions;

2. the path of particles following substorm injections, particularly those which lead to injection deep into the magnetosphere.

With this understanding, and assuming that the electrons in the outer radiation belt come from outside geosynchronous orbit, our information about conditions in that orbit could be extrapolated to the inner regions.

The data which are needed cannot be obtained without new missions and new observations. The most useful missions would be satellites in low-inclination, elliptical orbits which reach geosynchronous orbit i.e. geostationary transfer orbits. A number of small missions spaced in local time would be extremely valuable. In addition all operational missions should be encouraged to carry radiation detectors in order to provide the direct experience which will ultimately be the key to understanding the spacecraft performance. Such data will be much more valuable and will allow measurements in different orbits to be related to each other if the radiation detectors provide pitch angle distributions.

Acknowledgements. It is a pleasure to acknowledge the help of D.J. Rodgers and S. Szita of Mullard Space Science Laboratory in the preparation of this report. I have also been grateful for advice and comments of J. Lemaire of IASB throughout our joint work on the radiation belts and for the continuing support and stimulation of E.J. Daly of the European Space Technology Centre of ESA.

REFERENCES

Baker, D.N., J.B. Blake, L.B. Callis, J.R. Cummings, D. Hovestadt, S. Kanekal, B. Klecker, R.A. Mewaldt, and R.D. Zwickl, Relativistic electron accleration and decay time scales in the inner and outer radiation belts: SAMPEX, *Geophys. Res. Lett., 21,* 409–412, 1994.

M.H. Johnson, and J. Kierein, Combined Release and Radiation Effects Satellite(CRRES): spacecraft and mission, *J. Spacecraft and Rockets, 29,* 556–563, 1992.

D.J. Rodgers, A new empirical electron model, Proceedings of the Workshop on Radiation Belts: Models and Standards, Brussels 1996.

S. Szita, Ph.D. thesis, University of London, 1996.

Vampola, A.L., J.V. Osborn, and B.M. Johnson, CRRES magnetic electron spectrometer, AFGL-701-5a (MEA), *J. Spacecraft and Rockets, 29,* 595–595, 1992.

J.I. Vette, The AE-8 trapped electron model environment, NSSDC WDC-A-R&S 91-24, 1991.

A.D. Johnstone, Mullard Space Science Laboratory, Dept. of Space and Climate Physics, University College London (e-mail: adj@mssl.ucl.ac.uk)

DISCUSSION

Q: J. Albert. There are, in fact, physical models for the pitch-angle distribution of radiation belt electrons, based on interaction with whistler hiss.
A: A.D. Johnstone. Yes, and this knowledge should be folded into models, but more information is needed.
Q: M. Lauriente. How good is the instantaneous proton flux data in AP-8?
A: A.D. Johnstone. Never looked at it. Haven't the slighted opinion.
C: G. Ginet. If you are interested in forecasting radiation belt dynamics, measurements of the electromagnetic fields in the magnetosphere will be necessary. This is likely a more difficult job than particle measurements.
A: A.D. Johnstone. Predictions involve much more than measurement of the environment. If the predictions are being developed for users, then it is essential to direct the predictions towards providing the specific information which they can use. My view is that predictions are less useful than reliable data on the environment and its variability.
Q: J.B. Blake. Low altitude spacecraft such as SAMPEX measure in a qualitative way the evolution of the energetic electrons in the outer zone.
A: A.D. Johnstone. The qualitative information is extremely useful in building up the picture of events. However, ultimately a quantitative assessment of the height variation in the omnidirectional flux is going to be required. The height range from 600 km to 20,000 km is likely to be particularly important for the new generation of communications spacecraft.
Q: A.L. Vampola. The engineers are now sufficiently sophisticated that dose and surface charging are no longer a major problem—designs are in use. But the sporadic events that cause outages due to thick dielectric charging are a very serious problem and cannot yet be predicted.
A: A.D. Johnstone. I agree that deep dielectric charging is important and that the criteria which control the occurrence of discharging events are poorly known. Further study is required and perhaps could result in conditions which could be folded into radiation belt models.
Q: D.N. Baker. How far along are the designs and fabrication of new, relatively low altitude communication satellites (e.g., Iridium)? If they are quite far along, what can our community do to help avoid major economic losses from radiation effects?

A: A.D. Johnstone. I understand that some systems, especially Iridium, are already in production. It is probably time for the radiation belt community to become more active in approaching potential users.

Q: A.L. Vampola. The MEA angular response is well-enough determined that the convolution of the loss cone can be done even at geosynchronous altitude. But, it is very man-power intensive.

A: A.D. Johnstone. It will always be difficult to address the question of the altitude variation at high latitude and low altitude, using data from a near-equatorial orbit. The only reliable way is to make direct measurements in the region.

Q: D.N. Baker. The accumulated dose in the slot region for 6 months after the March 91 event was several orders of magnitude greater than before the event, so it too is non-negligible as an unpredicted hazard to spacecraft.

A: A.D. Johnstone. This demonstrates the need for spacecraft manufacturers to adopt a more mature approach to the space radiation environment, with scientific help of course.

C: X. Li. The comment is that the slot region is filled up often, not just during big events like the March 24, 1991 event. The slot region is a dynamical region, it is filled up even during moderate geomagnetic events, and the flux there decays (much faster than in the inner belt), and is filled up again.

Q: M. Walt. Why did you not include spacecraft charging as a hazard to satellites? This phenomena depends on energetic electron fluxes as well as thermal plasma densities?

A: A.D. Johnstone. The list of hazards was not intended to be comprehensive but to give examples of hazards for which penetrating particles are primarily responsible.

First results and perspectives of monitoring radiation belts

M.I. Panasyuk and E.N. Sosnovets

Skobeltsyn Institute of Nuclear Physics, Moscow State University, Moscow, Russia

O.S. Grafodatsky, V.I. Verkhoturov and Sh.N. Islyaev

NPO PM Spacecraft Corporation, Krasnoyarsk, Russia

The space environment monitoring programme (SEMP) is carried out on widely used communication and navigation satellites built at NPO PM Spacecraft Corporation. The unified scientific equipment for SEMP has been designed at INP MSU. The data delivery network has been also developed for the programme. Many computer codes for automated data processing and some sort of analysis have been elaborated as well. In addition to the description of the flying and ground-based equipment, the present report gives some examples of the numerous data sets obtained over the last years.

1. MONITORING PROGRAMME

The necessity of space environment monitoring can be understood for several reasons. First is the obvious inconsistency of the existing models with the real conditions in orbit. In particular, this inconsistency takes place at low altitudes and also at high altitudes (in the outer radiation belt). Another reason is an inability to examine the partial roles of the various sorts of space irradiation, that impact the spacecraft, without doing in-situ measurements.

The space environment monitoring programme (SEMP) is implemented by a cooperation of industrial and research organizations gathered by NPO PM Spacecraft Corporation [*Reshetnev et al., 1993*]. Major aims of the programme may be itemized as follows:

- to perform a realistic assessment of the role that the space environment plays in the degradation of a spacecraft;
- to examine the existing models of the radiation belts and to elaborate the renewed ones;
- to collect the representable scope of the experimental data, which provides an advantage in solving the problems of the Earth's magnetosphere physics.

1.1. Orbits and satellites

Table 1 presents the orbits that are explored by the communication and navigation satellites built at NPO PM. There are four different orbits. Their variety permits to cover almost all structural formations of the inner magnetosphere including the areas with the most intensive fluxes and the highest energy densities.

It is planned to have several satellites of each series in orbit. Also the tentative life time of a satellite should be greater than 5 years. All this gives a good opportunity to take the full coverage of the magnetosphere and even to carry out real multi-spacecraft measurements over all its areas.

In fact, realization of the programme started in 1991–1992, when the first GORIZONT and GLONASS satellites equipped with the SEMP instruments were launched. The first spacecraft from the new GALS series has already been launched in 1995. Other satellites of the MOLNIYA-3 and MUSSON-2 series are planned to be launched in 1996–1998.

Table 1. List of the spacecraft and orbits explored within SEMP

Satellite name	Operation time	Orbit
GORIZONT	1991–present	g/s (long.=E 80°, E 90°, E 103°, E 130°)
EXPRESS	1994–present	g/s (long.=W 15°)
GLONASS	1992–present	$i = 65°$, $r = 20,000$ km (circular)
GALS	1995–1996*	g/s
MOLNIYA-3	1996–1997	$i = 65°$, $A = 39,600$ km (elliptical)
MUSSON-2	1997–1998	$i = 99°$, $r = 1000$ km (circular)

*The first satellite of the GALS series was launched on November 23, 1995.

Table 2. Measured parameters

Method	Energy range
Electrostatic analyzer	E_e=1 keV (or 0.1–13 keV)
Windowed Geiger counter	$E_e > 40$ keV, $E_p > 1$ MeV
SSD	$E_e \sim 1$ MeV, $E_p > 13$ MeV
Cherenkov detector	$E_e > 5$ MeV $E_{p,z} > 500$ MeV/nucl
Dosimeter	≥ 1 g/cm^2 ≥ 2 g/cm^2

1.2. *Measured parameters*

Table 2 gives the parameters of the space environment chosen to be measured first. The reason of this choice is a necessity to assess three major effects of the space environment: degradation due to absorbed radiation dose, charging and single event upsets.

The number and type of measured parameters is supposed to be varied from one spacecraft series to another to provide the best investigation of the areas that a satellite flies through. In particular, protons of tens and hundreds MeV (inner radiation belt) will be measured on the MOLNIYA-3 and MUSSON-2.

1.3. *Instrumentation packages*

There were two packages of devices designed for SEMP named ADIPE and DIERA just to fit the different technical requirements from the different series of spacecraft. Comprehensive descriptions of these packages and the measured parameters are presented in [*Vlasova et al.*, 1993; *Ivanova et al.*, 1993; *Panasyuk and Sosnovets*, 1993].

The ADIPE package is dedicated to carry out the spectrometric measurements of the various sorts of irradiation in a wide range of energy: from magnetospheric plasma up to the galactic cosmic rays.

The DIERA measures in a more restricted number of data channels, i.e. for 8–14 parameters only. It has lower weight and smaller dimensions. This device has been assigned as a basic monitoring system due to its portability. It can be easily adapted to the wide variety of the telemetry systems used onboard.

All used spacecraft have controlled orientation in space. It is convenient to choose the main axis of an instrumentation package to be colinear to the Earth's radius directed to the satellite. This direction is approximately perpendicular to a field line of the magnetic field at the geostationary orbit.

The time resolution is chosen to be from 20 to 360 seconds depending on the orbit and operation mode.

1.4. *Data delivery system*

All the existing on-ground receiving stations, space flight centres, links between them and the organizing structure as a whole were unable to provide proper delivery of the data acquired in SEMP. A special data delivery system has been developed to solve this problem. The system uses PC-compatible computers equipped with a special telemetry-input adapter and the appropriate software packages. Such a computer is installed in a ground based station to receive the data from space monitors and to re-send them to the analysing centres. Before re-sending, the data are extracted from wider telemetry flow and then are packed. Full data transfer account is provided as well as checking for delivery losses. Direct modem connections and public networks are used as media of data transportation.

The data analysing centres are equipped with the relevant routines to receive a file, to quit a successful receipt or to request re-sending. All operations in both transmitting and receiving sites are quite automated. The system is cheap and reliable. By now, one on-ground receiving station and three analysing centres are equipped with the hardware and software tools mentioned above. Two other stations are planned to be equipped with the same technique in the near future.

2. DATA EXAMPLES

Figures 1 and 2 give the examples of the data obtained at geostationary and GLONASS (20,000 km circular) orbits. Considerable gaps take place because of the satellites' mode of rare downloads. The memory capacity has been extended in modern monitoring packages and now they can keep information up to 30 days without losses.

The data from three geostationary satellites, GORIZONT-

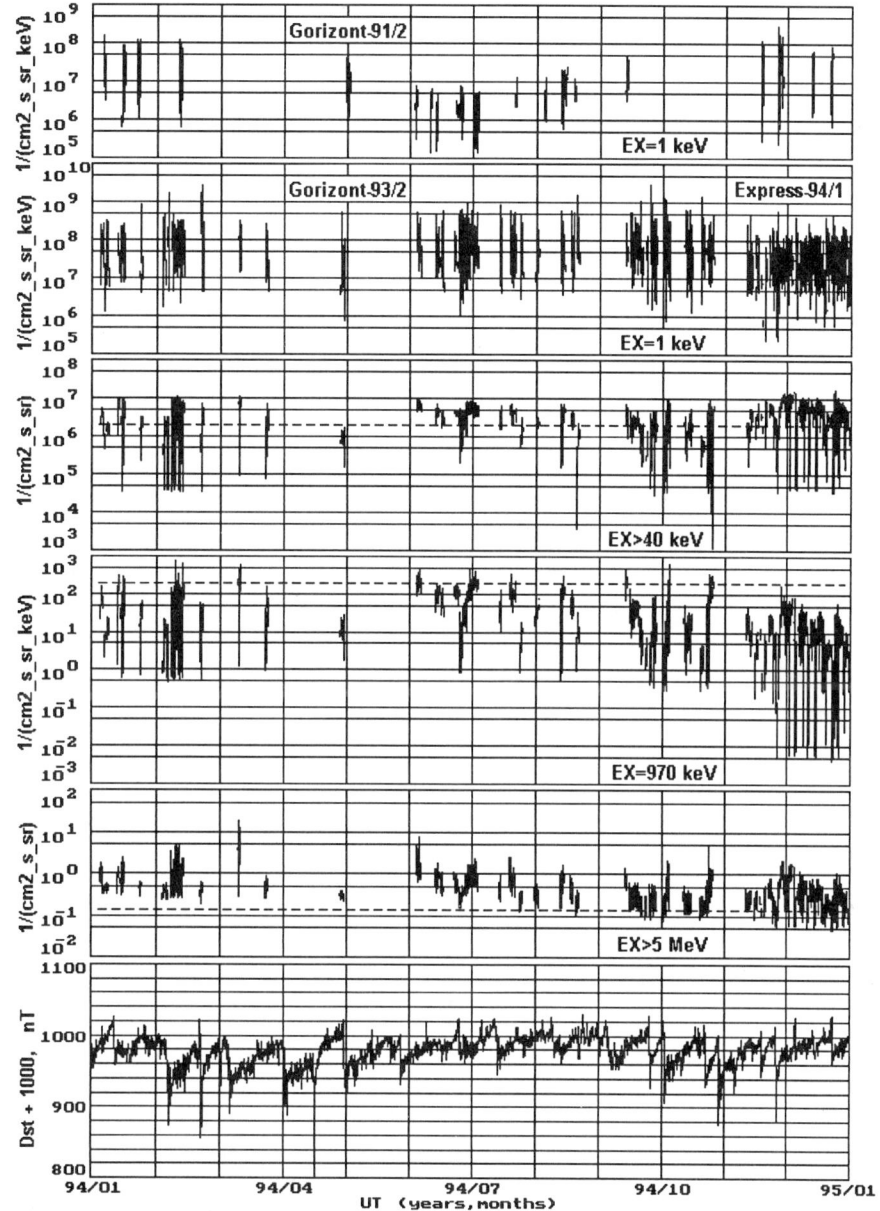

Figure 1. Intensities of the electron fluxes measured in some channels from 1 keV to 5 MeV in 1994. Three geostationary satellites are presented: GORIZONT-91/2, GORIZONT-93/2 and EXPRESS-94/1. Dashed lines represent the estimations from the INP-91 and AE-8 models.

91/2, GORIZONT-93/2 and EXPRESS-94/1, are presented in Figure 1. The magnitude of D_{st} (bottom panel) is artificially shifted by adding 1000 nT. Only the 1 keV channel perpendicular to the magnetic field line direction is taken from the GORIZONT-91/2 data set.

The plots on panels 2–5 (~ 1 keV, ≥ 40 keV, ~ 970 keV and > 5 MeV electrons) are compiled from the data of two spacecraft: GORIZONT-93/2 (01/01/94–30/10/94) and EXPRESS-94/1 (01/11/94–31/12/94). The merged plots demonstrate good consistence with each other because, in particular, the satellites carry the identical monitoring packages DIERA. The dashed lines give the model calculations. The INP-91 model [*Getselev et al.*, 1991] has been used for electrons with $E_e > 40$ keV, the AE-8 model [*Vette*, 1991] for electrons with $E_e > 5$ MeV.

Figure 1 illustrates "classical" behavior of the energetic electron intensity during a geomagnetic storm: decrease on an initial phase and 1–3 day delayed increase up to the level

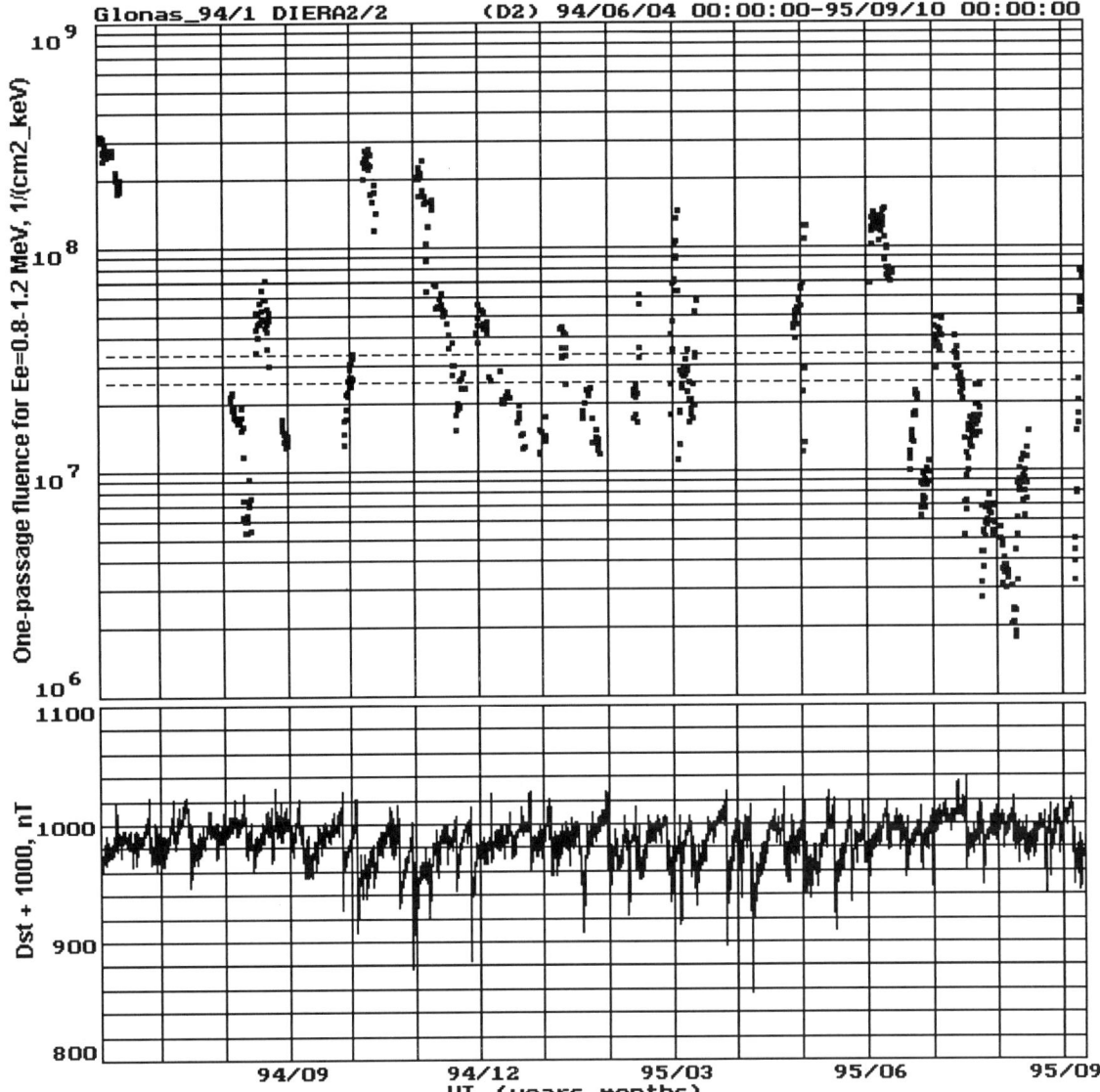

Figure 2. The 0.8–1.2 MeV electron flux as measured on GLONASS-94/1. The data is represented in the terms of fluence per a passage through the belt. The respective model estimates give the set of points located between the two dashed lines. INP-91 and AE-8 model are used. D_{st} is presented in bottom.

which sometimes may be higher than one before a storm. The significant annual variation in the 1 keV electrons can also be noted (top panel).

Figure 2 shows the flux of 0.8–1.2 MeV electrons as measured on GLONASS-94/1. The total amount of measured particles (fluence) per passage is represented in the plots. The model estimates were also made using tracing along a real orbit. Both the INP-91 and AE-8 models of the radiation belts were used. The model points are scattered in the region indicated by the two dashed lines. The differences in (B, L) values between odd and even orbits follow the scattering. The bottom panel represents $D_{st} + 1000$ nT.

Geomagnetic disturbances cause the major variation in the GLONASS orbit as well as in the geostationary orbit. The steady decrease is also seen in Figure 2, especially for 1995. Maybe we see an 11-year variation of the fluxes in the radiation belt which relates with the solar activity (the year 1995 is known as a year of minimum of solar activity). The observed increase of the intensity of electron fluxes for 1992–1994 co-relates well with the measured radiation dose on GLONASS.

Figure 3 shows diurnal values of radiation dose inside the spacecraft shell ($> 2\,\mathrm{g/cm^2}$ Al) from the two satellites: GLONASS-92/1 and GLONASS-94/1. The dashed line gives the model calculations for $\sim 3\,\mathrm{g/cm^2}$ Al (the INP-91 model

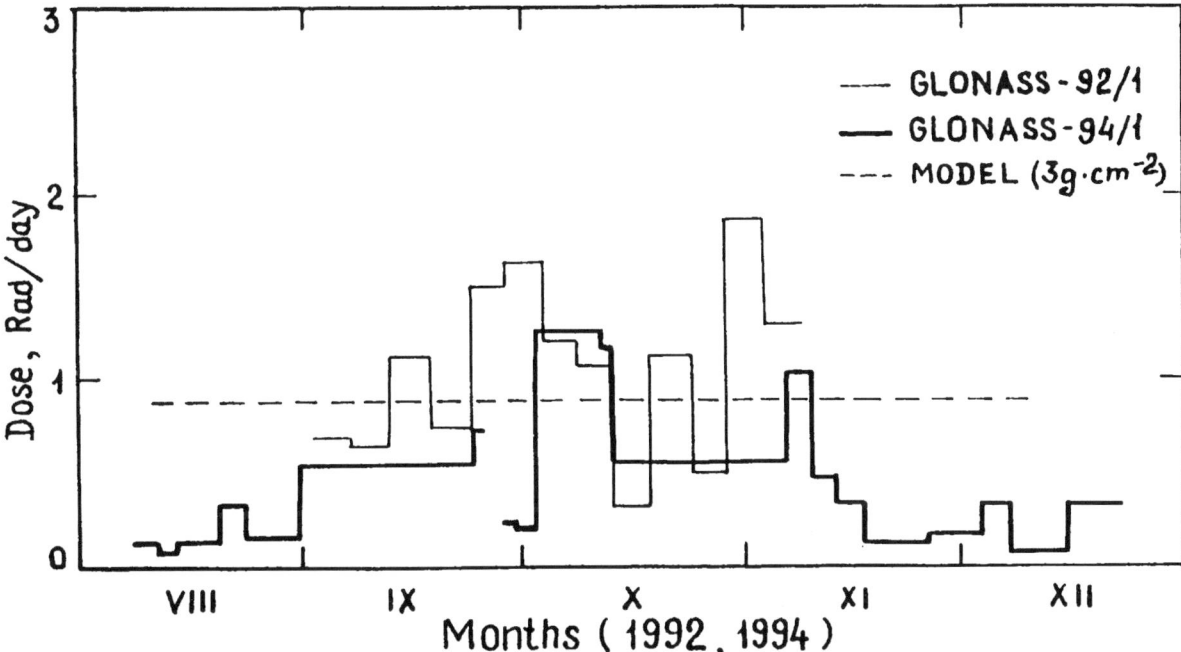

Figure 3. Diurnal values of radiation dose inside the spacecraft shell ($> 2\,\mathrm{g/cm^2}$ Al) from the GLONASS-92/1 and GLONASS-94/1 measurements

is used [*Getselev et al.,* 1991]). Although it is difficult to assess correctly a substance distribution around each dosimeter, they both were placed in just the same places on both satellites. It allows to compare their data directly. The mean value of the radiation dose from GLONASS-92/1 is approximately two times greater than the same from GLONASS-94/1 ($\sim 1.0\,\mathrm{rad/day}$ and $\sim 0.5\,\mathrm{rad/day}$, respectively).

A detailed analysis of the spatial and energy variations of the electron fluxes in the radiation belts and in the plasma sheet is given in [*Ivanova et al.,* these proceedings].

3. CONCLUSIONS

The following major results have been obtained to date by realizing SEMP:

- In-situ measurements make it possible to produce a realistic assessment of the space environment nearby a spacecraft. The assessment can be done for both radiation and charging effects. This knowledge may also be used when some decisions on spacecraft operating have to be issued.

- The collected data may provide a good basis for further development of the models of radiation belts. The performed comparison of the experimental data with the model estimations demonstrates the various consistences. Whether the experimental data is consistent with the model or not depends on what period and what energy ranges we consider. For geostationary orbit it follows from the different properties of electron fluxes in different energy ranges ($\sim$ keV, ten to hundred keV and few MeV).

All features of the spatial and energy distributions of the electron fluxes discovered by this time show how difficult is the problem of radiation environment modelling for the outer magnetosphere. However, the great attention paid to the problem presently, together with the huge sets of experimental data collected now in many countries from many satellites give good hope for progress in this field. We are ready to collaborate with all organizations and individuals having interest in this activity. We hope, for instance, to open access for the international scientific community to some sets of our data through the Internet soon.

REFERENCES

Getselev, I.V., A.N. Gusev, L.A. Darchieva, N.A. Kabashova, T.I. Morozova, A.V. Pavlov, M.I. Panasyuk, G.I. Pugacheva, S.Ya. Reizman, O.I. Savun, E.N., Sosnovets, L.V. Tverskaya, G.A. Timofeev and B.I. Yushkov, Model of spatial and energy distributions of charged particle (proton and electron) fluxes in the Earth's radiation belts, Preprint INP MSU-91-37/241, Moscow, 1991 (in Russian).

Ivanova, T.A., A.V. Zolotukhin, T.I. Morozova, I.A. Rubinstein, E.N. Sosnovets, M.V. Teltsov, V.I. Shumshurov, V.I. Verkhoturov, O.S. Grafodatsky, Sh.N. Islyaev and S.A. Maslov, Developing the methodology and establishing a global patrol service for monitoring of space environment parameters, Proceedings of the International Conference on Problems of Spacecraft/Environment Interactions, Novosibirsk, Russia, June 1992, ed. G. Drolshagen, ESA/ESTEC, p. 42, 1993.

Ivanova, T.A., Yu.V. Kutuzov, B.V. Marjin, N.N. Pavlov, I.A. Rubinstein, E.N. Sosnovets, M.V. Teltsov, L.V. Tverskaya and N.A.

Vlasova, Some characteristics of hot magnetospheric plasma at geostationary orbit, these proceedings.

Panasyuk, M.I., E.N. Sosnovets, The availability of data in the former Soviet Union, part 2, Gorizont, *STEP International, 3*, N2, 3, 1993.

Reshetnev, M.F., A.G. Kozlov, Sh.N., Islyaev, V.I. Verkhoturov, O. S. Grafodatsky and S.A. Maslov, Global patrol service: its goals, concepts, construction, Proceedings of the International Conference on Problems of Spacecraft/Environment Interactions, Novosibirsk, Russia, June 1992, ed. G. Drolshagen, ESA/ESTEC, 31, 1993.

Vette, J.I., The AE-8 trapped electron model environment, NSSDC/-WDC-A-R S91-24, 1991.

Vlasova, N.A., M.F. Goryainov, Yu.V. Kutuzov, B.V. Marjin, T.I. Morozova, I.A. Rubinstein, B.I. Savin, E.N. Sosnovets, L.V. Tverskaya, M.V. Teltsov, V.I. Verkhoturov, O.S. Grafodatsky, Sh.N. Islyaev and S.A. Maslov, ADIPE complex experiment on the study of space environment factors at synchronous orbit, Proceedings of the International Conference on Problems of Spacecraft/Environment Interactions, Novosibirsk, Russia, June 1992, ed. G. Drolshagen, ESA/ESTEC, 45, 1993.

M.I. Panasyuk, E.N. Sosnovets, Skobeltsyn Institute of Nuclear Physics, Moscow State University, Moscow 119899, Russia

O.S. Grafodatsky, V.I. Verkhoturov, Sh.N. Islyaev, NPO PM, Krasnoyarsk 660026, Russia

DISCUSSION

Q: J.B. Blake. What data from the Russian missions are available? How?

A: E.N. Sosnovets. We have rather large amounts of data from geostationary and Glonass satellites. To get a list of the data available please contact me or our director via e-mail.

Q: D.N. Baker. What is the typical time resolution of your available data? Do you have magnetic field data?

A: E.N. Sosnovets. Typical time resolution is:

- at geostationary orbit: 2–6 m for energetic particles and 4–8 s for plasma particles;
- at circular and elliptical orbit: 3 m;
- at circular polar orbit: 20 s.

We do not have magnetic field data on board.

Current and future data available in Japan

T. Kohno

Institute of Physical and Chemical Research (RIKEN)

The current status of radiation measurement with spacecraft in Japan is reviewed. Starting by the first GMS in 1978, efforts of radiation measurements in space were continued with both scientific and application satellites in Japan. The measured orbits were geostationary orbit, near earth orbit, highly eccentric orbits and far geomagnetic tail region. Protons, electrons and alphas have long been major target except for heavy ions with recent observation with GEOTAIL. The future plans of radiation observations in Japan will also be presented.

1. INTRODUCTION

The observation of energetic particles in space had been a hot topic at the early phase of space development by United States and Soviet Union (at that time) in the 60s [*Van Allen et al.*, 1959; *Williams and Mead*, 1965; *Vernov et al.*, 1967; *Lanzerottti et al.*, 1967]. But after the mid 70s the main stress of energetic particle observation shifted from the radiation belt particle to elemental and isotopic composition of Galactic Cosmic Rays, Solar Energetic Particles and Anomalous Cosmic Rays.

The first succeful launch of a Japanese scientific satellite was in 1971. A very preliminary particle measurement was tried by this satellite [*Takeuchi et al.*, 1974]. The first long duration observation of energetic particles in space was started by the first 'HIMAWARI' (GMS: Geostationary Meteorological Satellite). In Japan there are two organizations related to space development, National Space Development Agency of Japan (NASDA) and Institute of Space and Astronautical Science (ISAS). The NASDA and ISAS are mainly in charge of plactical space application and space science, respectively.

The satellite names used for radiation measurement in Japan so far are HIMAWARI-1–4 (GMS1–GMS4), 'OHZORA' (EXOS-C), 'AKEBONO' (EXOS-D), 'ETS-V' (KIKU-5), 'ETS-VI' (KIKU-6), GEOTAIL and IML-2. The measured energy regions are around a few MeV for electrons, about 1 to a few hundred MeV for protons and from a few to several tens of MeV/nuc for alphas. The details of their measurements will be described here.

2. CURRENT DATA

The space radiation observations performed so far in Japan will be reviewed here.

2.1. HIMAWARI (GMS) series

The first Japanese application satellite, HIMAWARI, was launched in 1977 into the geostationary orbit at 140° E using NASA's launch facility. This satellite series has been operated by the Japan Meteorological Agency. The Space Environment (energetic particle) Monitoring (SEM) by the GMS (Geostationary Meteorological Satellite) series continued until GMS-4, which was replaced by the current GMS-5 in March of 1995 where the SEM is no longer onboard. There is continuous data for about 16 years and the data of GMS-4 is still being received by Comunications Research Laboratory now. A data book of HIMAWARI observation from February 1978 to June 1986 was published [*WDC-C2*, 1986].

The detector arrangement of GMS/SEM is shown in Figure 1. The most simple five combination with one silicon detector and one moderator was used. The geometric factors of D1-D2 and D3-D5 are 0.042 and 0.389 $cm^2 sr$, respectively. An example of solar flare observation by GMS is shown in Figure 2.

2.2. ETS-V (KIKU-5)

An Engineering Test Satellite (ETS-V) was launched into the geostationary orbit at 150° E in 1987. This satellite had a TEchnical Data Acquisition (TEDA) subsystem on board. This subsystem includes a simple dose monitor consisting

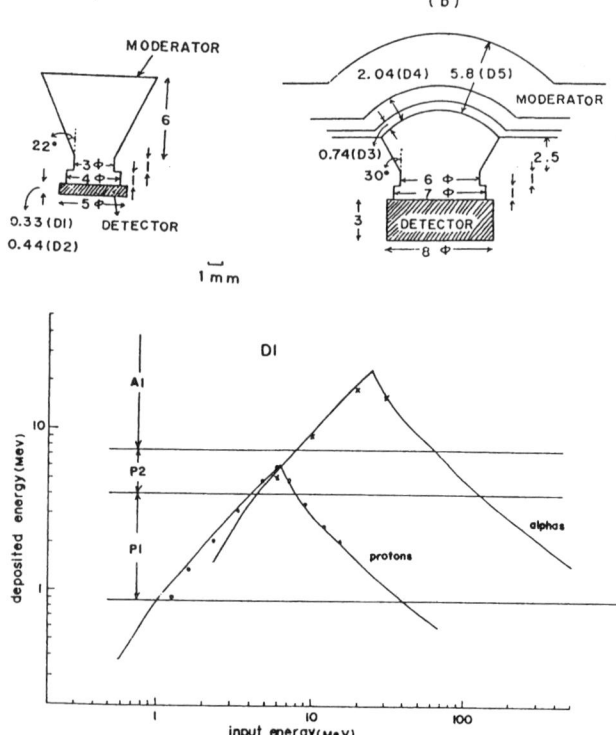

Figure 1. Configuration of each combination of a detector and a moderator (upper panel). The field of view of D1 and D2 (a) is narrower than those of D3-D5. Dome shaped moderators were used for D3-D5 (b) so that minimize the pathlength straggling in the moderator. An example of the responce curve in case of D1 is also shown (lower pannel). Crosses and circles show the results of accelerator calibrations.

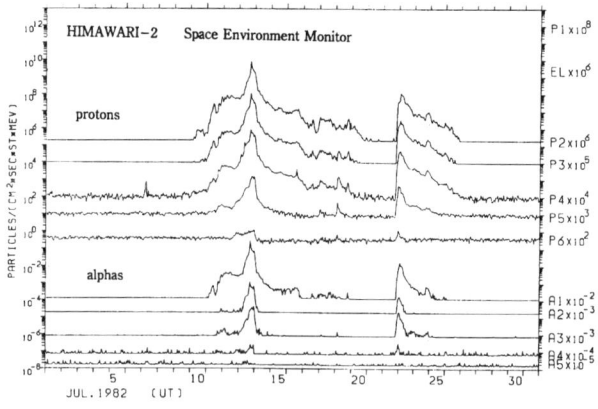

Figure 2. An example of solar flare originated particle observations by GMS-2

Table 1. Energy bands of GMS observations

channel	detector	particle	energy (MeV)
P1	D1	proton	1.2–4
P2	D1	proton	4–8
P3	D2	proton	8–16
P4	D3	proton	16–34
P5	D4	proton	34–80
P6	D5	proton	80–200
P7	D5	proton	200–500
A1	D1	alpha	9–70
A2	D2	alpha	30–70
A3	D3	alpha	65–170
A4	D4	alpha	130–250
A5	D5	alpha	320–370
EL	D3	electron	> 2

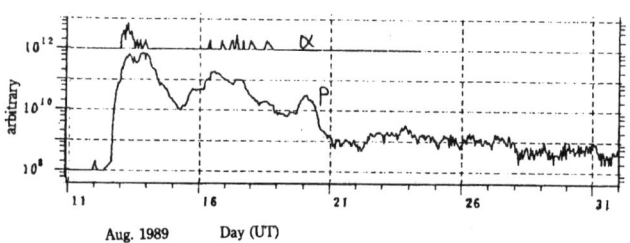

Figure 3. An example of observation by ETS-V for the large solar flare event of Aug. 1989

of two silicon detectors. The energy ranges observed is > 400 keV for electrons, 8–60 MeV for protons and > 60 MeV for alphas. This data is still being received now. An example of observed data is shown in Figure 3.

2.3. *ETS-VI (KIKU-6)*

The ETS-VI was launched by NASDA in 1994 using a H-II rocket intending a geostationary orbit. But due to a failure of the apogee engine, the resultant orbit was a highly elliptical transfer orbit with a perigee of about 8,000 km, apogee of about 38,000 km and an inclination of about 13°. The attitude is generally three axis stabilized except for the contingency mode with spinning stabilized. The DOse Monitor (DOM) could observe a wide space region as a result. The detailed observational results are presented at this workshop [*Goka et al.*, 1995]. A cross sectional view of the telescope is shown in Figure 4.

2.4. *OHZORA (EXOS-C)*

OHZORA was a scientific satellite developed by ISAS for the international program of MAP (Middle Atmosphere Program) with an orbit with a perigee of 350 km, an apogee of 850 km and an inclination of 75°. The HEP (High Energy Particle) instrument consists of two identical telescopes. The

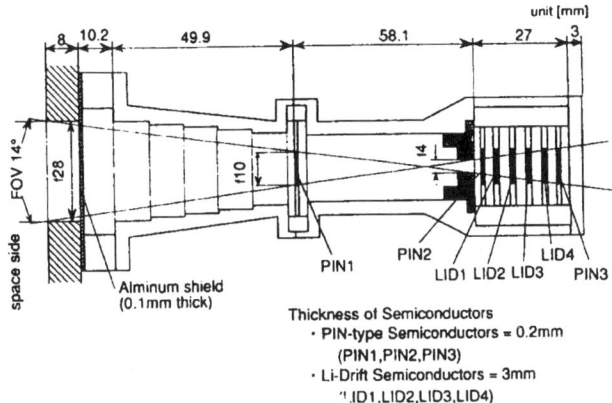

Figure 4. Cross sectional view of the ETS-VI/DOM telescope. The geometric factor is 0.003 cm^{2}sr.

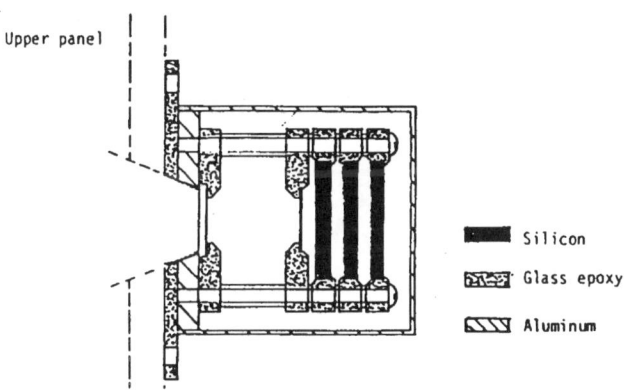

Figure 5. Cross sectional view of the OHZORA telescope. The geometric factor for the coincidence mode is 0.14 cm^{2}sr.

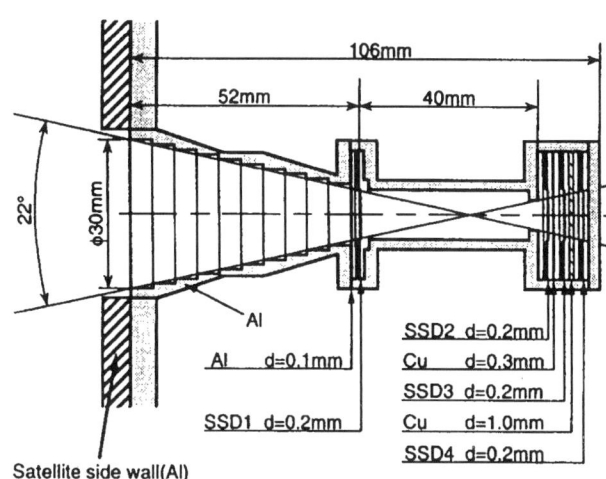

Figure 6. Cross sectional view of the AKEBONO telescope. In order to extend the observable energy region, two sheets of Cu absorber are used. The geometric factors for the coincidence mode (electrons and protons) and D1 only mode (alphas) are 0.023 cm^{2}sr and 0.19 cm^{2}sr, respectively.

direction of the center of the field of view is anti-parallel to the spin axis which is in the sun direction and the direction of another one is perpendicular to the spin axis [*Nagata et al.*, 1984]. The structure of the telescope is shown in Figure 5. The time period of data coverage of this satellite is about three years including solar minimum from February 1984 to March 1987. The intensity maps for electrons and protons below the radiation belt observed by this satellite are reported by [*Kohno et al.*, 1990]. The observed dynamic He behavior and electron injection are also reported [*Kohno et al.*, 1995; *Gusev et al.*, 1995]. The global distribution of trapped He is reported at this workshop [*Hasebe et al.*, 1995]. The raw data of this satellite is also under processing at the National Space Science Data Center [*Shing Fung*, private communication].

2.5. AKEBONO (EXOS-D)

A particle telescope named Radiation Monitor (RDM) was on board the scientific satellite 'AKEBONO' which was launched in February 1989 into an orbit of perigee of 270 km and apogee of 10,500 km with inclination of 75.1° [*Takagi et al.*, 1993]. The number of telescope is only one with the field of view of perpendicular to the spin axis which is in the sun direction. The cross sectional view of the telescope is shown in Figure 6. The observed energy ranges are three channels for electrons of 0.25–0.7, 0.7–2.0, > 2.0 MeV, three channels for protons of 6.3–15, 15–29, 29–38 MeV and one channel for alphas of 16–52 MeV. Different from OHZORA, the large value of its apogee make it possible to survey a wider space than OHZORA. The huge dataset of this satellite made it possible to study the dynamic structure of the radiation belt [*Yukimatsu et al.*, 1995]. The data from this satellite begins at 1989 and is still being taken now. The coverage period is over six years including solar maximum.

2.6. GEOTAIL

The GEOTAIL launched in 1992 is a Japanese side scientific satellite which is a part of the International Solar Terrestrial Physics Program. There are some sophisticated particle telescopes made of Si solid state detector arrays. The structures of the three telescopes of MI-1, MI-2 (A,B) and HI are shown in Figure 7. The total geometric factor of the four telescopes is about 100 cm^{2}sr. The energy regions to be observed by them are 2.3–55 MeV/nuc for He, 5.1–139 MeV/nuc for Ne and 13–230 MeV/nuc for Fe for example. The orbit is $8R_E \times 200R_E$ in the equatorial plane. GEOTAIL was launched in July 1992 and is still fully operational as of January 1996. In Figure 8 a part of the results of GEOTAIL observations of Anomalous Cosmic Rays is shown [*Hasebe et al.*, 1994].

2.7. IML-2

The real time monitoring of space environment radiation

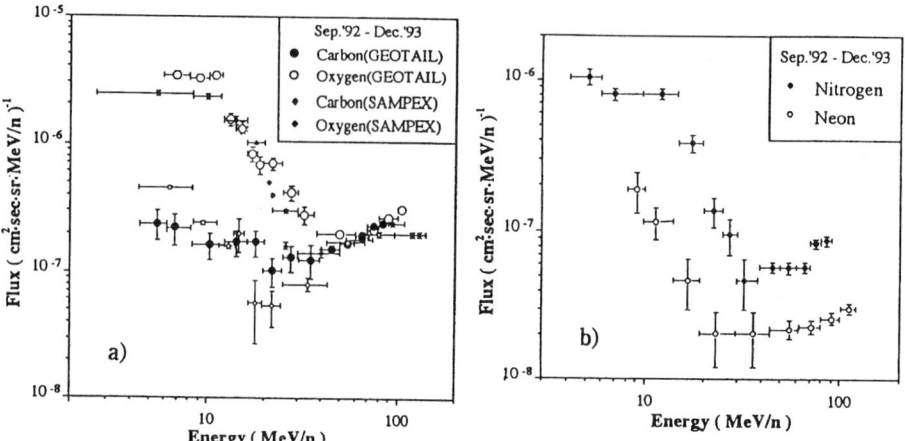

Figure 8. (a) Differential energy spectra for C and O during quiet times from September 1992 to December 1993 from the GEOTAIL satellite. C and O spectra taken by SAMPEX are also shown. (b) Differential energy spectra for N and Ne during the same period from GEOTAIL.

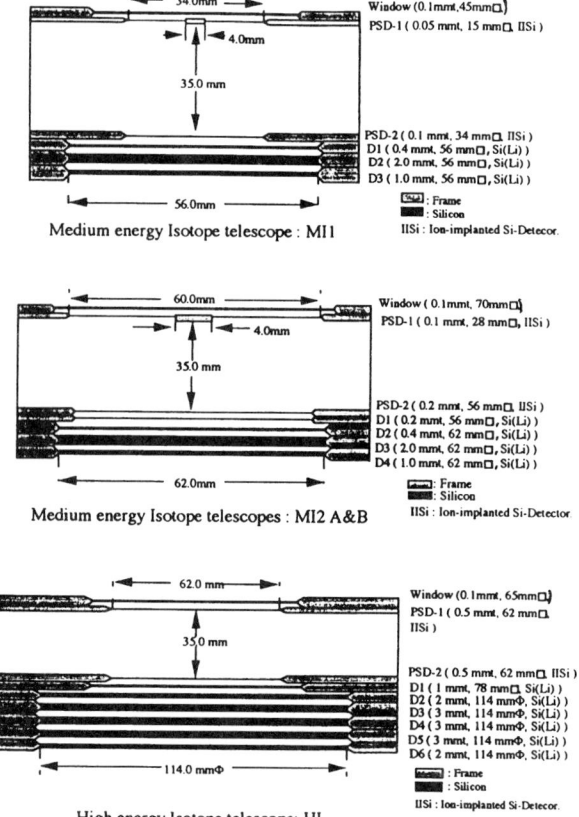

Figure 7. Schematic drawing of the telescopes of GEOTAIL mission: the Medium energy Isotope telescope MI-1 (top), the Medium energy Isotope telescopes MI-2A and MI-2B (middle), the High energy Isotope telescope HI (bottom)

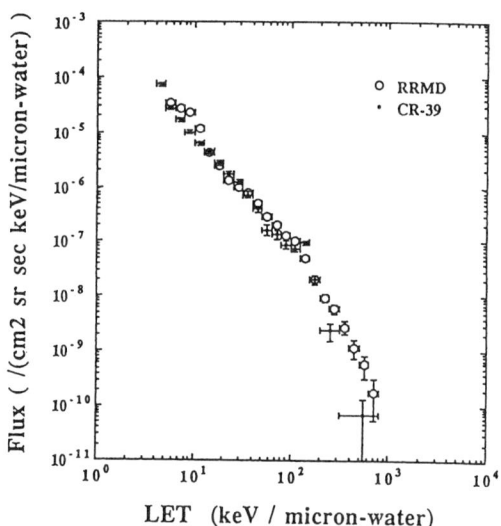

Figure 9. Differential LET distributions obtained by RRMD and CR-39 track detectors in the IML-2

has been achieved in the Space-Lab of STS-65 (second International Microgravity Laboratory: IML-2). A silicon solid state detector telescope (RRMD; Real time Radiation Monitoring Device) which is essentially identical to the HI telescope of the GEOTAIL satellite and CR-39 track detectors were used [*Doke et al.*, 1995]. The inclination of the orbit was 28.5° with an altitude of 300 km. The temporal variation of rates of particle flux, together with the dose equivalent and the LET distribution at three locations in the Space-Lab were reosonably given in real time. The LET distribution obtained by the RRMD and CR-39 is shown in Figure 9.

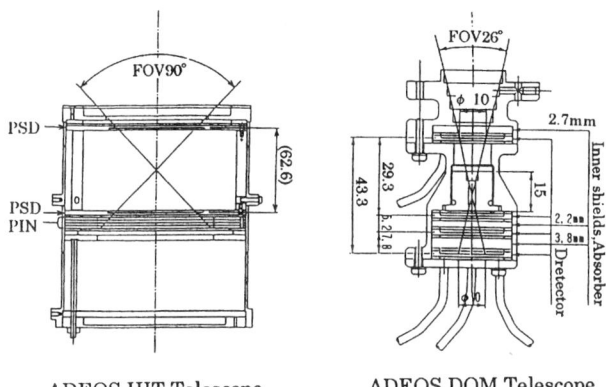

Figure 10. Cross sectional view of the ADEOS HIT and DOM telescopes. Two position sensitive detectors are used in HIT to know the incident direction of the particle. In order to extend the energy region to more than 100 MeV, two passive absorbers are used for the DOM telescope.

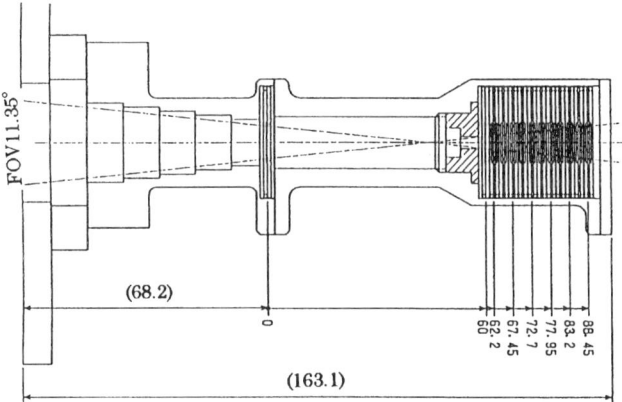

Figure 11. Cross sectional view of the ADEOS HIT and DOM telescopes. Two position sensitive detectors are used in HIT to know the incident direction of the particle. In order to extend the energy region to more than 100 MeV, two passive absorbers are used for the DOM telescope.

3. FUTURE PLANS IN JAPAN

There are some programs of radiation measurements in future at NASDA.

3.1. ADEOS

The ADvanced Earth Observing Satellite is going to be launched in August 1996. The planned orbit is sun-synchronous, subrecurrent with an altitude of 800 km and an inclination of 98.6°. There are two telescopes of the DOM and HIT type. The DOM observes protons with three channels of 50–70, 70–100 and > 100 MeV and the HIT measures heavy ions of energy ranges of 10–23 MeV/nuc for Li, Be, B, 15–36 MeV/nuc for C, N, O and 31–76 MeV/nuc for Fe. By using position sensitive detectors and PIN type energy detectors, the isotope separation can be expected. The geometric factor of the HIT telescope is about $25\,cm^2sr$. The structure of the telescopes is shown in Figure 10. There is a program of ADEOS-II, which is a successor of the first ADEOS and planned to be launched in September 1998. There is also a telescope identical to DOM of ETS-VI above described.

3.2. COMETS

There is also a DOM telescope on board the COMunications and Broadcasting Engineering Test Satellite (COMETS) which will be launched in summer of 1997 into the geostatiobary orbit at 121° E. The measurement range of the DOM of COMETS is 15 channels in 0.4–4.6 MeV for electrons, 9 channels in 7.5–46 MeV for protons, 12 channels in 7–35 MeV/nuc for alphas and 16 channels in > 140 MeV for particles heavier than alphas. These measurements will be able to be successive observations at geostationary orbit after GMS series, ETS-V and partially ETS-VI. A schematic view of the COMETS/DOM telescope is shown in Figure 11.

3.3. JEM

All items described above (ADEOS, ADEOS-II and COMETS) are already authorized and running now. But the following items are not yet authorized but under discussion.

It is well known that Japan will participate in the internationl Space Station program. There are various plans for space radiation observation at the Japanese Experiment Module (JEM). The following items are under discussion now:

1. protons and alphas: energy region from 5 Mev/nuc to 10 GeV/nuc using solid state detectors and Cherenkov detectors is considered.

2. Heavy particles: energy regions of 15–150 MeV/nuc for C and 30–330 MeV/nuc for Fe with position sensitive detectors are planned. Isotopic observation is also possible. Elemental observation of the ultra high energy heavy particles of up to 10^{16} eV/nuc using emulsion chamber technique is also proposed.

3. Neutrons: Neutron detectors using scintilator tube type and phoswitch type for the energy range of 5–100 MeV and bonner ball type for the energy region below 50 MeV are proposed.

4. γ rays: observation of 0.1–10 MeV γ rays is considered with ionization chamber type detectors.

There are many other environmental measurements on JEM proposed and under discussion.

Acknowledgements. The author is grateful to NASDA members who gave him valuable data about past and future programs. He is also indebted to many staffs of ISAS especially cordial collegues of OHZORA, AKEBONO and GEOTAIL group.

REFERENCES

Doke, T., T. Hayashi, J. Kikuchi, N. Hasebe, S. Nagaoka, M. Kato and G.D. Badhwar, Real Time Measurement of LET Distribution in the IML-2 Space-Lab (STS-65), *Nucl. Instr. Meth. Physics Res., A 365*, 524–532, 1995.

Goka, T., H. Matsumoto and T. Fukuda, Measurement of Radiation Belt Particles on ETS-6 Onboard Dosemeter, these proceedings, 1996.

Gusev, A.A., T. Kohno, I.M. Martin, G.I. Pugacheva, A. Turtelli, Jr., A.J. Tylka and K. Kudela, Injection and fast radial Diffusion of Energetic Electrons into the Inner Magnetosphere, *Planet. Space Sci., 43*, 1131–1134, 1995.

Hasebe, N. et al., Rapid Recovery of Anomalous Cosmic Ray Flux at 1 AU in Solar Cycle 22, *Geophys. Res. Letters, 21*, 3027–3030, 1994.

Hasebe, N., A. Ryowa, M. Kobayashi, K. Kondo, J. Hamada, Y. Mishima, K. Nagata, T. Kohno, J. Kikuchi and T. Doke, Global Distributions of Trapped He Fluxes Observed by OHZORA During the Geomagnetically Quiet Period of 1984–1987, these proceedings, 1996.

Kohno, T., K. Munakata, K. Nagata, H. Murakami, A. Nakamoto, N. Hasebe, J. Kikuchi and T. Doke, Intensity Maps of MeV Electrons and Protons below the Radiation Belt, *Planet. Space Sci., 90*, 483–490, 1990.

Kohno, T., A.A. Gusev, I.M. Martin and G.I. Pugacheva, The Trapped He Flux Dynamics Observed on the OHZORA Satellite during 1984–1987, *Geophys. Res. Letters, 22*, 877–880, 1995.

Lanzerotti, L.J., C.S. Roberts and W.L. Brown, Temporal Variations in the Electron Flux at Synchronous Altitudes, *J. Geophys. Res., 72*, 5893–5902, 1967.

Nagata, K., T. Kohno, H. Murakami, A. Nakamoto, N. Hasebe, T. Takenaka, J. Kikuchi and T. Doke, OHZORA High Energy Particle Observations, *J. Geomag. Geoelectr., 37*, 329–345, 1985.

Takagi, S., T. Nakamura, T. Kohno, N. Shiono and F. Makino, Observation of Space Radiation Environment with EXOS-D, *IEEE Trans. Nuc. Sci., 40*, 1491–1497, 1993.

Takeuchi H., T. Imai, S. Kumagaya, M. Wada and Y. Miyazaki, The Spatial Distribution of Quasi-Trapped Energetic Electrons Observed aboard the Satellite 'Shinsei', *Space Research, 14*, 309, 1974.

Van Allen, J.A., E.C. McIlwain and G.H. Ludwig, Radiation Observations with Satellite 1958e, *J. Geophys. Res., 64*, 271–286, 1959.

Vernov S.N., E.V. Gorchakov, P.I. Shavrin and K.N. Sharvina, Radiation Belts in the Region of the South Atlantic Magnetic Anomaly, *Space Sci. Rev., 7*, 490–533, 1967.

WDC-C2, Energetic Particle Intensity at Geostationary Orbit, in Space Environment Monitor Data from HIMAWARI, World Data Center C2 for Cosmic Rays, Cosmic Ray Laboratory, RIKEN, Japan, 1986.

Williams, D.J. and G.D. Mead, Night Side Magnetosphere Configuration as Obtained from Trapped Electrons at 1100 Kilometers, *J. Geophys. Res., 70*, 3017–3029, 1965.

Yukimatsu, A.S., M. Ejiri, T. Nagai, S. Takagi, A. Konno, T. Terasawa, T. Kohno and F. Makino, AKEBONO Observation of Radiation Belt Particles, these proceedings, 1996.

T. Kohno, Institute of Physical and Chemical Research (RIKEN), Japan

DISCUSSION

Q: D.N. Baker. Are the datasets you have described available (for collaborative studies) in a convenient form?

A: T. Kohno. I cannot say it is "convenient". Almost all data are basically open to general users. We will try to make a system for convenient data usage from outside.

UARS PEM contribution to radiation belt modelling

J.R. Sharber, J.D. Winningham, R. Link and R.A. Frahm

Southwest Research Institute, Department of Space Science, San Antonio, Texas

D.L. Chenette and E.E. Gaines

Research and Development Division, Lockheed Martin Missiles and Space, Palo Alto, California

The Upper Atmosphere Research Satellite (UARS), launched September 12, 1991, into a 585 km, 57° inclination orbit, carried a particle instrument, the Particle Environment Monitor (PEM), to assess the effects of solar and magnetospheric particle energy on the global atmospheric system. PEM consists of particle detectors, an X-ray imager, and a science magnetometer. It has operated since October, 1991, and thus provides a suite of observations made through the decrease of Solar Cycle 22 covering a wide range of activity levels, solar conditions, and local times. The objective of this paper is to provide familiarity with the PEM data set. A description of the observations and suggestions for the use of the data in the modelling and study of the radiation belts at low altitudes are presented.

1. INTRODUCTION

The Upper Atmosphere Research Satellite (UARS) embodies a program of global research stressing energy balance, dynamics, and chemistry, conducted by remote sensing from space. The satellite observatory orbits at an altitude of 585 km in a 57° inclination orbit. The sensors measure the energy radiated by the atmosphere, the energy absorbed or scattered from sunlight passing through the atmosphere, and particles incident on the upper atmosphere. Analysis furnishes detailed information on chemical constituents, temperature, atmospheric winds and dynamics, and the effects of solar radiation and particle energy input. The Particle Environment Monitor (PEM) has responsibility for assessing the influence of particle energy input on the atmospheric system.

The overall goal of PEM is to determine the magnetospheric and solar particle energy inputs to the atmospheric system (thermosphere, mesosphere, and stratosphere). Specific PEM objectives are:

1. to obtain a quantitative understanding of global energy deposition into the atmosphere over the solar cycle, and its variations with geophysical and solar activity;
2. to develop a self-consistent empirical and theoretical understanding of the coupling between the upper and lower atmosphere through energetic particles (composition, winds, heating electrodynamic coupling).

2. INSTRUMENTATION

In carrying out the above goal PEM makes both remote and *in situ* observations. Four instruments make up the complement: the Atmospheric X-ray Imaging Spectrometer (AXIS), the High Energy Particle Spectrometer (HEPS), the Medium Energy Particle Spectrometer (MEPS), and the Vector Magnetometer (VMAG). AXIS provides global-scale images and spectral measurements of bremsstrahlung X-rays over the energy range from 3 to 100 keV. HEPS and MEPS measure the energy spectra of precipitating electrons (5 eV to 5 MeV) and protons (5 eV to 150 MeV) at selected pitch angles. Measurements are made at a maximum of 63 energies for each species. The science magnetometer references the particle measurements to the magnetic field direction and yields magnetic field perturbations from which field aligned and horizontal ionospheric currents may be deduced. An outline of the meas-

Radiation Belts: Models and Standards
Geophysical Monograph 97
Copyright 1996 by the American Geophysical Union

Table 1. Particle Environment Monitor (PEM) Instrument Complement

Instrument	Description	Measurement Range Temporal-Spatial Resolution	Coverage
Atmospheric X-ray Imaging Spectrometer (AXIS)	16 element X-ray camera providing global multispectral imaging of bremsstrahlung; X-radiation resulting from electrons incident on the atmosphere; table lookup algorithms are used to obtain the incident electron spectrum	X-rays, $\sim$ 3–100 keV 8.19 s, 57.8 km (along track at 100 km)	field of view at 100 km altitude is $\sim$ 1230 km either side of UARS subsatellite track
High-Energy Particle Spectrometer (HEPS)	solid state detector telescopes providing differential spectral measurements of electrons and protons	electrons: 30 keV to 5 MeV 4.10 s, 31.1 km positive ions: 70 keV to 150 MeV 16.4 s, 125 km	look directions (w.r. spacecraft upward vertical); zenith (electrons and ions), $-15°$, $+15°$, $+45°$, $+90°$; nadir (electrons only), $-165°$, $+165°$
Medium Energy Particle Spectrometer (MEPS)	eight electrostatic energy analyzers providing differential spectral measurements of electrons and positive ions	electrons and positive ions: 5 eV to 32 keV 2.05 s, 15.6 km	look directions (w.r. spacecraft upward vertical); zenith (electrons and ions), $-23.7°$, $+6.3°$, $+21.3°$, $+36.3°$, $+66.3°$ nadir (electrons only), $-158.7°$, $+126.3°$, $+156.3°$
Vector Magnetometer (VMAG)	boom-mounted triaxial flux-gate magnetometer providing aspect information for the particle instruments and measurements of magnetic field perturbations used to calculate field-aligned currents	DC field: $-65,000$ nT to $+65,000$ nT 0.205 s, 1.56 km AC field: 5–50 Hz; δB_x and δB_z: 0–100 nT; δB_y: 0–10 nT 5 s, 38 km	local measurement of B_x, B_y, B_z and δB_x, δB_y, δB_z

urements provided by the PEM instrumentation is shown in Table 1. The PEM instrument paper [*Winningham et al.*, 1993] discusses each instrument in detail, and the reader is referred to that paper for specifics of designs, capabilities and calibrations of each instrument.

3. PEM OBSERVATIONS

To illustrate the capabilities of the PEM energy measurements we select measurements made during the large geomagnetic storm of November 8–9 of 1991. The period was characterized by a large magnetic disturbance which began with a sudden impulse at 0647 UT on November 8, reached a maximum excursion (-354 nT) between hour 0100 and 0200 UT of November 9, and was followed by a recovery phase that lasted until approximately noon on November 11. UARS, orbiting at 585 km in its circular orbit of $57°$ inclination, passed through the expanded auroral zone and into the polar cap during portions of several successive orbits.

We begin with the X-ray image, shown in Figure 1, made during the main phase of the storm between 1057 and 1143 UT on November 9. The image is produced from the 16 fields of view of the AXIS camera, which sees a $\sim$2500 km-wide path at the 100 km level as UARS moves over the Pacific Ocean and across the United States. The intensities of the X-ray fluxes are indicated by the scale to the left of the image. For reference the $Q = 3$ auroral oval [*Feldstein and Sarkov*, 1967], circles of constant geomagnetic latitude at $60°$ and $65°$, and the terminator at 100 km are shown. We observe an expanded oval with X-ray intensities of 10^4 photons/cm^2 s sr eV in the post-midnight sector.

The PEM electron measurements from HEPS and MEPS are shown as energy-time spectrograms in Figure 2. The lower panel shows electron observations of the MEPS sensor mounted at $36°$ with respect to the zenith while the upper two panels are HEPS electron data; the middle panel shows the precipitating component ($15°$), and the upper panel shows the trapped ($90°$) component. The energy range of HEPS is 30 keV to 5 MeV. These electron data were taken simultaneously with the X-ray image of Figure 1, in which the position of the foot of the UARS field line is indicated as the solid line through the image.

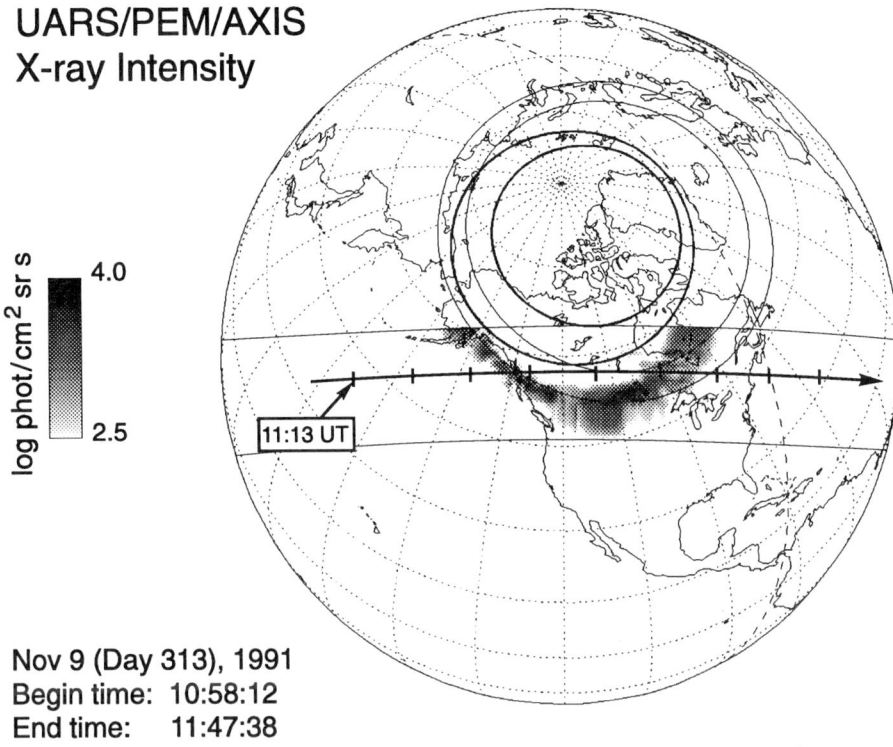

Figure 1. Incident X-ray intensity on November 9, 1991 between 1058 and 1148 UT showing integrated X-ray photon flux within a ~2500 km wide band measured by AXIS. The foot of the UARS field line is shown as a solid line; tick marks are three minutes apart. The $Q = 3$ auroral oval [*Feldstein and Starkov*, 1967], circles of constant geomagnetic latitude at 60° and 65°, and the terminator are shown for reference.

Looking first at the lower (MEPS) panel, the auroral fluxes are indicated by the increase in intensity and the spectral hardening at ~ 1118:30 UT and are observed until about 1132:00 UT. Several regions of enhanced flux are observed, with the most intense fluxes occuring at energies in the ~ 200 eV to ~ 2 keV range. With the aid of the HEPS data, shown in the upper panels, it appears very likely that UARS passed into the polar cap during this orbit. The polar cap was encountered at between 1122 and 1124:10 UT. The HEPS 15° telescope (middle panel) observed low fluxes in this region and MEPS encountered polar cap arc structures at the same time. The arc structure seen at about 1125:10 UT may be an intrusion of auroral precipitation to relatively high-latitudes.

The PEM energy range and measurement capabilities make it very useful for studies dealing with the relationships between the auroral and outer radiation belt regions. For example, examination of the upper panels of Figure 2 reveals that the high energy electrons precipitate over a much wider range of latitudes than do those at low energies. The populations are associated with three regions: the auroral region, the outer belt, and the inner belt. In the midnight sector auroral electrons are seen between invariant latitudes (ILAT) of 53° and 60.9°, the outer belt precipitation region overlaps the auroral region to some extent, then extends down to an invariant latitude of ~32° (~ 1112 UT). The population at low latitudes seen in the center panel between ~ 11:06 and ~ 11:12 UT includes locally trapped inner belt electrons since the upward looking HEPS detectors sample a portion of that population at low field-line inclination angles.

The complete PEM dataset will make possible the investigation of how these relationships vary with such geophysical phenomena as storms, substorms, relativistic electron precipitation events (REP's), solar particle events (SPE's), and moderate auroral conditions.

4. IONIZATION PRODUCTION

Precipitating electron spectra obtained from both HEPS and MEPS are used as inputs to compute rates of ionization production and energy deposition as functions of altitude. A sample electron spectrum is shown in Figure 3. The data were taken during the interval 1120:28-32 UT on November 9. The MEPS part of the spectrum is a 2-sweep average, while the HEPS portion is a single 4-second unaveraged accumulation. Because of the high fluxes encountered, the lower HEPS energy channels were saturated and are not shown. The spectrum shows a small peak at about 400 eV, probably due to parallel electric field acceleration of the low-energy electrons, a quasithermal population in the several keV range, and a power law high-energy tail. At energies above a few keV,

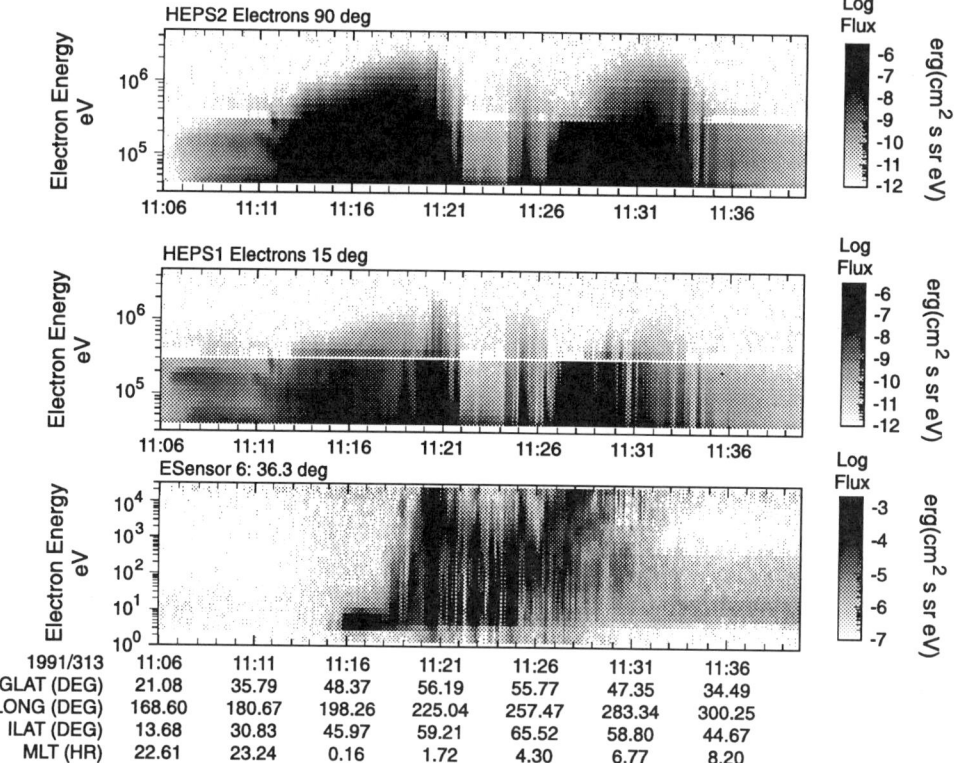

Figure 2. Electron energy-time spectrograms obtained from the PEM high and medium energy particle spectrometers (HEPS and MEPS) on November 9, 1991 between 1106 and 1139 UT. Top and middle panels show measurements from the 90° (HEPS-1) and the 15° (HEPS-2) sensor heads, respectively; the lower panel shows the MEPS 36° measurements. All panels show differential energy flux (erg/cm^2 s sr eV).

the spectrum may be conviently represented by a κ distribution [*Christon et al.*, 1991] having a thermal peak of 3350 eV, a density of 2.02 cm^{-3}, and a κ of 5.6. Many of the input electron spectra can be fitted by κ distributions, but this is not always the case in the PEM dataset. A spectral characterization study is now underway to determine the most representative functional forms of the spectra.

The ionization rate profile from the spectrum measured at 1120:28-32 UT is shown in Figure 4. It is computed using the CEPXS/ONELD multistream discrete ordinates code [*Lorence*, 1992], which solves the coupled electron-photon Boltzmann transport equations over the energy range 1 keV–100 MeV. The ionization production rate peaks at a value of 3×10^5 cm^{-3}s^{-1} at an altitude of 102 km. A secondary ledge on the curve is seen starting at about 50 km. This is the contribution from the absorption of forward scattered bremsstrahlung X-rays produced by the incident electrons. As a reference marker, we include in the figure the rate of ionization production at the 50° geographic latitude resulting from galactic cosmic rays during solar maximum conditions [*Brasseur and Solomon*, 1986]. It is clear that the precipitating energetic auroral electrons are the dominant source of ionization at all altitudes above 30 km.

5. UARS ENERGY INPUT MODEL

One of the PEM objectives is to develop a global energy input model to be used as input to other models such as the NCAR global circulation model [*Roble and Ridley*, 1987]. Our model has similarities to low-altitude models already in existence [*Hardy et al.*, 1985, 1987; *Fuller-Rowell and Evans*, 1987], which are statistical models based on large satellite databases and keyed to auroral activity. However, they do not at present fully include the effects of the high-energy (> 30 keV) particles. Our approach will include this population explicitly. The PEM model will provide spectral parameters and/or energy deposition rates. Due to the changing nature of the magnetospheric energy input, the model is designed to provide these quantities as functions of local time, latitude, and geomagnetic activity. The model is statistical and uses the particle observations (HEPS + MEPS) taken within the loss cone over a specified pixel size (viz., 5° latitude × 5° longitude). The model will thus provide spectral parameters and energy inputs from the auroral oval and radiation belt regions.

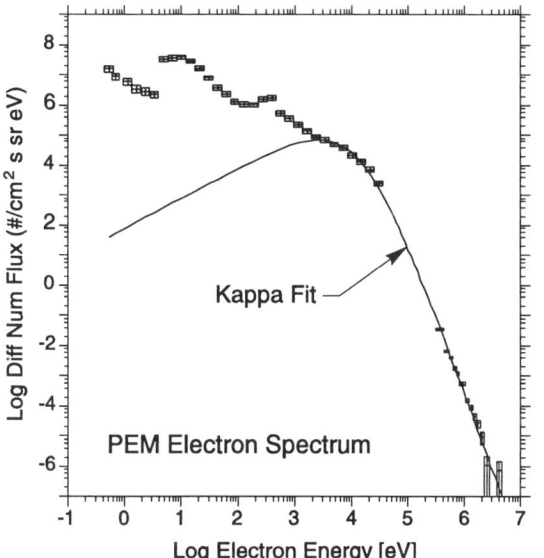

Figure 3. Electron differential number spectrum measured during 1120:28–32 UT of November 9, 1991. The spectrum shows a small peak at about 400 eV, a quasithermal population in the several keV range, and a power law high-energy tail. At energies above a few keV, this spectrum may be conviently represented by a κ distribution [*Christon et al.*, 1991].

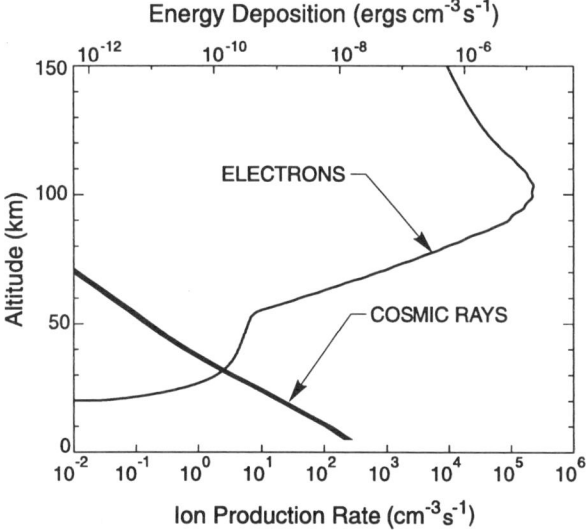

Figure 4. Ionization rate profile from the spectrum measured at 1120:28-32 UT computed using the CEPXS/ONELD multistream discrete ordinates code [*Lorence*, 1992]. The ionization production rate peaks at a value of $3 \times 10^5 \, cm^{-3} s^{-1}$ at an altitude of 102 km. The secondary ledge starting at about 50 km is the contribution from the absorption of forward scattered bremsstrahlung X-rays produced by the incident electrons. Shown for comparison is the rate of ionization produced by galactic cosmic rays at 50° geographic latitude during solar maximum conditions [*Brasseur and Solomon*, 1986].

6. RADIATION BELT MODELLING STUDY

UARS was launched in September of 1991 and has operated nearly continuously until April/May, 1995, after which time the solar collector had to be "parked" in an optimal collecting orientation because of a failing clutch mechanism. PEM coverage after that time has been considerably reduced as duty cycling became the normal UARS operating procedure. As a result, the primary PEM database is a high-quality, nearly continuous set of particle and X-ray data accumulated over an interval of $\sim$ 3.5 years over a wide range of energies, local times and geophysical conditions. The data interval spans a major part of the declining phase of Solar Cycle 22.

The PEM energy range and measurement capabilities make it particularly well-suited for studies dealing with the relationships between the auroral and outer belt regions. Event studies designed to investigate the morphology of the low altitude radiation belt component and its variation in both intensity and location with respect to the auroral precipitation region are an obvious starting point. An understanding of the changes in the morphology with geophysical events such as storms, substorms, relativistic electron precipitation events (REPs), solar proton events (SPEs), and even moderate auroral activity are the likely result of these kinds of studies. Similar studies, with emphasis on the precipitating component, have already been carried out by UARS investigators [*Sharber et al.*, 1993; *Chenette et al.*, 1993; *Gaines et al.*, 1995] or are ongoing, as with the November 1993 storm initiative in which the PEM team is a supplying particle data, X-ray images, and field-aligned current observations.

The UARS project point of view is necessarily focused on the effects on the atmosphere rather than magnetospheric or radiation belt processes. Such processes become relevant to UARS science if they enable an improved understanding of the atmospheric effects. This is the basis for both the event studies and the PEM statistical model. But PEM makes measurements at a variety of viewing directions with respect to the local zenith; i.e., measurements are made within and outside of the atmospheric loss cone (see Table 1). Therefore, developing a radiation belt model using PEM observations is feasible. Such a model would employ techniques similar to the PEM energy input model described above and already under development, but it would contain a measurement of the trapped radiation component (and spectra) as well as the the precipitating component. Among its other features, the model could include an anisotropy indicator (coming straight from the measurements) from which an assessment of particle "dumping" from the radiation belts over a range of particle energies, locations, and geophysical conditions would be possible.

7. SUMMARY POINTS

- UARS was launched September 12, 1991. To date we have 3.5 years of nearly continuous data. During the spring of 1995, a reduction is spacecraft available power reduced PEM on-time to a few days each month. To date, that condition continues.
- The PEM energy range and measurement capabilities make it particularly well-suited for studies dealing with

the relationships between the auroral and outer belt regions. This will make possible the study of how these relationships vary with geophysical events such as storms, substorms, REPs, SPEs, and more moderate auroral activity.

- We have begun a program of modelling the particle energy input into the upper and middle atmosphere. Similar techniques with the same data set can be employed to produce a model of the radiation belt as seen by PEM at 585 km altitude.
- Based on our UARS experience, it is clear that future work should place emphasis on the importance of loss of particles from the radiation belt population and the impact of these losses on the atmosphere.

Acknowledgements. Funds for PEM instrument development and operations provided by NASA contract NAS5-27753 and NASA grant NAG5-3148 to Southwest Research Institute are gratefully acknowledged. We thank C.A. Gurgiolo for numerous helpful discussions and A.O. Sawka for technical assistence.

REFERENCES

Brasseur, G. and S. Solomon, *Aeronomy of the Middle Atmosphere*, D. Reidel, Dordrecht, Holland, 1986.

Chenette, D.L., D.W. Datlowe, R.W. Robinson, T.L. Schumaker, R.R. Vondrak and J.D. Winningham, Atmospheric energy input and ionization by energetic electrons during the geomagnetic storm of 8–9 November 1991, *Geophys. Res. Lett., 20,* 1323, 1993.

Christon, S.P., D.J. Williams, D.G., Mitchell, C.Y. Huang and L.A. Frank, Spectral characteristics of plasma sheet ion and electron populations during disturbed geomagnetic conditions, *J. Geophys. Res., 96,* 1–22, 1991.

Feldstein, Y.I. and G.V. Sarkov, Dynamics of auroral belt and polar geomagnetic disturbances, *Planet. Space Sci., 15,* 209, 1967.

Fuller-Rowell, T.J. and D.S. Evans, Height-integrated Pedersen and Hall conductivity patterns inferred from the TIROS-NOAA satellite data, *J. Geophys. Res., 92,* 7506–7618, 1987.

Gaines E.E., D.L. Chenette, W.L. Imhof, C.H. Jackman and J.D. Winningham, Relativistic electron fluxes in May 1992 and their effect on the middle atmosphere, *J. Geophys. Res., 100,* 1027–1033, 1995.

Hardy, D.A., M.S. Gussenhoven and E. Holeman, A statistical model of auroral precipitation, *J. Geophys. Res., 90,* 4229–4248, 1985.

Hardy, D.A., M.S. Gussenhoven, R. Raistrick and W.J. McNeil, Statistical and functional representations of the pattern of auroral energy flux, number flux, and conductivity, *J. Geophys. Res., 92,* 12275–12294, 1987.

Lorence, L.J., Jr., CEPXS/ONELD version 2.0: A discrete ordinates code package for general one-dimensional coupled electron-photon transport, *IEEE Trans. Nucl. Sci., 39,* 1031, 1992.

Sharber, J.R., R.A. Frahm, J.D. Winningham, J.C. Biard, D. Lummerzheim, M.H. Rees, D.L. Chenette, E.E. Gaines, R.W. Nightingale, W.L. Imhof, Observations of the UARS Particle Environment Monitor and computation of ionization rates in the middle and upper atmosphere during a geomagnetic storm, *Geophys. Res. Lett., 20,* 1319, 1993.

Roble, R.G. and E.C. Ridley, An auroral model for the NCAR thermospheric general circulation model (TGCM), *Annales Geophys., 5A,* 369–382, 1987.

Winningham et al., The UARS Particle Environment Monitor, *J. Geophys. Res., 98,* 10,649–10,646, 1993.

J.R. Sharber, J.D. Winningham, R. Link and R.A. Frahm, Southwest Research Institute, 6220 Culebra Road, Department of Space Science, San Antonio, Texas 78238, U.S.A.

D.L. Chenette and E.E. Gaines, Research and Development Division, Lockheed Martin Missiles and Space, Org. 91–20, Bldg. 252, 3251 Hanover Street, Palo Alto, California 94304, U.S.A.

DISCUSSION

Q: M. Lauriente. Did you experience any SEU's?

A: J.R. Sharber. Yes, some. We can provide you a log of these events.

C: J. Albert. Regarding your comments on the importance of losses to the atmosphere, I would like to point out that some losses are caused by pitch-angle diffusion resulting from lightning induced waves.

Radiation Belt Observations From CREAM and CREDO

C. Dyer and A. Sims

Space Department, DRA Farnborough

C. Underwood

CSER, University of Surrey, Guildford

These experiments are designed to measure protons, cosmic rays and accumulated dose. Through the variety of missions employed they have now achieved wide coverage of the magnetosphere as well as a significant portion of a solar cycle. The LEO observations have shown the westward drift of the South Atlantic Anomaly, new regimes of trapped protons in the region of $L = 2.6$ following solar flare events in March 1991 and October 1992, and an altitude dependence of trapped protons which is at variance with AP-8. Total dose monitors deployed at high inclination or high altitude show the extreme time variability of the outer radiation belt, while the total dose is significantly less than AE-8 predictions. The need for both further developments in the models and a comprehensive programme of flight experiments is emphasised.

1. INTRODUCTION

While models of the space radiation environment are extensively employed for design purposes, comparisons between the predicted and observed doses and upset rates show a mixed rate of success with discrepancies up to orders of magnitude. It is not clear whether these discrepancies arise from errors in the particle environment or in the characterisation of the devices. The CREAM and CREDO experiments have been designed to redress this deficiency by providing measurements of the relevant aspects of the environment. Since 1988 the coverage of these instruments has expanded to cover the entire atmosphere and magnetosphere as well as a large portion of the solar cycle, thus providing a unique data set. In this short paper we will highlight the results of relevance to models of the trapped radiation. For fuller discussion and a more complete set of results, including cosmic ray effects, the reader is referred to recent review papers [*Dyer*, 1995; *Dyer et al.*, 1996a].

Radiation Belts: Models and Standards
Geophysical Monograph 97
Published in 1996 by the American Geophysical Union

The Cosmic Radiation Environment and Activation Monitor (CREAM) has flown on six Shuttle flights between September 1991 and February 1995, covering the full range of inclinations as well as altitudes between 210 and 550 km. Meanwhile the Cosmic Radiation Environment and Dosimetry experiment (CREDO) has operated continuously on UOSAT-3 in 800 km, 98.7° orbit since April 1990. Similar detectors were launched on KITSAT-1 (1330 km, 66° inclination) in August 1992 and POSAT-1 (790 km, 98.7° inclination) in September 1993. Since the summer of 1994, CREDO-II versions have been operating on APEX in an eccentric orbit (350×2486 km) at 70° inclination, and on STRV in geostationary transfer orbit (298×35953 km, 7° inclination).

2. THE CREAM AND CREDO EXPERIMENTS

The CREAM and CREDO detectors are designed to monitor those aspects of the space radiation environment of concern for electronics, i.e. charge deposition spectra, linear energy transfer spectra and total dose. In these instruments the SEU environment is monitored by means of pulse-height analysis of the charge-deposition spectra in ten pin diodes, each 1 cm^2 in area and 300 μm in depth. In the Shuttle version the channel thresholds range from a charge deposition of 0.021 pC,

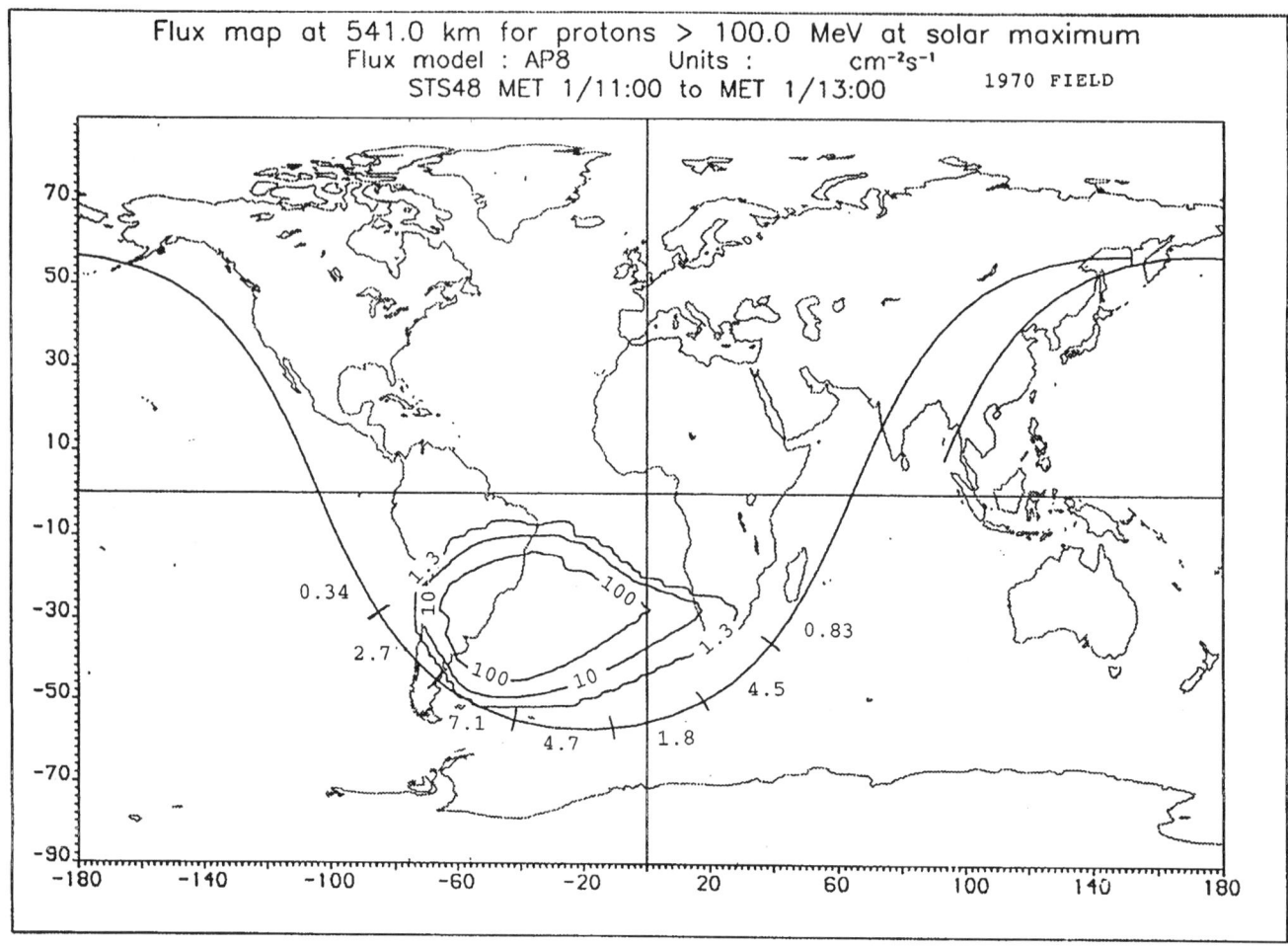

Figure 1. Ground track of orbit 23 for STS-48 is shown with respect to proton flux contours from AP-8 and 1970 field. With this field the orbit just misses the SAA.

equivalent to a normally incident particle with a linear energy transfer (LET) of $6.8\,\mathrm{MeV}/(\mathrm{g\,cm}^{-2})$, for channel 1 to 21 pC, equivalent to $6808\,\mathrm{MeV}/(\mathrm{g\,cm}^{-2})$, for channel 9. For the CREDO version on UOSAT-3 the range of thresholds is 32.2 to $6430\,\mathrm{MeV}/(\mathrm{g\,cm}^{-2})$. Data are accumulated into preset time bins, 5 minutes for STS-48 and 2 minutes for STS-53. A variety of locations is employed to investigate the influence of shielding, ranging from the sleep station wall for minimum shielding to the airlock ceiling for maximum shielding. The shielding at the wall has a minimum value of $1.3\,\mathrm{g\,cm}^{-2}$ (proton threshold 32 MeV) in certain directions with 50% of solid angles having shielding less than $6\,\mathrm{g\,cm}^{-2}$ (proton threshold 75 MeV). The shielding determines the energy threshold and eliminates trapped heavy ions and electrons so that channel 1 serves as a proton monitor. The CREDO versions utilise RADFETs to obtain total dose information. These MOSFET devices have a thick gate oxide ($> 1,000$ Å) designed to trap positive charges and shift the threshold voltage.

The CREDO-II version was developed for flights on the APEX (Advanced Photovoltaics and Electronics Experiment) spacecraft and STRV (Space Technology Research Vehicle) and represents a significant advance over previous versions as it employs a telescope technique to detect coincidences between parallel planes of pin diodes in order to provide directional information and define the pathlength of particles through the diodes to within 25%. A high area detector is used for cosmic rays while a low area detector using two $1\,\mathrm{cm}^2$-diodes is used to monitor the high rates of protons in the inner belt. The pulse-height analysis technique is also applied to "non-coincident" events, such that each telescope array also acts as an omnidirectional detector of area $8\,\mathrm{cm}^2$. In the APEX version the time resolution is fixed at 290 s for compatibility with the telemetry system. The detector box walls are designed to eliminate all but the most energetic of the trapped electrons which are further eliminated by the LET thresholds. Total dose is measured at ten locations on APEX and three locations on STRV-1a using RADET dosimeters.

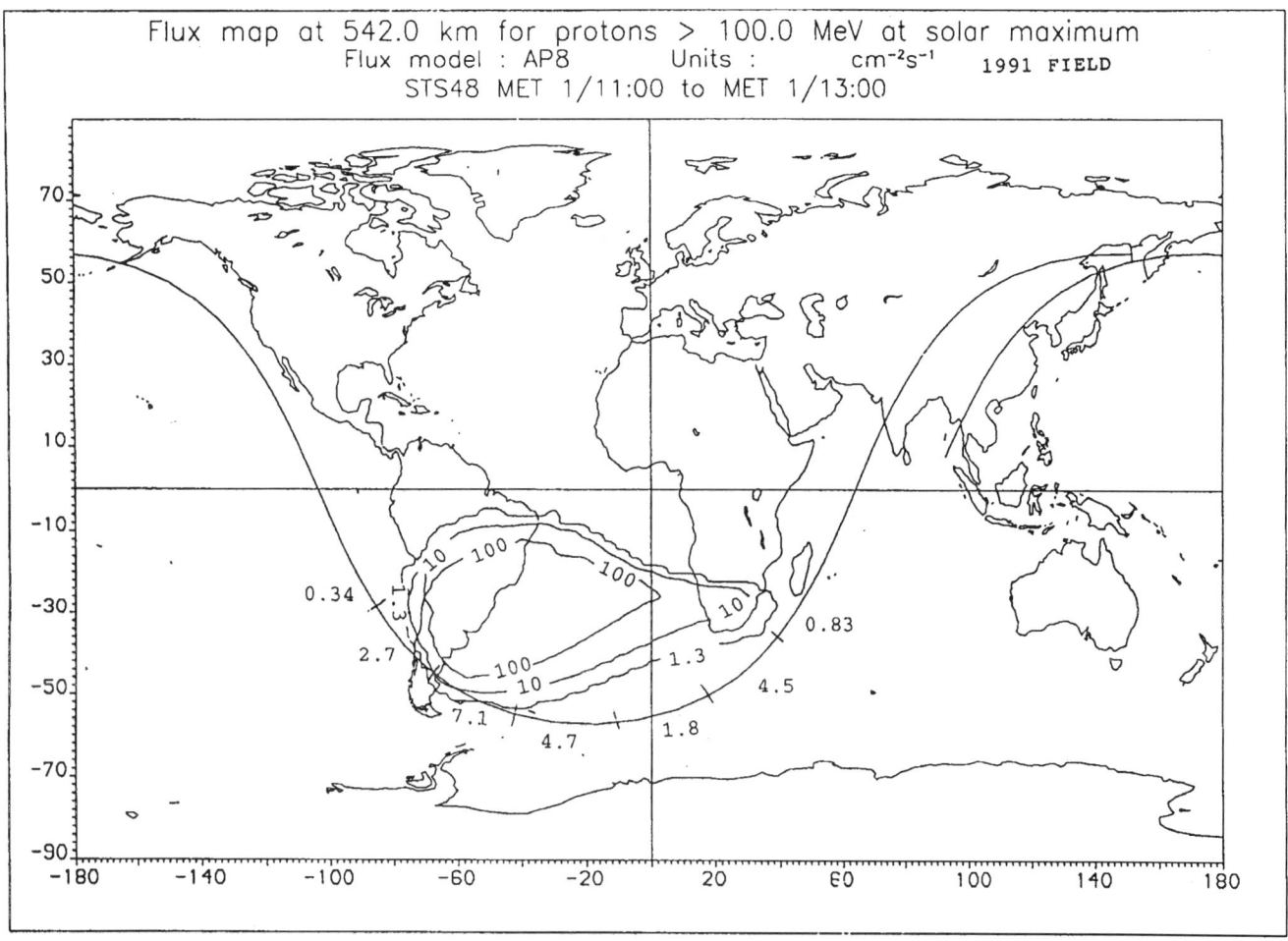

Figure 2. Ground track of orbit 23 for STS-48 is shown with respect to proton flux contours (> 100 MeV) from AP-8 and 1991 field. With the updated field the orbit intersects the SAA. An additional peak is seen off of South Africa due to the new radiation belt created in March 1991.

3. TRAPPED PROTON RESULTS

The CREAM experiment on Shuttle has now flown on missions 48, 44, 53, 56, 68 and 63 over the time period September 1991 to February 1995. Data from the first 4 missions have been extensively analysed and presented in [*Dyer et al.*, 1992a, 1993, 1996b]. The reader is referred to these papers for a complete discussion of the results, which include observations of the westward shift in the position of the South Atlantic Anomaly, a new regime of trapped protons, modulation of cosmic rays, induced radioactivity, enhancements by secondary particles, and discrepancies in LET spectra compared with predictions. Some of the highlights are given below.

It was noted that certain SAA passes (e.g. orbit 23 of STS-48 and orbit 40 of STS-44) were not predicted when using the recommended technique of employing the field model pertaining to the data from which the models were created (i.e. 1970). However, use of the 1991 geomagnetic field does predict peaks for these orbits as it accounts for the steady drift of the SAA contours to the west due to evolution of the geomagnetic field. This is illustrated in Figures 1 and 2 where the ground track of orbit 23 for STS-48 is shown with respect to the SAA contours obtained using the 1970 and 1991 fields respectively. It can be seen that the orbit just clips the contours to the southwest for the 1991 contours but misses for the 1970 case. For this orbit there is a second peak observed off of South Africa which is not predicted by either field model. This region is where the $L = 2.5$ shell intersects this altitude orbit and the high fluxes are due to the second proton belt observed by CRRES to be created by the solar flare event of 23 March 1991 [*Mullen et al.*, 1991]. Careful analysis of the STS-53 data again showed a small enhancement in this region when cosmic ray contributions are carefully subtracted [*Dyer et al.*, 1993, 1996b]. This was originally believed to be the remnants of the March 1991 event but evidence from UOSAT-3 (see below) now points

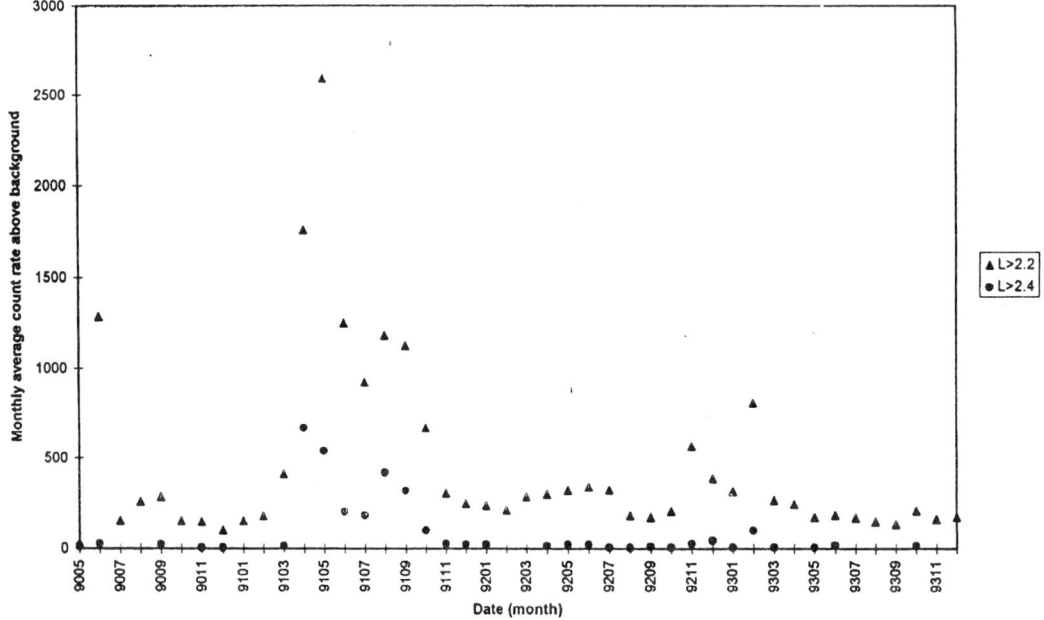

Figure 3. Monthly-averaged count rates at $L > 2.2$ and 2.4 from UOSAT-3 with cosmic-ray background subtracted show new regimes of trapped radiation following flare events in March 91 and October 92.

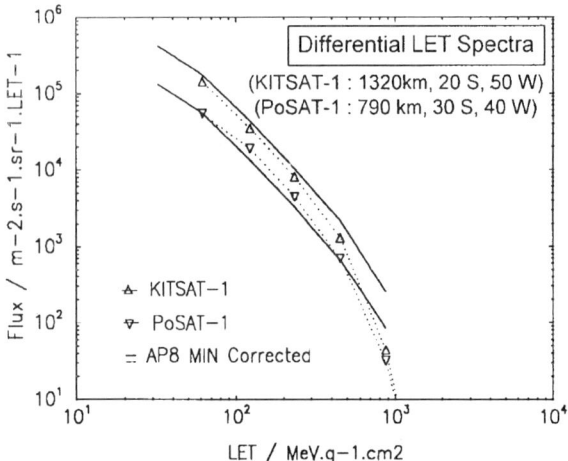

Figure 4. Energy-deposition spectra in the SAA from CRE on POSAT and KITSAT during May 1994 show less altitude dependence than predicted by AP-8.

towards a second enhancement, possibly associated with a flare in October 1992. The westward movement of the SAA and creation of the new regime off of South Africa in March 1991 have been confirmed by health physics monitors on Shuttle missions [*Konradi et al.*, 1994].

The UOSAT series of microsatellites (50–60 kg) has been developed by the University of Surrey and are in low earth orbit with altitudes between 700 and 1300 km. The later spacecraft in the series have included the radiation monitors CREDO, provided by DRA, and CRE (Cosmic Ray Experiment) produced at Surrey. The former is described above while the latter comprises a single diode of area 9 cm^2. Pulse-height analysis is employed to give the energy-deposition spectra of the radiation. RADFETs are also included to record the accumulated total dose at a number of locations. Results on the environment have been reported in [*Dyer et al.*, 1991, 1992b, 1993, 1996b]. The CREDO detector on UOSAT-3 has the advantage of continuous coverage, although the orbit gives only short duration passages through the regime of the new radiation belt. The UOSAT data have been carefully examined by mapping the count-rates into (B, L) space following subtraction of cosmic-ray contributions by means of fits to cosmic-ray counts obtained at identical geomagnetic latitudes outside of the belts. In addition days containing direct solar-flare particles have been excluded based on data from the GOES spacecraft. The remaining counts taken over the (B, L) region of the new belt accessible to UOSAT have been averaged on a monthly basis and the resulting time variations for L values greater than 2.2 and 2.4 are plotted in Figure 3 to show the time history of this region of the radiation belts. The marked increase at March 1991 and the decay through to October 1991 are clearly seen. There appears to have been a second increase in November 1992, possibly arising from the proton flare of 31 October 1992, and this was probably responsible for the enhancement seen by STS-53. There is also a hint of an enhancement early on following the May 1990 solar flare. Clearly the slot region is highly dynamic.

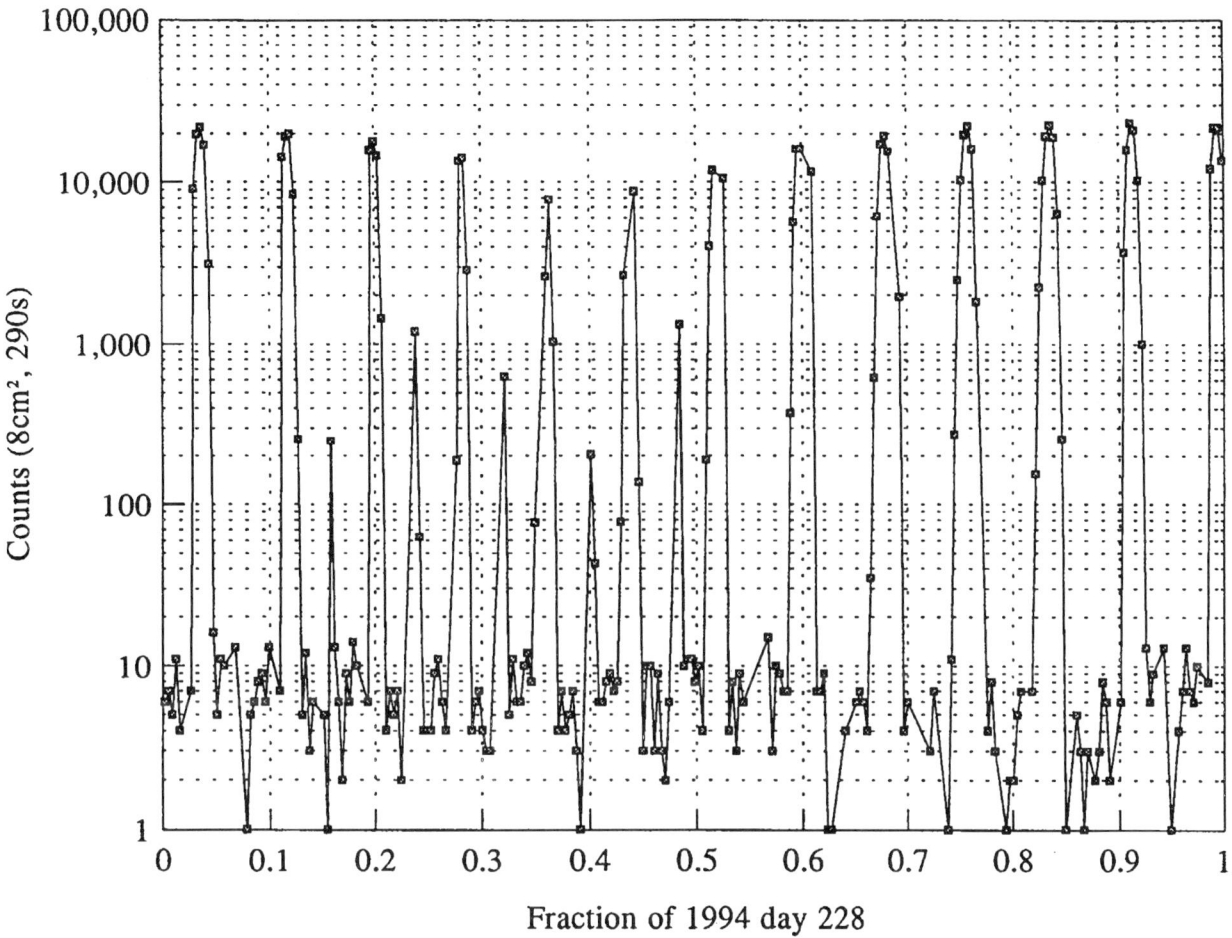

Figure 5. Count-rate profile for a typical day from the CREDO-II high area detector on APEX. Frequent intense inner-belt passes are seen for this eccentric orbit with apogee 2486 km.

KITSAT-1 was launched into 1330 km altitude, 66° inclination orbit alongside the TOPEX/Poseidon mission in August 1992 enabling the investigation of upsets and environments at somewhat higher altitude. The KITSAT spacecraft include the CRE experiment to yield collateral information on the radiation environment, while intercalibration of the CRE and CREDO experiments was made possible by the launch of CRE on POSAT-1 into 790 km, 98° orbit in September 1993. Results on the environment have been presented in [*Underwood et al.*, 1994]. Encouraging agreement is obtained between the three experiments despite the different design. For the region of the SAA, the LET spectra observed from KITSAT and POSAT have been compared with each other and with predictions based on the AP-8 model for solar minimum, allowing for protons stopping and slowing in the detector. Figure 4 shows that agreement is good considering that uncertainties of a factor two are inherent in allowing for spacecraft shielding and particle anisotropies. The influence of the higher altitude in affording about a factor five increase can be seen and this is also reflected in the total dose data from the RADFETs which show 0.7 to 0.9 rads per day on POSAT and 4 to 6 rads per day on KITSAT. The observed increase with altitude is less than that predicted by AP-8.

The Advanced Photovoltaics and Electronics Experiment spacecraft contains a number of experiments concerned with space environmental effects on solar arrays and electronics. The 250 kg spacecraft was launched on a Pegasus, aircraft-launched rocket on 3 August 1994 into an eccentric orbit of 352×2486 km, 70° inclination, which takes it into the intense regimes of the inner proton belt as well as giving significant exposure to cosmic rays, outer-belt electrons and solar particles at high latitudes. The Cosmic Ray Upset Experiment (CRUX) is provided by NASA/GSFC [*Adolphsen et al.*, 1995ab] and is a test package of SRAMs and HEXFETs. The Cosmic Radiation Environment and Dosimetry experiment is provided by DRA Farnborough to obtain the complementary environment data on cosmic rays, trapped protons and dose, which are required to interpret the CRUX observations. This version is the CREDO-II telescope described above. Total dose is measured at ten locations using RADFET dosimet-

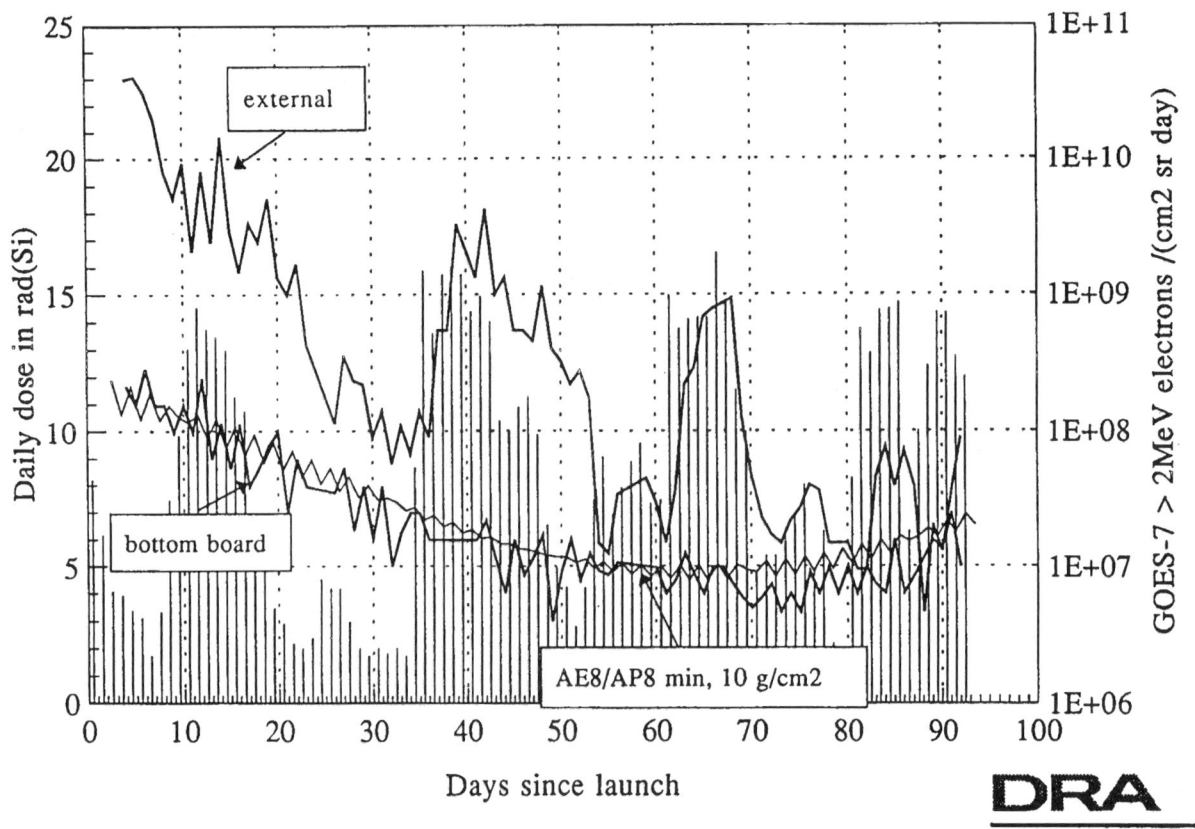

Figure 6. Observed vs. predicted daily dose rate during the first 90 days of the APEX mission. The flux of > 2 MeV electrons measured by GOES-7 is shown as a histogram and is seen to correlate with periodic surges in the daily dose rate observed in the lightly shielded dosimeter. The smooth downward trend in the dose rate is due to the precession of apogee. These outer belt electron enhancements are also seen by STRV (see paper by Wrenn and Sims).

ers. Eight such dosimeters are at various positions within the CREDO box, one is placed within the CRUX experiment and one is located immediately outside the CREDO box. Fuller results are given in [Sims et al., 1995]. Figure 5 shows data from the charged particle monitor for 1994 day 228. The plot shows raw counts in a single channel which corresponds to protons of energy greater than 90 MeV external to the spacecraft. The larger peaks (near 10,000 counts) occur when the spacecraft moves up into the main inner proton belt near apogee; the intermediate peaks (100 to 1,000 counts) correspond to traversals of the South Atlantic Anomaly region. The background count rate (near 10 counts) is due to the galactic cosmic ray background, and is modulated by latitude, although this is partially obscured by statistical fluctuations. Upsets in the memory devices tested by CRUX are dominated by proton interactions and are showing good correlation with the proton measurements from CREDO.

4. OUTER BELT ELECTRON RESULTS

For APEX the most exposed RADFET is showing an annual dose of 4.3 krad, while dose-rate enhancements are seen at the same time as dose-rate and energetic electron enhancements observed by STRV in GTO and GOES-7 in GEO. This is illustrated in Figure 6 and shows that enhancements in the outer radiation belt are observable at low altitude in the high latitude "horn regions".

The two Space Technology Research Vehicle microsatellites have been built at the Defence Research Agency Farnborough and contain a set of experiments designed to monitor various aspects of the space environment. They were launched on 17 June 1994 into GTO orbit of inclination 7°. From the radiation effects viewpoint the orbit is ideal, as both

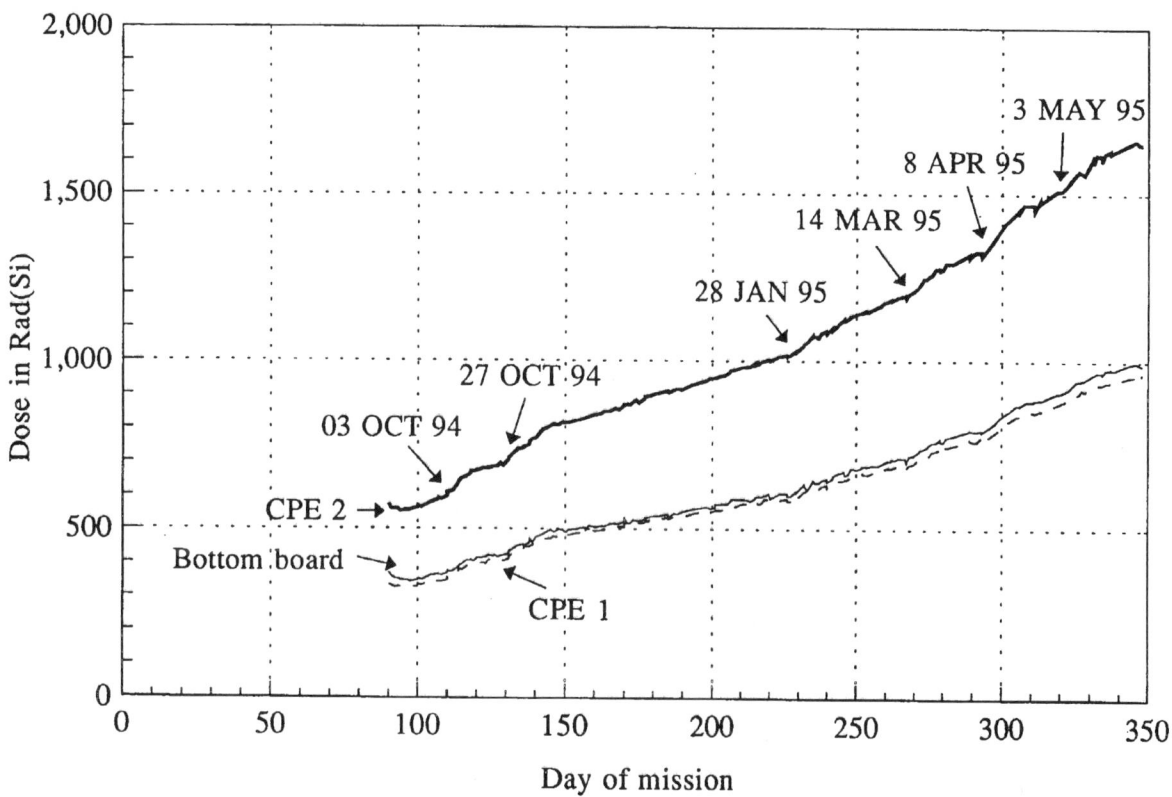

Figure 7. Dose accumulated with time for RADFETs within CREDO on STRV-1a. Enhanced rates correspond to electron increases seen by STRV and GOES.

inner and outer radiation belts are experienced and much of the orbit is fully exposed to cosmic rays. Three complementary radiation monitors are flown. The CREDO-II unit is on STRV-1a and measures energetic protons and cosmic rays together with dose. On STRV-1b the Radiation Environment Monitor, provided by ESTEC and PSI Switzerland [*Daly et al.*, 1992], monitors electrons and protons, while the RADMON detector [*Buehler et al.*, 1993], provided by JPL, monitors the upset and dose environment by microdosimetry. In addition there is a suite of instruments designed to measure spacecraft charging and the plasma environment and comprising a cold ion detector, Langmuir probe and surface charge detector. The cold ion detector has the beneficial property of having a background channel which detects electrons with energies greater than 1 MeV. Data from CREDO are presented in [*Dyer et al.*, 1995] while data from REM and the cold ion detector are also reported in these proceedings [*Bühler et al.*, 1996; *Wrenn and Sims*, 1996]. Three RADFETs are located in the CREDO box and accumulated doses vs. time are shown in Figure 7. These levels are quite low, ranging from 1 to 1.5 krad per year but show enhanced rates coincident with electron increases seen by the cold ion detector, illustrating the highly dynamic nature of the outer electron belt. The electron enhancements are highly significant as they can produce deep dielectric charging events which are an alternative source of anomalies to single event upsets.

5. DISCUSSION

Results from the CREAM and CREDO experiments flown on a wide range of platforms allow the following conclusions:

1. Low earth orbits are dominated by the SAA which is moving westwards due to the evolving geomagnetic field.

2. The increase in SAA protons with altitude is less than predicted by the AP-8 model.

3. Long term trapping of protons can occur in the slot region at $L = 2.6$ following certain solar particle events (e.g. March 1991, October 1992) and for low earth orbits this is experienced off of South Africa.

4. The outer zone region of trapped electrons is highly dynamic.

The above observations illustrate the need for modifications to the trapped radiation models and continued observations. As radiation-hardened components inevitably become less available and increasing reliance is placed on commercial components for future, high performance space systems, the need for accurate models and space test data will become more acute.

Acknowledgements. This work has been supported at the Defence Research Agency by the UK Ministry of Defence. Many of these flights are the result of international collaboration and the support of BMDO on CREAM and STRV is gratefully acknowledged, as is the support of the USAF/Space Test Programme on CREAM and APEX.

REFERENCES

J. Adolphsen, J. Barth, E. Stassinopoulos, K. LaBel, T. Gruner, M. Wennersten and C. Seidleck, SEP data from the APEX Cosmic Ray Upset Experiment: Predicting the performance of commercial devices in space, IEEE Proceedings of the RADECS 95 Conference, 1995a (in press).

J. Adolphsen, E. Stassinopoulos, J. Barth, K. LaBel and C. Seidleck, Single event upset rates on 1M and 256k memories: CRUX experiment on APEX, *IEEE Trans. Nuc. Sci.*, 42, 6, 1964–1974, 1995b.

M. Buehler, B. Blaes, G. Soli and G. Tardio, On-chip p-MOSFET dosimetry, *IEEE Trans. Nuc. Sci.*, 40, 6, 1442–1449, 1993.

P. Bühler, L. Desorgher, A. Zehnder, L. Adams and E. Daly, Exploring the radiation belts with the radiation environment monitor REM, these proceedings.

E. Daly, L. Adams, A. Zehnder and S. Ljungfelt, ESA's radiation environment monitor and its technological role, IAF 92-0779, 1992.

C.S. Dyer, "In-flight experiments", Short course lecture notes from 3rd European symposium on Radiations and their effects on Components and Systems, RADECS 95, Arcachon, France, 18 Sept. 1995.

C.S. Dyer, A.J. Sims, J. Farren, J. Stephen and C. Underwood, Radiation environment measurements & single event upset observations in sun-synchronous orbit, *IEEE Trans. Nuc. Sci.*, 37, 6, 1700–1707, 1991.

C.S. Dyer, A.J. Sims, J. Farren, J. Stephen and C. Underwood, Comparative measurements of the single event upset and total dose environments using the CREAM instruments, *IEEE Trans. Nuc. Sci.*, 39, 3, 413–417, 1992a.

C.S. Dyer, A.J. Sims, P.R. Truscott, J. Farren and C. Underwood, Radiation measurements on Shuttle missions using the CREAM experiment, *IEEE Trans. Nuc. Sci.*, 39, 6, 1809–1816, 1992b.

C.S. Dyer, A.J. Sims, P.R. Truscott, J. Farren and C. Underwood, The low earth orbit radiation environment and its evolution from measurements using the CREAM and CREDO experiments, *IEEE Trans. Nuc. Sci.*, 40, 6, 1471–1478, 1993.

C.S. Dyer, A.J. Sims, P.R. Truscott, C. Peerless and C. Watson, Measurements of the radiation environment from LEO to GTO using the CREAM & CREDO experiments, *IEEE Trans. Nuc. Sci.*, 42, 1975–1982, 1995.

C.S. Dyer, A.J. Sims and C. Underwood, Measurements of the SEE environment from sea level to GEO using the CREAM & CREDO experiments, Special Issue of *IEEE Trans. Nuc. Sci.* on Single Event Effects and the Space Radiation Environment, Vol. 43, No. 2, 1996a.

C.S. Dyer, A.J. Sims, P.R. Truscott, C. Peerless and C. Underwood, Temporal variations in the new proton belt created in March 1991 observed using the CREAM and CREDO experiments, *Adv. Space Res.*, 17, 2, 159–162, 1996b.

A. Konradi, G.D. Badhwar and L.A. Braby, Recent Space Shuttle observations of the South Atlantic Anomaly and the radiation belt models, *Adv. Space Res.*, 14, 10, 911–921, 1994.

E. Mullen, M. Gussenhoven, K. Ray and M. Violet, A double-peaked inner radiation belt: cause and effect as seen on CRRES, *IEEE Trans. Nuc. Sci.*, 38, 6, 1713–1717, 1991.

A. Sims, C. Dyer, C. Watson and C. Peerless, Measurements of the radiation environment on the APEX spacecraft, IEEE Proceedings of the RADECS 95 Conference, 1995 (in press).

C. Underwood, D. Brock, P. Williams, S. Kim, R. Dilao, P. Santos, M. Brito, C. Dyer and A. Sims, Radiation environment measurements with the cosmic ray experiments on-board the KITSAT-1 and POSAT-1 micro-satellites, *IEEE Trans. Nuc. Sci.*, 41, 6, 2353–2360, 1994.

G. Wrenn and A. Sims, Internal charging in the outer zone and operational anomalies, these proceedings.

C. Dyer and A. Sims, Space Department, DRA Farnborough
C. Underwood, CSER, University of Surrey, Guildford

Los Alamos Geosynchronous Space Weather Data For Radiation Belt Modeling

G.D. Reeves, R.D. Belian, T.C. Cayton, M.G. Henderson, R.A. Christensen, P.S. McLachlan and J.C. Ingraham

Los Alamos National Laboratory, Mail Stop D 436, Los Alamos, New Mexico

This paper presents an overview of a database of Los Alamos geosynchronous energetic particle data and the tools available to access and analyze those data. Los Alamos geosynchronous energetic particle measurements began in 1976 and the on-line data coverage currently begins in 1979. Typically data are available simultaneously from three geosynchronous satellites. Two generations of instruments have flown-the Charged Particle Analyzer (CPA) and Synchronous Orbit Particle Analyzer (SOPA). Both instruments measure electrons and ions with energies from tens of keV to tens of MeV. The data which have been made available on-line consist of 1-spin (approximately 10-second) averages or 1-minute (approximately 6-spin) averages. This paper includes a brief description of the data holdings, instructions for accessing digital data and summary plots, and instructions for accessing other reference material related to the data via the World Wide Web.

1. INTRODUCTION

As magnetospheric physics begins to include programmatic applications the need for timely access to key data sets has become more acute. Nowhere is this more apparent than in the area that has come to be known as "Space Weather". Broadly, Space Weather refers to being able to specify the plasma environment in space and the effects of that environment on the spacecraft that operate there. Naturally, a key focus of Space Weather applications is on the region near geosynchronous orbit where hundreds of military and civilian spacecraft operate. Satellites that operate in geosynchronous orbit can be affected by surface charging, deep dielectric charging, single event upsets, high detector or electronic backgrounds, rapid changes in the local electric or magnetic fields, and other interactions with the environment. Each of these can be related to one or more physical processes that affect the geosynchronous environment. For example surface charging can be enhanced by the injection of hot plasma into geosynchronous orbit during substorms. Deep dielectric charging is more common during relativistic electron enhancements produced by the interaction of the magnetosphere with high-speed solar wind streams from coronal holes.

Geosynchronous orbit is an interesting region for space weather applications not only because of the large number of spacecraft that operate there but also because of the large number of magnetospheric processes that affect it. Geosynchronous orbit lies near the outer edge of the radiation belts but still within the region of stable trapping for electrons and ions with energies greater than several tens of keV. Geosynchronous orbit is the source region for injection of ions into the ring current and for the inward radial diffusion of both electrons and ions into the radiation belts. Geosynchronous orbit is also near the inner edge of the thermal plasma sheet and is the region where dispersionless injections of electrons and ions during substorms are most commonly observed. Geosynchronous orbit is also a good place to measure the temporal behavior of relativistic electron enhancements, magnetopause compressions, and solar energetic particle events.

Because of the importance of geosynchronous orbit, both as a prime location for satellites and as an interesting region of the magnetosphere, Los Alamos National Laboratory has flown a series of energetic particle detectors on geosynchronous satellites beginning in 1976 and continuing through the present. This paper describes those data and an on-going project to make the data available on-line both as digital data and as summary plots and data synthesis products. The primary access to data, plots, and other information is currently the "World Wide Web"—a graphical interface to the internet. One can view this document as a summary of what is available on "the Web" and a set of instructions for accessing more detailed information.

Radiation Belts: Models and Standards
Geophysical Monograph 97
Copyright 1996 by the American Geophysical Union

2. THE DATA

Geosynchronous orbit is a circular orbit located at a geocentric distance of approximately 6.62 R_E (42,000 km) where the orbital period is approximately 24-hours. In that orbit a spacecraft will stay above a particular geographic longitude. A spacecraft at the geographic equator can be up to ±11° off the magnetic equator due to the tip of the earth's dipole with respect to its spin axis. Thus geosynchronous satellites at different geographic longitudes will be at slightly different magnetic latitudes and therefore slightly different L-shells. In addition the asymmetries and temporal variation of the earth's magnetic field can also make a geosynchronous satellite sample different magnetic L-shells. However, the variation in L is typically quite small and, compared to an elliptically orbiting satellite like CRRES, geosynchronous satellites are essentially fixed at $L = 6.6$.

As part of an ongoing program the geosynchronous satellites which carry Los Alamos energetic particle instruments are referred to by their International Satellite Designator Numbers (ISDN). An example is satellite 1989-046. The first four digits refer to the year of launch. A given satellite such as 1989-046 might be operated at a single geographic longitude for its entire lifetime or it might be moved to a different longitude according to the needs of the mission. In general though, one satellite has operated near 70° W. Longitude, one has operated between 130° and 170° E. Longitude, and one has operated between 30° and 70° E. Longitude. Other longitudes have been covered at various times and for various amounts of time [ep_locations.html][1].

Los Alamos has flown two generations of energetic particle detectors at geosynchronous orbit. The Charged Particle Analyzer (CPA) instrument was flown on satellites from 1976 to 1987 and one or more CPA-equipped satellites operated through 1995. The Synchronous Orbit Particle Analyzer (SOPA) was flown on satellites beginning in 1989. Four SOPA-equipped satellites have been launched so far. Typically data are received from three or four satellites simultaneously. Nominal data coverage is 24-hours per day but data gaps do exist. Frequently a data gap on one satellite is due to switching ground receivers from that satellite to another satellite in the constellation (see Figure 1). Between 1989 and 1996 data are typically available from both CPA-equipped satellites and SOPA-equipped satellites.

Although the CPA and SOPA instruments are similar there are some differences. CPA measures electrons from 30 keV to 2 MeV in 12 energy channels. It measures protons from approximately 75 keV to approximately 200 MeV in 26 energy channels. The energy thresholds for protons are "approximate" because there is some variation from one spacecraft to another. For example the nominal lowest energy proton threshold varies from 70 keV on spacecraft 1984-037 to 147 keV on spacecraft 1977-007. Six "low energy" electron channels are measured with five telescopes at angles of 0°, ±30°, and ±60° from the spin plane while the remaining

---

[1] Throughout this document if a full URL is not given the prefix http://leadbelly.lanl.gov/lanl_ep_data/ should be assumed. For example the full URL for lanl_ep.html is http://leadbelly.lanl.gov/lanl_ep_data/lanl_ep.html.

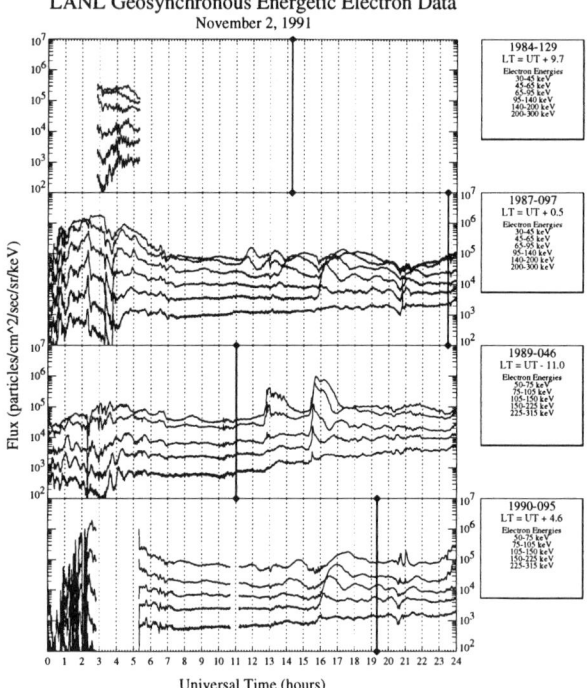

Figure 1. A typical electron summary plot. These plots are viewable as GIF images with any Web browser. There is one plot for each from 1979 to the present. Additional hypertext buttons allow the user to page through plots, to select other dates, or to get additional information about the plots and the data.

measurements are made with single-look direction telescopes mounted at 0° with respect to the spin plane (referred to as the "belly band"). More detailed information on the CPA can be found in *Higbie et al.* [1978] or in the LANL Energetic Particle bibliography [ep_publications.html].

The SOPA instrument measures electrons from 50 keV to greater than 1.5 MeV in 10 energy channels and protons from 50 keV to 50 MeV in 12 channels (in addition there are ten channels for heavy ions including alpha particles, carbon, nitrogen, oxygen, and others). Protons and electrons are measured together using three telescopes mounted at 0°, 30°, and −60° with respect to the spin plane (belly band). More detailed information on the SOPA can be found in *Belian et al.* [1992] or in the LANL Energetic Particle bibliography [ep_publications.html].

Spacecraft carrying both generations of instruments (CPA and SOPA) are actively controlled such that the spin axis of the satellite points continuously toward the center of the earth. Therefore the nominal dipole magnetic field direction is approximately perpendicular to the spin axis. In that configuration complete pitch angle coverage is obtained for all electrons and ions each spin of the spacecraft (about 10.24 seconds). When the field becomes inclined and is no longer perpendicular to the spin axis excellent pitch angle coverage is still obtained for all SOPA channels (from 3 telescopes) and from the six "low energy" CPA channels (from 5 telescopes) while the other CPA measurements are limited in pitch angle according to the inclination of the field. This limitation should be remembered when analyzing spin-averaged data.

In addition, satellites carrying SOPAs also carry the Energy Spectrometer for Particles (ESP) instrument [*Meier et al.,* 1996] which measures electrons from 0.7 to 26 MeV in 6 channels and protons from 11 to greater than 20 MeV in three channels. Those data are not yet included in our on-line database and will not be discussed further in this publication.

3. ON-LINE ACCESS

Digital data have been stored on-line using a SUN workstation and 12 GB of hard disk storage. The name of the workstation is `leadbelly.lanl.gov` and its internet node number is 128.165.207.108. Leadbelly is named after the great blues pioneer Huddie Ledbetter, better known as Leadbelly [`http://leadbelly.lanl.gov/leadbelly.html`].

A single data file is stored for each satellite for each day. By convention files are named with the date (Universal Time) and International Satellite Designator Number. The date is in YYMMDD format so a file for November 2, 1991 for satellite 1989-046 will be called `911102_1989-046`. A file extension may be added to indicate how the data were processed. For example, `911102_1989046.flux.sum`. The data are stored as ASCII text files which have then been compressed with the `gzip` utility. Each file includes ephemeris information. The first column is universal time (in decimal hours), followed by geographic latitude ($-90°$ to $+90°$), geographic longitude ($-180°$ to $+180°$), geocentric radius (in R_E), and count rates (counts/second) for each energy channel. The data may be processed to extract only certain energy channels, to convert count rates to flux, or to sum sets of energy channels.

Two sets of data files are archived. From 1989 to the present the raw telemetry data were stored on optical platters. Those data have been reprocessed to produce 1-spin ($\approx$ 10-second) averages. From 1979 to 1989 1-minute averages were stored along with the raw telemetry data on magnetic tape. Those data have been reprocessed to produce 1-minute average data files in the same format as the 10-second data files. Over 6,000 of the original magnetic tapes are being reprocessed to produce 10-second averages to replace the 1-minute averages and to fill in 1976 to 1979 when no averages were archived. Currently 10-second averages have been produced for 1986 (the PROMIS period which includes coverage by the Viking auroral imager) and for 1979 (which includes ISEE 1 and 2 tail coverage) in addition to all the data from 1989 onward.

We believe this data set represents a unique resource for space plasma physics in general and for radiation belt studies and modeling in particular. It is one of the longest and most continuous sets of satellite data. Multiple satellites provide good coverage in local time. The instruments have broad spectral and pitch angle coverage. And, the data are available on-line and are acquired in real time.

3.1. *The Energetic Particle Home Page [lanl_ep.html]*

All of the data and summary plots described here are available electronically over the internet. We have chosen to use the graphical interface and server protocol known as the World Wide Web (or simply the Web) as the primary means for accessing those products. The web server is also known as `leadbelly.lanl.gov` and the Universal Resource Locator (URL) is `http://leadbelly.lanl.gov/`.

The Los Alamos Energetic Particle "Home Page" is located at `lanl_ep.html`. The LANL EP Home Page is a common point of reference for finding other pages. From there the user can:

1. request digital data as described below,
2. access summary plots of the data—also described below, or
3. obtain supplemental information about the data or about related topics.

Supplementary information includes information about the satellites, the CPA and SOPA instruments, and about the database itself. It also includes information about the energetic particle team, a bibliography of publications which use the LANL energetic particle data, and on-line collaborative projects.

3.2. *Requesting Digital Data [ep_request.html]*

Digital data are stored as compressed text files with ephemeris and count rates. Typically the data need to be processed before they are useful to the average user. Therefore we have established a request system. The first part of the request system is a World Wide Web form [`ep_request.html`]. To request data you specify information such as your name and E-mail, the date and times you want data for, what satellites, what energy channels, the time resolution required, and whether you want flux units or count rates. The request form generates an input file to a program that actually processes the data. Processed data are put in a unique directory for each request. One data file is produced for each satellite and each day requested. The data are provided as text files (not compressed) and include ephemeris as well as fluxes or count rates. The requester is notified by E-mail when the data are ready and can download the data by anonymous FTP or through a Web interface.

The data are received at Los Alamos in real time but are processed daily. Therefore data are typically available within 24 hours of when they were acquired. The system is currently optimized to request as little as a few minutes of data or as much as several weeks worth of data. Long-term surveys (months or years worth of data) currently require too much processing power for the on-line system. In the future, hourly and daily averages will be produced and put on-line for long-term studies.

3.3. *Viewing Summary Plots [summary_plot_chose.html]*

While access to digital data is often essential for a study or as input to a model, it is often more convenient to quickly view a summary plot of the data. Summary plots can be useful to determine what satellites were providing data at a particular time, where they were located, and whether something interesting was happening. A quick check of a summary plot can let you know whether it is even worth while requesting digital data.

One key to making summary plots useful is that it must be quick and easy to view them, to find the date you are interested in, and to page through plots as one would with hard copy. For this to be practical, the plots were pre-generated and saved as GIF images which are viewable by a web browser. Currently only one type of summary plot has been produced. An example is shown in Figure 1. These summary plots highlight substorm injection activity. The plot shows 30–300 keV (for CPA) or 50–315 keV (for SOPA) electron fluxes over 24-hours of universal time. The plot has stacked panels with one panel for each satellite. The time at which the satellite passed midnight is indicated with a vertical bar. Substorm injections show up most clearly in the electron fluxes when the satellite is near midnight or in the dawn sector (e.g. in Figure 1, $\approx$ 1530 UT, for satellite 1989-046). Drifting injections of electrons can be seen at other local times (e.g. in Figure 1, $\approx$ 1530 UT, for satellites 1987-097 and 1990-095).

Using buttons the user can page through plots, forwards or backwards, from one day to another or by using a form [summary_plot_choose.html] the user can chose the year, month, and day of interest. Additional information and links to the other Los Alamos energetic particle web pages are also provided.

As with the digital data summary plots are produced on a daily basis and are generally available 24-hours after the data were acquired. Other useful summary plots are envisioned. Summary plots of "low energy" protons in a format similar to those already available for electrons will be available in the near future. For higher energies monthly and/or yearly summary plots will be made available for relativistic electron enhancements (e.g. > 2 MeV) and for solar energetic particle events (e.g. > 10 MeV). Several data synthesis products are being developed. A Geosynchronous Electron Flux (GEF) index has been developed and is available for testing. The GEF index is a single-variable time series of 50–300 keV electrons from whatever satellite is closest to midnight. This index is useful for comparison with other indices such as AE or D_{st} and for more complex analyses that are not amenable to input from multiple energy channels and from multiple satellites. A complementary Global Geosynchronous Synthesis model which interpolates between satellites for full local time coverage is also being developed.

4. CONCLUSIONS

The development of new models of the radiation belts will be made easier by ready access to energetic particle data from geosynchronous orbit. Better theoretical models will benefit from a better physical understanding of the processes occurring at geosynchronous orbit and from data that provide boundary conditions and constraints on those models. Empirical models always require data for input. In the simplest cases the models represent conditions given by an average over a set of data. New models, particularly those which will be developed for space weather applications, will require that the models be driven in near-real-time by the actual conditions measured in space.

To aid all these efforts the Los Alamos energetic particle team has developed an on-line database of energetic particle data from geosynchronous orbit which is accessible over the internet. Digital data can be requested and downloaded on-line. Summary plots can also be viewed on-line. In addition much of the information needed to properly interpret the data can also be found on-line. Only a portion of that information could be included in this brief introduction to the data system. We also note that, while much of the data system can be considered complete, many more useful features will be added as they are developed. Readers are encouraged to browse the database and web pages for themselves and to provide us with comments and suggestions.

REFERENCES

Belian, R.D., G.R. Gisler, T. Cayton and R. Christensen, High Z energetic particles at geosynchronous orbit during the great solar proton event of October, 1989., *J. Geophys. Res., 97*, 16,897, 1992.

Higbie, P.R., R.D. Belian and D.N. Baker, High-resolution energetic particle measurements at 6.6 R_E. 1, Electron micropulsations, *J. Geophys. Res., 83*, 4851, 1978.

Meier, M.M., R.D. Belian, T.E. Cayton, R.A. Christensen, B. Garcia, K.M. Grace, J.C. Ingraham, J.G. Laros and G.D. Reeves, The energy spectrometer for particles (ESP): Instrument description and orbital performance, Proc. Taos Workshop on Earths Trapped Particles, Taos, NM, in press, 1996.

R.D. Belian, T.C. Cayton, R.A. Christensen, M.G. Henderson, J.C. Ingraham, P.S. McLachlan and G.D. Reeves, Los Alamos National Laboratory, Mail Stop D 436, Los Alamos, NM 87545, USA (E-mail: reeves@lanl.gov)

Outer Zone Relativistic Electron Flux Variations Observed By SAMPEX During Nov. 1–8, 1993

X. Li[1], D.N. Baker[1], M. Temerin[2], J.B. Blake[3] and S.G. Kanekal[4]

A drastic change in the outer zone energetic electron distribution in the magnetosphere during Nov. 3–4, 1993 has been measured by instruments on board the Solar, Anomalous, and Magnetospheric Particle Explorer (SAMPEX), which has a low-altitude (520×675 km) and nearly polar orbit (inclination $82°$). There was an overall flux drop after the strong activity associated with a large southward interplanetary magnetic field and strong solar wind pressure enhancement late on Nov. 3. In particular, the most energetic electrons (> 3 MeV) were completely lost for $L = 3$–8 down to the cosmic ray background of the detector which corresponds to a three order of magnitude decrease for about 12 hours starting at the beginning of Nov. 4. There was no clear sign of enhanced electron precipitation which could lead to the total loss of the electrons at low L-shells ($L < 5$). The outer zone electron fluxes recovered to a higher level than before and moved to lower L-shell ($L < 3$) over a time span of a day. Several mechanisms contributing to the loss of the electrons are discussed in this paper.

1. INTRODUCTION

The outer radiation belt ($L > 3$) consists dominantly of high-energy electrons and is characterized by large temporal flux fluctuations. The relativistic electron component is of practical importance because of its harmful effect on spacecraft subsystems Gussenhoven et al., 1987. It has also been suggested that energetic electrons can play a role in affecting the atmosphere since they penetrate to lower altitudes than most other magnetospheric particles Baker et al., 1987; Callis et al., 1991.

In this paper, we report the electron flux variations in response to a major solar wind disturbance which occurred late on 3 November, 1993 and discuss physical mechanisms likely contributing to the loss of the relativistic electrons. Around 2330 UT of 3 November, 1993, instruments on board IMP-8 measured a large southward interplanetary magnetic field (IMF) component and strongly enhanced solar wind pressure. The observations showed that the pressure in the solar wind increased over an order of magnitude mainly due to the enhancement of the solar wind density [John Foster, Private Comm., 1995]. At that time, IMP-8 was on the night side, $(X, Y, Z) \sim (-30, 23, 5) R_E$ in GSE coordinates. There was no solar wind measurement available up stream of the magnetosphere during this time period.

2. INSTRUMENTATION

The electron data used for this study are from two instruments on board SAMPEX. This spacecraft is in the designed orbit altitude of 520×675 km with an orbital inclination of $82°$ and an orbital period of 100 minutes Baker et al., 1994. The Proton/Electron Telescope (PET) measures 0.4–30 MeV electrons (see Cook et al., 1993 for details). One of the PET channels (P1) has a threshold of 400 keV. (This channel also responds to > 4 MeV protons, but other measurements show that electrons almost always dominate the P1 rate in the outer zone). The P1 data used here were acquired at 6 s resolution. The other sensor used in this study is the Heavy Ion Large Telescope (HILT). HILT is a large ion drift chamber-proportional counter with solid-state detector (SSD) elements at the back of the telescope (see Klecker et al., 1993 for details). Although these detectors are also sensitive to ≥ 4 MeV protons, at mid-latitudes, the HILT SSDs responds almost ex-

[1]LASP, University of Colorado, Boulder, Colorado
[2]Space Sciences Lab, University of California, Berkeley, California
[3]The Aerospace Corporation, Los Angeles, California
[4]NASA/GSFC, Greenbelt, Maryland

Radiation Belts: Models and Standards
Geophysical Monograph 97
Copyright 1996 by the American Geophysical Union

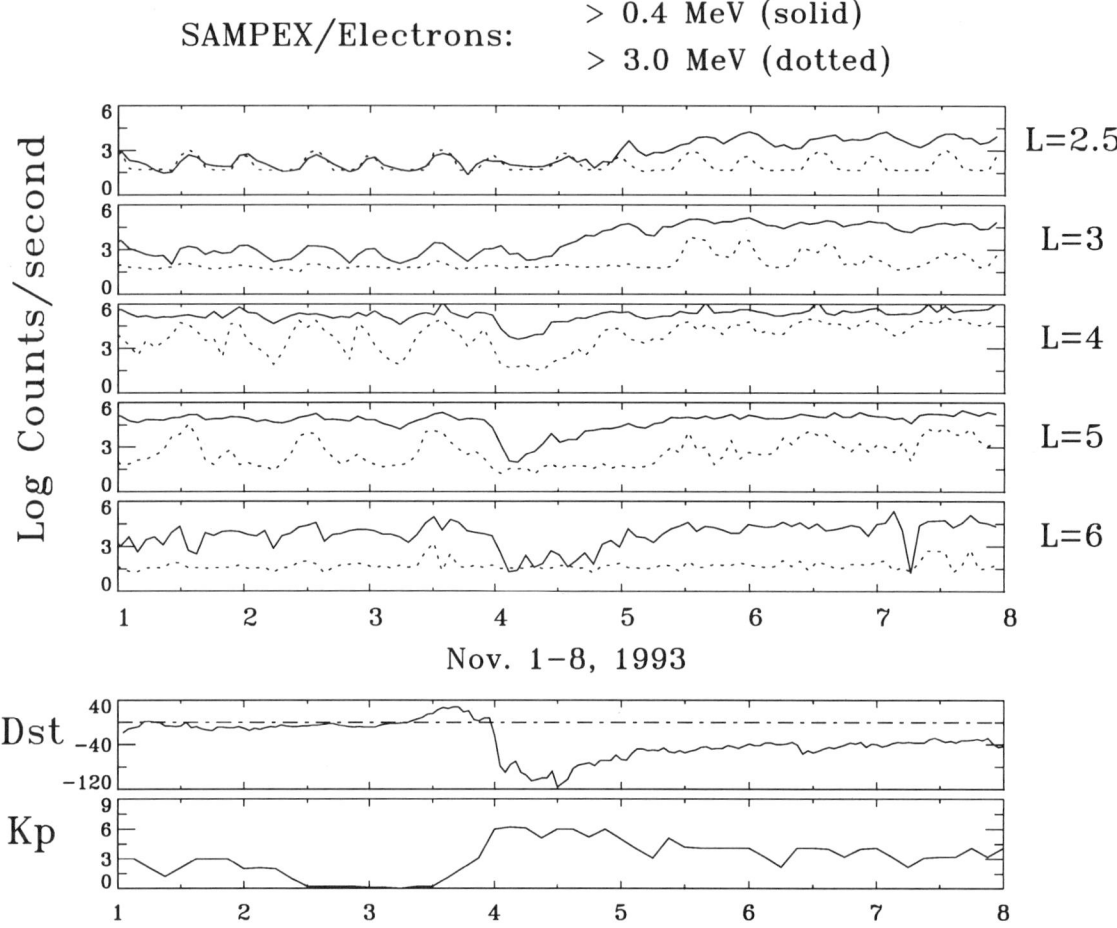

Figure 1. Energetic electrons measured by SAMPEX during Nov. 1–8, 1993. The orbitally-averaged countrate is plotted vs. time for different L-shells. Also plotted (lower panels) are the D_{st} and K_p indices during the same time period.

clusively to energetic electrons. Two thresholds ($E > 1$ MeV and $E > 3$ MeV) are set for HILT SSDs. The sixteen solid-state (SiLi) detectors are placed in a square array; each row is connected to a separate preamplifier-amplifier chain with two discriminator thresholds of 1 MeV sampled 10 times per second and 3 MeV sampled every 6 seconds.

Because of the relative positioning of the four rows and the magnetic field, the relative count rates of the four rows usually can be used to infer if the loss cone is filled or nearly filled. If the loss cone is filled, the electrons are isotropic over the HILT aperture and as a result all four detector rows show the same count rate (see [*Blake et al.*, 1996]).

3. OBSERVATIONS

Figure 1 shows orbitally-averaged electron countrates at different L-shells and D_{st} and K_p indices plotted vs. time for Nov. 1–8, 1993. Due to the satellite's orbit, either precipitating electrons, or trapped electrons with their mirror points lower than the spacecraft are measured. One may notice the peaks with a periodicity of about 12 hours in the countrate, more evident in the lower L-shells. These are due to the South Atlantic Anomaly. In the South Atlantic region, the Earth's magnetic field is much weaker (due to the offset, tilted dipole) and more particles have their mirror points lower than the spacecraft. From these regular variations (prior to Nov. 4), we can tell that the energetic electrons are in a more or less steady state. But right after the strong compression by the solar wind late on Nov. 3, these regularities disappeared for more than a day, which means that the energetic electron environment was greatly altered. Other features we can learn from Figure 1 are:

1. electron fluxes at all energies dropped quickly after the solar wind disturbance, then recovered and moved to lower L-shells;

2. the less energetic electrons recovered first and the more energetic electrons recovered later;

3. the D_{st} temporal profile resembles the flux, which suggests there is some correlation between them (this point will be further discussed later).

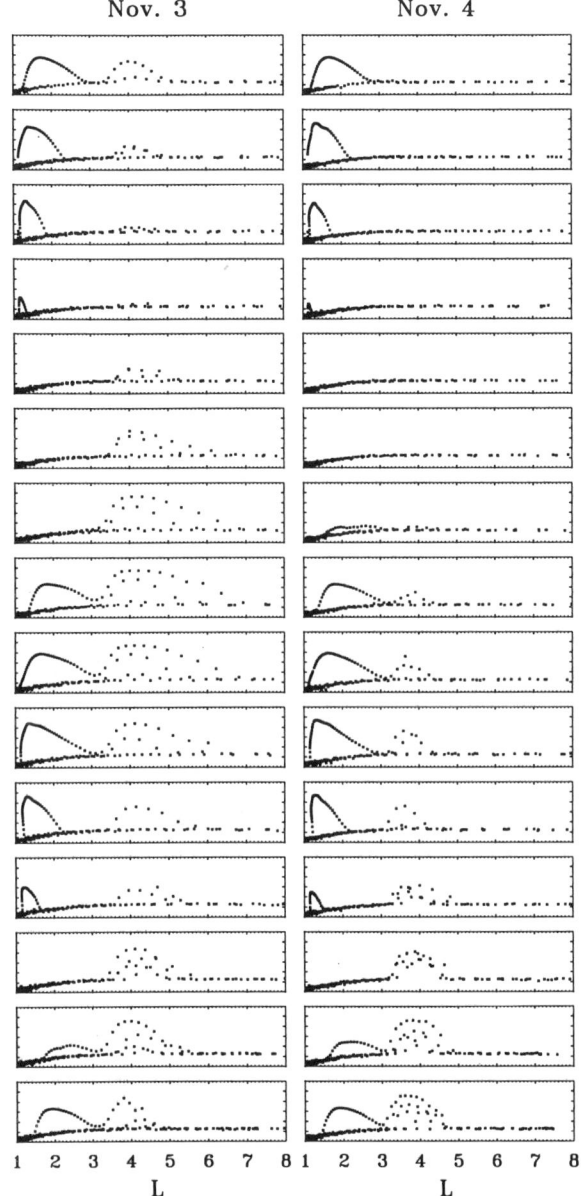

Figure 2. Countrate scaled from 10^0/s to 10^6/s plotted vs. L for electrons with energy > 3 MeV. Each panel is for one orbit (~ 100 minutes).

L-shells. The left set of panels is for Nov. 3 and the right is for Nov. 4. Each panel is for one spacecraft orbit, during which the spacecraft passes a given L-shell four times. We can see several things from Figure 2:

1. There is a stable background measurement at all L-shells from the galactic cosmic ray background, which decreases at lower L-shells, indicating a cut off boundary for the galactic cosmic ray penetration into the inner magnetosphere.

2. Generally there are two peaks due to the radiation belts corresponding to the inner and the outer belt. The inner belt is mostly contaminated by very energetic protons ($E > 100$ MeV), which are very stable compared with the outer belt electrons. The variation of the inner peak from panel to panel is mostly due to the orbital effect (i.e., South Atlantic Anomaly), which also applies somewhat to the outer belt.

3. After the strong solar wind compression at the end of Nov. 3, all electrons with energy > 3 MeV disappeared from the outer belt region for about 12 hours, then slowly recovered.

The GPS satellites with a circular orbit of 4.2 R_E and an inclination of 53° and Los Alamos Geosynchronous spacecraft observed similar features during this same time period [*Tom Cayton and Geoff Reeves*, Private Comm., 1995]. So this effect is not simply due to measurements near the loss cone by SAMPEX. Results from high time resolution (0.1 s, > 1 MeV) count rates of HILT (not shown here) show that enhanced precipitation (when count rates from all four detector rows are the same) started at larger L about 1800 UT of Nov. 3, and moved to lower L with time. However, count rates from row 1 and row 4 at lower L (< 5) are still clearly different even early on Nov. 4 when the overall flux dropped significantly, which means the loss cone at lower L is not filled. This implies that the electron loss due to precipitation into the atmosphere cannot account for all of the oberved drops in the flux at lower L. A natural question arises: where did these energetic electrons go?

While the electrons with energy > 3 MeV had such a dramatic change during the storm, the electrons with less energy also changed significantly. The energy spectral index is a good parameter to describe the relative flux in different energy ranges. We attempt to estimate this index by plotting the flux instead of the countrate. However, it is not trivial to exactly calculate the flux from the countrate. The problem lies in the determination of the actual response of the detectors to the electrons with given energy and some unwanted but inevitable response to other contaminating particles (e.g. > 100 MeV protons). Nonetheless, it is worthwhile to estimate the flux and see the relative change of the energy spectra before and after the storm. During the main phase of the storm, there were rapid changes in all energies and rapid variation in relative countrate. Also right after the main phase, the countrate of electron with energy > 3 MeV was down to (likely below) the background for about 12 hours. Thus, spectral fits at these times are not so meaningful.

Figure 3 presents the results of Monte Carlo calculations of the efficiencies/geometric factors of the HILT solid-state detectors, the upper and lower curves are for electrons with

One may also notice that the countrate from P1 remained almost flat when it reached 10^5 or higher, this is because of the saturation of the detector at higher flux levels.

Another feature that should be noted during this storm is that the most energetic electrons (> 3 MeV) appear to be depleted for $L = 3$-8 down to the cosmic ray background of the detector which corresponds to a three order of magnitude decrease. In Figure 2, the countrate for electrons with energy > 3 MeV from HILT are plotted at intervals of 6 seconds vs.

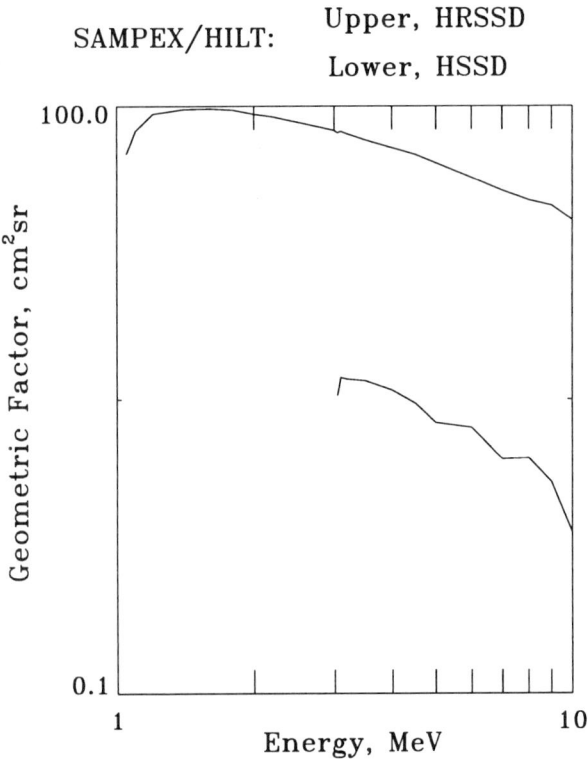

Figure 3. The results of Monte Carlo calculations of the efficiencies/geometric factors of the HILT solid-state detectors.

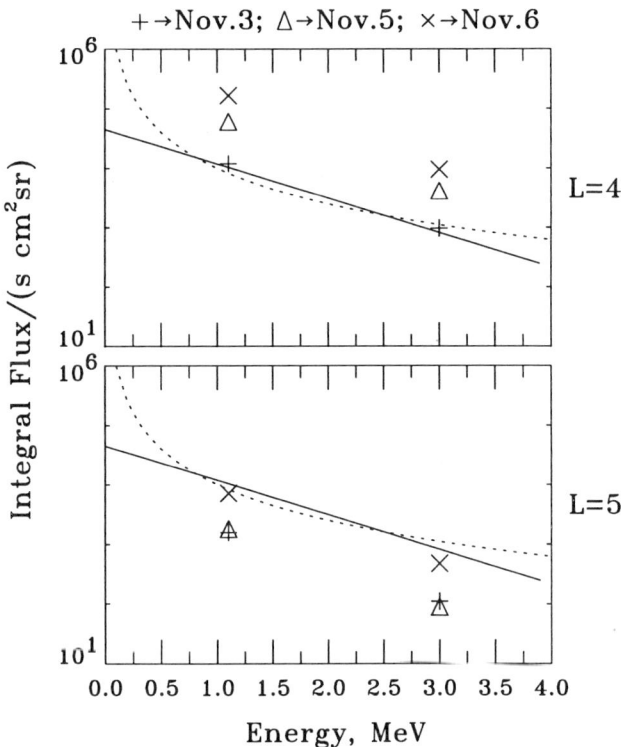

Figure 4. Daily averaged integral energy flux for Nov. 3, Nov. 5, and Nov. 6 of 1993. The solid curve and dotted curve are from an exponential fit and power law fit, respectively, for Nov. 3 at $L = 4$. See text for details.

energy $> 1\,\mathrm{MeV}$ and $> 3\,\mathrm{MeV}$, respectively. The results are for the sums of the all four detector rows; for an isotropic flux each row responds identically. Using a "bow-tie analysis" Baker, 1974, we can obtain the integral energy flux corresponding to a certain energy threshold from the countrate.

Figure 4 shows the daily-averaged integral energy flux from HILT for Nov. 3, Nov. 5 and Nov. 6 represented by $+$, $\triangle$, and $\times$, respectively. Results are shown only for $L = 4$ and $L = 5$, the region of the outer zone. The solid line is from an exponential fit for Nov. 3 at $L = 4$, Flux $= \mathrm{const}_E \times \exp(-E/E_0)$, with $E_0 = 0.75\,\mathrm{MeV}$ and the dotted line is from a power law fit also for Nov. 3 at $L = 4$, Flux $= \mathrm{const}_P \times E^{-\gamma}$, with $\gamma = 2$. There are no significant changes in the daily-averaged integral energy spectral index for Nov. 3, 5, and 6. It is also evident that the overall flux increased after Nov. 5. We should keep in mind that the electrons measured by instruments on board SAMPEX are only part of the whole particle distribution (only the electrons with mirror points lower than the spacecraft). Also the geometric factors used for determining the integral flux are obtained assuming an isotropic flux and it is assumed that each row of the HILT detector responds identically. The particle pitch angle distribution inferred from different countrate ($E > 1\,\mathrm{MeV}$) from detectors in different rows of the HILT (see Blake et al., 1995 for details) indicates (not shown here) that the particle pitch angle distribution was pancake-like before the strong compression and became isotropic right after the strong compression. About 17 hours later it returned to pancake-like again.

4. SUMMARY OF OBSERVATIONS

There was an overall flux drop following the strong activity associated with large negative IMF B_z and high solar wind pressure, and in particular there was a depletion for longer than 12 hours in the SAMPEX measurements of the electrons with energy $> 3\,\mathrm{MeV}$. During the recovery, less energetic electrons recovered first and more energetic electrons recovered later. The overall fluxes recovered to levels higher than before and moved to lower L-shell over a time span of a day. The GPS satellites and Los Alamos Geosynchronous spacecraft observed similar features during this same time period [Tom Cayton and Geoff Reeves, Private Comm., 1995]

The estimated daily-averaged integral energy spectra for electrons with energy around $1\,\mathrm{MeV}$ to $3\,\mathrm{MeV}$ can be best represented by an exponential fitting, Flux $= \mathrm{const}_E \times \exp(-E/E_0)$, with $E_0 = 0.75\,\mathrm{MeV}$, or a power law fitting with the power law index=2, on Nov. 3 at $L = 4$. The energy spectra did not change significantly.

5. DISCUSSION

The general features of this event, which has its own characteristics, are typical of some other magnetospheric storms in which the flux of the electrons drops at the onset of intense magnetospheric activity and recovers in a few hours to days after the initial drop, with the higher energy electrons taking a longer time than the lower energy electrons to recover or surpass their initial levels. Neither the initial drop nor the subsequent recovery in the electron flux is completely understood. It is important to understand the basic electron acceleration and loss mechanisms in the magnetosphere. Regarding the question where the energetic electrons went at the start of the storm, at least three different mechanisms may have contributed to the initial flux drop.

The most straightforward mechanism is the adiabatic effect (conserving all three invariants) due to the change in the magnetic field from the injection of the ring current. An important consequence of the ring current is a decrease of the magnetic field in the inner magnetosphere. This can be measured on the ground and is usually quantified by the D_{st} index. In particular, if the third adiabatic invariant (Φ) is conserved then the electron drift orbit around the Earth must move outward. Conservation of the first two adiabatic invariants (μ and J) then implies that the electron loses energy. If the electron distribution is steeply falling with energy, as it usually is, a fixed-energy detector will see a drop in the flux. The resemblance of the D_{st} temporal profile to the electron flux variation (see Figure 1) suggests a correlation between them. The magnitude of the D_{st} effect can be estimated given the initial radial profile of the electron spectrum and the D_{st} index. Such estimates made from data from the CRRES satellite Rinaldi et al., 1994 show that the D_{st} effect can usually only explain part of the drop.

A second possible mechanism is that the electrons precipitate into the ionosphere as a result of pitch angle diffusion from interaction with VLF waves e.g., Kennel and Petschek, 1966. This is recognized as a usual loss mechanism for radiation belt particles. However, for this event SAMPEX did not measure such enhanced precipitation at lower L-shells ($L < 5$). Enhanced precipitation was measured at higher L-shells. Although electrons at lower L-shell may still be subject to the precipitation loss if they can be moved outward to larger L-shell by other means, it is not clear that this can be the leading loss process for the electrons at lower L-shells.

A third possibility is that the electrons drift out to the magnetopause and then out of the magnetosphere. Early on Nov. 4, 1993, the magnetopause was observed to come inside of geosynchronous orbit [*G. Reeves and M. Thomsen*, private comm., 1995]. The magnetic field configuration was severely distorted and the electron's drift orbit may have been altered to bring it to the magnetopause. Our preliminary modeling work using a guiding center particle code suggests the feasibility of this process.

The above is a qualitative discussion, in order to understand more accurately where the energetic electrons went, a more precise modeling effort is necessary, which is being undertaken.

Acknowledgements. We acknowledge useful discussion with some other SAMPEX team members, namely, Mark Looper, Richard Selesnick, and Dick Mewaldt. This work is supported by NASA grant NAG5-2681 and NAGW-1098.

REFERENCES

Baker, D.N., Energetic particle fluxes and spectra in the Jovian magnetosphere, Ph. D. thesis, University of Iowa, Iowa City, 1974.

Baker, D.N., J.B. Blake, D.J. Gorney and P.R. Higbie, Highly relativistic magnetospheric electrons: a role in coupling to the middle atmosphere?, *J. Geophys. Res. Lett.* **14**, 1027, 1987.

Baker, D.N., J.B. Blake, L.B. Callis, J.R. Cummings, D. Hovestadt, S. Kanekal, B. Blecker, R.A. Mewaldt and R.D. Zwickl, Relativistic electron acceleration and decay time scales in the inner and outer radiation belts: SAMPEX, *J. Geophys. Res. Lett.* **21**, 409, 1994.

Blake, J.B., M.D. Looper, D.N. Baker, R. Nakamura, B. Klecker and D. Hovestadt, New high temporal and spatial resolution measurements by SAMPEX of the precipitation of relativistic electrons, *Adv. Space Res.*, in press, 1996.

Callis, L.B., et al., Precipitating relativistic electrons: Their long term effect on stratospheric odd nitrogen levels, *J. Geophys. Res.* **96**, 2939, 1991.

Cook, W.R., et al., PET: A proton/electron telescope for studies of magnetospheric, solar, and galactic particles, *IEEE Trans. Geosci. Rem. Sensing* **31**, 565, 1993.

Gussenhoven, M.S., E.G. Mullen, R.C. Filz, D.H. Brautigam and F.A. Hanser, New low-altitude dose measurements, *IEEE Trans. Nuc. Sci.* **34**, 676, 1987.

Kennel, C. and H. Petschek, Limit on stably trapped particle fluxes, *J. Geophys. Res.* **71**, 1, 1966.

Klecker, B., et al., HILT: A heavy ion large area proportional counter telescope for solar and anomalous cosmic rays, *IEEE Trans. Geosci. Rem. Sensing* **31**, 542, 1993.

Rinaldi, M.A., W. Nightingale, Y.T. Chiu and M. Schulz, Short-term response of outer-belt Short-term response of outer-belt relativistic electrons to D_{st} variations, *AGU, Eos*, Nov. 1, 545, 1994.

D.N. Baker and X. Li, LASP, University of Colorado, 1234 Innovation Drive, Boulder, CO 80303, USA.

M. Temerin, Space Sciences Laboratory, University of California, Berkeley, CA 94720, USA

J.B. Blake, Space Sciences Department, The Aerospace Corporation, Los Angeles, CA 90009-2957, USA

S.G. Kanekal, NASA GSFC Code 690 and HSTX, Greenbelt, MD 20771, USA

ISEE Measurements for Radiation Belt Modeling

R.H.W. Friedel, E. Keppler, G. Loidl and A. Korth

Max-Planck-Institut für Aeronomie, Katlenburg-Lindau, Germany

ISEE-2 electron data (the KED instrument) from November 1977 to October 1987 is available at the MPAe. These data provide integral energy pitch angle information and differential energy spin-averaged information. About half this dataset has been processed from the raw telemetry (to March 1982). The electron measurements of this data are to be part of the database for the TREND-3 radiation belt modeling effort. In order to parameterize the data according to magnetospheric coordinates such as L, B_0/B_m or α_0, magnetic field models have to be used which are capable of providing these parameters not only as a function of satellite position, but also pitch angle. Since we have access to magnetic field data on ISEE-2, particular care was taken to compare model output to in-situ measurements. Using the UNIRAD software package developed at BIRA, a comparative study of several magnetic field models was undertaken. In general, no magnetic field model is equally good at every L, and even those providing input parameters (such as K_p, AE or solar wind parameters) cannot follow fast field variations. We also include a survey of energetic particle data from near-earth satellite missions available at the MPAe.

1. INTRODUCTION

ISEE (International Sun-Earth-Explorer) was a joint NASA-ESA project consisting of three satellites: ISEE-1 and ISEE-2 which flew in tandem around the earth (on a deep-tail orbit), and ISEE-3 which was put at the libration point $235\,R_E$ from earth.

ISEE-1 and ISEE-2 were launched together on 22 October 1977 on almost identical elliptical orbits, with a perigee at 438 km and an apogee at $22.7\,R_E$. The orbital period was 57.5 hours. The energetic particle instrument on ISEE-1 became non-functional after January 1980, while the instrument on ISEE-2 continued functioning until the end of mission in September 1987. It is this almost 10-year (one solar cycle) data set that is of interest for radiation belt modeling. Data until February 1982 has been processed, while data until September 1987 is still in the raw telemetry format. Resurrection of the old programs to produce the Level-1 Master Science Files from unprocessed data is difficult but progressing at the MPAe.

However, the dataset is not continuous: due to the long orbital period ISEE-2 only passing through the inner magnetosphere once every two and a half days, traversing from $L = 10$ back out to $L = 10$ in around 9 hours ($L = 6$ back to $L = 6$ in around 3 hours), at speeds near 8 km/s. This, together with the non-continuous data coverage, significantly reduces the amount of data available for the inner magnetosphere. It is still one of the longest datasets for energetic particles in the inner magnetosphere, and every effort will be made to make all of it available for radiation belt modeling.

2. INSTRUMENTATION

The instrument used for this study is the KED (Keppler-Daughter) subsystem of the MEPE (Medium Energy Particle Experiment) package on ISEE-2. For details see *Keppler et al.* [1978] and *Williams et al.* [1978].

KED consists of two subsystems: a Wide Angle Spectrometer (WAPS) and four Narrow Angle Spectrometers (NAPS), which provide measurable fluxes only within the closed field line region.

The operational modes for integral and differential data are different, with the latter having a generally lower time

Radiation Belts: Models and Standards
Geophysical Monograph 97
Copyright 1996 by the American Geophysical Union

Table 1. Differential data channels (keV) for KED on ISEE-2

channel	1	2	3	4	5	6	7	8	9	10	11	12	13
low	17	28	38	48	62	80	104	134	174	225	292	484	807
high	28	38	48	62	80	104	134	174	225	292	484	807	1000

Table 2. Variables in Model Data File. BLXTRA is the magnetic model module from UNIRAD.

Variable	Definition	Origin
iy, id	year, day of year	
model	flag for internal magnetic field model	BLXTRA namelist
mmoflg	flag for B-value at earth's surface 0: $M = 0.311653$ Gauß R_E 1: $M = M(\text{BLTIME})$ shows the used magnetic value on earth's surface	BLXTRA namelist
outer	flag for external field model	BLXTRA namelist
ih, im, sec	hours, minutes, seconds	
dlonm, dlatm, radim	longitude, latitude (degrees), radius (km)	
bm	measured magnetic field, $bm = \sqrt{bxm^2 + bym^2 + bzm^2}$ (nT)	
flux	flux for 18 pitch angles (0–10, 10–20, ..., 170–180)	
spec	normalized energy spectra for 12 energy ranges (kev): (17.5–28.0, 28.0–37.6, 47.6-61.5, 61.5–79.5, 79.5–103.5, 103.5–133.1, 133.1–172.5, 172.5–223.3, 223.3–289.5, 289.5–480.5, 480.4–801.0, 801.0–1000.0)	
iokp	own fitted K_p. Search for best fitting value: Compares measured values of magnetic field to model field values for all possible model input values of K_p. Only for Tsyganenko T89.	
value_kp	geomagnetic activity index K_p	OMNI data base
oni	density p of solar wind	OMNI data base
ofs	velocity V of solar wind	OMNI data base
iodst	geomagnetic activity index D_{st}	OMNI data base
b	B_{model} total (Gauß)	BLXTRA
ly, ldy, lh, lm, ls	local time, year, day of year, hours, minutes, seconds	
bmir	B-value at mirror point for 9 pitch angles	BLXTRA
lval	corresponding L-values for 9 pitch angles (to get all 18 values: 18=1, 17=2, 16=3, ...)	BLXTRA
inlat	invariant latitude (degree)	
qf	quality flag (see Sect. 5)	

resolution.

WAPS measures only in the spin plane of the satellite and thus provides limited pitch-angle information, while the NAPS provide four look-directions and reasonably cover the unit sphere.

Integral data is obtained from all detectors with a resolution of at least 4 sectors per spin, while spectra are taken from all detectors in turn, but are always spin-averages. There is thus NO pitch angle information for differential data. The differential data has the channels assignment as shown in Table 1.

3. CHOICE OF DATA

The requirements for radiation belt electron modeling are

Table 3. List of radiation belt data available at the MPAe

Mission	Instr.	Data availability	Elec. keV	Ions keV	Mass Media	Processing Status	Contact Person
GEOS 1	S321	27.04.77–02.11.77[+]	16–300	28–3300	tape	Vax, accessible	A. Korth
GEOS 2	S321	26.07.78–31.01 83[#]	16–300	28–3300	tape	Vax, accessible	A. Korth
ISEE 1		02.11.77–mid 82[+]	22–1200	24–2081	DAT	Vax, accessible	E. Keppler
ISEE 2	KED	02.11.77–23.09.87[+]	17–1251	25–800	DAT	Vax, accessible, some high level, some raw telemetry	E. Keppler
AMPTE	CHEM	01.09.84–01.07.88[+]	—	1–320[*]	tape	Vax, accessible	B. Wilken
Viking	MICS	01.03.86–01.12.86[+]	—	10–320[*]	DAT	UNIX, current, all high level	B. Wilken
CRRES	MEB	28.07.90–12.10.91[+]	21–285	37–3200	DAT	Vax, UNIX, current, all high level	A. Korth
CRRES	MICS	28.07.90–12.10.91[+]	—	40–2000[*]	DAT	Vax, UNIX, Fortran, current	B. Wilken

[*] ion composition measurements available
[+] magnetic field measurements available
[#] magnetic field measurements available 78–79

to obtain pitch angle / energy matrices with as high a time resolution as possible. With the limitations imposed by the instrument the following strategy was adopted:

- Only NAPS integral data were used since they yield pitch-angle information. The data was collected into 18 pitch-angle bins over 16 spins yielding a data resolution of just over one minute.
- Differential data from all NAPS was averaged over the same period. This spectrum was then normalised and is used to split up the integral data at the various pitch angles. The assumption here is that the spectrum does not vary greatly with pitch angle for the time resolution used.
- For each data point a magnetic field model (UNIRAD) was used to calculate model field, and pitch-angle dependent L-value.
- Other sorting parameters of interest (D_{st}, K_p, solar wind parameters) are extracted from the OMNI data base.
- A final data frame containing all the information is written for each data time. Data is organised into ASCII files per day.

Particles with different pitch angle measured at a given point in space reside on different L-shells: the UNIRAD package provides the calculation of pitch-angle dependent L-values.

4. MODEL DATA FILE

Files are ASCII, one file per day. Only those times for which L-values could be calculated are included (within magnetopause) so file length varies, average is about 200 kB with a maximum of around 600 kB.

Header:
```
iy, idy, model, mmoflg, outer
```
Records:
```
ih, im, sec, dlonm, dlatm, radim, bm
flux(M), M=1,18
spec(M), M=1,12
iokp, value_kp, oni, ofs, iodst, b
bmir(M), M=1,10, ly, ldy, lh, lm, ls
lcal(M), M=1,10
inlat(M), M=1,10, QF
```

Model L-values (`lcal`) and B-mirror (`bmir`) values are only given for pitch angle bins 0–90° as they are symmetrical around 90°. The variables used above are explained in Table 2.

5. ORDERING PARAMETERS: A MAGNETIC FIELD STUDY

Most data in radiation belt studies are ordered by L-value, which is obtained by models. Normally no attempt is made to check the validity of such models against the data at hand: up to 20% of data used is ordered by L-values which are incorrect at that given time: either the field was too disturbed or the satellite was no longer within closed field lines even though the model still yielded an L-value.

On ISEE-2 magnetic field data was available and is used in a study to check the validity of model data. Using various field models of the UNIRAD package yielded some interesting

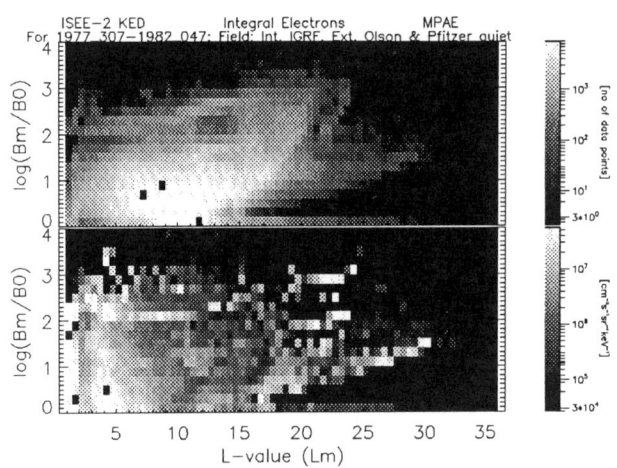

Figure 1. Final flux map for integral data for all the processed data available.

statistics on model validity. In general, no static model is equally good at all L-values, and in general no model agrees with the data (1 standard deviation) for more than 70% of the data points.

Using tuneable models such as T89 with input parameters such as K_p yields better results with the best result being achieved when the input parameter was treated as a free variable which was adjusted to obtain the least-error-fit with data: up to 80% good fits. Such a pseudo K_p shows much larger variability compared to the published K_p since it no longer is a global parameter, although the general trend of the 3-hour K_p is reproduced.

Care must be taken to ensure valid L values are used when constructing a model radiation belt data base, to ensure that input errors into the model are minimised. To this end the ISEE-2 data base of electron data was extended by one parameter which can be used in the later process of assembling a flux-map to choose which data to use. This parameter (a byte) is used as a 8-bit flag where each bit has the following meaning if set:

no bit valid data point—model values O.K., no problems

bit 0 local magnetic field measurement differs from model by more than 5%

bit 1 local magnetic field measurement differs from model by more than 10%

bit 2 local magnetic field measurement differs from model by more than 20%

bit 3 local magnetic field measurement differs from model by more than 50%

bit 4 local magnetic field measurement differs from model by more than 100%

bit 5 flux below magnetospheric threshold (on open field lines)

bit 6 no spectral data

The details of this magnetic field study will be the subject of a separate paper.

6. FIRST RESULTS FROM ISEE-2

Final flux maps have been assembled for all of the processed data up to February 1982. As this work is part of a larger modeling effort (TREND-3, sponsored by ESA) the current maps conform with the agreed standards and are analogous to the maps used by MSSL for the new CRRES radiation belt model.

Magnetic field model used is the Olson-Pfitzer quiet model, no input parameters. A sample output plot for the ISEE data is shown in Figure 1.

L Values are calculated using the BLXTRA part of UNIRAD. The other ordering parameter is the log of the ratio of magnetic field strengths at the mirroring point to the equator. The top panel shows the number of data points in each bin, and the bottom panel the average flux in that bin.

7. RADIATION BELT DATA AT THE MPAE

Missions covering the radiation belts are shown in Table 3. The plan is to make the majority of these data available through NSSDC in a similar format as described in this paper, suitable for radiation belt modeling.

Acknowledgements. Original ISEE-2 raw data tapes have been obtained courtesy of Ted Fritz. Lorne Matteson and Judy Stevenson at NOAA, Boulder have been of great help in resurrecting the original ISEE-2 processing chain. Many thanks to Daniel Heynderickx of BIRA, Belgium for help with his UNIRAD package.

REFERENCES

Keppler, E., B. Wilken, G. Umlauft, H. Fischer, K. Richter, E. Bubla and K. Fischer, Ein Spektrometer für geladene Teilchen mittlerer Energien—Experiment (KED-ISEE), *BMFT-Forschungsbericht*, W 78-19, 1978.

Williams, D.J., E. Keppler, T.A. Fritz, B. Wilken and G. Wibberenz, The ISEE 1 and 2 medium energy particles experiment, *IEEE Trans. Geosci. Electron.*, GE-16, 270–280, 1978.

R.H.W. Friedel, E. Keppler, G.D. Loidl and A. Korth, Max-Planck-Institut für Aeronomie, Postfach 20, 37189 Katlenburg-Lindau, Germany

Measurement of Radiation Belt Particles with ETS-6 Onboard Dosimeter

T. Goka, H. Matsumoto and T. Fukuda

National Space Development Agency of Japan, 2-1-1 Sengen, Tsukuba-shi, Ibaraki, 305, Japan

S. Takagi

Mitsubishi Research Institute, Inc., 2-3-6 Otemachi, Chiyoda-ku, Tokyo, 100, Japan

Radiation belt particles (electrons, protons and alpha particles) have been measured with a dosimeter mounted on board ETS-VI (Engineering Test Satellite VI) from September 1994. The ETS-VI was launched into an orbit with an apogee of about 38,000 km, a perigee of about 8,000 km and an inclination of about 13°, by the National Space Development Agency of Japan (NASDA) on August 28, 1994. Altitude distributions of proton and electron fluxes and spectra are obtained and compared with the NASA AP-8 MIN and AE-8 MIN models. The proton fluxes as measured in space are about 2 to 10 times larger than the model estimated values, and the measured electron energy spectra are harder than the spectra calculated with the NASA models. Variations of energetic charged particle fluxes at $L = 4$–6 near the geomagnetic equator have been observed. The flux reached its highest value a few days after an enhancement of geomagnetic activity. A strong correlation between the trapped particle fluxes and the geomagnetic activity has been found.

1. INTRODUCTION

For space missions, dose estimation in the space environment is quite important for protection of electronic devices and astronauts. Electronic devices damaged by the radiation belt particles cause serious trouble for the satellite operations. For manned space missions, the space vehicles must be designed to decrease the radiation exposure for astronauts.

New observations have been continuing on the space radiation environment by using the Japanese ETS-VI satellite onboard dosimeter. The orbit of ETS-VI is nearly a geostationary transfer orbit passing through the outer zone of the radiation belt.

2. SATELLITE AND INSTRUMENTS

ETS-VI was launched on August 28, 1994, with a Japanese H-II rocket, and has been placed in an orbit with an apogee of 7.1 R_E (Earth radius = 6371.2km), a perigee of 2.2 R_E, an inclination of about 13°. The perigee of ETS-VI was changed to 2.3 R_E in November, 1994, to decrease the proton dose of the satellite.

ETS-VI carries an equipment called TEDA (Technical Data Acquisition Equipment), and the aim of this equipment is to examine environmental effects of devices used for space and find the cause of anomalies in spacecraft. TEDA consists of a dosimeter (DOM) and a magnetometer (MAM).

The dosimeter (DOM) is mounted to observe the space radiation environment. This instrument consists of semiconductor detectors, and the energies and counts of protons, electrons, alpha particles and heavy ions are measured using a simple E-ΔE technique. A cross sectional view of the instrument is shown in the paper in this workshop by [*Kohno*, 1996]. ETS-VI is three axis stabilized, and the view direction of the dosimeter is always nearly perpendicular to the magnetic field line. The specifications of DOM are shown in

Table 1. DOM telescope specifications

Energy Range	electron	0.5 MeV–5.8 MeV
	proton	8 MeV–45 MeV
	α particle	20 MeV–187 MeV
	heavy ion	>13 MeV/u
Geometric Factor	0.0029 cm^{2}sr	
Sampling Rate	32 sec	

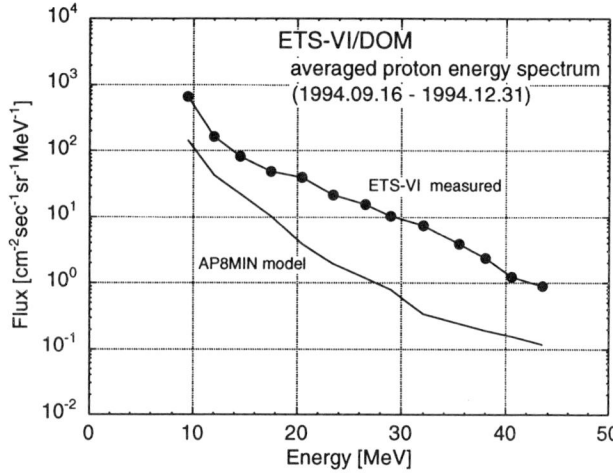

Figure 1. Proton energy spectrum measured by ETS-VI compared to that calculated from the AP-8 MIN model. The fluxes are integrated along all data points in the orbit.

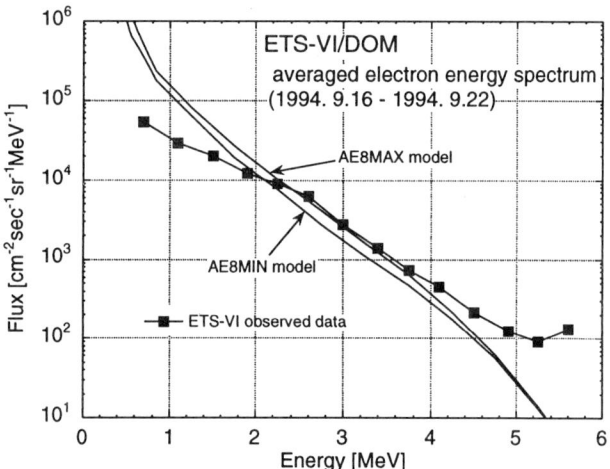

Figure 2. Electron energy spectrum measured by ETS-VI compared to that calculated from AE-8 MIN and AE-8 MAX model. The fluxes are integrated along all data points in the orbit.

Table 1. This type of monitor has good sensitivity for charged particles, but it is difficult to separate protons from electrons clearly. Because electrons are major in the outer zone of the radiation belt ($L > 3$), there is a possibility to contaminate the proton channel count with electrons.

The magnetometer (MAM) is mounted on ETS-VI and observed the variation of the geomagnetic field [*Nagai et al.*, to be published]. This instrument consists of a triaxial fluxgate magnetometer and measures the variation of the geomagnetic field in the satellite local coordinates, X_{sc}, Y_{sc} and Z_{sc}. The satellite X_{sc} axis is parallel to the satellite velocity vector, the Z_{sc} component is directed to the center of the earth, and the Y_{sc} axis is $(-X_{sc} \times Z_{sc})$ in the Cartesian coordinate system pointing nearly northward. The magnetometer is operated in two resolution modes: 32 nT in Range-L and 0.125 nT in Range-H. The dynamic ranges of these two modes are 65,536 nT in Range-L and 256 nT in Range-H, respectively.

3. ENERGY SPECTRA AND SPATIAL DISTRIBUTIONS OF CHARGED PARTICLES

The operation of the DOM telescope was started on September 16, 1994. The observed proton and electron energy spectra are shown in Figures 1 and 2, compared to the spectra calculated from the AP-8 MIN and AE-8 models. The observed fluxes are averaged along all points of the orbit where the data have been acquired at the ground stations, and similarly the model estimated fluxes are integrated along all points of the ETS-VI orbit. These data include both the inner and outer zones of the radiation belt particle fluxes. The measured proton fluxes are about 2 to 10 times larger than the model estimated values, but this difference between the observed proton fluxes and the models does not exceed the acceptable error range of the models. The observed energy spectral shape is harder than that of the models, while the absolute values of the fluxes are in good agreement with the model.

To consider details of the radiation belts, Figures 3 and 4 show the proton and electron energy spectra at each L-value. The spectra of the AP-8 MIN and AE-8 MIN models are also plotted and compared to the measured data. The spectral shapes of the measured proton fluxes at low L values give good agreement with the models, but the fluxes are about 10 times larger than the model. While the models have no proton fluxes at $L = 3.8$, the observed data of the DOM telescope indicates that protons exist in that region. And the fluxes at $L = 3.8$ are about two orders of magnitude smaller than those at $L = 2.2$. This result suggests that a small rate of the outer zone electrons are counted in the proton channels. But the effects of the contaminations of electrons to the proton channel counts are negligible in the inner zone (low L value), because the electron contaminated counts are not large and the proton fluxes in that region are quite high. The fluxes in the region $L < 3$ are dominant for the fluxes averaged on the whole orbit of ETS-VI, so the effects of the electron counts can also be ignored. On the other hand, the observed electron fluxes at $L < 3$ are larger than those of the model, while the measured fluxes in the low energy range are smaller than those of the model data at high L-values. This result indicates that the slot region observed with ETS-VI is more indistinct than than the models.

Spatial distributions of proton and alpha particles as a function of L-value are shown in Figures 5 and 6. AKEBONO observed data [*Takagi et al.*, 1993] are also plotted. It seemed

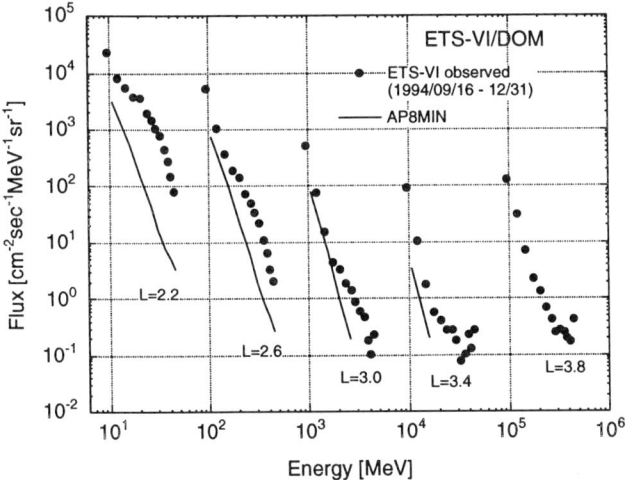

Figure 3. Proton energy spectrum at each L-value measured by ETS-VI and that from the AP-8 MIN model. The spectra are offset in energy.

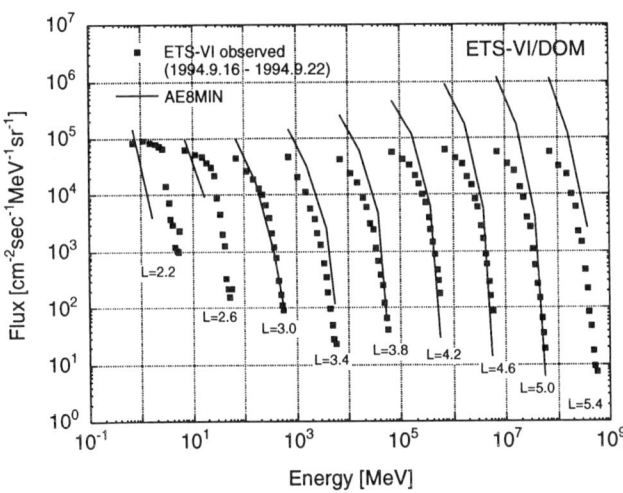

Figure 4. Electron energy spectrum at each L-value measured by ETS-VI and that from the AE-8 MIN model. The spectra are offset in energy.

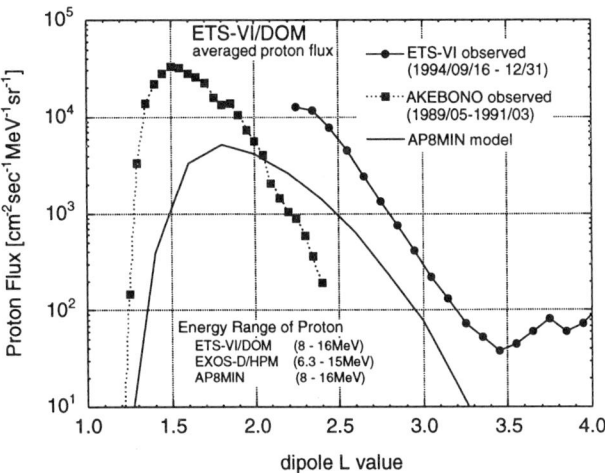

Figure 5. Proton distribution measured by ETS-VI as a function of L-value, compared to the AP-8 MIN model and the AKEBONO data.

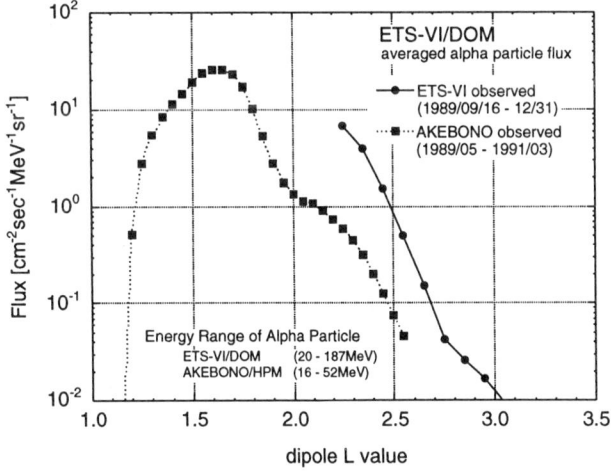

Figure 6. Alpha particle distribution measured by ETS-VI as a function of L-value compared to the AKEBONO measured data.

that the proton and alpha particle fluxes measured by ETS-VI are higher than those measured by AKEBONO, but the distribution shapes of protons and alpha particles are similar. And the discrepancies of the fluxes measured with ETS-VI and AKEBONO are too large for the electron contaminations, therefore these differences are related to the difference of the observed periods, AKEBONO data corresponding to the solar maximum in 1990, while the ETS-VI data was taken during solar minimum. In addition, these differences are also due to the different methods of L-value calculation between these two satellites.

4. THE CORRELATION BETWEEN THE CHARGED PARTICLES AND THE GEOMAGNETIC ACTIVITY

Figures 7 and 8 show the time profile of fluxes at $L < 3$ and $L = 4$–5 compared to the sum of daily K_p indices, during September 16 to December 31, 1994. While the flux was stable at low altitude ($L < 3$), the observed flux at $L = 4$–5 fluctuated, and a strong correlation between the flux and the geomagnetic activity has been found. It is likely that the proton channels of the DOM telescope are contaminated with the outer zone electrons.

Figure 9 shows the time profiles of charged particle flux at $L = 4$–5 and of the magnetic field observed by the on-board magnetometer. The magnetic field is presented in the dipole VDH coordinate system: H is parallel to the di-

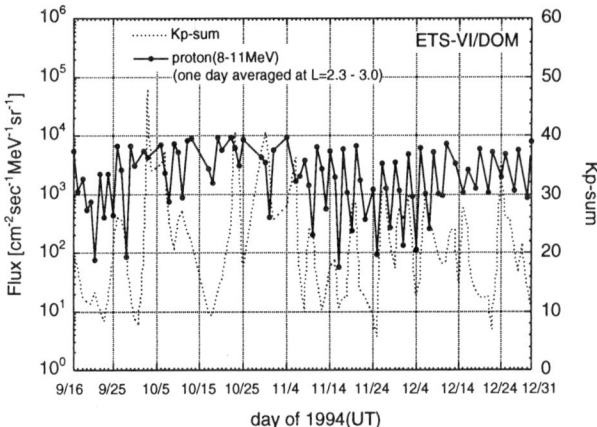

Figure 7. Time profile of energetic charged particle flux at $L < 3$ measured by ETS-VI. The fluxes are averaged over a day. Daily sums of K_p indices are plotted.

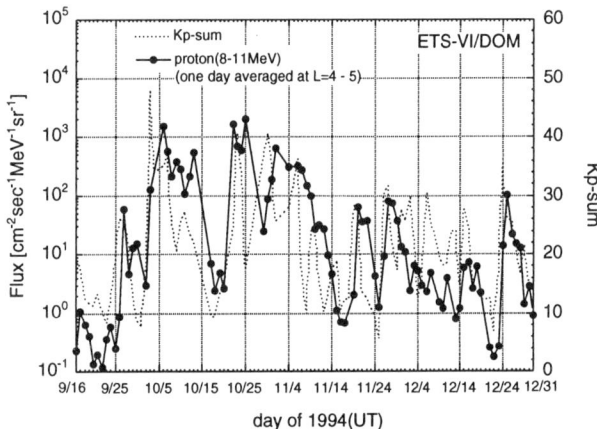

Figure 8. Time profile of energetic charged particle flux at $L = 4$–5 measured by ETS-VI. The fluxes are averaged over a day.

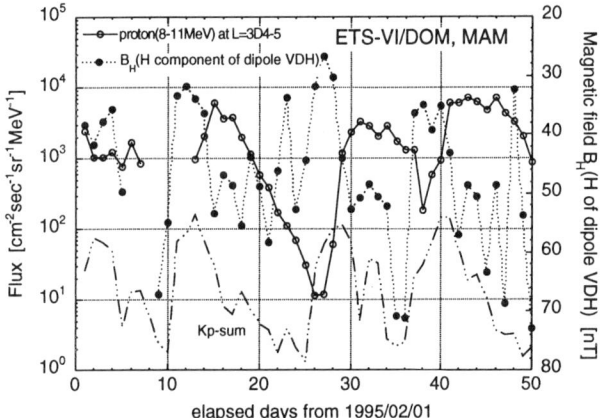

Figure 9. Time profile of energetic charged particle flux at $L = 4$–5 and observed magnetic field B_H (H component in the dipole V, D, H coordinates). B_H is averaged over all data in a day in the nightside and $L > 6$.

pole axis pointing northward, V is orthogonal to H and to a radius vector pointing outward, V completes the Cartesian coordinate system pointing nearly eastward. In Figure 9, the H component of the magnetic field averaged over the time period when the satellite is in the nightside and $L > 6$ are plotted and compared to the sum of K_p indices for the day. Since the observed magnetic field data are averaged, the detailed structure of the magnetic field is not examined from this figure, the long term variation can be seen. In the short term variation, the substorm associated magnetic field variations of near synchronous orbit are well known (e.g., *Nagai* [1991]), the field configuration becomes more tail-like prior to onset and becomes dipole-like after the onset. Therefore, the H component of the magnetic field decreased before the onset, and increased after the onset. The variation of the averaged H component of the magnetic field in Figure 9 is caused by the growth of the ring current, and there is a strong correlation between this magnetic field data and the geomagnetic index. The flux reached its highest value a few days after when the H component decreased and reached its lowest value, and the flux decreased slowly while the H component increased rapidly.

5. CONCLUSION

The observations of charged particles in the radiation belts have been continued with the dosimeter on ETS-VI. The observed proton flux averaged over about three months is about 2 to 10 times larger than the NASA radiation belt models, but this difference between the averaged results of the observations and the models does not exceed the acceptable error range of the models. However, the energy spectra of the observed electron fluxes at various L values are harder than those of the models. The spatial distribution of alpha particles is also acquired. A strong correlation between the charged particles at $L = 4$–5 and the K_p index is observed, and the time profile of the flux in that region has a time lag behind the night-side geomagnetic disturbance.

Acknowledgements. We are grateful to Dr. T. Nagai for advice and suggestion of the MAM observed data, and to Dr. T. Kohno and Mr. T. Imai for designing and calibrating the DOM telescope.

REFERENCES

Kohno, T., Current and Future Data Available in Japan, these proceedings, 1996

Nagai, T., et al., submitted to J. Geomag. Geoelectr.

Nagai, T., An empirical model of substorm-related magnetic field variations at synchronous orbit, in Magnetospheric Substorms, edited by J.R. Kan, T.A. Potemra, S. Kokubun, and T. Iijima, *Geophysical Monograph, 64*, AGU Washington DC, 91–95, 1991.

Takagi, S., Nakamura, T., Kohno, T., Shiono, N. and Makino, F., Observation of space radiation environment with EXOS-D, *IEEE Trans. Nucl. Sci., 40*, 6, 1491–1497, 1993.

T. Goka, H. Matsumoto and T. Fukuda, National Space Development Agency of Japan, 2-1-1 Sengen, Tsukuba-shi, Ibaraki, 305, Japan

S. Takagi, Mitsubishi Research Institute, Inc., 2-3-6 Otemachi, Chiyoda-ku, Tokyo, 100, Japan

Global Distributions of Trapped He Fluxes From OHZORA Satellite During the Geomagnetically Quiet Period of 1984–1987

N. Hasebe[1], A. Ryowa[2], M. Kobayashi[2], K. Kondoh[2], J. Hamada[2], Y. Mishima[2], K. Nagata[3], K. Kohno[4], J. Kikuchi[5] and T. Doke[5]

We present the global distributions of trapped He (4.8–37 MeV) fluxes in the low altitude region under quiet-time conditions ($|D_{st}| < 30$ nT) of geomagnetic activity. Observations were obtained by the OHZORA satellite during 1984–1987 in the altitude of 350–850 km. We have compared the global distributions of He fluxes with those for trapped electrons and protons at the same altitudes. Auroral zones observed from energetic He are not so prominent as compared with those from electron and proton fluxes. It is found that the anomalous distributions of He fluxes are localized at the South Atlantic Anomaly, while the anomalous distributions of electron and proton fluxes are clearly seen to extend from South Atlantic region to Southern Anomaly. Further, the L-distributions of trapped He fluxes are also examined during quiet-time conditions. It is found that the distributions for lower energies (4.8–13 MeV) have two peaks at $L = \sim 1.4$ and ~ 2.7, while the distribution for higher energies (13–37 MeV) has a single peak at $L \sim 1.5$ and does not have the peak at $L = \sim 2.7$.

1. INTRODUCTION

It has been recognized that radiation belts are important in space flight because they can cause an interference with scientific measurements, a damage to materials and electronic devices, and serious hazards to human bodies. The radiation belts at low altitudes are particularly interesting because the space station and space shuttles have low-altitude orbits [*Miah et al.*, 1992]. Then accurate and detailed observations for energetic particle fluxes are necessary to evaluate radiation exposure for spacecraft with different orbits.

The polar orbiting satellite SAMPEX has observed trapped heavy ions, predominantly N, O, and Ne at energies in a narrow L-shell region near $L = 2$ [*Cummings et al.*, 1993]. The trapped population is primarily located in the low-altitude radiation belts near the South Atlantic Anomaly (SAA) [*Blake and Friesen*, 1993]. The SAMPEX findings have renewed an interest in the interaction of the ACR with the radiation belts.

The formation of a new proton radiation belt was observed at $L \sim 2.5$ following the Storm Sudden Commencement (SSC). The observations of the L-shell structures related to the SSC and ACR provided a growing interest to study the dynamics of the low-altitude radiation belt.

The distribution of particle fluxes during solar minimum and solar maximum are quite essential as a baseline data for the particle environment. Global maps of proton and electron fluxes at low altitude from OHZORA data were previously presented by *Nagata et al.* [1988] and *Kohno et al.* [1990]. The global distributions of He flux had not yet been presented before. We present them here for the first time with their L-structure as observed by OHZORA at the altitude of 350–850 km during the quiet-time period of 1984–1987.

2. INSTRUMENT AND OBSERVATION

The EXOS-C (OHZORA) satellite was launched on 14 Feb. 1984 into an orbit with an inclination of 75°, an apogee of 850 km and a perigee of 350 km. The misson ended in Feb. 1987.

---

[1] Faculty of General Education, Ehime Univ., Matsuyama, Ehime 790, Japan
[2] Dept. of Phys., Ehime Univ., Matsuyama, Ehime 790, Japan
[3] Faculty of Engineering, Tamagawa Univ., Machida, Tokyo 194, Japan
[4] Inst. of Phys. and Chem. Res. (RIKEN), Wako, Saitama 351-01, Japan
[5] Adv. Res. Center for Science and Engineering, Waseda Univ., Shinjuku, Tokyo 169, Japan

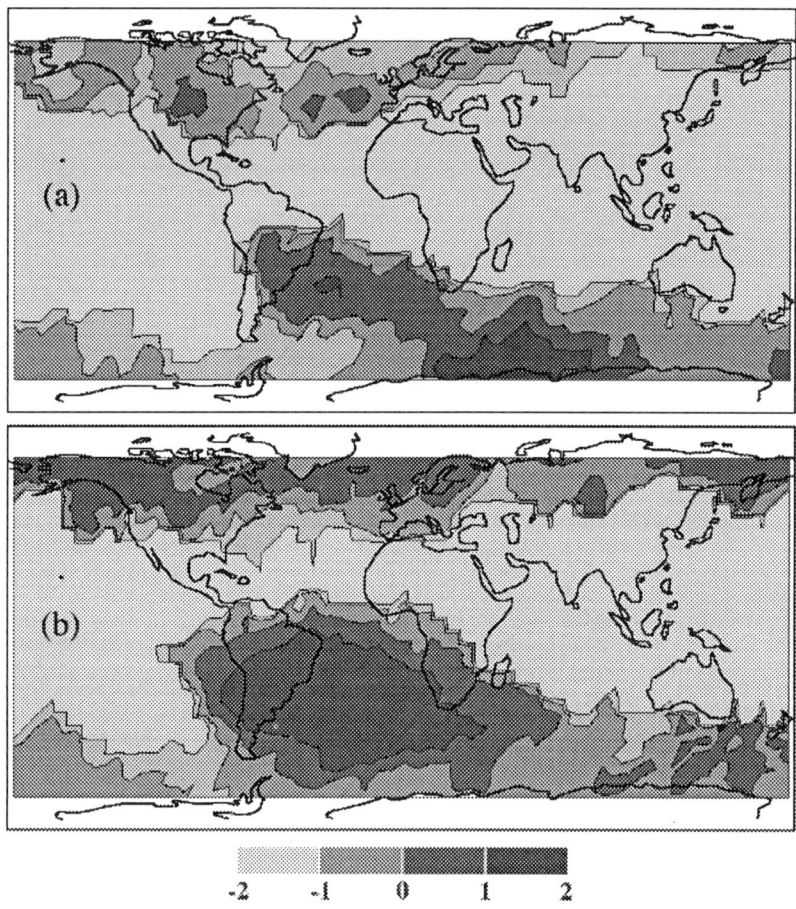

Figure 1. The global distributions of the fluxes for He ions with energies 4.8–13 MeV obtained from the OHZORA observation during the period Feb. 1984–Feb. 1987. The particle data are sampled in the altitude range of (a) 350–475 km, and (b) 725–800 km.

The High Energy Particle (HEP) instrument [*Nagata et al.*, 1985] aboard the OHZORA consists of two identical telescope (Sensor-1 and Sensor-2), each of which employs two surface barrier Si detectors and three Si (Li) detectors. The telescopes measure proton, helium and electron fluxes using the well-established $\Delta E \times E$ method. A thin aluminum foil (6 μm) is placed in front of the telescopes to shield sun light. One of the telescopes (Sensor-1) looks in the anti-sunward direction and the other (Sensor-2) is mounted perpendicular to the Sensor-1. The geometric factor for each telescope is 0.14 cm^{2}sr for coincidence-mode and 0.84 cm^{2}sr for single-detector-mode. The acceptance angle for each telescope is ±25°. Energies covered by the HEP instrument are 0.19–3.2 MeV for electron, 0.64–35 MeV for proton and 4.8–37 MeV for He. Further details about the sensors have been given in *Nagata et al.* [1985].

3. RESULTS AND DISCUSSION

Fluxes of the trapped ions in the radiation belts change with the geomagnetic activity and their variations differ with the altitude and location. In order to study global distributions of the He fluxes during the quiet-time period of geomagnetic activity during the solar minimum, the quiet time intervals were selected such that $|D_{st}| < 30$ nT. We divided all the data into two altitude segments (350–475 km, and 725–850 km) in order to examine the altitude-dependence. The global maps and the L-value distributions for He ion in the SAA during the quiet-time period of 1984–1987 are presented in the following sections.

3.1. Global Maps

In order to obtain the global maps of particle fluxes, we divided the whole globe into 6° × 6° meshes for both latitude and longitude. For regions with no data, we calculated particle flux in a given mesh by averaging the fluxes in four adjacent meshes.

The global maps of electron and proton fluxes have already been published by *Kohno et al.* [1990] using OHZORA data. They found that there are high flux regions in the auroral zones and the SAA. The distributions of integral He fluxes

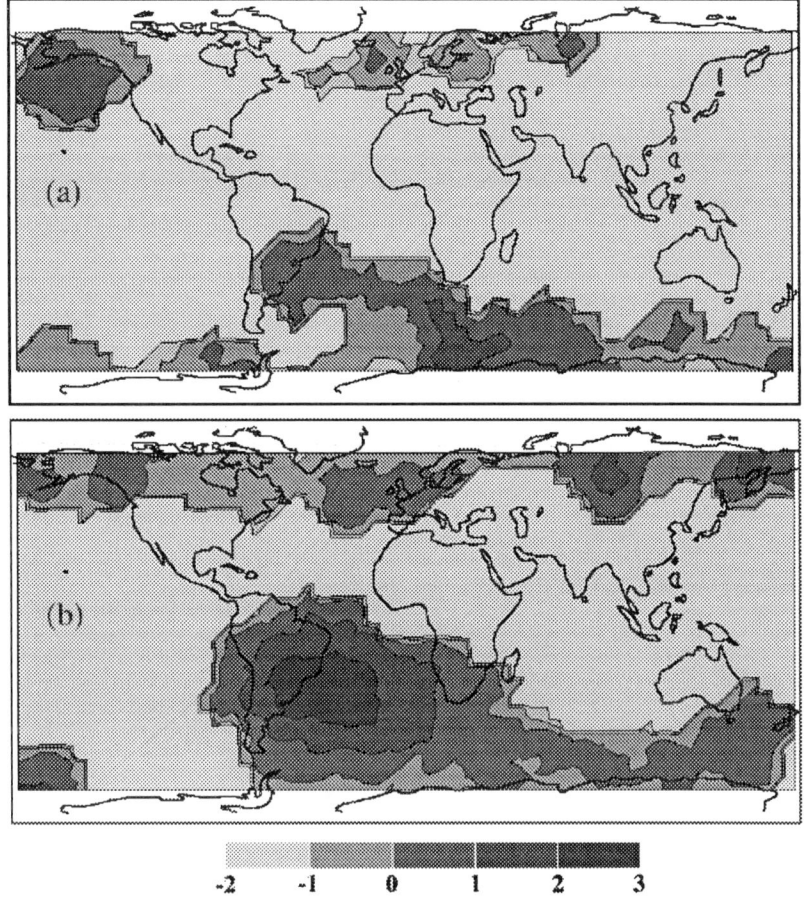

Figure 2. The global distributions of the fluxes for He ions with energies 13–37 MeV obtained from the OHZORA observation during the period Feb. 1984–Feb. 1987. The particle data are sampled in the altitude range of (a) 350–475 km, and (b) 725–800 km.

in the two energy ranges are now available and are discussed below.

Figures 1a and 1b show the global maps of the He fluxes in the energies 4.8-13 MeV at the two altitude (350–475 km and 725–850 km). The distributions of He fluxes are different from those for p and e. The He fluxes are smaller by about 4 orders of magnitudes than those for p and e [*Kohno et al.*, 1995]. The auroral zones where high fluxes are observed for p and e don't exist for He fluxes. High intensities of the fluxes are localized only in the SAA, and the highest flux is at most $\sim 10^1$ counts/cm^{2}sr sec in the SAA. The anomaly region for He expands with increasing altitude.

Figures 2a and 2b show the global maps of the He fluxes in the energy range 13–37 MeV at the two different altitudes. Auroral zones cannot be clearly identified. The region with the high He flux is seen only in the SAA but not in the auroral zones.

3.2. *L-Value Distributions*

Recently, the behavior of transiently trapped He ions in L-shell during and just after substorms has been studied with observations from the SAMPEX, CRRES, AKEBONO and OHZORA satellites [*Cummings et al.*, 1993; *Chen et al.*, 1994; *Takagi et al.*, 1993; *Kohno et al.*, 1995]. They showed that energetic He ions are injected into the region at $L = 2$–3 and trapped in these L-shells during substorms. But they have not investigated the L-distributions under the quiet-time condition long after the substorms.

We have investigated the long-term distribution for He in the SAA under quiet-time conditions. Figure 3a shows these distributions for both energy intervals 4.8–13 MeV and 13–37 MeV at low altitudes (350–475 km). Figure 3b shows the same distributions for higher altitudes (725–850 km). It is found that in the lower energy range the distributions have two peaks, one at $L \sim 1.4$ and one at ~ 2.7, and that their peak intensities increase with altitude. In the higher energy range the distributions have a single peak at $L \sim 1.5$ and the peak at $L \sim 2.7$ is not present.

The peak at $L \sim 2.7$ for the 4.8–13 MeV energy interval at the altitudes 725–850 km, suggests that part of the He ions injected during substorms can be trapped for a long time after

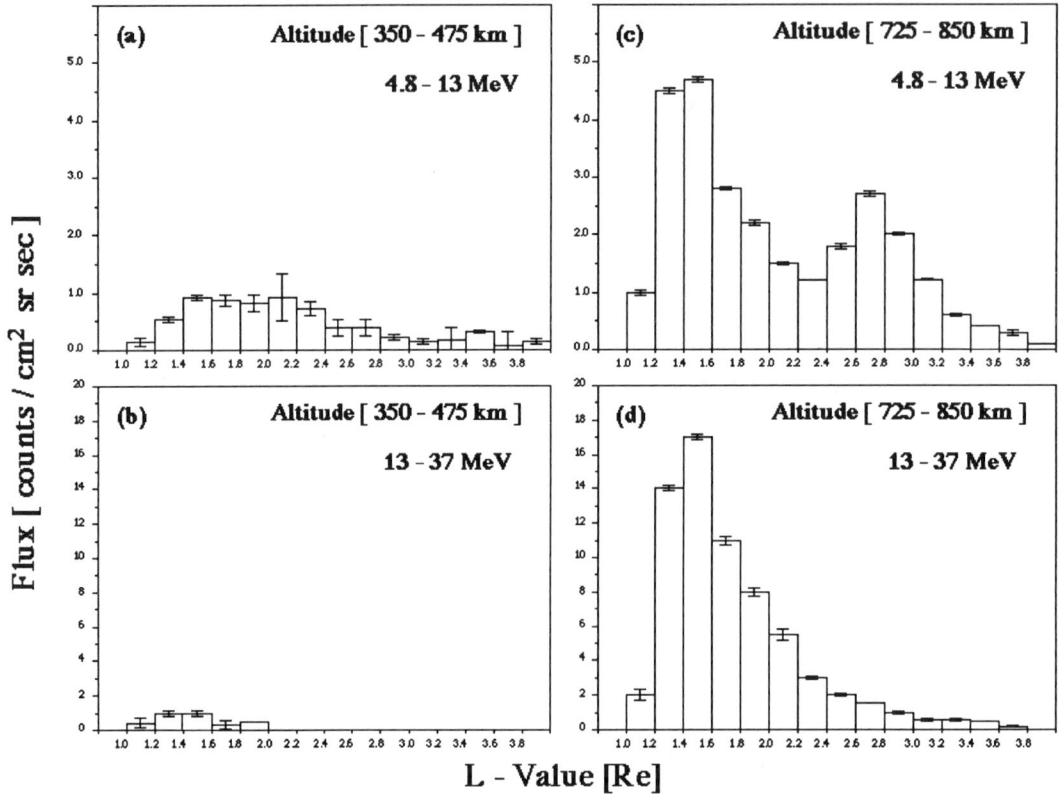

Figure 3. The distributions of He fluxes versus L-value in the altitude range 350–475 km (panels a and b), and 725–850 km (panels c and d) in the South Atlantic Anomaly. Figures (a) and (c) are for 4.8–13 MeV, and Figures (b) and (d) for 13–37 MeV.

substorms (see Figures 3a–d). To make the motion of transiently trapped He ions in radiation belts clear, it is necessary to investigate the time evolution of the L-distributions of ions associated with substorms.

REFERENCES

Blake, J.B. and L.M. Friesen, A technique to determine the charge state of the anomalous low-energy cosmic rays, *P. 15th ICRC (Provdiv)*, *2*, 341, 1977.

Chen, J., G.T. Guzik, Y. Sang, J.P. Wefel and J.F. Cooper, Energetic helium particles trapped in the magnetosphere, *Geophys. Res. Lett.*, *21*, 1583, 1994.

Cummings, J.R., A.C. Cummings, R.A. Mewaldt, R.S. Selesnick and E.C. Stone, New evidence for geomagnetically trapped anomalous cosmic rays, *Geophys. Res. Lett.*, *20*, 2003, 1993.

Kohno, T., K. Munakata, K. Nagata, H. Murakami, A. Nakamoto, N. Hasebe, J. Kikuchi and T. Doke, Intensity maps of MeV electrons and protons below the radiation belt, *Planet. Space Sci.*, *38*, 483, 1990.

Kohno, T., A.A. Gusev, I.M. Martine and G.I. Pugacheva, The trapped He flux dynamics observed on the OHZORA satellite during 1984–1987, *Geophys. Res. Lett.*, *22*, 877, 1995.

Miah, M.A., K. Nagata, T. Kohno, H. Murakami, A. Nakamoto, N. Hasebe, J. Kikuchi and T. Doke, Spatial and temporal features of 0.64–35 MeV protons in the space station environment: EXOS-C observations, *J. Geomag. Geoelectr.*, *44*, 591, 1992.

Nagata, K., T. Kohno, H. Murakami, A. Nakamoto, N. Hasebe, T. Takenaka, J. Kikuchi and T. Doke, OHZORA high energy particle observations, *J. Geomag. Geoelectr.*, *37*, 329, 1985.

Nagata, K., T. Kohno, H. Murakami, A. Nakamoto, N. Hasebe, J. Kikuchi and T. Doke, *Planet. Space Sci.*, *36*, 591, 1988.

Takagi, S., T. Nakamura, T. Kohno, N. Shiono and F. Makino, Observation of space radiation environment with EXOS-D, *IEEE Trans. Nucl. Sci.*, *40*, 1491, 1993.

N. Hasebe, Faculty of General Education, Ehime Univ., Matsuyama, Ehime 790, Japan.

A. Ryowa, M. Kobayashi, K. Kondoh, J. Hamada and Y. Mishima, Dept. of Phys., Ehime Univ., Matsuyama, Ehime 790, Japan.

K. Nagata, Faculty of Engineering, Tamagawa Univ., Machida, Tokyo 194, Japan.

K. Kohno, Inst. of Phys. and Chem. Res. (RIKEN), Wako, Saitama 351-01, Japan.

J. Kikuchi and T. Doke, Adv. Res. Center for Science and Engineering, Waseda Univ., Shinjuku, Tokyo 169, Japan.

Energetic Particle Data Archived at IEP SAS

K. Kudela and M. Slivka

Institute of Experimental Physics, Slovak Academy of Sciences, Košice, Slovakia

Low and high apogee satellites in which the Institute of Experimental Physics, Slovak Academy of Sciences (IEP SAS) participated in the framework of the former Intercosmos Programme, yielded a relatively large amount of flux measurements which have been archived. Some of these data can be useful for multidisciplinary groups interested in the study of magnetosphere and its surroundings. We review the data available and what they can be useful for. Two data sets, namely those obtained from low altitude measurements from Active (IK-24) and those from the high apogee satellite Prognoz-10 are briefly discussed.

1. INTRODUCTION. HISTORICAL OVERVIEW

The activities of the Space Physics Group (SPG) of the Institute of Experimental Physics, Slovak Academy of Sciences (IEP SAS) at Košice, Slovakia began in 1970 with the low altitude satellite Intercosmos-3 launched in Russia, and continued on the IK-5 and IK-13 satellites in 1972–1975. The analysis of the energetic particles collected by these low altitude satellites was devoted to the study of particle precipitation and redistribution during strong geomagnetic disturbances. The devices for particle measurements were constructed at Charles University, Prague, according to an IEP physical design. The data were stored on film and paper tape. After 1974, the first devices for satellite measurements were constructed by IEP SAS. The first experiment, SK-1, prepared jointly with PTI in St. Petersburg, Russia, was devoted to the study of neutron and gamma ray flux on board IK-17 with circular orbit 500 km [*Efimov et al.*, 1983]. The system was previously tested on two balloon flights in Russia [*Dubinsky et al.*, 1977]. Since 1980 the SPG of IEP SAS has been more oriented towards the study of particles of lower energies, between a few tens keV and several MeV. This energy region is interesting because a variety of physical processes in the magnetosphere are affected by particles in this range of energy. Among them are the process of acceleration at the plasma discontinuities in the interplanetary medium, at the planetary shocks, interactions with hydromagnetic waves in magnetospheric plasmas, energization during the geomagnetic storms, etc. The measurements were obtained from the high apogee Prognoz-type satellites by the new series of scientific instrumentations DOK. The low altitude measurements of particles have been continuing as well. The data obtained are archived and can be used by other scientists. In the next two sections we illustrate their potential use.

2. SPE-1 DATA

Measurements of particle fluxes were obtained with the SPE-1 instrumentation onboard the low-altitude satellite Active (IK-24), built with silicon surface barrier detectors. The SPE-1 apparatus contained 3 pairs of detectors. The satellite was oriented with one axis to the zenith and another one along the velocity vector velocity. The axes of the detectors were oriented 99°, 69° and 39° with respect to the zenith axis of the satellite. In each pair the first detector with a thickness of 100μ for measurement of protons and the second one with thickness of 300μ for electrons were used. The full acceptance angle of each detector was 20°. All electron detectors were covered by a Mylar foil to stop protons above 700 keV. The proton detectors had a magnetic filter which reject electrons above to 650 keV from the acceptance cone. The diameter of the detectors was 8 mm and the geometric factor was $0.03\,cm^2 sr$. Active detector cooling was used for noise reduction. The energy range for protons (all ions are detected) is 25–800 keV and for electrons it is 20–400 keV. These intervals were divided quasi-logarithmically in 7, 15 or 31 energy channels according to the mode of operation. The high inclination of the satellite (82.6°) and eccentric orbit with perigee 500 km and apogee 2500 km together with the orbit evolution allowed the measurement of electrons and protons in all local time sectors within a 3 months inter-

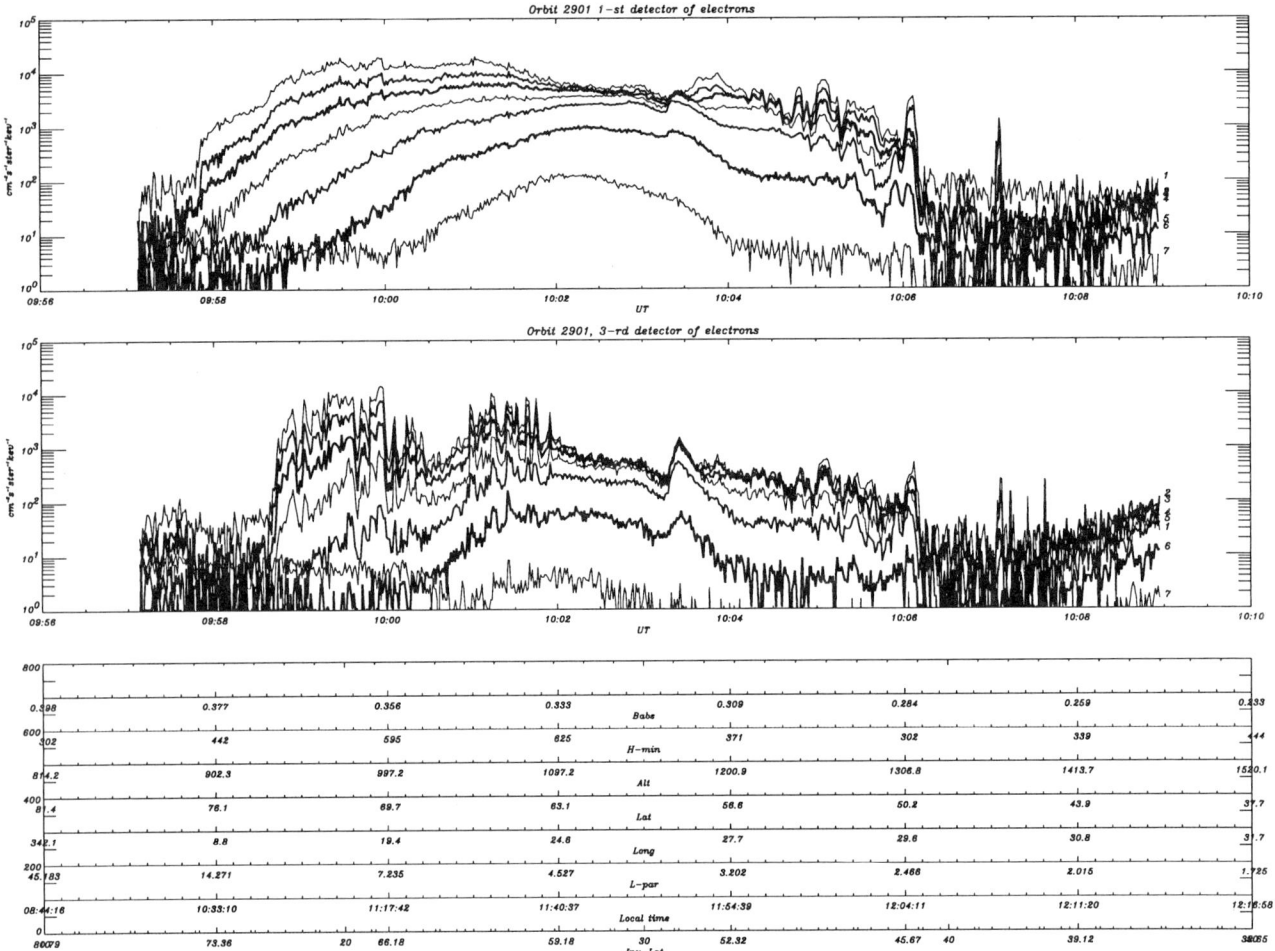

Figure 1. The time dependences of electron fluxes for the first detector of the SPE-1 intrumentation (top panel) and for the third detector of the SPE-1 instrumentation (bottom panel). Both detectors were working in the operation mode A with registration of seven energy channels. The different curves in the top panel correspond to energies 17.7–27.9 keV, 27.9–44.2 keV, 44.2–69.9 keV, 69.9–111 keV, 111–175 keV, 175–277 keV and > 277 keV respectively. The different curves in the bottom panel correspond to energies 20.6–32 keV, 32–49.6 keV, 49.6–76.9 keV, 76.9–119 keV, 119–185 keV, 185–287 keV and > 287 keV, respectively. Pitch angles of different detectors are shown on the top of both panels. On the bottom of this figure other parameters (the absolute value of the interplanetary magnetic field vector, the minimum altitude of the mirror point, altitude, latitude and longitude of the satellite position, the L-parameter, the local time and the invariant latitude) are shown for orbit 2901 of the Activ satellite (May 19, 1990).

val. The SPE-1 instrumentation was developed by IEP SAS Košice in collaboration with Space Research Institute, Moscow, Russia. A short description of the device is given by *Kudela et al.* [1991]. There were two telemetry systems on this satellite (RTS and STO). RTS used the special memory instrument onboard the satellite to collect experimental data. It had four subsystems with different time steps of measurement. SPE-1 apparatus measured in 3 modes of operation. The modes differ in total number of energy steps (7, 15 or 31 energy channels). Better energy resolution lead to worse temporal resolution and vice versa. The STO telemetry used data transfer from the satellite during its radiovisibility from the ground point of connection. Most of the data are obtained from Panská Ves station. The advantage of STO is relatively fine temporal resolution (8 energy channels from each detector obtained each 0.2 s). The time resolution of all modes of operation of SPE-1 can be found in *Shuiskaya et al.* [1994]. The data archived from SPE-1 (both for RTS and STO telemetries) are ready on 4 mm DAT tapes. They contain the time, the count rates of all available energy channels of 6 detectors, the position of the satellite in geographic and geomagnetic coordinates, as well as the orientation of the detector axes with respect to the geomagnetic field vector. In RTS telemetry the data cover the time period from October 1989 to

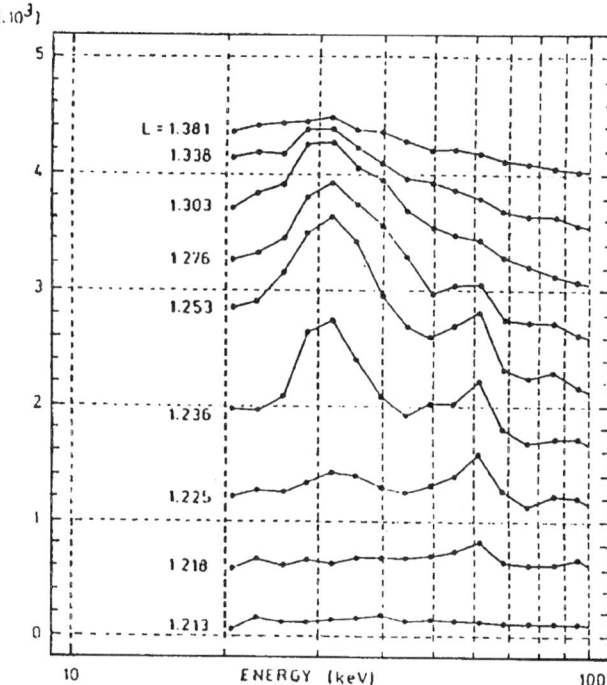

Figure 2. Differential energy spectra of electrons in the inner zone measured by detector 3 on November 11, 1989, in a mode with 31 energy channels. Of these the first 17 are displayed. L values are indicated for the corresponding curves. The lowest curve is labelled in $(cm^2 s\, sr\, keV)^{-1}$. Each successive curve, at higher L, is shifted upward by a factor of 500.

February 1993 and for STO telemetry the data go from October 1989 to September 1992. Figure 1 is showing one second averages of the count rates of electrons of different energies at two different sets of local pitch angles (two detectors with different orientation with respect to the zenith are displayed in the upper and lower panel of Figure 1). Strong fluctuations are seen especially at high local pitch angles, within the local loss cone, in the subauroral morningside region as well as in the middle latitudes. Figure 2, taken from *Kudela et al.* [1992a], shows the differential energy spectra of electrons in the inner zone. It was measured by third detector on November 11, 1989 in a mode with 31 energy channels by RTS telemetry. The lowest 17 energy channels are displayed. L values are indicated for the corresponding curves. The lowest curve is labelled in $(cm^2 s\, keV)^{-1}$. Each successive curve, at higher L, is shifted upward by factor of 500. The first spectrum in the time of highest L ($L = 1.381$) corresponds to altitude 860 km and ratio $B/B_{eq} = 2.20$. The spectrum of lowest L ($L = 1.213$) is at an altitude of 1129 km and is close to the equator ($B/B_{eq} = 1.11$). The satellite Active covers the longitude $268°$–$270°$E in the time interval displayed on this figure. This measurement extended the results obtained earlier by other authors indicating the existence of peaks in the differential energy spectra of electrons (cited in referenced paper) to lower energies.

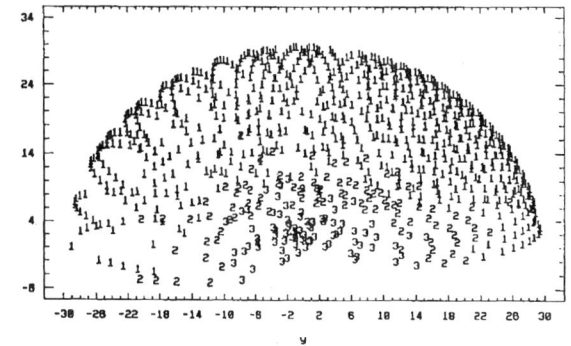

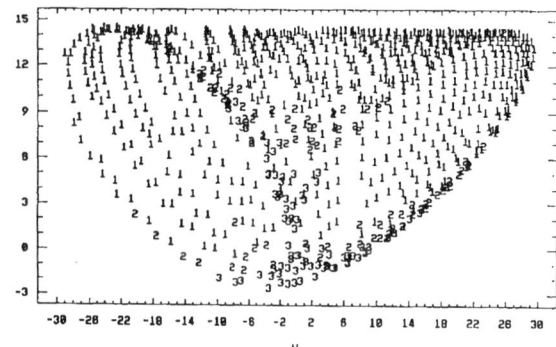

Figure 3. Plots of the Prognoz-10 orbit in GSE coordinates. The top panel shows a projection in the X, Y plane and the bottom panel shows a projection in the Y, Z plane. A region identifier is shown every 100 min, where 1 denotes solar wind; 2, magnetosheath; 3, magnetosphere.

3. DOK-1 DATA

The DOK-1 instrument, constructed at SPG IEP SAS, in the cooperation with Space Research Institute, Moscow, was intended to measure the energy spectra and angular distribution of electrons with energy 10–180 keV and protons with energy 10 keV–2 MeV (see *Fischer et al.* [1985]). It was installed on board of the Prognoz-10 satellite launched in the ecliptic orbit with apogee 200,520 km, perigee 421 km, inclination 65° and period 5785 minutes. The measurements are available from April 26 until November 5, 1985. The satellite measured basic parameters of plasma, energetic particles and waves. DOK-1 instrumentation had three pairs of Si surface-barrier detectors, with sensitive layer 100 μ, active surface with diameter 3 mm and geometric factor $0.01\, cm^2 sr$. All detectors had passive cooling, due to the orientation of Prognoz type satellites. The axes of these detectors were parallel in each pair and oriented 50°, 90° and 180° with respect to the axis x (spinning axis oriented towards the Sun, the spin period was 120 s). A magnetic filter deflecting low energy electrons from the viewing cone was situated in front of one detector of each pair (proton detector). Two telemetries worked on

Prognoz-10 satellite: the first one provided six integral count rates (for each of the detectors, their threshold energies were different in the interval 10–20 keV), the second one (BROD) was working during the time intervals of crossings of the magnetospheric boundary regions and DOK-1 was measuring in each of the detector's 4 energy channels and better temporal resolution was obtained. Data from DOK-1 in both telemetry modes are archived at IEP SAS. The reviews of 2 min integral count rates in graphical form are in *Slivka* [1988] and *Kudela and Slivka* [1989ab]. The 10 sec data from DOK-1, as well as 2 min averages of count rates of protons and electrons, their anisotropies, magnetic field and its fluctuations, position of the satellite in GSE and geometric parameters of the connection to the model bow shock and the magnetopause, are available on 4 mm DAT tapes. The coverage of the archived data provides the opportunity of statistical studies of particle occurence in the magnetosheath and in the nearby region upstream from the bow shock. Two sets of latitudes (low latitudes at inbound passes, medium latitudes at the outbound passes) as well as a wide range of longitudes from dawn through noon till the dusk region, are covered (Figure 3, taken from *Kudela et al.* [1994]). The plots of the Prognoz-10 orbit in GSE coordinates are displayed in two projections in GSE. A region identifier (1 for solar wind, 2 for magnetosheath, 3 for magnetosphere) is shown each 100 minutes. The data from the Prognoz-10 satellite were used to study the sources of medium energy particles in the magnetosheath and in the upstream region. In *Kudela et al.* [1990] a multipoint case study of energetic particle observations of upstream events (Prognoz-10) on June 7, 1985, using magnetospheric CCE and geosynchronous satellite particle data, indicated the high probability that the observed burst of upstream energetic particles for that particular case is due to particle leakage from the magnetosphere during the disturbed period. The result was similar to that obtained by *Sarris et al.* [1978] with another set of data and another period. Although the case studies indicate the leakage from the magnetosphere is a very promising mechanism of the upstream particle increases, the statistical studies are showing also the importance of the contribution of shock accelerated particles to the medium energy particles in the magnetosheath and in the upstream region. The detailed statistical studies, based on the available electron and ion data from Prognoz-10, confirm the importance of both types of processes, i.e. the acceleration at the bow shock and magnetospheric leakage [*Kudela et al.*, 1992b; *Kudela et al.*, 1994].

4. FINAL REMARKS

Recently, after the launch of the Interball-tail satellite, SPG IEP SAS started to archive the available data from the DOK-2 experiment working on the main satellite and from its simplified version, DOK-2S on the subsatellite Magion-4. The description of the DOK-2 and DOK-2S instruments is in *Lutsenko et al.* [1995] and of the frames of the measurements in *Kudela et al.* [1995]. The DOK-2 measures the fluxes and energy spectra of electrons (20–600 keV) and protons (20–1,500 keV) using two pairs of Si detectors. There are several improvements of the DOK-2 and DOK-2S instruments, constructed in SPG IEP SAS in cooperation with Space Research Institute in Moscow and Demokritos University of Xanthi in Greece, with respect to DOK-1: the energy spectra is for both electron and ion measurements much more detailed (56 energy channels), the discrimination between electrons and ions is much better, and the measurement is highly flexible (allowing to obtain very detailed temporal evolution of nonstationary processes yielding in the bursts of particles in the magnetosheath, in the geomagnetic tail and in the upstream region). The processing before the archiving needs the accumulation not only of DOK 2 and DOK 2S measurements, but also magnetic field and orientation data. After these data will be available, the archiving will be done similarly as in the cases of DOK-1 and SPE-1, and thus they can be of relevance to a larger community of magnetospheric physicists. The high resolution energy spectra are promising to contribute in much more detail to the question of relative importance of two mechanisms (magnetospheric leakage, bow shock acceleration) of the ion population in the magnetosheath and upstream region than it was in the case of Prognoz-10. First results show the inverse velocity dispersion, i.e. the appearance of the low energy ions earlier than the higher energy ones and the consecutive hardening of the spectra. These observations support the Fermi acceleration mechanism as proposed for instance by *Terasawa et al.* [1981]. This confirms again that the medium energy ions are mainly accelerated in the bow shock region.

Without a large amount of work in processing data before their archiving, the stage characterized by a relative lack of obtaining physical results, it would be impossible to bring the data to the form suitable for extensive physical analysis, especially for statistical studies, and for the comparison with other data sets important for their use by other researchers. For those who are interested in physical analysis of the data shortly described here, we suggest to contact any of the authors, best by E-mail (kkudela@kosice.upjs.sk or slivka@kosice.upjs.sk).

Acknowledgements. The data archived were obtained from the instruments which were constructed by the technical group of SPG. We would like to express our thanks especially to J. Rojko, as well as to J. Balaz and other colleagues from the group and to cooperators from other institutions (L. Michaelli, P. Opatrny). Our thanks are directed also to P. Triska and his colleagues working on the Magion subsatellites, to S. Fischer with whom the data analysis was often discussed, and to V. Lutsenko and F.K. Shuiskaya from Space Research Institute in Moscow (Co-PI, Co-I of the instruments), whose work was especially substantial for obtaining the data before and just after the launch of the satellites and who contributed significantly to the design of the instrument. Finally, we are thankful to J. Štetiarova for her help in data handling and VEGA agency for grant 1353 support.

REFERENCES

Dubinsky, J., K. Kudela, Yu.E. Efimov, Yu.A. Chichikalyuk, L. Michaeli and T. Vasek, Apparatus for Balloon Measurements of the Neutron Flux, *Bull. Astronom. Inst. Czechoslovakia*, 28, 241–243, 1977.

Efimov, Yu.E., K. Kudela, L. Michaeli, J. Rojko and Yu.A. Chichikalyuk, Nauchno-kosmocheskoje priborostrojenije (in Russian), 1, Moscow, *Metallurgija*, 76–79, 1983.

Fischer, S., K. Kudela, V. Lutsenko, J. Matišin and J. Rojko, DOK-1

Instrument, in Intershock Project, ed. by S. Ficher, Publication of the Astronomical Institute of the Czechoslovak Academy of Sciences No. 60, 166, 1985.

Kudela, K. and M. Slivka, Time dependences of low energy protons and electron fluxes measured on board of the Prognoz 10 Intercosmos satellite, Part II. Preprint of IEP SAS, UEF-03-89, Košice, 1989.

Kudela, K. and M. Slivka, Time dependences of low energy proton and electron fluxes measured on board of the Prognoz 10 Intercosmos satellite, Part III. Preprint of IEP SAS, UEF-04-89, Košice, 1989.

Kudela, K., D.G. Sibeck, R.D. Belian, S. Ficher and V. Lutsenko, Possible Leakage of Energetic Particles From the Magnetosphere Into the Upstream Region on June 7, 1985, *J. Geophys. Res., 95*, 20,825, 1990.

Kudela, K., J. Matišin, F.K. Shuiskaya, O.S. Akentieva and T.V. Romantsova, Inner Zone Electron Peaks Observed by the Active Satellite, *J. Geophys. Res., 97*, 8,681, 1992.

Kudela K., D.G. Sibeck, M. Slivka, S. Fischer, V.N. Lutsenko and D. Venkatesan, Energetic Electrons and Ions in the Magnetosheath at Low and Medium Latitudes: Prognoz 10 Data, *J. Geophys. Res., 97*, 14,849, 1992.

Kudela, K., D.G. Sibeck and M. Slivka, Prognoz-10 Energetic Particle Data: Leakage from the Magnetosphere versus Bow Shock Acceleration, *J. Geophys. Res., 99*, 23,461, 1994.

Kudela, K., M. Slivka, J. Rojko and V.N. Lutsenko, The Apparatus DOK-2 (Project INTERBALL): Output Data Structure and Modes of Operation, Preprint IEP SAS, UEF-01-95, 1995.

Lutsenko, V.N., J. Rojko, K. Kudela, T.V. Gretchko, J. Baláz, J. Matišin, E.T. Sarris, K. Kalaitzides and N. Pachalidis, Energetic Particle Experiment DOK-2 (INTERBALL Project), in Interball, Mission and Payload, French Space Agency Publication, 249–255, 1995.

Sarris, E.T., M. Krimigis, C.O. Bostrom, and T.P. Armstrong, Simultaneous multispacecraft observations of energetic proton bursts inside and outside the magnetosphere, *J. Geophys. Res., 83*, 4,289, 1978.

Shuiskaya, F.K., O.S. Akentieva, T.V. Romantsova, K. Kudela, J. Matišin, and J. Štetiarova, *Kosmicheskije issledovania, 32*, 70, 1994.

Slivka M., and K. Kudela, Time dependences of low energy proton and electron fluxes measured on board of the Prognoz 10 Intercosmos satellite, Part I. Preprint of IEP SAS, UEF-03-88, Košice, 1988.

Terasawa, T., Energy spectrum of ions accelerated through Fermi process at the terrestrial bow shock, *J. Geophys. Res., 86*, 7595, 1981.

K. Kudela and M. Slivka, Institute of Experimental Physics, Slovak Academy of Sciences, Watson str. 47, 04353 Košice, Slovakia

Monitoring of the Radiation Belts with the Radiation Environment Monitor REM

P. Bühler, L. Desorgher and A. Zehnder

Paul Scherrer Institute, CH-5232 Villigen

L. Adams and E. Daly

ESA/ESTEC, NL-2200 AG Noordwijk

The Radiation Environment Monitor has been built for the monitoring of spacecraft radiation environments. It measures fluxes of high energetic protons and electrons with some amount of energy resolution. In 1994 two of these instruments were successfully launched and provide information on the behaviour of the earth radiation belts. In this paper we give an introduction to the instrument and present some examples of data.

1. INTRODUCTION

The Radiation Environment Monitor REM has been developed at the Paul Scherrer Institute in collaboration with the Compagnie Industrielle Radioelectrique SA under an ESA contract. In the year 1994 two REM instruments were launched into space. One instrument was launched on June 17, 1994 with the UK small satellite STRV-1B and one was fixed in September 1994 on the outside of the Russian MIR spacestation. The STRV-1B satellite orbits the Earth on a Geostationary Transfer Orbit. It passes repeatedly through the radiation belts and thus is an excellent platform for studying the radiation environment through a range of altitudes. The Low Earth Orbit of the MIR spacestation allows study of the South Atlantic Anomaly and the precipitation of electrons at high latitudes. STRV-REM is planned to be operational for at least until July 1996. MIR-REM forms part of the joint ESA/Russian EUROMIR-95 mission and will be operational until February 1996.

2. THE INSTRUMENT

REM accumulates energy transfer spectra of charged particles in silicon detectors. It is sensitive to protons in the energy range between 30 and 600 MeV and to electrons in the energy range between 1 and 10 MeV [*Bühler et al., 1996*]. The REM instruments consist of two independent silicon detectors. Both detectors are covered with a spherical dome of aluminium. One detector has an additional layer of tantalum. Whereas the detector without tantalum (e-detector) sees electrons as well as protons, the extra tantalum reduces the penetration of electrons and makes this detector better at measuring protons (p-detector). The response of the instrument to incident particles is mainly defined by the shielding and the differential energy release in silicon. Due to the variation of the energy loss of protons in silicon in the energy range between 10 and 1000 MeV the incident energy of the protons is measured. On the other hand, the energy-loss curve of electrons in silicon is practically flat between 1 and 10 MeV and thus the energy of the electrons is only poorly determined. The detector response to protons and electrons has been measured and numerically simulated for various energies. The results are used to deduce the energy spectra of the particles from the measured pulse height histograms.

3. STRV-REM MEASUREMENTS OF THE OUTER BELT ELECTRONS

The measured intensity and shape of the outer belt is highly variable. In Figure 1 the count rates at $L = 5.5$, detected in the e-detector of the STRV-REM, are plotted versus time. The vertical lines are equally spaced in time, with a period of 26.5 days and help to guide the eye.

Two essential properties of the count rate curve can be

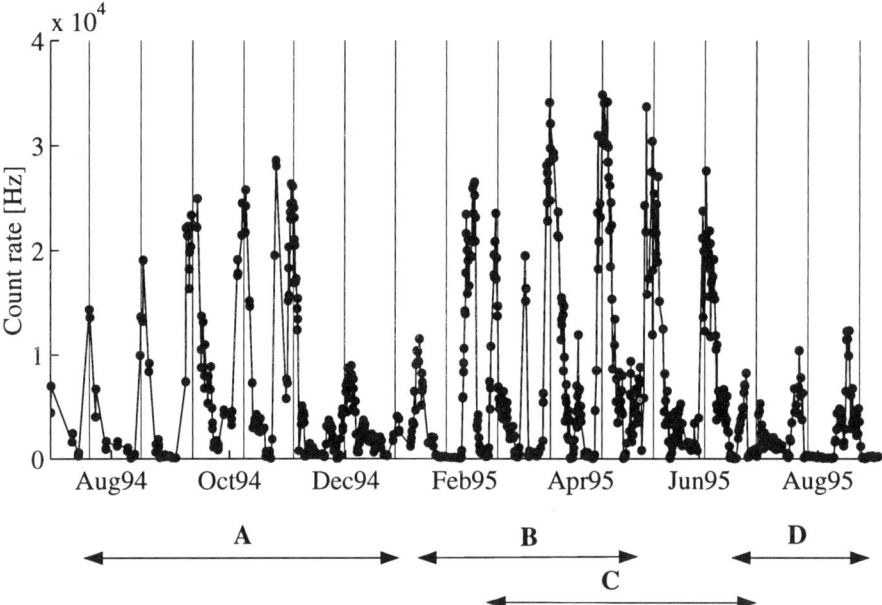

Figure 1. STRV-REM e-detector count rates versus time at $L = 5.5$. A, B, C, and D mark different groups of peaks with constant delay between contiguous maxima. The vertical lines are equally spaced in time with a period of 26.5 days.

noted. First we note that the count rates periodically increase by up to a factor 100. The count rate peaks can be grouped into different series (A, B, C, D in Figure 1) with constant delay of approximately 26.5 days between contiguous maxima. This period fits well with the mean solar rotation period. An obvious reason for the periodic variations is the interaction of fast solar wind streams with the earth's magnetosphere [*Simon and Legrand*, 1992]. Solar coronal holes, from where fast wind streams originate, have lifetimes of typically a few solar rotations. Thus the different series of count rate peaks can be understood as a reaction of the magnetosphere to fast solar wind streams originating from specific solar coronal holes.

There is one large peak at the end of October 94 which does not fit into this scheme and can be associated with the occurrence of a rather strong solar proton event on October 19 [*Solar-Geophysical Data Prompt Report*, 1994].

As a second property we note a seasonal variation with higher electron fluxes in fall 1994 and spring 1995 and low rates during winter 1994/1995 and summer 1995. Seasonal variations of the geomagnetic activity are known [*Russell and McPherron*, 1973]. The orientation of the IMF with respect to the earth's magnetic axis plays an important role in the coupling of the solar wind with the magnetosphere. Southward IMF clearly favours geomagnetic activity. As the earth's magnetic axis is tilted with respect to the ecliptic plane, the IMF orientation varies from season to season. This causes a seasonal modulation of the geo-effectiveness.

The increase of the electron fluxes normally happens with a timescale of one to two days. Figure 2 shows a series of passages through the outer belt dating from around the first of April, 95. On March 28 the count rates are very low. They increase first at large L-values. After a maximum is reached, the peak "moves" towards smaller L-values where it remains, only slowly decreasing, for a few days. By April 7 the peak has disappeared. This general behaviour is typical for the whole data set. However, the position and shape of the belt in L-space vary strongly.

4. MIR-REM MEASUREMENTS OF THE FOOTPRINTS OF THE RADIATION BELTS

Due to the inclination of 55°, the circular MIR spacestation orbit covers nearly the same L-values as the STRV satellite does, from 1 to 6.5. But whereas the STRV-REM on its near equatorial orbit experiences particles with a wide range of equatorial pitch angles, MIR-REM only sees particles with equatorial pitch angles small enough to penetrate to the low MIR altitudes.

As the dipole axis of the magnetic field is tilted and shifted relative to the earth's rotation axis, the penetration depth of trapped particles is a function of geographic position. Thus at a given altitude the particle fluxes depend not only on L but also on geographic position.

At the MIR altitude this effect is most pronounced for the protons of the inner belt. This is demonstrated in Figure 3 where MIR-REM count rates, calculated magnetic field strength, and L-value are plotted as a function of time. Peaks showing up in both detectors are protons and peaks only showing up in the e-detector are electrons. The measurements marked with the two vertical lines have the same L-value of 1.5. But whereas in the first case the proton flux is large, no protons are detected in the second case (note the difference in the magnetic field strength). Plotting count rates versus geographic coordinates shows that the high proton count rates fall into the region of the South Atlantic Anomaly.

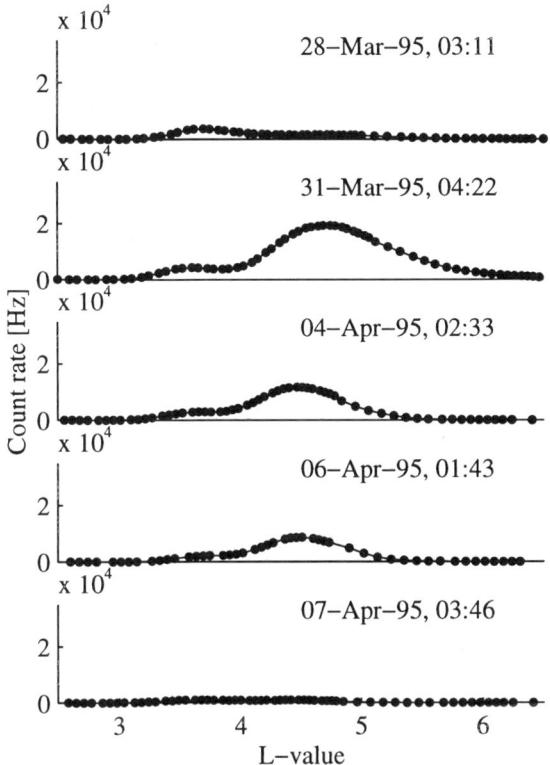

Figure 2. STRV-REM e-detector count rates versus L-value for a series of outer radiation belt passages. Enhancements tend to start at large L-values and "move" to lower L-values.

5. CONCLUSIONS

Since the middle of the year 1994 two REM instruments are measuring the high energy particle fluxes on particular orbits: STRV-REM on a Geostationary Transfer Orbit and MIR-REM on a Low Earth Orbit. Both orbits pass repeatedly through the inner and outer radiation belts.

The nearly equatorial STRV-1B orbit provides good spatial and temporal resolution and thus is ideal for studying variations of the radiation belt.

The particle environment on the MIR orbit is dominated by inner belt protons in the South Atlantic Anomaly. During a few orbits per day the MIR space station passes through the outer radiation belt at high magnetic latitudes. Future plans include a study comparing the MIR-REM electron fluxes with the ones measured simultaneously on STRV-1B.

Acknowledgements. This research is supported by ESA/ESTEC/-WMA contract No. 11108/94/NL/JG(SC).

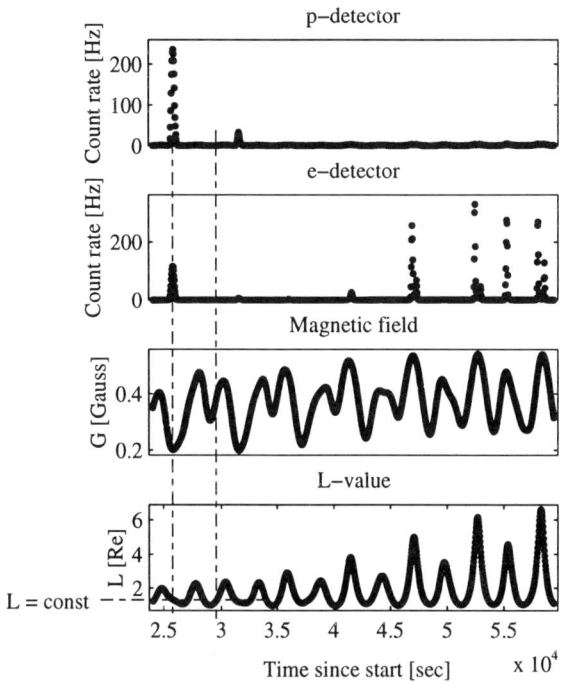

Figure 3. MIR-REM data taken on April 5, 1995. The first two panels show the count rates in the p- and e-detector, respectively. Panel three shows the calculated magnetic field strength, and panel four the corresponding L-values.

REFERENCES

Bühler, P., S. Ljungfelt, A. Mchedlishvili, N. Schlumpf, A. Zehnder, L. Adams, E. Daly and R. Nickson, Radiation Environment Monitor, *Nucl. Instr. and Meth. in Phys. Res., A 368*, 825, 1996.

Russell, T.Ch. and R.L. McPherron, Semiannual variation of geomagnetic activity, *J. Geophys. Res., 278*, 92, 1973.

Simon, P.A. and J.P. Legrand, The cyclic behaviours of the two interplanetary sources of geomagnetic activity and of their relevant solar sources, in Proceedings of Solar-Terrestrial Predictions IV, Ottawa, Canada, May 18–22, 1992.

Solar-Geophysical Data Prompt Report, National Geophysical Data Center, Boulder, Colorado, October 1994.

P. Bühler, L. Desorgher and A. Zehnder, Paul Scherrer Institute, CH-5232 Villigen

L. Adams and E. Daly, ESA/ESTEC, NL-2200 AG Noordwijk

Some characteristics of hot magnetospheric plasma at geostationary orbit

T.A. Ivanova, Yu.V. Kutuzov, B.V. Marjin, N.N. Pavlov, I.A. Rubinshtein, E.N. Sosnovets, M.B. Teltsov, L.V. Tverskaya and N.A. Vlasova

Skobeltsyn Institute of Nuclear Physics, Moscow State University, Moscow, Russia

The measurement of hot magnetospheric plasma is a part of the space environment monitoring programme (SEMP). This programme has continued on Russian communication satellites of the GORIZONT series since 1991 and, more recently, on the EXPRESS spacecraft. This report presents an overview and preliminary analysis of the spectrometric data for 0.1–13 keV electron plasma. An approximation by a 2-Maxwellian distribution function has been used to determine the temperatures and densities for each spectrum. Temperatures, estimated over the course of a year, are found to be very stable. The distribution of densities in local time and specific profiles of some daily density plots allow us to conclude that the high-temperature population seems to be provided by the temporal-spatial structures bound with the geomagnetic tail.

1. EXPERIMENTS

The instruments for investigation of hot magnetospheric plasma have been developed at INP MSU. The method of electrostatic deflection with microchannel-multiplier detection is applied to measure the flux of charged particles. Table 1 describes characteristics of the satellites and onboard instruments [*Vlasova et al.*, 1993]. As is seen, the most comprehensive electron and ion spectrometers measuring in two different directions are on the satellite with the longest life time. Three operating devices of this instrument, denoted as EX, EZ and IZ, are described below.

2. DATA ACQUISITION AND PROCESSING

Since the monitoring of radiation belts is routine work, data acquisition and processing is as automated as possible. The specific telemetry-input adapter for PC compatible computers and software packages for data transportation and processing have been developed to equip both a receiving ground station and the scientific centres with the aim of most effective data acquisition. The interactive graphic browser with the data cutting facility should be mentioned among the specific program tools for the data preprocessing. It was used extensively for rectifying the data. Some codes for more advanced data processing, like spectrum approximation, were also developed and included in the list of automatically invoked programs to allow prompt analysis.

Figure 1 gives all data of the ~1 keV channel from the EZ instrument on the GORIZONT-91/2 in a very compressed form. This figure may be considered as a chart of the data availability as well as a demonstration of the range of measured intensity variations. Numerous gaps in the data array occur because of the relatively small internal memory of the instrument and the problem with organizing the sessions for data download. The upper panel in Figure 2 presents the one-day data fragment of the EZ measurements in details.

3. DATA ANALYSIS

For understanding the spectrometric data we have used the well known fact that the magnetospheric plasma of 50 eV–80 keV can be satisfactorily approximated by a 2-Maxwellian distribution function [*Garrett et al.*, 1990]. The term "a 2-Maxwellian distribution" means that two plasma components with different temperatures are measured in the same direction—in contrast to "a bi-Maxwellian distribution", for which the plasma components are defined as measured in the perpendicular direction [*Garrett and DeForest*, 1979].

The computer code developed for this analysis uses a method which differs from the technique of Garrett. It fol-

Table 1. List of the spacecraft/instruments measuring hot plasma at geostationary orbit

Satellite	Operation time	Longitude	Detector is directed to	Energy range	Index
GORIZONT-91/2	1992, Feb.–present	E 80	radially from the Earth	$E_e = 0.1$–$13\,\text{keV}$ (15 ch.)	EX
			to geogr. south	$E_e = 0.1$–$13\,\text{keV}$ (30 ch.)	EZ
			to geogr. south	$E_i = 0.1$–$13\,\text{keV/n}$ (30 ch.)	IZ
GORIZONT-93/2	1993, Nov.–1994, Oct.	E 130	radially from the Earth	$E_e = 1\,\text{keV}$	41
EXPRESS	1994, Oct.–present	W 15	radially from the Earth	$E_e = 1\,\text{keV}$	ED

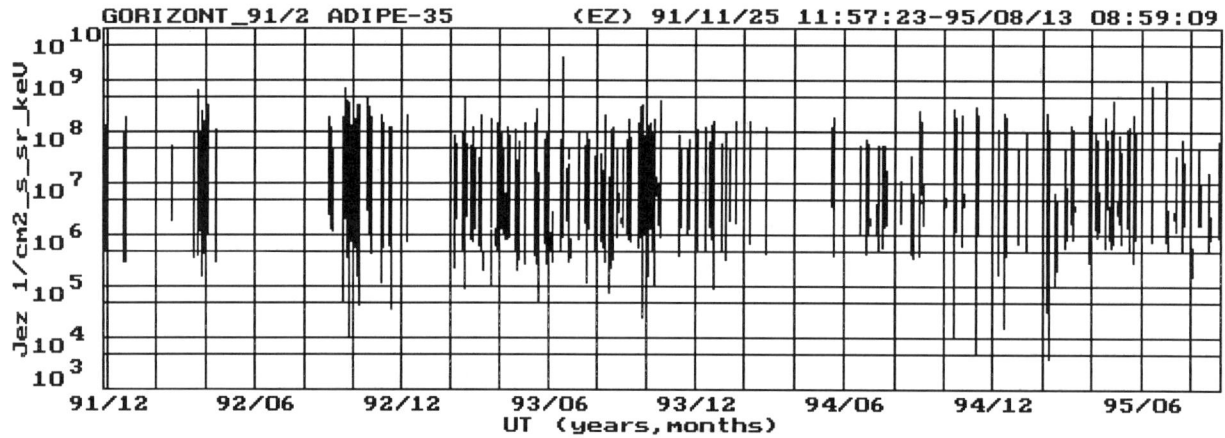

Figure 1. An overview of all available data from the device EZ ($\sim 1\,\text{keV}$ channel)

lows, with some improvements, as reported in [*Novikov and Mileev*, 1989]. The code fits every spectrum using a weight function to equalize both low-energy and high-energy parts of a spectrum. The correlation factor C_{cor} is used as a criterion of fitting quality.

Only the measurements of hot magnetospheric plasma for one year—1993—have been chosen to be analyzed in this report. The first results of fitting gave the best approximation quality (the distance between C_{cor} and 1) for the electrons from the EX device. The quality of EZ approximation was worse and IZ demonstrated the worst approximation quality. We discuss only electron plasma below.

The next restriction concerns the level of geomagnetic activity. We have decided to investigate the behavior of electron plasma for quiet geomagnetic periods first, with some optional testing of more disturbed ones. We expected to find and study the mean values of the determined parameters of a 2-Maxwellian approximation for relatively quiet times, when the quality of approximation would be supposed to be the best. In fact, we have used the spectra with C_{cor} greater than a specified value to proceed the assessments (although involving of K_p or AE would be reasonable).

An example of a 2-Maxwellian approximation applied to a set of one-day spectra from EZ is given in Figure 2. T_1, T_2, n_1 and n_2 denote determined temperatures and densities of two Maxwellian populations of electron plasma. C_2 is C_{cor} of the approximation. C_1 is the C_{cor} calculated for the best-fit single Maxwellian distribution function.

Figure 3 presents LT-distributions of approximated plasma parameters for all the data for 1993. Mean values and standard deviations are given for temperatures and densities. Summarized results derived from the figure are presented in Table 2. Only the spectra with $C_{\text{cor}} > 0.90$ are included in both the figure and table.

The determined EZ parameters are seen not to be so smooth as those of EX since they are strongly affected by the moving solar-cell battery which crosses the sensitive cone of a detector twice a day. These periods (around 12 and 02 hours UT in Figure 2 and around 06 and 18 hours LT in Figure 3) should be carefully corrected (or excluded) in future studies.

4. DISCUSSION

The average temperatures and densities determined for hot electron plasma measured at geostationary orbit demonstrate rather good consistency with the average values published in *Garrett et al.* [1990] except T_2 which is 3–4 times smaller. As a possible explanation of this discrepancy we can note the different geomagnetic latitudes of compared satellites, differences in the periods of observation and in the periods of data averaging.

The following features of the reported LT-distributions of plasma parameters may be noted:

- The LT-binned temperatures and densities for directions EX and EZ are consistent with each other with

Table 2. Total characteristics of electron fluxes from the EX and EZ spectrometers for 1993

	EX				EZ			
	T_1 (keV)	n_1 (cm^{-3})	T_2 (keV)	n_2 (cm^{-3})	T_1 (keV)	n_1 (cm^{-3})	T_2 (keV)	n_2 (cm^{-3})
Mean value	0.24	0.84	2.4	0.24	0.18	0.67	2.6	0.46
Stand. deviation	0.060	0.47	0.49	0.19	0.059	0.39	1.7	3.0

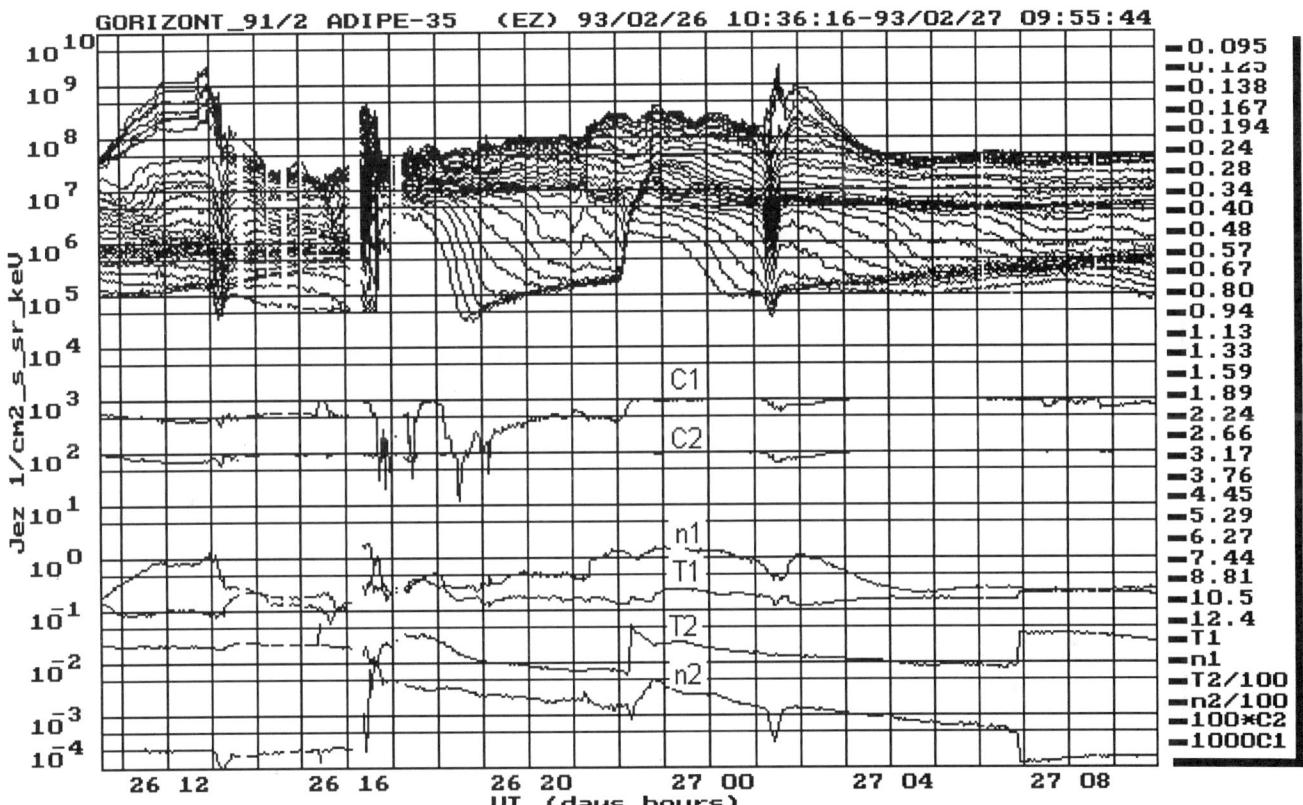

Figure 2. An example of a one-day data set from the device EZ. Approximated temperatures, densities and correlation coefficients are given in lower part of the graph. Note the offset of the four curves as given in the legends.

an accuracy not worse than of a factor 2. The most remarkable differences are in the lowest densities around local noon and also in the $\sim 25\%$ deficiency of T_1 for EZ over all LT bins.

- The values of T_1 and T_2 for both directions show remarkable stability: quasi-constant levels over all LT bins and relatively low standard deviations (except some peculiarities in EZ which may be caused mostly by the moving solar-cell batteries). T_1 demonstrates also very slight diurnal variation which has minimum near local noon and maximum after midnight.

- The diurnal variation of n_2 for the direction EX shows an asymmetric form: n_2 rises more steeply within 18–05 hours LT than it falls within 06–18 hours LT.

- n_1 demonstrates significantly different behavior in comparison with $n2$, for EZ it tends to be closer to n_2 than for EX.

Quasi-equality of the plasma parameters for EX and EZ directions confirms the well-known quasi-isotropy of plasma at the geostationary orbit giving also more detailed information about the same distribution of both relevant populations (T_1, n_1) and (T_2, n_2).

Very flat and meaningful (due to small standard deviations) distribution of T_2 in LT says that the high-temperature component of plasma, having been thermalized somewhere in a collisionness source, is only placed into the region of observation without significant changes in its spectrum.

The LT-distribution of n_2, presented here, looks like the LT-distribution of the electron flux of the lowest energy chan-

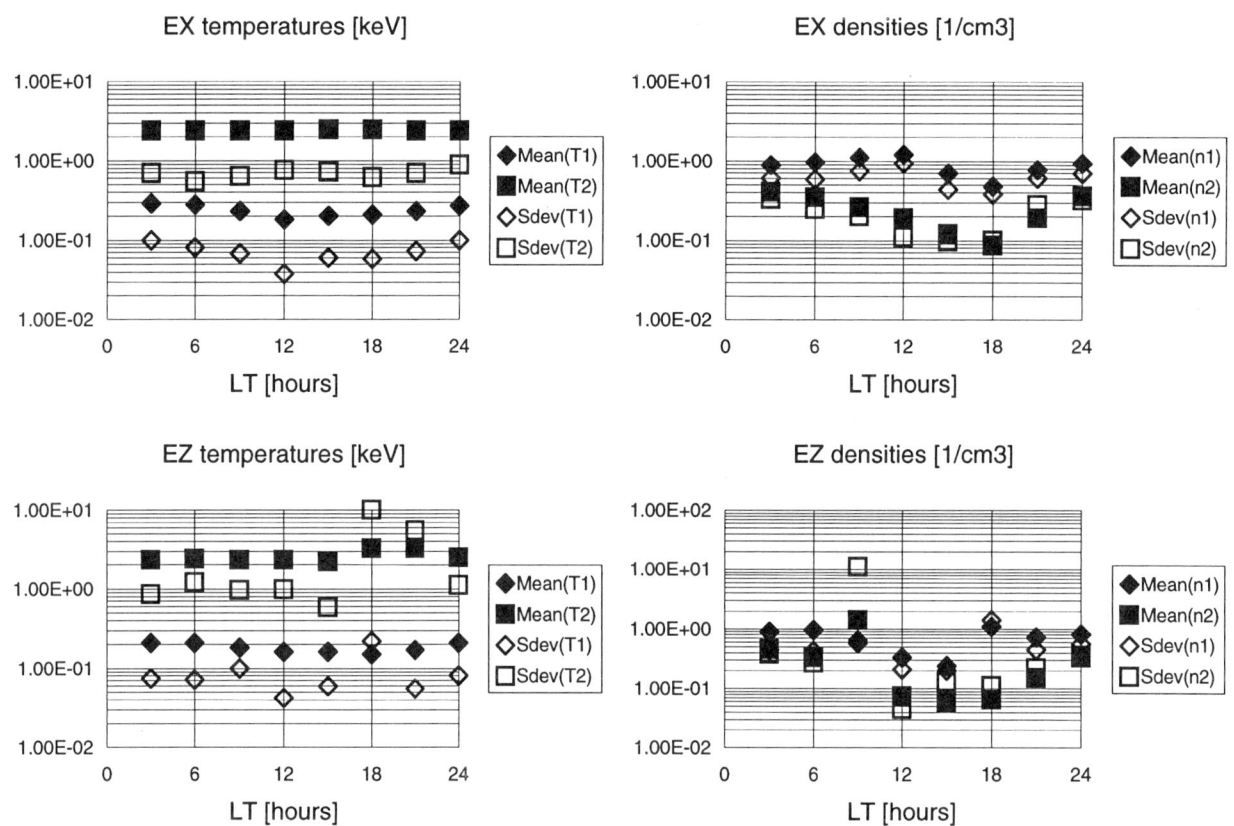

Figure 3. LT-distributions of T and n (with their standard deviations) determined from 2-Maxwellian approximations. All spectra in 1993 for which $C_{cor} > 0.90$ are used.

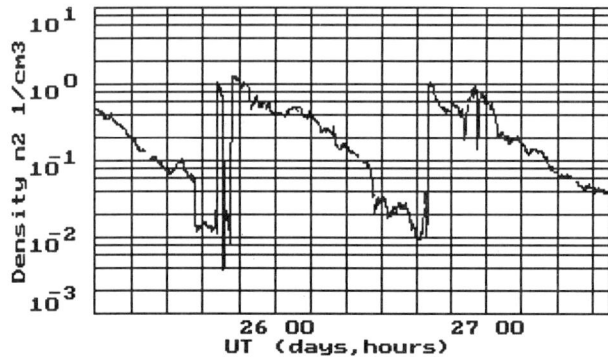

Figure 4. Profile of calculated density n_2 for the direction EX in September 1993

nel given in *Rodgers and Johnstone* [1994]. However, we wish to give a somewhat different interpretation of this curve. While *Rodgers and Johnstone* [1994] consider the asymmetric maximum near 5 hours LT as a result of the short-time injection events that dominate in this location, we often see sequences of long-term profiles that remain for several days. Such curves are obviously responsible for the final asymmetric LT-distribution. An example of such a sequence is given in Figure 4. The repeating of n_2-profile from one orbit to the next with approximately the same LT-location may support the conclusion that we observe spatial structure instead of a series of short-time injection events. We suppose that the satellite crosses the region that has been pumped (or is being pumped) with one or a series of injections or possibly it is filled with the plasma of T_2 temperature by another mechanism most probably related to magnetic storms. Under this assumption the region has a certain spatial distribution of plasma density which is seen in the curve in Figure 4 while a satellite crosses the region. That distribution most probably exists as a dynamic structure. As is obvious from the location of this structure, the geomagnetic tail plays a major role in its formation.

5. SUMMARY

Some statistical characteristics of hot electron plasma at the geostationary orbit for relatively quiet periods, based on a 2-Maxwellian spectrum representation are determined in this paper. Also some conclusions concerning the spatial character of temporal structures of hot electron plasma in the night side of the outer magnetosphere have been derived from this analysis. All reported results are still preliminary and they could be corrected or confirmed after more detailed study.

Acknowledgements. The authors thank Dr. Elise Antonova for the fruitful remarks and Prof. Alan Johnstone for the interest to this study.

REFERENCES

Novikov, L.S. and V.N. Mileev, A physics-mathematical model of satellite charging in geostationary and high-elliptic orbits, in *Studies on geomagnetism, aeronomy and solar physics, 86*, NAUKA, 64–98, 1989 (in Russian).

Garrett, H.B., The Geosynchronous Plasma Environment, AIAA-90-0289, 1990.

Garrett, H.B. and S.E. DeForest, Time-Varying Photoelectron Flux Effects on Spacecraft Potential at Geosynchronous Orbit, *J. Geophys. Res., 84*, 2083–2088, 1979.

Rodgers, D.J. and A.D. Johnstone, Statistics of the outer radiation belt, Proceedings of Taos workshop on the Earth's trapped particle environment, 1994, in press.

Vlasova, N.A., Goryainov, M.F., Kutuzov, Yu.V., Marjin, B.V., Morozova, T.I., Rubinstein, I.A., Savin, B.I., Sosnovets, E.N., Tverskaya, L.V., Teltsov, M.V., Verkhoturov, V.I., Grafodatsky, O.S., Islyaev, Sh.N. and S.A. Maslov, ADIPE complex experiment on the study of space environment factors at synchronous orbit, Proceedings of the International Conference on Problems of Spacecraft/Environment Interactions, Novosibirsk, Russia, June 1992, ed. G. Drolshagen, ESA/ESTEC, 45, 1993.

T.A. Ivanova, Yu.V. Kutuzov, B.V. Marjin, N.N. Pavlov, I.A. Rubinshtein, E.N. Sosnovets, M.B. Teltsov, L.V. Tverskaya, N.A. Vlasova, Skobeltsyn Institute of Nuclear Physics, Moscow State University, Moscow 119899, Russia

Internal Charging in the Outer Zone and Operational Anomalies

G.L. Wrenn and A.J. Sims

Defence Research Agency, Farnborough, England

In order to show how temporal variations of relativistic electrons populating the outer belt jeopardise spacecraft operations, two case studies are outlined. The first describes phantom commands caused by internal charging, the second monitors the background noise in sensitive detectors. Some lessons for radiation belt modelling studies are inferred.

1. INTRODUCTION

This poster attempts to represent the interests of one class of end-user, the operators of communications satellites. It might be argued that suppliers of GEOsynchronous COMSATs have not been well served by the radiation belt modelling community. In the early days there were many surface charging problems but thanks to excellent work with ATS, GEOS and SCATHA, it was possible to produce design guidelines [*Purvis et al.*, 1984] which effectively overcame these problems. Nevertheless, the models do not address substorm injections which are the important ingredient. In respect of total dose, the situation is rather better but there must be concern that money has been wasted on over-protection because a recommended model (AE-8) [*Vette*, 1991] is too conservative. However, things have taken a turn for the worse in the last three years. Operators have been plagued by hundreds of phantom commands and upsets (including some expensive failures) which are due to internal charging from frequent transient enhancements of the outer electron belt—again features which are totally ignored in existing models.

It is clear that temporal variations of the outer belt have more than academic interest, as the host of communications satellites in GEO are regularly exposed to enhanced fluences of relativistic electrons. Their fluxes often increase by 2 or 3 orders of magnitude for several days, these enhancements being most pronounced near solar minimum. The intensifications of the penetrating electrons can cause:

- accelerated total dose effects;
- internal charging, leading to ElectroStatic Discharge and operational anomalies;
- increased noise in sensitive detectors.

Here we present specific examples to show the observed characteristics of:

1. phantom commands disturbing a GEO spacecraft referred to as DRA δ;
2. background noise level in a detector aboard STRV-1a in Geosynchronous Transfer Orbit.

2. INTERNAL CHARGING ANOMALIES

In recent years DRA δ has suffered over 120 identical status switching anomalies, but fortunately these have had little operational impact; they have been described by *Wrenn* [1995]. Without exception these switchings occur at times of enhanced fluence, as monitored by electron detectors on other geostationary satellites. Figure 1 records all the switches logged during 1994 in relation to the measured 2-day electron fluence (event day + previous day) for energies greater than 2 MeV (GOES-7) and greater than 200 keV (METEOSAT-3). The variations in the daily geomagnetic activity index A_p are also shown. The threshold fluence for switching fits the internal charging process which is represented schematically in Figure 2, in this case the shielding is determined to be equivalent to 0.2 mm of Aluminium or less.

Figure 3 describes the distribution of the phantom switches with respect to local time and season. This pattern is significantly different to those appropriate to anomalies caused by surface charging [*Wrenn and Sims*, 1993], it seemingly reflects the asymmetry of the geomagnetic field and the lower L values sampled near noon by a satellite in GEO. However, Figure 4 establishes that the minimum time between successive switchings is some 30 hours. This justifies the selection of 2-day fluence as a suitable metric but it rather suggests

276 INTERNAL CHARGING IN THE OUTER ZONE

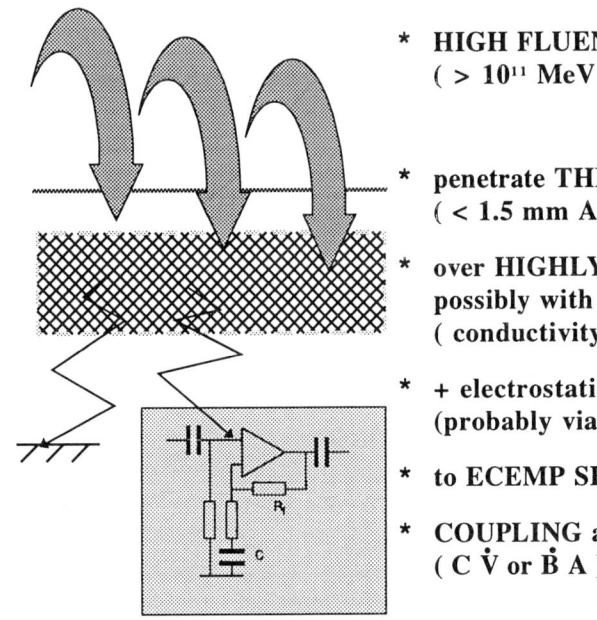

* **HIGH FLUENCE of .5 - 5 MeV ELECTRONS**
  (> 10^{11} MeV cm^{-2})

* penetrate **THIN SHIELDING**
  (< 1.5 mm Al equivalent)

* over **HIGHLY INSULATING** dielectric,
  possibly with embedded conductor
  (conductivity < 10^{-14} ohm^{-1} m^{-1})

* + electrostatic **DISCHARGE PATHS**
  (probably via ground lines or structure)

* to **ECEMP SENSITIVE CIRCUITS** with

* **COUPLING** as to generate **CRITICAL** pulses
  (C $\dot{V}$ or $\dot{B}$ A)

Figure 2. The internal charging process which produces anomalies

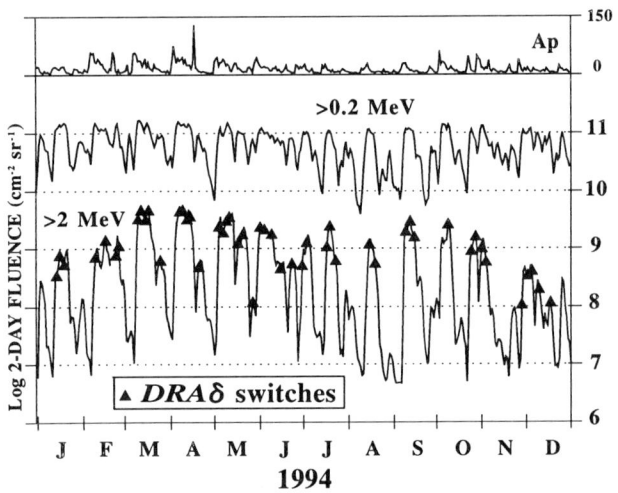

Figure 1. DRA δ anomalies during energetic electron enhancements

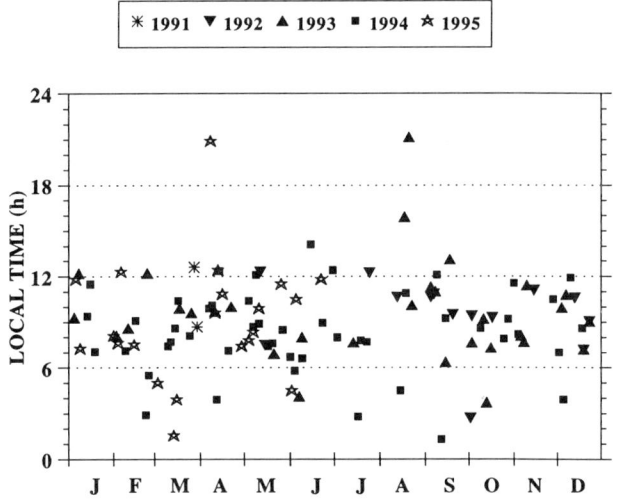

Figure 3. The distribution of DRA δ anomalies in local time and season

that some further explanation is required for the pre-noon clustering of the switches.

In order to quantify the correlation between the switches and relativistic electron flux each day is categorised as LOW, MODERATE or HIGH risk depending upon the daily fluence of > 2 MeV electrons measured at GOES-7, see Table 1.

Figure 5 shows the number of such days in 1993 and 1994 and how many exhibited switches. Although the analysis is rather crude, it well demonstrates that most of the switches occurred on 'high risk' days with very few on 'low risk' days. The frequency of 'high risk' days increased between 1993 and 1994 as the solar cycle approached minimum. This effect is further explored in Figure 6 where the number of switches per calendar month is compared with the number of days per month for which the GOES-7 flux exceeded $10^8 \text{cm}^{-2}\text{day}^{-1}\text{sr}^{-1}$ between May 1992 and May 1995. It is clear that the switches occurred preferentially during intervals when the flux remained high for many days. Such intervals were most frequent in 1994, with extreme values in May and June, but the pattern may yet be changed as the declining phase of the solar cycle continues.

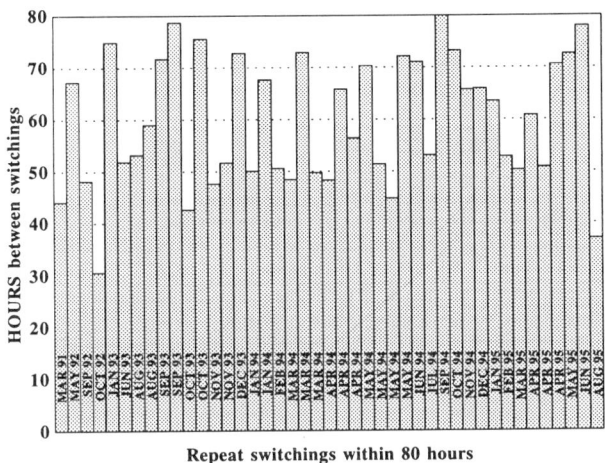

Figure 4. Intervals (< 80 hours) between successive DRA δ switches

Table 1. Fluence conditions defining risk category of each day

Risk		Fluence ($cm^{-2} sr^{-1} day^{-1}$)	
		Day	Previous Day
GREEN	Low	$< 8.5 \times 10^7$	$< 8.5 \times 10^7$
AMBER	Moderate	$> 8.5 \times 10^7$ $< 2 \times 10^8$	$< 2 \times 10^8$
RED	High	$> 4.5 \times 10^8$ $> 2 \times 10^8$	$> 2 \times 10^8$

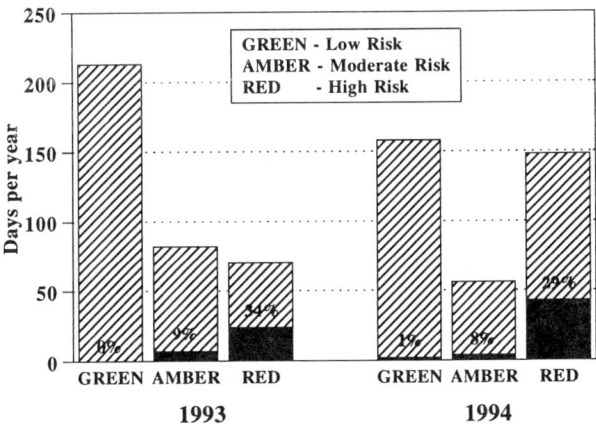

Figure 5. Probability of DRA δ switching for low, moderate and high risk days

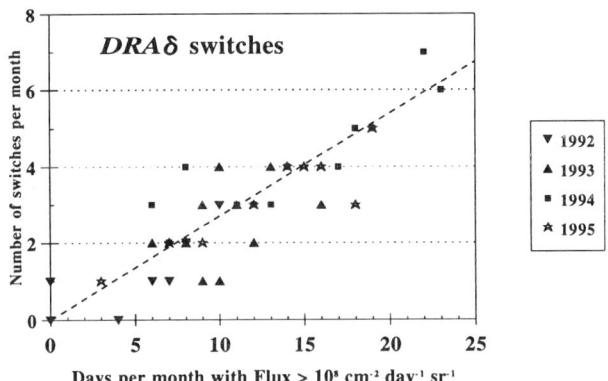

Figure 6. Relationship between frequency of DRA δ switches and high flux days

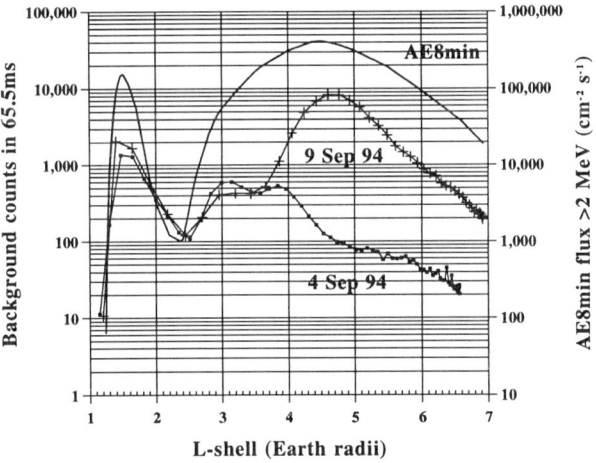

Figure 7. Background count rate profiles for STRV-1a Cold Ion Detector

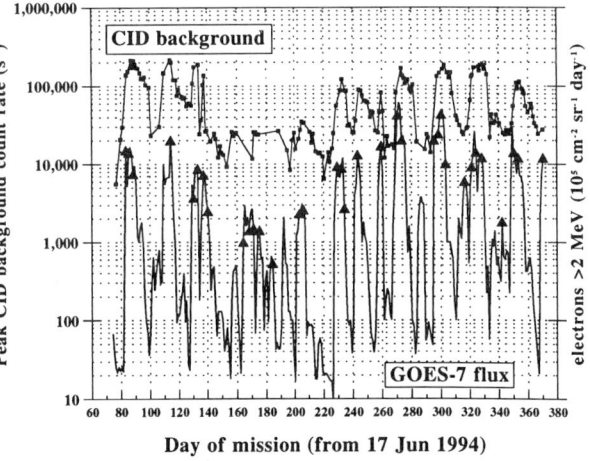

Figure 8. Variation of CID background count compared to GOES-7 electron flux, DRA δ switches are shown as in Figure 1.

3. NOISE IN MICROCHANNEL PLATE DETECTOR

The Cold Ion Detector on STRV-1a (launched into GTO on 17th June 1994) employs microchannel plate detectors with limited shielding ($\sim 2\,\text{mm}$ of Aluminium) within an external sensor. Protons $> 20\,\text{MeV}$ (inner belt) and electrons $> 1\,\text{MeV}$ (outer belt) can reach the plates and register as a background count rate. The background channel thus provides a day-to-day image of the position and shape of the outer belt and tracks its temporal variability. Figure 7 compares the outbound profiles for two days, 4th September 1994 and 9th September 1994 to illustrate a typical flux enhancement through the zone. Figure 8 plots the daily peak values of the background count rate and also the GOES-7 $> 2\,\text{MeV}$ electron daily fluence values, for the period 1 September 1994 to 22 June 1995. The matching of the periodic variabilities is sufficient to establish that the enhancements seen at GEO are symptomatic of the solar-driven evolution of the whole outer zone. Detailed measurements from the Radiation Environment Monitor on STRV-1b are provided by *Bühler et al.* [1996] in these proceedings.

4. CONCLUSIONS

- Energetic electrons are a key element in space weather prediction.
- Existing models such as AE-8 do not specify short-term temporal variations.
- The hazard of internal charging has been underestimated in the past.
- Optimised shielding protection must address effects due to internal charging as well as total dose.
- Understanding of the physical processes which generate outer belt enhancements is critical to the production of realistic models.
- More on-board detectors are needed for spacecraft health monitoring and analysis of inevitable operational anomalies.
- Real-time solar observatory and upstream solar wind monitors will permit environment forecasts to aid the planning of special operations and to activate circumvention procedures when required.
- Future work on radiation belt models should be driven by 'user requirements'.

Acknowledgements. The GOES-7 data were supplied by Dave Speich of the NOAA Space Environment Center, Boulder. The Meteosat-3 and STRV-1a data were provided by UCL Mullard Space Science Laboratory.

REFERENCES

Purvis, C.K., H.B. Garrett, A.C. Whittlesey, and N.J. Stevens, Design guidelines for assessing and controlling spacecraft charging effects, NASA Technical Paper 2361, 1984.

Vette, J.I., The AE-8 trapped electron model environment, NSSDC Report 91-24, NASA/GSFC November 1991.

Wrenn, G.L., Conclusive evidence for internal dielectric charging anomalies on geosynchronous communications spacecraft, *J. Spacecraft and Rockets* **32**, pp. 514–520, 1995.

Wrenn, G.L. and Sims, A.J., Surface charging of spacecraft in geosynchronous orbit, in The Behavior of Systems in the Space Environment, ed. by R.N. DeWitt, D.P. Duston and A.K. Hyder, Kluwer, Dordrecht, The Netherlands, pp. 491–511, 1993.

Bühler, P., L. Desorgher, A. Zehnder, L. Adams and E. Daly, Exploring the radiation belts with the Radiation Environment Monitor REM, these proceedings, 1996.

G.L. Wrenn and A.J. Sims, Space Department, Q134, DRA Farnborough, Hants GU14 6TD, UK (E-mail: G_L_Wrenn@scs.dra.hmg.gb)

Missions and Data Acquisition
Report of Discussion Group D

Reporter: R. Friedel

Max-Planck-Institut für Aeronomie, Postfach 20, 37189 Katlenburg-Lindau, Germany

Participants: A.D. Johnstone (chair), T. Kohno (co-chair), R. Friedel (reporter), D.N. Baker, J.B. Blake, M.I. Panasyuk, G.D. Reeves, J. Sharber

- Need to start with User Requirements:
  - Users generally don't care about science
  - Increased use of simple on-board measurements e.g. (GGS, ESA) (Russian SSI).
  - Move toward short term forecasting (e.g. data from libration point L_1).

- Data availability:
  - There is enough data already coming in to do a good job of near-realtime monitoring.
  - A standard data package needs to be defined:
    * format (ASCII?);
    * products (relativistic e^-, high energy p^+, ...?);
    * standard access (ftp via WWW).

- Need for near realtime models:
  - pilot project to assemble flux map of radiation belt based on current data (max. 1 week delay);
  - provide this as on-line service.

- New missions:
  - increased need for pitch-angle information;
  - standard instrumentation;
  - CRRES-type orbit under-sampled;
  - use of micro-satellites in geostationary transfer orbit.

We have learned at the meeting that:

1. the need for accurate radiation belt data has increased with new satellite systems being operated in different regions of space, and with increasing sophistication increasing their susceptibility to radiation damage;
2. users need statistical surveys of the radiation belts to assess the risks to their spacecraft and up-to-date reports to diagnose the causes of any anomalies experienced;
3. the variability of the radiation belts is much more dramatic, in all regions of space, and at all time scales down to seconds, than had been suspected a few years ago;
4. there are already many data sets covering the vital regions of space being collected regularly, but not being fully exploited for radiation monitoring;
5. all the major national and international agencies are involved in data collection.

We therefore recommend that:

1. a global radiation belt database be compiled from current data collections using modern computing and networking technology to achieve a near-realtime response;
2. spacecraft operators be encouraged to place standard radiation monitoring equipment on all spacecraft as is already being done in Russia;
3. such equipment should cover hot plasma, energetic electrons in the outer belt and energetic protons;
4. include such data in the database to provide more comprehensive spatial coverage especially where it is needed and the longterm coverage to properly assess variability;
5. bearing in mind that present orbital coverage of operational spacecraft and the minimal capability of suitable standard monitoring equipment, the quality of the data would be enhanced by the deployment of four spacecraft in geostationary transfer orbit, equally spaced in local time, carrying instruments capable of detailed pitch angle resolution. Such spacecraft could be microsatellites (~ 50 kg).

As first steps in this programme we suggest that:

1. the major agencies be requested to make the important data sets freely available in the interests of global coverage, available to all;
2. an international working group be formed to pursue these objectives and to seek the support necessary for an initial pilot study to collect and correlate the current data sets.

R. Friedel, Max-Planck-Institut für Aeronomie, Postfach 20, 37189 Katlenburg-Lindau, Germany

DISCUSSION

Q: M.K. Hudson. We need magnetometers for physical models, not just organizing particle data, eg. there were no dayside spacecraft with $\vec{B}$ data for the March 91 event.

A: A.D. Johnstone. Magnetometers are non-invasive pieces of equipment for most communication satellites to carry. This should certainly be expected for scientific satellites. While magnetometers are not demanding on spacecraft resources (mass, power) accurate interpretation of the data requires a magnetically clean spacecraft. In general this is expensive to achieve and requires specialist knowledge by the constractors. In that sense it is difficult to include magnetometers in a standard monitoring equipment to operational spacecraft.

Q: S.F. Fung. The suggestion of having environment monitoring particle and magnetic field detectors is intriguing. For physical modeling, one would like to have the highest possible resolution data, comparable to those from scientific instruments. Flying such high quality monitoring instruments may not be feasible, in view of their requirements or resources.

A: A.D. Johnstone. The idea of a standard monitoring package is to keep it small and low-powered so that it can be carried on as many spacecraft as possible. The hope is for wide deployment—hundreds of satellites?—and the data even if crude would provide a fantastic level of information about the radiation belt dynamics and the magnetosphere. The challenge for instrument builders is to make the scientific return as sophisticated as possible within the constraints.

Global Imaging by Energetic Neutral Particles

T. Beutier and J.-A. Sauvaud

Centre d'Etude Spatiale des Rayonnements, Toulouse, France

D. Boscher and S. Bourdarie

CERT-ONERA, Toulouse, France

The use of energetic neutral atoms seems to be a very promising technique to provide a global view of the magnetospheric cavity. Simulation results of neutral images produced by a dynamic convection-diffusion model of ions in the internal magnetosphere are presented. In particular, a series of pictures of the auroral precipitation of ion and the formation of the azimuthal asymmetry of the ring current ions is shown.

1. INTRODUCTION

The detection of energetic neutral atoms (ENA) produced by charge exchange between the thermal neutral population of the geocorona and energetic ions can provide global monitoring of the 3-D structure of the plasma reservoirs forming the magnetospheric cavity. As the geocoronal neutral content is relatively stable and as the trajectories of the energetic hydrogen atoms produced by charge exchange are essentially straight lines, the processing of the 3-D imaging of the plasma reservoir is a relatively easy task. Depending on the energy range of the ENA detectors, the monitoring of the radiation belts, the ring current, the inner plasma sheet and of the auroral precipitation can be performed [*Roelof et al.*, 1985; *Roelof*, 1987].

The ENA flux dynamics is a good application to trapped particle simulation codes. We have, recently, developed new diffusive-convective codes able to simulate the dynamics of electron and proton radiation belts. A first set of codes consists of the determination of the distribution function built in a phase space where the "active" dimensions are the adiabatic invariant and where the three other dimensions (i.e. the phases) are averaged (see *Beutier et al.* [1978], for the proton case). These kinds of models, which use a dipolar field,
are able to reproduce accurrently the main spatial and temporal features of belts using a small set of physical processus. However, we are limited to energy greater than 50 keV and for characteristic times greater than 10 hours.

So we have developed a second set of codes using a four dimensional phase space formalism where the longitude plays now an active role in addition to the three adiabatic invariants [*Bourdarie et al.*, 1995ab]. These kinds of models, using an asymetric time dependent magnetic field, are now able to simulate the radiation belt dynamics during substorm injections for energies above 1 keV and time scales equal to about 10 minutes.

In Section 2 we describe the calculation of ENA fluxes. We give the density and cross section values used in the calculation. The results are given in Section 3 and conclusions are presented in Section 4.

2. ENA FLUX CALCULATION

We start from the proton distribution function $f(M,J,\Phi,\varphi)$ in each point of a quadri-dimensional space where the four dimensions are the three action variables of trapped particle motion, J_1, J_2, J_3 (so called the magnetic moment, M, the longitudinal invariant, J, and the magnetic flux, Φ) and the third phase variable, φ_3 (so called longitude). The distribution function is averaged over the two first phase variables. From these values, we interpolate to calculate the proton flux for various positions (r, θ, φ), energies and pitch angles. The external boundary is fixed at 6 Earth radii of altitude and we can perform calculations for all energies between 1 keV and 300 MeV and all pitch angles between 0° and 90°.

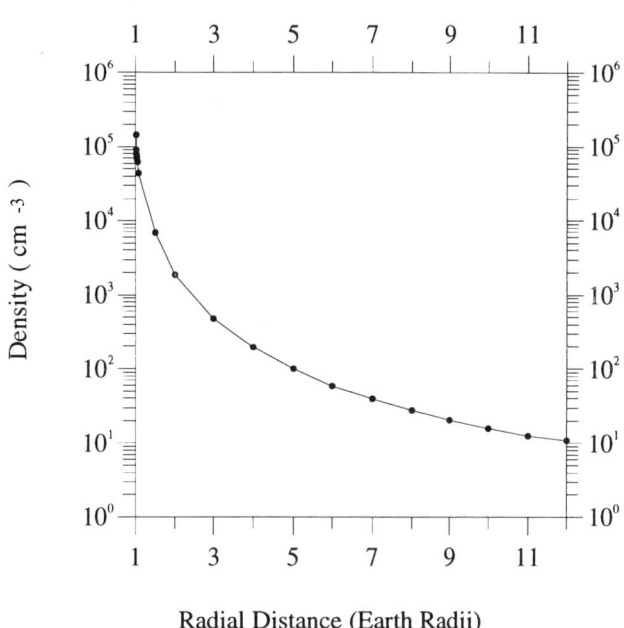

Figure 1. Geocoronal atomic hydrogen density as a function of radial distance.

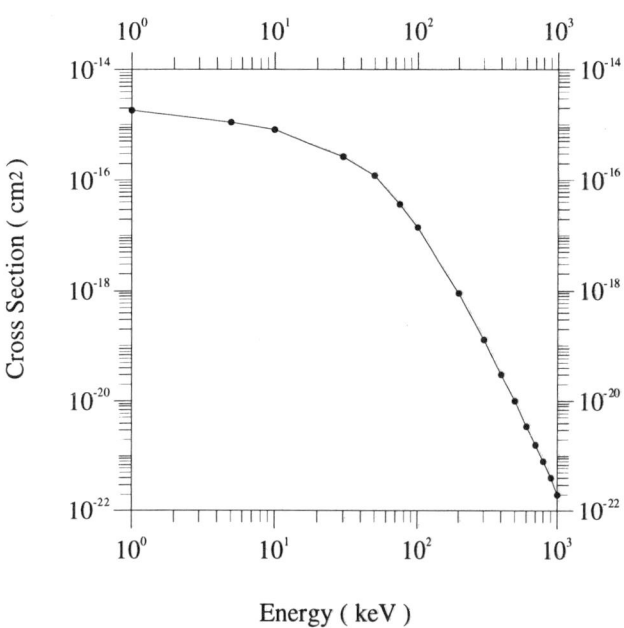

Figure 2. Charge exchange cross section. The horizontal axis corresponds to the initial proton energy nearly identical to the resulting atomic hydrogen energy.

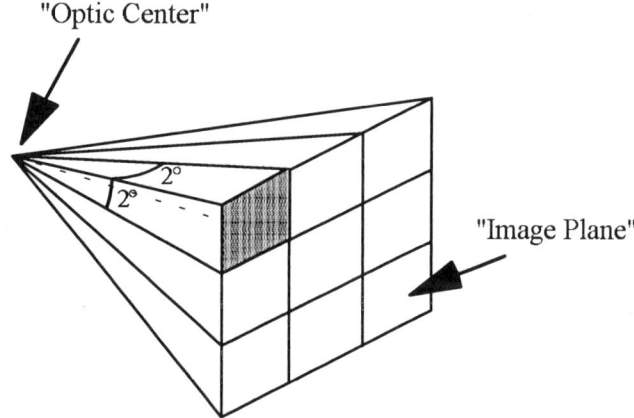

Figure 3. Simplified view of neutral imager

From these distributions, we can easily obtain the number, τ, of energetic neutral hydrogen atoms with a kinetic energy higher than E_0 produced per unit volume [centered in (r, θ, φ)], time and solid angle. To simplify, we assume that neutral atoms are isotropic in each volume units (i.e., all ion phases exist when the charge exchange event takes place). We have:

$$\tau(r, \theta, \varphi) = \frac{n_H(r, \theta, \varphi)}{4\pi}$$
$$\times \int_0^{\frac{\pi}{2}} \left[\int_{E_0}^{\infty} p^2 f(r, \varphi, E, \alpha_e) \sigma(E) \, dE \right] \sin \alpha_e \, d\alpha_e \,, \quad (1)$$

where r is the radial distance, θ the colatitude, φ the longitude, p the particle momentum, E the kinetic energy, α_e the equatorial pitch angle, $\sigma(E)$ the proton-hydrogen charge exchange cross section and n_H the geocoronal hydrogen density in (r, θ, φ).

For the cold atomic hydrogen density, shown in Figure 1, values are taken from *Rairden et al.* [1986]. These are the classical values issued from the model of *Chamberlain* [1963], adjusted to the geocoronal observations of the Dynamic Explorer 1 spacecraft. They correspond to a gas temperature of 1050 K.

We have extracted the hydrogen-proton charge exchange cross section, shown in Figure 2, from results compiled by *Claflin* [1970].

The measured ENA flux corresponds to the integral along the sight line passing by the "optic center" and the center of the given pixel. The number of energetic neutral hydrogen atoms ($E > E_0$) seen by the pixel ij per unit surface, time and solid angle is:

$$J_{ij} = \int_{E_0}^{\infty} \tau(l) dl \,, \quad (2)$$

where l is the location along the sight line and $\tau(l)$ is the production rate of ENA defined by Eq. (1).

We use an idealized camera of 49×49 pixels with $2°$ aperture each. Figure 3 gives a schematic view of this camera.

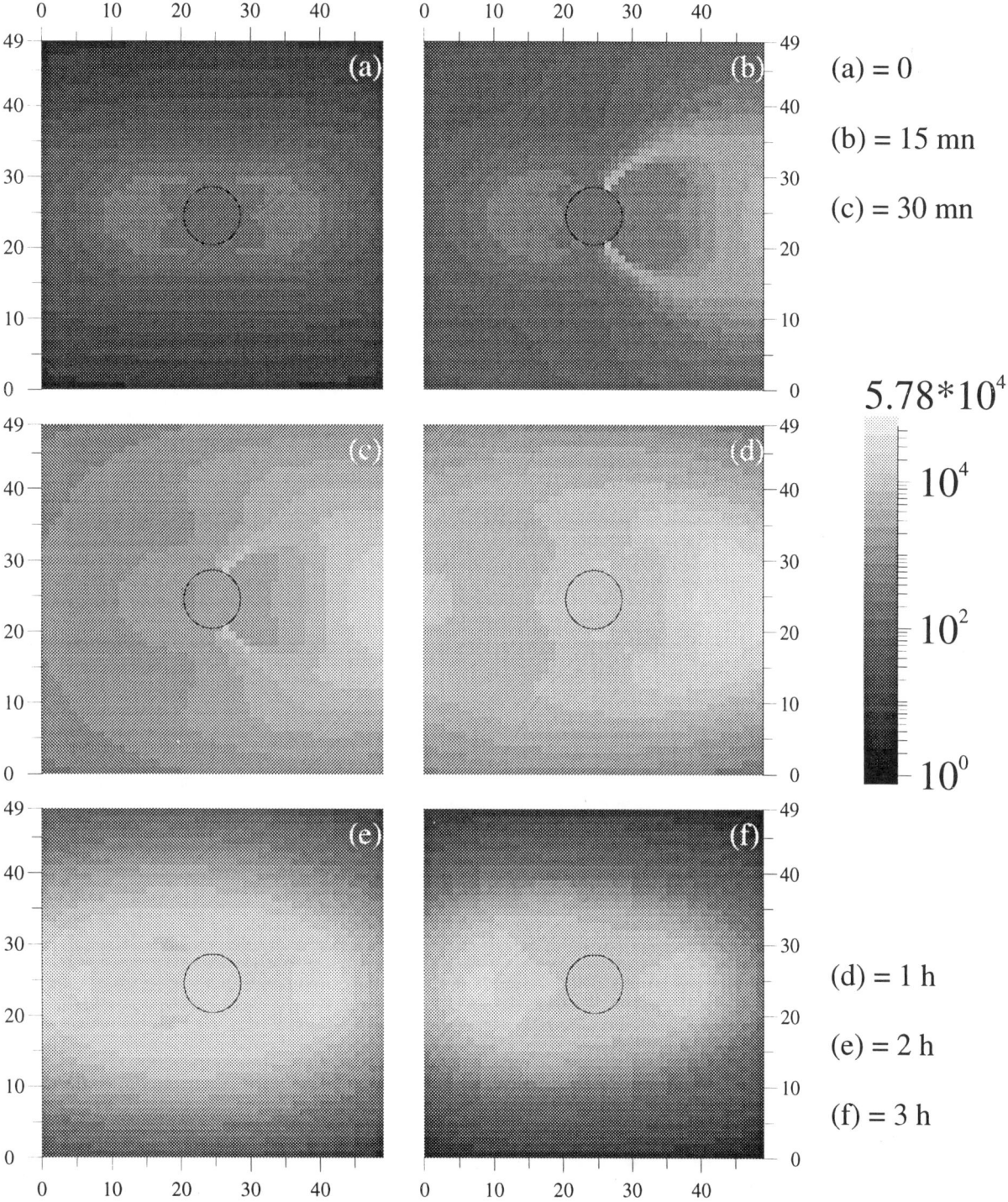

Figure 4. Energetic neutral hydrogen fluxes ($cm^{-2}s^{-1}sr^{-1}$, $E > 30\,keV$) seen by an idealized neutral imager located at $6R_E$ altitude, in the equatorial plane, at 18:00 LT. The panels (a), (b), (c), (d), (e) and (f) correspond, respectively, to the initial state ($t=0$), to the end of injection ($t = 15$ mn), and 30 mn, 1 h, 2 h, and 3 h after the start of the calculation. The axes are in pixel indices. Noon is on the left and the dark circle corresponds to the Earth's location.

3. RESULTS

Figure 4 gives the flux temporal variations of energetic neutral hydrogen per cm^2, per second and per steradian with energies higher than 30 keV seen by a camera located at an altitude of 6 R_E, in the equatorial plane, at 18:00 LT.

As initial conditions we use the distribution function deduced from the NASA AP-8 model. In order to gain an idea of radiation belt dynamics we have injected a strong distribution function (i.e. the AP-8 one multiplied by 1,000) during 15 mn at the external boundary ($L = 7$) for all pitch angles and for local time between 23:00 and 01:00. The injected spectrum and field models used in this simulation are identical to those given by *Bourdarie et al.* [1995b]. Note that in these preliminary calculations, the injected distribution function is arbitrary and taken sufficiently strong to distinctly see it from initial state.

On all panels, noon is on the left and the dark circle shows the Earth's location. The X and Y axes indicate the pixel indices. The fluxes in neutral hydrogen are given by a grey scale.

At the end of the injection (b) we can see the precipitation along the field line connected with the injection region. Drift and radial motions of particles are quite visible. Ring current increasing can also be seen, in particular 3 hours after the start of the injection (f).

4. CONCLUSIONS

These preliminary calculations show the accuracy of radiation belt models to simulate ENA fluxes seen by a neutral imager. However, some improvements seem necessary, in particular the use of a realistic injection distribution function and the necessity for taking into account the pitch angle (non isotropy of the flux).

The radiation belt modeling is a useful tool to prepare future missions of global monitoring of the magnetosphere.

REFERENCES

Beutier, T., D. Boscher, and M. France, SALAMMBO: A Three-Dimensional Simulation of the Proton Radiation Belt, *J. Geophys. Res., 100*, 17181–17188, 1995.

Bourdarie, S., D. Boscher, and T. Beutier, Dynamic physical modeling of trapped particles for satellite survey, *Proc. Workshop on Radiation Belt: Models and Standards*, Brussels, Oct. 17–20, 1995a.

Bourdarie, S., D. Boscher, and, T. Beutier, Modélisation du transport de particules chargées dans la magnétosphère terrestre interne, *GdR Plasmae*, Tournon, France, 1995b.

Chamberlain, J.W., Planetary coronae and atmospheric evaporation, *Planet. Space Sci., 11*, 901–960, 1963.

Claflin, E.S., Charge-exchange cross sections for hydrogen and helium ions incident on atomic hydrogen: 1 to 1000 keV, *U.S. Air Force Rep. SAMSO-TR-70-258*, Space Phys. Lab., Los Angeles, California, 1970.

Rairden, R.L., L.A. Frank, and J.D. Craven, Geocoronal imaging with dynamics explorer, *J. Geophys. Res., 91*, 13613–13630, 1986.

Roelof, E.C., D.G. Mitchell, and D.J. Williams, Energetic neutral atoms ($E \sim 50$ keV) from the ring current: IMP 7/8 and ISEE 1, *J. Geophys. Res., 90*, 10991–11008, 1985.

Roelof, E.C., Energetic neutral atom image of a storm-time ring current, *Geophys. Res. Letters, 14*, 652–655, 1987.

T. Beutier and J.-A. Sauvaud, Centre d'Etude Spatiale des Rayonnements (CESR), 9 Avenue du Colonel Roche, 31029 Toulouse Cedex, France

D. Boscher and S. Bourdarie, Centre d'Etudes et de Recherches de Toulouse (CERT-ONERA / DERTS), 2 Avenue Edouard belin, 31055 Toulouse Cedex, France

Global Imaging and Radio Remote Sensing of the Magnetosphere

S.F. Fung and J.L. Green

NASA Goddard Space Flight Center, Greenbelt, Maryland

Many space agencies are at various stages of planning for a new type of magnetospheric spacecraft carrying instruments capable of imaging various magnetospheric plasma regimes. These potential missions are benefiting from recent developments in sensors, optics, electronics and signal processing techniques. When combined together, the imaging and remote sensing instruments would make possible an exciting capability to view directly the global distribution, transport and energization of both cold and hot magnetospheric plasmas. Global magnetospheric imaging will greatly extend our knowledge drawn from in-situ sampling of the vast magnetospheric plasma regions over the past three decades. Global imaging can be accomplished on time scales varying from a few to tens of minutes, allowing the observations to be easily placed in the storm and substorm context. For example, while the energetic neutral atom imaging will yield information directly on the energetic particles found in the ring current and radiation belts, the radio plasma sounding technique will monitor the variations in the geomagnetic field caused by the storm-time ring current, which in turn reflects the radiation belt dynamics. Therefore, long term magnetospheric imaging will lead to new insight for understanding and modeling of the structure and dynamics of both the high and low energy magnetospheric plasmas, such as the radiation belts.

1. INTRODUCTION

Imaging and remote sensing measurements have long served the advancement of astronomy and astrophysics. In magnetospheric physics, the importance of global imaging and remote sensing has also been demonstrated by the successes in satellite imaging of the aurora (e.g., *Frank and Craven* [1988]; *Murphree et al.* [1990]). In the report, "Space Science in the Twenty-First Century: Imperatives for the Decades 1995 to 2015," the Solar and Space Physics Task Group under the Space Science Board of the National Academy of Sciences has in fact identified magnetospheric imaging as an innovative and exciting initiative for the study of magnetospheric dynamics. In response to this report, NASA formed the Magnetosphere Imager (MI) mission team to examine the feasibility of various types of instruments for magnetospheric imaging. The results of the MI mission team have recently been published in *Armstrong and Johnson* [1995] and *Armstrong et al.* [1995].

To embark on the new thrust of global imaging science, NASA has recently selected the Imager for Magnetopause-to-Aurora Global Exploration (IMAGE) mission as the first of a new series of mid-size explorers (MidEx) to be flown in 1999 (NASA press release no. 96–68, April 10, 1996). IMAGE will carry a number of global imaging instruments to study the global response of the magnetosphere to the changes in the solar wind. Information on IMAGE can be obtained via the world wide web at:
http://bolero.gsfc.nasa.gov/~image/IMAGE.html.

Many of the techniques suitable for global magnetospheric imaging have been reviewed by *Williams et al.* [1992]. We will provide a brief summary below on those techniques and some of the global science objectives which can be achieved by them. Among the most innovative ones are perhaps the energetic neutral atom (ENA) [*Roelof*, 1987] and the radio plasma (RPI) [*Green et al.*, 1993; *Reiff et al.*, 1994ab] imaging

Radiation Belts: Models and Standards
Geophysical Monograph 97
This paper is not subject to U.S. copyright.
Published in 1996 by the American Geophysical Union

techniques. In the balance of the paper, we will discuss an innovative application of the radio sounding technique and in particular, the roles that the radio and energetic neutral imaging techniques can play in providing information on the dynamics of storm-time ring currents and their contributions to the structure of the radiation belts.

The Earth's magnetosphere is extremely dynamic, with large-scale changes in size and shape in response to interplanetary conditions, and major internal reconfigurations. Unlike single-spacecraft measurements, which allow only sampling of magnetospheric conditions in specific regions at any given time, global remote sensing techniques can provide simultaneous viewing of various plasma domains such as the plasmasphere, ring current, and radiation belts.

2. INSTRUMENTATION FOR MAGNETOSPHERIC IMAGING

As described by *Williams et al.* [1992], there are a number of remote sensing and imaging techniques which can produce global images of the magnetosphere. They include the measurements of (1) the fluxes of energetic neutral atoms resulting from charge exchange reactions between energetic ions and the cold geocorona, (2) the solar extreme-ultraviolet radiation resonantly scattered by He^+ and/or O^+, (3) the emissions in far ultraviolet, visible, and X-ray wavelengths caused by the precipitating auroral particles, and (4) the echoing of variable-frequency radio signals from different density levels within large-scale magnetospheric structures using a magnetospheric radio sounder.

Neutral atom imaging measures the neutrals resulting from the change-exchange reactions between the neutral hydrogen geocorona and the energetic particles in the inner magnetosphere. This technique will provide important information on storm-time ring-current dynamics and trapped radiation distributions. In addition, imaging the resonantly scattered solar emission by the cold He and O ions at 304 Å and O^+ at 834 Å will yield important information about how these cold ions are distributed in the plasmasphere.

Auroral imaging provides a measure of the energy deposition by precipitating particles as well as the global morphology of a substorm. In the far-ultraviolet (FUV) wavelength range the aurora is typically brighter than the dayside atmospheric background [*Rairden et al.*, 1986] and can be used to monitor the auroral oval during orbital night and day. Ultraviolet (UV) images can provide information on the mean energies and energy fluxes of the precipitating electrons and their relationship with the global structure of the aurora. In addition, imaging the x-ray radiation resulting the Bremsstrahlung emissions of more energetic (up to hundreds of keV) precipitating electrons will reveal information of the losses of energetic electrons from the ring current and radiation belts.

Finally, radio wave imaging involves the measurements of transmitted radio wave pulses which are reflected at propagation cutoffs in the structured magnetospheric plasma. Operating in the frequency range from 3 kHz to 1 MHz, a radio sounder can remotely measure a wide range of densities while determining the positions (i.e., ranges from the transmitter) of different critical plasma boundaries, such as the plasmapause and magnetopause, on time scales of a few minutes.

Clearly, the strength of a global imaging or remote sensing mission lies in combining different instrumentation techniques within a single mission. In the remainder of this paper, we will discuss the use of the radio imaging technique to measure various aspects of the dynamics of the high energy particle environment of the inner magnetosphere. More information on the neutral atom imaging technique can be found in the paper by Beutier et al. in this volume.

3. MAGNETOSPHERIC RADIO IMAGING

The use of radio sounding techniques for the study of the ionospheric plasma dates back to *G. Briet and M.A. Tuve* in 1926. Ground based sounders measure the electron number density (N_e) profile up to the peak in the F region of the ionosphere. These instruments provided the foundation for the success of the Alouette and International Satellites for Ionospheric Studies (ISIS) programs which pioneered the use of space borne, swept frequency sounders to obtain electron density N_e profiles of the topside ionosphere. Repeated measurements during the orbits produced orbital plane images which routinely provided density measurements accurate to within 10% (limited only by frequency resolution). The Alouette/ISIS experience also demonstrated that even with a high-powered transmitter (compared to the low-powered sounder possible today) a radio sounder can be compatible with other imaging instruments on the same satellite.

The feasibility of magnetospheric imaging and radio remote sensing using advanced radio sounding technique has been extensively studied by *Green et al.* [1993], *Calvert et al.* [1995] and *Green et al.* [1996]. Both the magnetopause and plasmasphere, as well as the cusp and boundary layers, can be observed by a radio sounder in a high-inclination polar orbit with an apogee greater than $6 R_E$. Radio imaging will provide measurements of magnetospheric densities with unprecedented precision and coverage in the plasmasphere, inner magnetosphere and magnetopause, such that the structure and dynamics of different magnetospheric plasma regions can be determined.

Like a radar, a radio sounder transmits and receives coded electromagnetic radio pulses. A basic radio sounder measures the time delay between the transmitted pulse and the echo. The time delay measurement is then converted into a distance. In a magnetized plasma the reflection location depends on the wave frequency and its mode (or polarization), the ordinary (O) or extraordinary (X) mode.

Reflection of the O mode occurs at the plasma cutoff where the sounder wave frequency equals the local electron plasma frequency f_p, which is determined by the local electron density, $f_p \approx 9 N_e^{1/2}$ kHz (with N_e in cm^{-3}). This condition forms the basis for measuring plasma density at a remote location. As the sounder frequency is increased, the waves penetrate to greater distances, into regions of larger plasma density, yielding echoes with successively larger delay times. From the echo delays as a function of sounding frequency, the electron density profile from the spacecraft can be determined.

For the X mode, the cutoff frequency is given by $f_x = (f_b/2) + [f_p^2 + f_b^2/4]^{1/2}$, where $f_b = 2.8 B$ MHz (where B

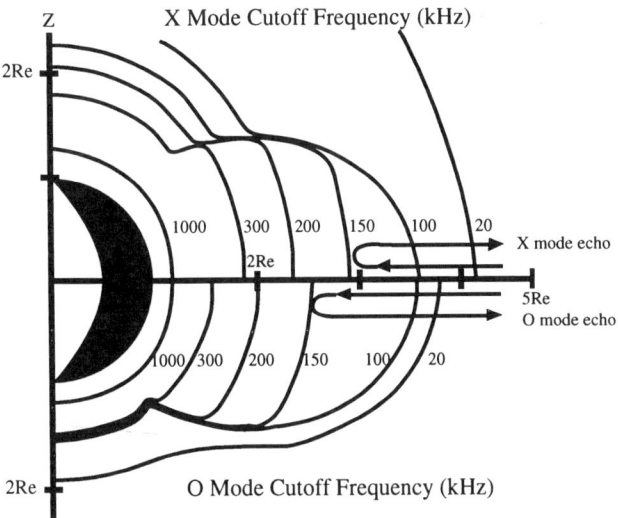

Figure 1. Model cutoff surfaces for the extraordinary (X) (upper half) and ordinary (O) (lower half) mode waves. The plasma and magnetic field modes have been adjusted for clarity in illustrating the spatial separation between the cutoffs in the two modes at a given frequency.

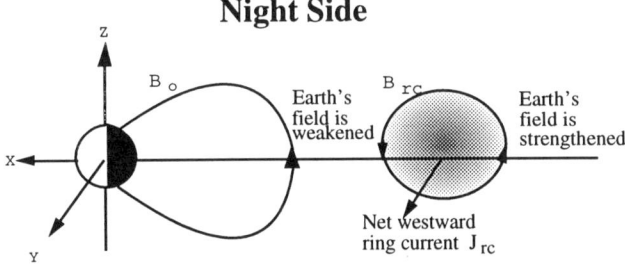

Figure 2. Earth's magnetic field B_0 is perturbed by the magnetic field B_{rc} due to the ring current J_{rc}.

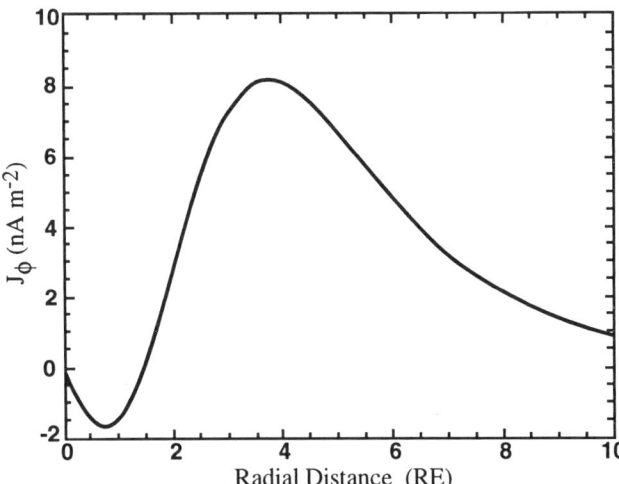

Figure 3. Ring current density profile produced by using the *Hilmer and Voigt* [1995] model, with the negative component flowing eastward and positive component flowing westward, respectively.

is the magnetic field strength in gauss). The O and the X modes can propagate freely at or near the speed of light when the wave frequencies exceed the plasma cutoffs and hence are called the free-space modes. It is the X mode that is the most important in measuring the effect of the ring current, as we will demonstrate in the next section.

Figure 1 illustrates the difference in the reflection locations of the O and X mode for a radio imager located on the magnetic equator outside of the plasmasphere with sounder frequency of 150 kHz. Contour plots of the f_x (top half) and f_p (bottom half) cutoff frequencies are shown in Figure 1. Radio waves propagating at frequencies above the cutoff frequencies in either of the two free-space modes will travel at nearly the speed of light in a straight line over most of their trajectories. Upon encountering a plasma cutoff (location where the wave frequency equals the f_p), as shown in the lower portion of Figure 1, the echoes will be specularly reflected. From knowledge of the local fp the plasma density N_e (in cm^{-3}) can be calculated. From knowledge of f_p in the vicinity of the f_x reflection point f_b can be calculated and hence the magnetic field can be determined.

4. SOUNDING OF RADIATION BELTS & THE RING CURRENT

Although the sounder waves (the X or O mode) are reflected mostly by the cutoffs of the cold electron component of the magnetospheric plasma, their echo signatures may be used to study the dynamics of the radiation belts as we describe in this section.

As motions of charged particles are controlled by the background magnetic field, much of the radiation belt and ring current dynamics can be elucidated by observing the variations in the global geomagnetic field caused by the rise and decay of the ring current. Figure 2 depicts the situation in which a net ring current flows westward in the nightside magnetosphere. The magnetic perturbations due to the current will decrease the geomagnetic field in the region between the earth and the ring current, while the earth's field is enhanced by the current's field in the region beyond the ring current L-shells.

Since the X-mode cutoff frequency depends on the magnitude of the local background magnetic field (and density), an increase or decrease in the background field will cause the X-mode cutoff surfaces (Figure 1) to shift farther away from or closer to Earth, respectively. To investigate the detectability of the changes in the geomagnetic field caused by the ring current, we have performed a ray-tracing modeling study.

For illustrative purposes, we have adopted for the ray tracing calculations a diffusive equilibrium plasma model based on those given by *Aikyo and Ondoh* [1971] and *Angerami and Thomas* [1964], and a background (unperturbed) magnetic field given by a dipole. Figure 3 shows the ring current density profile obtained by using the *Hilmer and Voigt* model [1995]. The gross features (the main positive and negat-

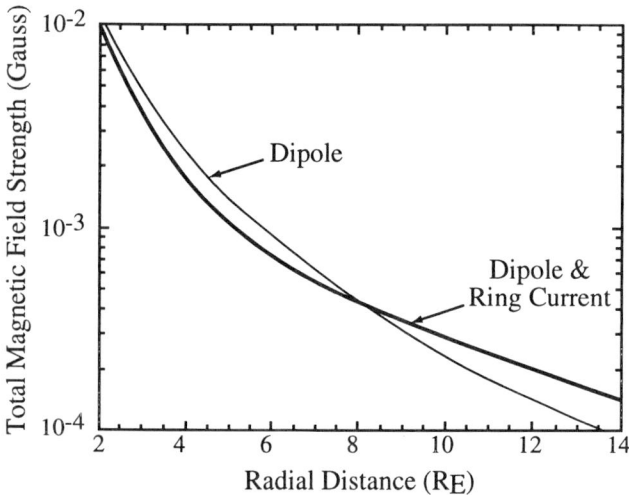

Figure 4. Magnetic field profiles of a pure dipole and the case of a dipole plus the perturbations of the model ring current in Figure 3.

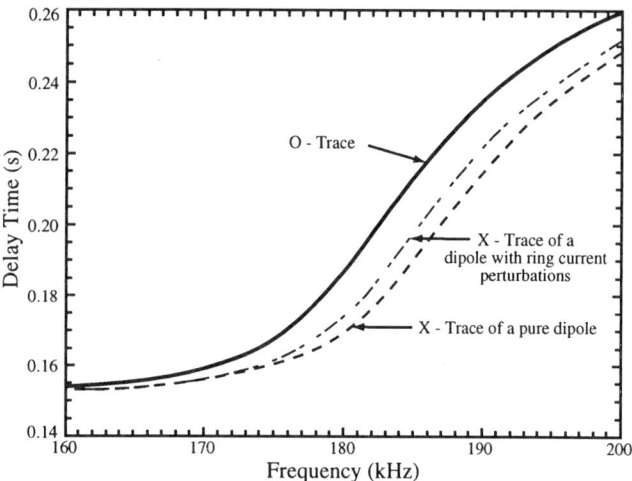

Figure 5. A plasmagram of O mode and X mode echo delay times as a function of the sounder wave frequency.

ive component currents) of this profile resemble those of a storm-time ring current observed on September 5, 1984 (see Figure 4 in *Lui et al.* [1987]).

Effects of the magnetic perturbations of the net westward ring current (c.f. Figure 2) are clearly seen in Figure 4, in which decreased and increased field regions (compared to the earth's dipole field) are observed. However, the presence of an eastward (negative) current (see Figure 3) at the lower L-shell range of the ring current region effectively causes the net westward (positive) current to center at higher L values (near $L = 8$).

Using the ray tracing modeling code developed by *Green* [1988], we have modeled the propagation of both the X and O mode waves with frequencies in the range $160 < f < 200$ kHz, launched earthward along the equator from a sounder located at $L = 10$. The X and O mode waves are reflected upon encountering their respective plasma cutoffs and return as echoes when they reach the sounder location. The total echo delay times for the various frequency waves are plotted in a plasmagram shown in Figure 5. The O mode echoes have longer delay times than the X mode echoes at a given frequency because the O mode cutoff surfaces are located closer to Earth and thus farther from the sounder at $10\,R_E$.

The X mode echoes for both the case of a pure dipole and the case of a dipole plus ring current perturbations are also shown in Figure 5. The X-mode calculations show that the echoes would suffer slightly longer time delays when the ring current is present because their cutoff surfaces have shifted earthward (farther away from the sounder). For a nominal time delay of 0.2 s, the change in the time delay is of the order of a few percent, about 25 ms. With advanced digital sounding techniques [*Calvert et al.*, 1995; *Green et al.*, 1996], the changes in the delay times should be readily measurable.

Although significant changes in the magnetospheric cold plasma density structure are known to occur during magnetic storms and substorms (e.g. *Carpenter et al.* [1993]; *Car-*

penter [1995]) (changes that are themselves important topics of study by the EUV photon imager and the radio sounder), the density model in our example was held constant so that the O mode trace in Figure 5 was unchanged. This can be justified by the fact that in a number of important observing situations, the density profile is not expected to depart significantly from its quiet time levels as the dilation of the geomagnetic field by the ring current proceeds. One such important situation should occur wihin the outer storm-time plasmasphere at longitudes where there has not been a significant storm-time loss of flux tube electron content and where during a storm the equatorial density profile preserves its quiet time property of varying roughly inversely with flux tube volume (e.g. *Chappell et al.* [1971]; *Carpenter and Anderson* 1992]). Within longitudinally localized sectors, depletion of flux tube electron content by a factor of up to 3 does occur within the outer storm-time plasmasphere (e.g. *Park* [1974]; *Carpenter et al.* [1993]), the occurrence of corresponding density profile changes could be identified by the sounder and EUV imager and an appropriately varying density model applied to the ring current detection problem.

5. DISCUSSION

In this paper we have summarized a number of potential techniques that are applicable to obtaining global images of the magnetosphere. These images will provide information on the global dynamics that affect the development of the storm-time ring current and the radiation belts.

In particular, both the energetic neutral atom and radio plasma imaging techniques will be important in that they provide complementary measurements of the ring current and the radiation belts. Energetic neutral particle imaging measures the previously trapped energetic ions which make up the bulk of the ring current, yielding information such as pitch angle distributions [*Fok et al.*, 1995]. On the other hand the radio imaging technique, while measuring the cold plasma component, will provide critical information on the

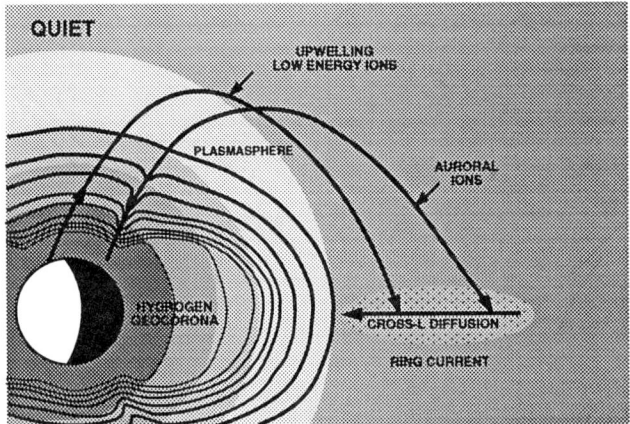

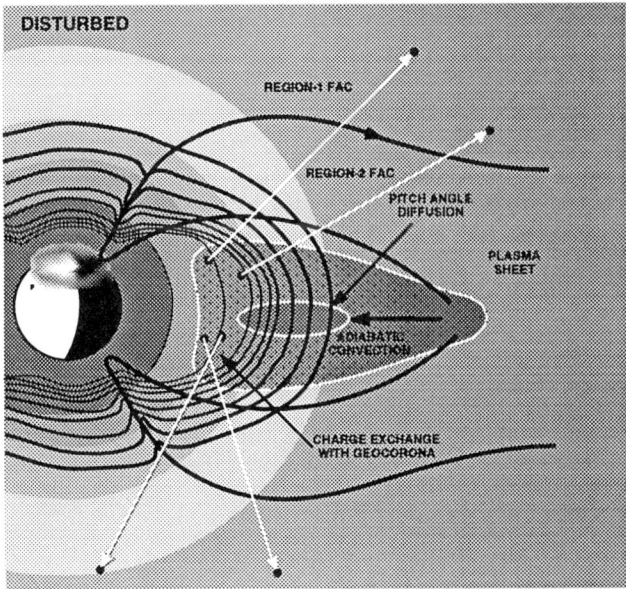

Figure 6. A schematic of the different magnetospheric configurations for quiet (upper) and storm (lower) times observable by an energetic neutral atom imager and a radio plasma imaging instrument.

changes in the geomagnetic field and in the plasmaspheric density structures as a function of the phases of a storm or substorm, and on the effects of those changes on the dynamics of the trapped particles in the radiation belts, as depicted in Figure 6. These techniques are therefore useful for investigating substorm injection boundaries and the inward motion as well as the rise and decay of the storm-time ring current.

In conclusion, global imaging measurements, while complementing in situ observations, will provide space physics research with new perspectives. Missions such as IMAGE should lead to significant advancement of our understanding of global magnetospheric structures and dynamics.

Acknowledgements. The authors gratefully acknowledge the early inspiration from the late Dr. S.D. Shawhan and useful discussions with the MI Science Definition Team. We also thank Dr. D.L. Carpenter for helpful discussions and insightful comments.

REFERENCES

Aikyo, K., and T. Ondoh, Propagation of nonducted VLF waves in the vicinity of the plasmapause, *J. Radio Res. Labs.*, *18*, 153, 1971.

Angerami, J.J., and J.O. Thomas, The distribution of ions and electrons in the Earth's exosphere, *J. Geophys. Res.*, *69*, 4537, 1964.

Armstrong, T.P., and C.L. Johnson, *Magnetosphere Imager Science Definition Team—Interim Report*, NASA Reference Publication 1378, Marshall Space Flight Center, Huntsville, Alabama, September, 1995.

Armstrong, T.P., D.L. Gallagher, and C.L. Johnson, *Magnetosphere Imager Science Definition Team—Executive Summary*, NASA Reference Publication 1379, Marshall Space Flight Center, Huntsville, Alabama, September, 1995.

Beutier, T., J.-A. Sauvaud, D. Boscher, and S. Bourdarie, Global imaging by energetic neutral particles: A scientific implement of a new readiation belt model, Workshop on Radiation Belts: Models and Standards, Brussels, 17-20, 1995 (this volume).

Breit, G., and M.A. Tuve, A test for the existence of the conducting layer, *Phys. Rev.*, *28*, 554–575, 1926.

Calvert, W., R.F. Benson, D.L. Carpenter, S.F. Fung, D.L. Gallagher, J.L. Green, D.M. Haines, P.H. Reiff, B.W. Reinisch, M.F. Smith, and W.W.L. Taylor, The feasibility of radio sounding in the magnetosphere, *Radio Science*, *30*, 5, 1577–1615, 1995.

Carpenter, D.L., Earth's plasmasphere awaits rediscovery, EOS, *Trans. Am. Geophys. Union*, *76*, 89, 1995.

Carpenter, D.L., and R.R. Anderson, An ISEE/whistler model of equatorial electron density in the magnetosphere, *J. Geophys. Res.*, *97*, 1097, 1992.

Carpenter, D.L., B.L. Giles, C.R. Chappell, P.M.E. Décréau, R.R. Anderson, A.M. Persoon, A.J. Smith, Y. Corcuff, and P. Canu, Plasmasphere dynamics in the duskside bulge region: a new look at an old topic, *J. Geophys. Res.*, *98*, 19,243, 1993.

Chappell, C.R., K.K. Harris, and G.W. Sharp, The dayside of the plasmasphere, *J. Geophys. Res.*, *76*, 7632, 1971.

Fok, M.-C., T.E. Moore, J.U. Kozyra, G.C. Ho, and D.C. Hamilton, Three-dimensional ring current decay model, *J. Geophys. Res.*, *100*, 9619, 1995.

Frank, L.A., and J.D. Craven, Imaging results from Dynamics Explorer 1, *Rev. Geophys.*, *26*, 249–283, 1988.

Green, J.L., R.F. Benson, W. Calvert, S.F. Fung, P.H. Reiff, B.W. Reinisch, and W.W.L. Taylor, A Study of Radio Plasma Imaging for the proposed IMI mission, NSSDC Technical Publication, February 1993.

Green, J.L., Ray tracing of planetary radio emissions, *Planetary Radio Emissions* II, Proceedings of the 2nd International workshop held at Graz, Austria, 1988.

Green, J.L., W.W.L. Taylor, S.F. Fung, R.F. Benson, W. Calvert, B.W. Reinisch, D.L. Gallagher, and P.H. Reiff, Radio remote sensing of magnetospheric plasmas, Proceedings of the Chapman Conference on Space Plasma Measurement Techniques, Santa Fe, NM, submitted, 1996.

Hilmer, R.V., and G.-H. Voigt, A magnetospheric magnetic field model with flexible current systems driven by independent physical parameters, *J. Geophys. Res.*, *100*, 5613–5626, 1995.

Lui, A.T.Y., R.W. McEntire, and S.M. Krimigis, Evolution of the ring current during two magnetic storms, *J. Geophys. Res.*, *92*,

7459, 1987.

Murphree, J.S., L.L. Cogger, and R.D. Elphinstone, Observations of distortions of optical features in the UV auroral distribution, *IEEE Trans. Plasma Sci., 17*, 109–115, 1990.

Park, C.G., Some features of plasma distribution in the plasmasphere deduced from Antarctic whistlers, *J. Geophys. Res., 79*, 169, 1974.

Rairden, R.L., L.A. frank, and J.D. Craven, Geocoronal imaging with Dynamics Explorer, *J. Geophys. Res., 91*, 13,613, 1986.

Reiff, P.H., J.L. Green, R.F. Benson, D.L. Carpenter, W. Calvert, S.F. Fung, D.L. Gallagher, B.W. Reinisch, M.F. Smith and W.W.L. Taylor, Radio Imaging of the Magnetosphere, *EOS, 75*, 129, March 15, 1994a.

Reiff, P.H., J.L. Green, R. Benson, D.L. Carpenter, W. Calvert, S.F. Fung, D.L. Gallagher, Y. Omura, B.W. Reinisch, M.F. Smith and W.W.L. Taylor, Remote sensing of substorm dynamics via radio sounding, in Substorms-2, Proceedings of the Second International Conference on Substorms, Ed. J.R. Kan, J.D. Craven, and S.-I. Akasofu, University of Alaska Press, Fairbanks, Alaska, pp. 281–287, 1994b.

Roelof, E.C., Energetic neutral atom image of a storm-time ring current, *Geophys. Res. Lett., 14*, 652, 1987.

Space Physics Strategy Implementation Study, The NASA Space Physics Program for 1995–2010, Vol. 1: Goals, Objectives, and Strategy, Vol. 2: Program Plan, April 1991.

Williams, D., E.C. Roelof, and D.G. Mitchell, Global magnetospheric imaging, *Rev. Geophys., 30*, 183–208, 1992.

S.F. Fung and J.L. Green, NASA Goddard Space Flight Center, Greenbelt, MD, USA

Artificial Neural Network (ANN) Forecasting of Energetic Electrons at Geosynchronous Orbit

G.A. Stringer, I. Heuten, C. Salazar and B. Stokes

Department of Physics, Southern Oregon State College, Ashland, Oregon

Levenburg-Marquardt (LM) backpropagation neural networks have been trained to predict hour-ahead Log(electron fluxes) with overall prediction efficiency up to 95% and rms error of 0.1 near the flux peaks. The recent history of Log(GOES-7 1-hour averaged electron fluxes), D_{st}, and K_{p} along with the magnetic local time were used as inputs. Training/testing strategies explore the effects of the number of hidden layer neurons, the selection of initial weights and biases, the number of training epochs, and the relative importance of each type of history on ANN performance.

1. INTRODUCTION

Space weather in the near Earth environment has caused disruptions of power distribution systems and electronic communication [*Siscoe et al.*, 1994; *Freeman*, 1994]. Sudden increases in trapped energetic particle fluxes can cause temporary disruption of service or even loss of expensive satellite equipment [*Baker et al.*, 1986; *Rostoker et al.*, 1995; *Wrenn*, 1995]. Highly reliable and precise space weather forecasting is an economic prerequisite to implementing a damage avoidance strategy such as temporary shutdown which would result in lost revenue.

Artificial Neural Networks (ANN) have been used to model and predict day-averaged Log(flux) (LF) for 3-5 MeV day-averaged geosynchronous electrons [*Koons and Gorney*, 1991; *Stringer and McPherron*, 1993]. These training strategies produced prediction efficiencies < 63%, far too low to motivate cost effective damage avoidance strategies. In addition, the 1-day time resolution of these models were too coarse to capture the sudden flux increases that are often observed.

A recent study of geosynchronous satellite anomalies [*Wrenn*, 1995] found different localized occurrence zones depending on the type of anomaly. Surface discharge effects occur pre-midnight to dawn while deep dielectric charging occurs more frequently from dawn to noon local time. Simultaneous observations taken by LANL geosynchronous electron instruments [*Reeves*, 1995] show very different patterns at widely separated local times. Thus successful ANN modeling of geosynchronous electron fluxes for purposes of anticipating anomalies requires:

1. higher time/space resolution than provided by 1-day averages;
2. ability to "nowcast" as well as "forecast" the electron flux so that reliable values can be projected on a time/space grid spanning all local times from the present into the near future.

This information would provide a basis for statistically reliable forecasts of satellite anomalies.

This paper describes initial efforts to implement such a system. An hour-averaged database was adopted. It includes energetic electron data along with correlated solar wind and magnetospheric parameters. Initially the GOES electron data were hour-averaged in MLT for the database. The NSSDC Common Data Format was used. The goals were to

1. improve prediction of peak fluxes,
2. determine optimal training parameters, and
3. learn how to extract some physics from trained ANNs.

The following improvements in the use of ANN's to study and predict geosynchronous Log(electron flux) were realized:

1. improved time resolution through the use of 1-hour averaged data;

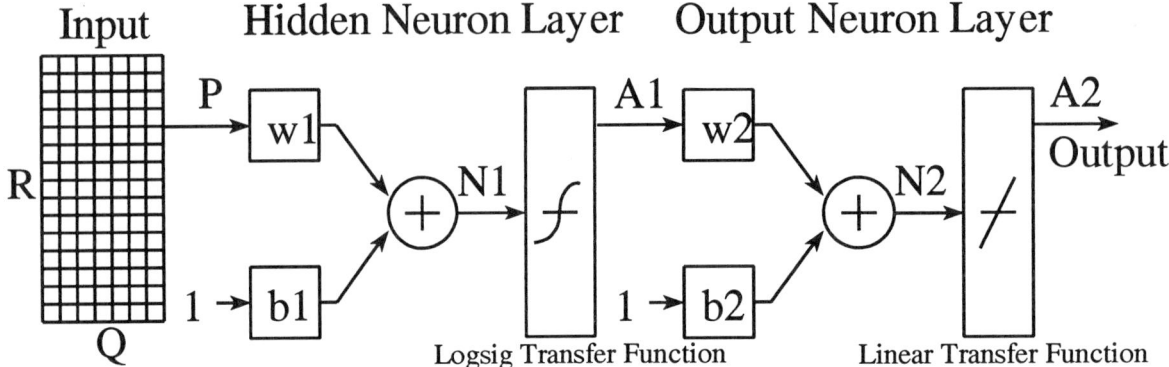

Figure 2. Detailed ANN schematic showing the structure of neurons in the hidden and output layers. Each neuron is characterized by a set of weights for each input, a bias, a summing node, a characteristic transfer function and an output.

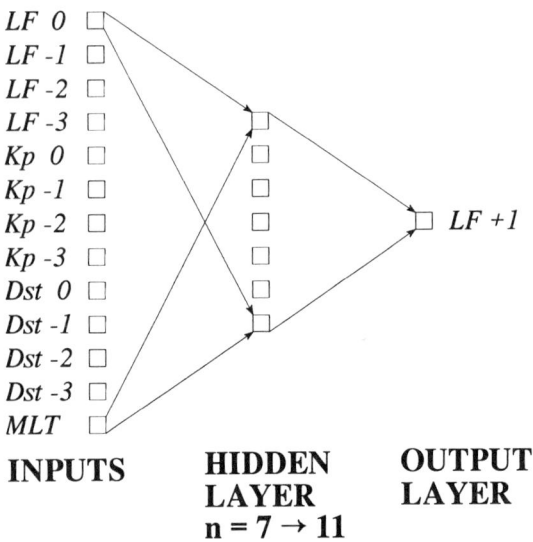

Figure 1. Schematic of an artificial neural network showing thirteen inputs, one hidden layer with 7-11 neurons, and one output neuron. Only a few of the many interconnecting paths between layers of neurons are shown.

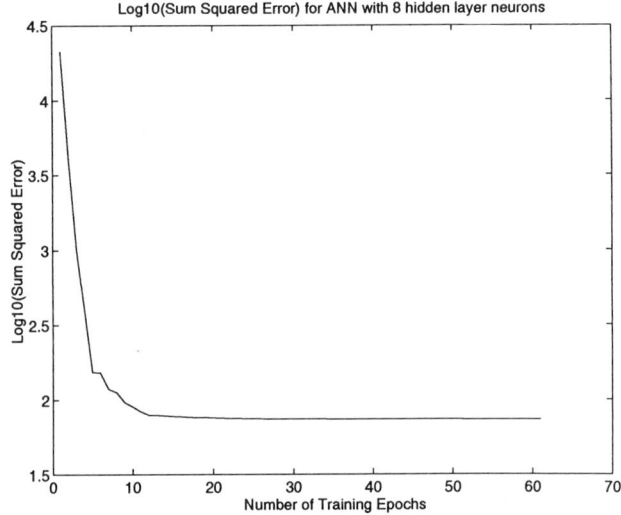

Figure 3. Sum-Squared-Error, SSE, versus number of training epochs. Illustration of the rapid convergence of the Levenberg-Marquardt, LM, backpropagation algorithm.

2. overall prediction efficiencies (PE) up to 95% for hour-ahead-Log(flux);

3. low rms error of 0.1 near the peaks gives reliable prediction of potentially damaging peak fluxes.

In addition, a technique was developed for using trained ANN's to explore the first order relationship between input training parameters and the output variables. This may shed light on the physical processes captured by a trained ANN model.

2. THE ARTIFICIAL NEURAL NETWORK

Artificial Neural Networks (ANNs) are software approximations of layered parallel arrays of biological neurons. The MATLAB Neural Network Toolbox [*Demuth and Beale*, 1994] running on a variety of PCs was used in this work. It allowed easy user modification and control of the source code which allowed integration of the training and testing process. It also supported good graphical display primitives for monitoring the training and testing process.

Figure 1 illustrates the relationships between the input data array, the internal neuron layers of the network, and the output vector. According to MATLAB semantics this diagram is a 2-layer network. It has a hidden layer containing a variable number of neurons and an output layer containing one neuron for each output. The number of neurons in the hidden layer is roughly analogous to the idea of polynomial degree used in normal curve fitting routines. If too many neurons are used, the network may reproduce the training data very well but poorly match other data. This explains why concurrent training and testing are necessary to detect this possibility (sometimes called overtraining) in an ANN.

A convenient set of input variables was selected for the first training/testing efforts. It consisted of the 4-hour histories of Log(flux), K_p and D_{st} along with the magnetic local time, MLT (since K_p is a 3-hour index we interpolated the 3 hour values to get 1-hour resolution). The output variable for training and testing was Log(flux) one hour later (LF+1). Thus there are 13 ANN inputs and one ANN output.

Figure 2 shows a more detailed schematic of the individual neurons used in each layer. A two layer network having a sigmoid transfer function in the first (hidden) layer and a linear transfer function in the second (output) layer was used. This type of ANN can be trained to approximate most functions arbitrarily well [*Hagan et al.*, 1996]. Individual neurons are characterized by:

1. a weight for each input,
2. a bias or offset,
3. summing points which combine all of the weighted inputs plus offset to provide a net neuron input N,
4. a transfer function which specifies the relationship between the neuron input N and the neuron output A.

Neural networks are characterized by multiple neurons arranged in layers with interconnections between all of the outputs in one layer and all of the inputs in the next layer.

ANN training proceeds generally as follows:

1. An array of Q facts, each with R variables, are applied to the input P and are fed forward through the ANN. This produces an output vector of length Q at output $A2$ of the second layer.
2. After each such epoch of training the ANN is tested to see if it has reached a predetermined training objective, e.g. Sum Squared Error (SSE) < 0.1. If not an algorithm that modifies the weights and biases is back propagated through the network and the training process is repeated.

The Levenberg-Marquardt (LM) algorithm was used [*Press et al.*, 1992]. The LM algorithm required 16 Megabytes of RAM to train the ANNs when 1,000 training/testing facts were used.

3. ANN TRAINING PROCEDURE

The training set was constructed from historical GOES-7 data and the OMNI data set from the National Space Science Data Center that had previously been added to the CDF database. The 1-hour averaged electron fluxes were computed from 5-minute corrected electron data contained in each hour of MLT. For example, all of the electron fluxes occurring between 0000 and 0100 hours MLT were averaged and assigned a value of 0 for MLT. Finally, all facts with an electron flux less than 20 counts per second were excluded from the training set. This forced the ANN to concentrate more on matching the peaks in the electron flux.

Training/testing proceeds by sequential and repeated application of these data sets to the input of the ANN. The training set included 1,000 selected facts starting with July 1,

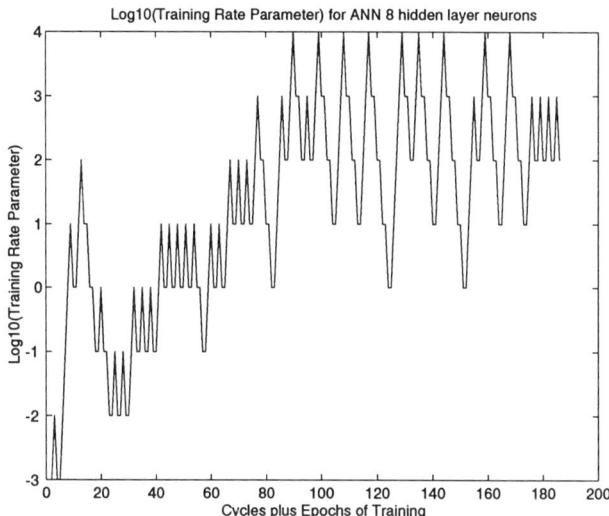

Figure 4. The LM training rate parameter, f, is adaptively adjusted during the LM training process. Changes occur several times per epoch. Large values occur near the end of training.

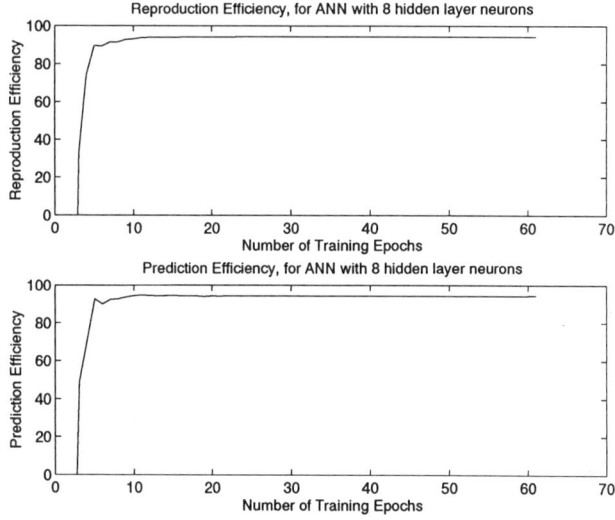

Figure 5. Reproduction Efficiency, RE, measures how well the trained ANN reproduces the training target. Prediction Efficiency, PE, measures how well the trained ANN matches data not used for training. The two curves are remarkably similar.

1989, enough to cover at least two solar rotation cycles. A similar testing set of 1,000 facts was constructed using data starting on January 1, 1990. This set of facts was not allowed to alter the ANN weights and biases. The ANN output was compared with the target vector using rms error and PE as figures of merit.

Figure 3 shows typical SSE as a function of the training epoch. The initial value for the training rate parameter, f, is set in the program but it is adaptively adjusted by the training algorithm depending on the average gradient of the

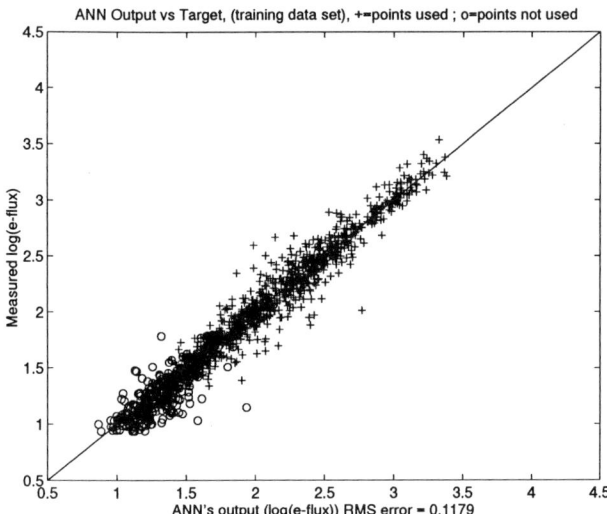

Figure 6. Scatter plot comparing ANN output with measured LF+1 in the training data set. Points shown with open circles were not used for training. RMS error was computed for Log(E-flux) greater than 1.9 to emphasize ANN performance near the peak fluxes.

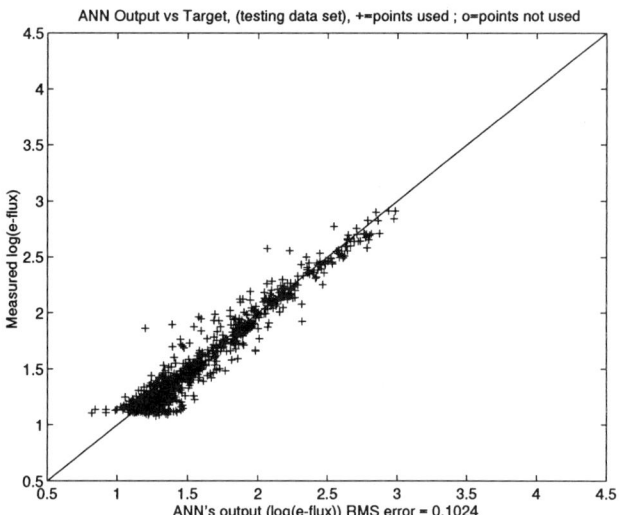

Figure 7. Scatter plot comparing ANN output with measured LF+1 in the testing data set. None of these data were used for training the ANN.

error surface. Figure 4 shows how f is adaptively adjusted by the LM algorithm and tends to increase as the local minimum of the error surface is found. The reproduction efficiency (RE) in the training set and the prediction efficiency (PE) in the testing set vary as shown in Figure 5. Again the LM algorithm leads to rapid convergence to the final value for RE and PE. There is little evidence that overtraining is a problem as indicated by the lack of a noteworthy relative maximum in the PE curve.

ANN performance is affected by the choice of several other variables. The initial weights and biases are selected by MAT-LAB from a pseudo-random number generator. The same pattern of numbers occurs each time a new MATLAB session is started. Sometimes these weights and biases evolve with training produce noteworthy ANN performance. Often the results are less desirable. Thus one is forced to rerun the training algorithm many times and save the good results. The number of neurons in the hidden layer also affects the training time and the ultimate performance of the network. Our training strategy varied the number of hidden layer neurons from 7 to 11. There is no way to separate the effects of these two variables on ANN performance nor is there a logical method for pre-selecting best values.

4. TRAINING/TESTING RESULTS

The scatter plot in Figure 6 compares measured LF+1 values (the target vector) with the trained ANN output. The scatter plot in Figure 7 illustrates the typical relationship between the output of a fully trained ANN and the testing target vector. Figure 8 shows the same information plotted as a time sequence. RMS error was computed only for Log(E-flux) values greater than 1.9 to emphasize ANN behavior near the peak fluxes. RMS errors from 0.1 to 0.2 were common.

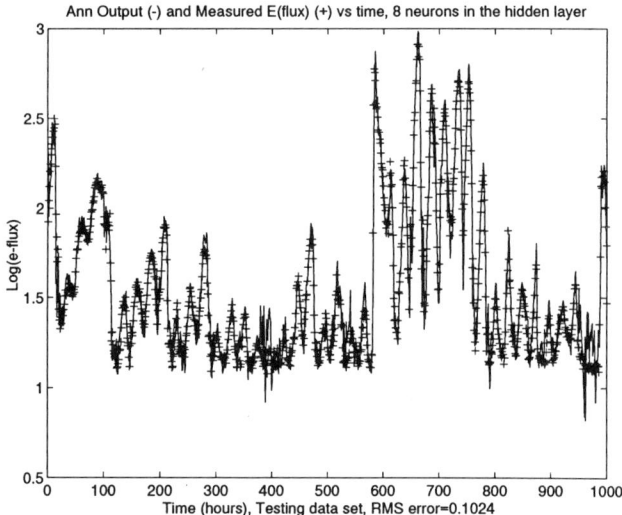

Figure 8. Time series plot comparing ANN output (solid curve) with measured values (+) for LF+1 in the testing data set. Lowest RMS error was obtained with 8 hidden layer neurons.

While these initial results appear quite good it is fair to note that the autocorrelation of Log(flux) at a lag of one hour is about the same as the values reached for RE and PE shown in Figure 5. This would seem to indicate that the ANN is not much better than linear extrapolation one hour into the future. Indeed, analysis of the relative importance of the four different types of input parameters on the prediction of LF+1 shows that, to first order, the output depends only on the previous history of LF. This, of course, isn't the complete story since Figures 7 and 8 clearly show many cases where the turn around of LF+1 near a peak is accurately predicted by the ANN immediately following a four hour monotonic rise. This could happen via a second order effect involving one or more of the other input variables. One possible indication that this does occur is the large bias values in the hidden layer. Because of their logsig transfer functions, large values of b_1 will inhibit the neurons ability to change its output until a very large input is applied. Although large biases are seen it is not yet clear how these are connected to each of the individual training parameters.

5. FUTURE PLANS

Many interesting avenues for further study of ANN models of geosynchronous electrons are now within reach. Electron data from GMS-4 and the LANL spacecraft will soon be added to the CDF database. This should provide concurrent data at six different MLT positions that can be used for training and testing. Rather than having a single output at a single location in MLT the ANN can then have multiple outputs trained for each position. With the use of upstream data from the solar wind using IMP-8 and WIND, the ANN model should be able to produce accurate forecasts 2-3 hours ahead. If ANNs can be taught to use pattern recognition strategies with YOHKOH or SOHO images concentrating near the West limb of the sun along with the GOES X-ray flux this should extend the prediction times still further. If ANNs can be taught to interpolate between satellite locations then a position/time matrix covering all 24 hours of MLT and times from the present to 3-4 hours into the future will be possible.

Acknowledgements. The support of the Murdock Charitable Trust, NASA grant NAG8-222 and the Oregon Space Grant Consortium is gratefully acknowledged.

REFERENCES

Baker, D.N., Belian, R.D., Higbie, P.R., Klebesadel, R.W. and Blake, J.B., Hostile Energetic Particle Radiation Environments in Earth's Outer Magnetosphere, AGARD Symposium, The Hague, Netherlands, 2-6 June 1986.

Demuth, H. and Beale, M., *Neural Network TOOLBOX for Use with MATLAB*, The Math Works, Inc., 1994.

Freeman, J.W., Jr., Storms in Space: A Fictionalized Account of "The Big One", EOS, Transactions, *American Geophysical Union, 74*, 412, 1994.

Hagan, M.T., H.B. Demuth and M. Beale, *Neural Network Design*, pp. 12-19, PWS Publishing, 1996.

Koons, H.C. and Gorney, D.J., A Neural Network Model of the Relativistic Electron Flux at Geosynchronous Orbit, *J. Geophys. Res., 96*, 5549, 1991.

Press, W.H., S.A. Teukolsky, W.T. Vetterling and B.P. Flannery, *Numerical Recipes in C*, p. 683, Cambridge University Press, 1992.

Rostoker, G., Baker, D.N. and Skone, S.H., Correlated Measurements of Relativistic Electrons at SAMPEX With ULF Measurements From the CANOPUS Magnetometer Chain, IUGG XXI General Assembly, Abstracts Week B, B 180, 1995.

Siscoe, G., et al., Developing Service Promises Accurate Space Weather Forecasts in the Future, *EOS, Transactions American Geophysical Union, 75*, 353, 1994.

Stringer, G.A. and McPherron, R.L., Neural Networks and Predictions of Day-Ahead Relativistic Electrons at Geosynchronous Orbit, Proceedings of the International Workshop on Artificial Intelligence Applications in Solar/Terrestrial Physics, Lund, Sweden, 139, 1993.

Wrenn, G.L., Conclusive Evidence for Internal Dielectric Charging Anomalies on Geosynchronous Communications Spacecraft, *J. Spacecraft and Rockets, 32*, 514, 1995.

G.A. Stringer, I. Heuten, C. Salazar and B. Stokes, Department of Physics, Southern Oregon State College, Ashland, OR 97520, USA

ESA Update of AE-8 Using CRRES Data and a Neural Network

A.L. Vampola

Space Environmental Effects, Vista, California

Under European Space Agency auspices, 100 to 1700 keV electron data from the Medium Electrons A spectrometer on CRRES is being used to update AE-8 beyond $L = 3$. The data are analyzed and then used to train a neural network to predict energetic fluxes using the K_p magnetic field index. After the network is trained, the entire K_p index from 1932 to 1993 is processed to obtain daily average fluxes at various energies and L values. Specific products will include long-term averages of fluxes, peak flux vs. energy and L, and flux vs. frequency-of-occurrence tables. The daily average data show that AE-8 has excessive fluxes at high energy at the highest L values.

1. INTRODUCTION

Historically, the outer zone energetic electron environment has been very poorly known due to a lack of measurements in the outer zone. This does not seem to have led to loss of satellite systems through dose damage, but satellite operations have been disrupted and some systems have been damaged by the energetic electrons through a mechanism known as "thick dielectric charging" [*Meulenberg*, 1978]. In this process, energetic electrons embed in dielectrics in circuit boards and cables, producing potentials in excess of the breakdown potential in the dielectric. The subsequent discharge can act as a spurious signal or in some cases can damage circuits. This mechanism can be prevented from occurring if the charging current due to the energetic electrons is prevented from exceeding the maximum discharge current from the dielectric. But in order to determine the amount of shielding required to reduce charging currents to safe levels, one must know what maximum energetic electron flux may occur. A table of flux vs. frequency-of-occurrence at various energies is needed in order to perform shielding trade-off studies. One may decide to accept an occasional non-damaging discharge in return for limiting the amount of shielding that is used.

Radiation Belts: Models and Standards
Geophysical Monograph 97
Copyright 1996 by the American Geophysical Union

2. AE-8

The best model of the energetic electron environment that is presently available is the AE-8 model issued by the National Space Science Data Center in 1983 [*Vette*, 1991]. It is a static average model, with versions for solar maximum and solar minimum. In the inner zone, it is an adequate representation, although it does predict excessive fluxes of electrons at high energy [*Vampola*, 1993; *Abel et al.*, 1994]. Because of the intense energetic proton fluxes in the same region, the error in the high energy portion of the electron spectrum is immaterial to almost all missions. However, in the outer zone, the picture is far different.

The AE-8 environment is an update of the old AE-4 model using geosynchronous data from ATS 1, 5, and 6 and extreme extrapolations up the field line of low altitude data from OV3-3, OV1-19, and AZUR. The AE-4 environment had been generated using a number of data sets, but those had been obtained by threshold detectors (with the exception of the OGO-1 and OGO-3 spectrometers). The OGO instruments (five-channel magnetic electron spectrometers similar in concept to the OV3-3, OV1-19 and CRRES spectrometers) provided highly reliable data, but due to the approach used (a pulsed electromagnet), the live-time was less than 1% and the data statistics were very poor. The other instruments had the typical problems of electron threshold detectors: very inaccurate energy determination and poor knowledge of detection efficiency. The AE-4 model was known to be inaccurate, but it was the best available.

Above 2 MeV the AE-8 model is not based on reliable data. AZUR had a single threshold above this energy, but the

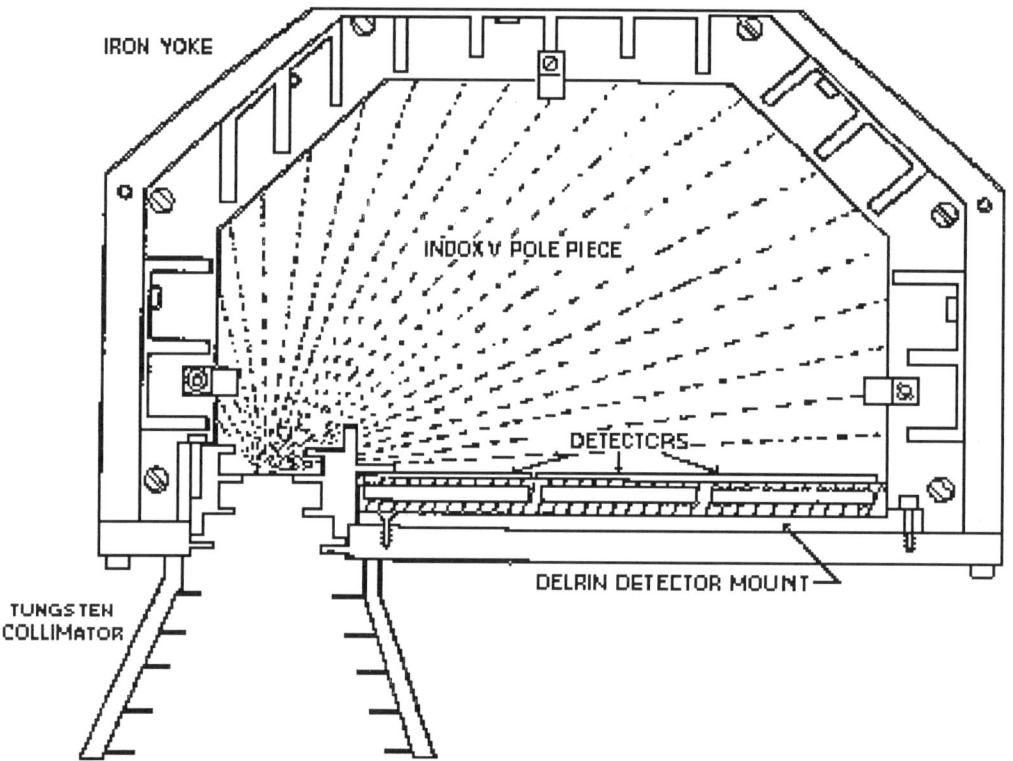

Figure 1. Layout of the CRRES MEA Magnetic Electron Spectrometer. Electrons incident through the collimator are bent through 180° by an 800 gauss transverse magnetic field and focused upon an 18-element ion-implanted silicon detector array. Both the external and internal collimators are tungsten. The detector element closest to the collimator is the background detector. Ridges on walls and the dashed lines indicated on the pole pieces are aluminum anti-scatter devices. This geometry insures that almost all off-angle electrons incident through the aperture are absorbed by a perpendicular wall rather than scattering forward off of a parallel wall. The use of aluminum minimizes bremsstrahlung generation.

efficiency as a function of energy was very poorly known. The OV1-19 had channels every 300 keV up to 5.1 MeV, but the raw data had been re-analyzed without consultation with the people who designed, calibrated, and flew the detector. Those doing this re-analysis mis-identified some of the channels, were not aware of calibration data taken with the instrument (including background responses), and used totally arbitrary "background corrections" to reduce the high energy fluxes to values with which they felt comfortable. The AE-8 model has been criticized for its deficiency in high energy electrons [*Baker et al.*, 1986]. In fact, above 2 MeV, the model is just an extrapolation of unknown validity. Several groups in the US and in Europe have been looking at the deficiencies in AE-8, but none is engaged in updating it.

The inaccuracy in AE-8 in the outer zone at high energy has not had serious effects on the longevity of satellites due to dose damage for the following reasons: on-orbit experience is used to guide the design of solar arrays; and, annealing of dose damage is not usually taken into account in the design of shielding for the electronics (which results in sufficiently robust shielding designs that they overcome the inaccuracies in AE-8). However, the lack of a proper description of the high energy electron environment continues to plague spacecraft in geosynchronous and other outer-zone orbits. When flux intensities are high, the buildup of embedded charge in dielectrics due to the high energy electrons, with subsequent breakdown, produces spurious operation of many spacecraft. These same high energy electrons produce noise in sensor and data transmission systems. Since the AE-8 model is an averaged model, it does not address the high flux rates which occasionally occur and which are the cause of these problems. One publication [*Vampola*, 1987] did address these maximal flux rates, but used the same spacecraft data sets as were used in generating AE-8. A second publication [*Vampola*, 1995] presents a maximum observed energetic electron spectrum at geosynchronous orbit, but does not provide similar information for other magnetospheric locations. No frequency-of-occurrence information has been published for energetic electrons in the outer zone.

3. THE CRRES MEA SPECTROMETER

Because of the known deficiencies in AE-8, especially at high energy in the outer zone, the United States Air Force funded a mission (designated RADSAT) to measure the particle environment and its effects on electronics in the geosynchronous

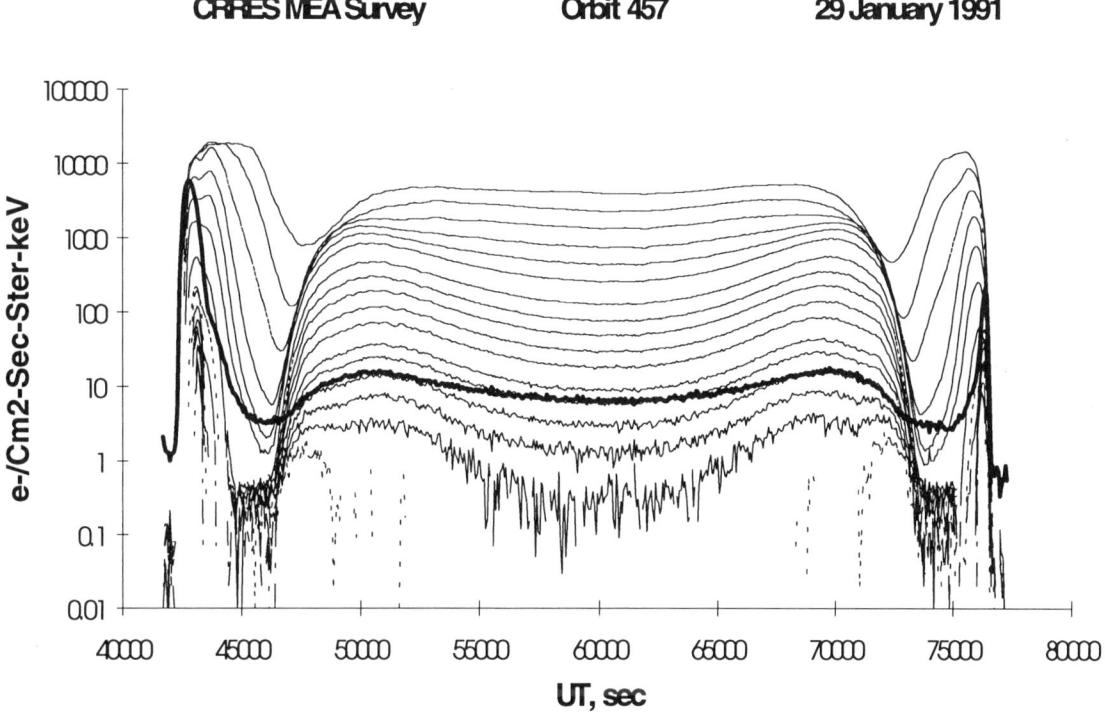

Figure 2. Standard survey plot of the CRRES MEA. The thin lines are the 1-minute averages (118 data samples) of each of the 17 electron detectors (center energies from 153 to 1582 keV). The heavy line is the background monitor counts/second. The satellite orbit starts at perigee beneath the stable trapping region, ascends up through the inner zone, the slot, and into the outer zone to apogee, then returns back through those regions to perigee. The fluxes measured on the downward leg differ from those of the ascending leg because the orbit inclination results in different magnetic latitudes being traversed.

transfer orbit. The RADSAT mission was combined with a NASA chemical release mission and flown on CRRES (Combined Release and Radiation Effects Satellite). The CRRES satellite was redundantly instrumented with ionizing particle detectors with the express intention of using the resulting data to update both the AE-8 and AP-8 models.

CRRES was launched on 25 July 1990 into a highly elliptical orbit (350×33584 km, $18.1°$ inclination) and provided almost continuous data until loss of communications occurred on 11 October 1991 [*Johnson and Kierein*, 1992]. Included in the instrument complement was the modified and refurbished backup instrument for the OV1-19 magnetic focusing electron spectrometer which supplied low altitude data for the AE-8 model. For the CRRES mission, the energy range was changed to 100 keV to 1.7 MeV, since this range is of primary interest to engineering applications and the RADSAT portion of CRRES was directed toward engineering. This instrument was identified as Medium Electrons A (MEA).

The MEA was one of a large set of magnetic-focusing electron spectrometers built in the 1964–1969 period and flown from 1965 (OV2-1, a launch failure) through 1990 (CRRES). Data from some of these were used in constructing the NSSDC environmental models. Magnetic focusing spectrometers use a transverse magnetic field to momentum analyze electrons and focus them upon a detector. Since the energy is known by virtue of the geometry of the instrument and its magnetic field, the energy deposit in the detector can be used to discriminate against various types of background and to enhance the detection efficiency to approximately 100%. Also, by virtue of the fact that each energy interval has its own detector and electronics, the instrument has approximately 100% live time for all energies. The final result is an instrument which provides excellent statistics and has a very good background rejection.

The CRRES MEA had an 18 element ion-implanted silicon detector array at its $180°$ primary focus, Figure 1. Seventeen channels were devoted to electron detection while the 18th detector was devoted to monitoring the penetrating background (cosmic rays, energetic protons, bremsstrahlung). The center energies for the 17 differential energy electron channels varied from 140 keV to 1.54 MeV at approximately 100 keV intervals with minor overlap. Counts from each channel were processed by a dedicated Data Processing Unit [*Koga et al.*, 1992], stored in 12-bit compression registers (9-bit resolution with 17-bit range) and read out each 0.512 seconds. Figure 2 is an example of the survey plots which have been generated from each orbit of data. A more detailed description of the instrument is available in the literature [*Vampola et al.*, 1992].

The MEA performed as-designed during the entire mission. Due to the large amount of high quality calibration data

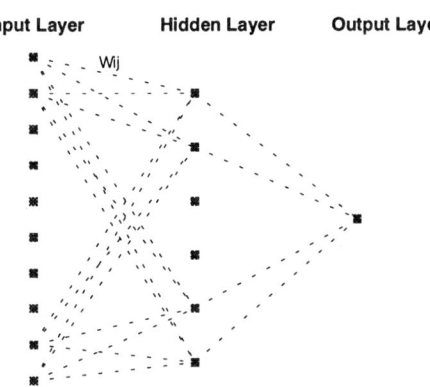

Figure 3. Diagram of a neural network. The input layer represents values of the independent variable. The output layer is the dependent variable. The W_{ij} are elements of the weight matrix. Each neuron in a layer is connected to every neuron in the next layer. Not all of the connections are shown.

obtained in its final flight configuration, including using the flight DPU for data acquisition, the geometric factors, energy responses, and efficiencies of each channel are known with very good precision. The on-orbit background response is well-defined. No failures or degradation in performance occurred in any channel during the entire mission.

4. NEURAL NETWORK ANALYSIS

A neural network, as used in the present context, is a software program which uses a trial-and-error algorithm to predict a relationship between a set of input parameters and a set of output parameters. Typically, three layers of "neurons" are used: an input layer to which the independent variable data set is presented, a "hidden" intermediate layer, and an output layer which consists of one or more "neurons" which acquire the value of the dependent variables (Figure 3). Connections exist between each neuron in one layer and all of the neurons in the next layer. Connection strengths are represented by a weighting matrix, W_{ij}. The output value is thus a combination of weighted values of the input values.

In practice, the network must first be "trained" by presenting it with sets of input values and with outputs which are known to be associated with those input values. The software program then modifies the elements W_{ij} of the weighting matrix, calculates an output, compares it with the known outputs, and then iterates on the W_{ij} values to produce new outputs which differ from the known outputs by a smaller amount than the result of the previous iteration. When the difference is sufficiently small and when usable results are also obtained when other sets of known input and output values are used, the network is considered to be trained. It is then used by presenting it with sets of input data for which the outputs are not known.

The neural network which was used for this study has been used previously [*Koons and Gorney*, 1991; *Koons et al.*, 1994]. The first study, using $\sum K_\mathrm{p}$ as an input variable and the fluence of $> 300\,\mathrm{keV}$ electrons at geosynchronous orbit as the output variable, was successful in predicting the energetic electron fluxes up to one day in advance (within one-half an order of magnitude on a variable with five orders of magnitude variation). This same neural network analysis approach indicates that the current particle models were generated from data obtained during an epoch in which fluxes were only about one-third of their normal value (averaged over the past 62 years).

Up until now, only geosynchronous orbit energetic electron data have been available for such a study. The present study provides the data required to extend such a study to the entire outer zone. In this study, the input is the $\sum K_\mathrm{p}$ magnetic index. The desired output will be the logarithm of the differential J_perp flux intensity. The CRRES MEA data base will be used for training and validating the neural network.

Let us use a relevant example to clarify the procedure: Presume that we have trained the network to provide to us the integral electron flux above 153 keV at $L = 6.0$ using the current day and the previous 9 days of $\sum K_\mathrm{p}$. We then use the trained network to provide the expected output for a number of cases, both typical and atypical. If the predicted outputs from the network are sufficiently close to the actual values obtained from the CRRES MEA data base, we will consider the network trained and then can use it and the K_p index to estimate the integral electron flux above 153 keV at $L = 6$ for any time period since 1932 (the beginning of the K_p index).

In this project, the CRRES MEA data were processed into 428 daily averages of fluxes at 153, 417, 782, 1178, and 1582 keV. The set was divided into two equal subsets, one of which is used to train the neural network and the other is used to check the trained network. The days are randomized within and between the two subsets.

5. PRODUCTS EXPECTED FROM THE STUDY

5.1. *CRRES MEA Data Set*

The following parameters will be derived from the CRRES MEA data:

- Average pitch-angle distributions as a function of L and local time (to enable one to extrapolate equatorial fluxes to positions lower on the field line).
- Mission average flux at 0.25 L intervals from $L = 3.0$ to $L = 8.0$, 17 energies between 153 keV and 1578 keV, to be compared with the AE-8 model.
- Peak observed intensities as a function of energy and L.
- Probability of observing a flux intensity vs. intensity as a function of energy and L.
- Determination of the average energy spectrum (shape) as a function of L.

Additionally, two data bases will be generated which will be made available for use by others:

- Data Base A. This data base will include a record for each passage of CRRES through each of the specified

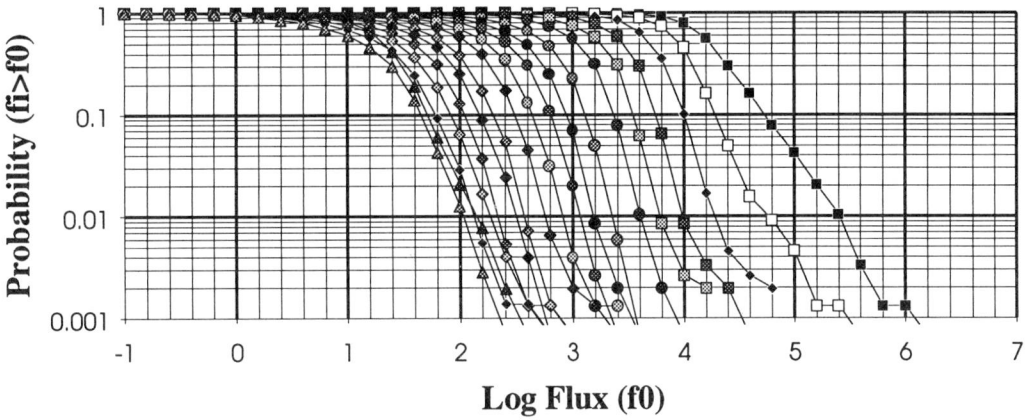

Figure 4. Probability of seeing a flux f greater than a given flux, f_0. Data are mission averages from the 17 channels of the CRRES MEA instrument. The highest intensity curve is for 153 keV electrons, the lowest is for 1582 keV electrons.

L intervals. The records will include time, ephemeris parameters, $D_{\rm st}$, $K_{\rm p}$, and $A_{\rm p}$, and fitting parameters for the average flux spectrum and average pitch-angle distribution.

- Data Base B. This data base will be similar to Data Base A, but will include the actual average fluxes and pitch-angle distributions instead of fitting parameters.

5.2. Neural network study

The products from the neural network study will be:

- A 62-year averaged flux at L intervals from $L = 3$, 4, 5, 6, 6.5, and 7, at five energies from 153 keV to 1578 keV, to supersede the AE-8 model.
- Probability of observing a flux intensity vs. intensity as a function of energy and L.
- Determination of the average energy spectrum as a function of L.
- Peak intensities as a function of energy and L.
- A software program which uses $\sum K_{\rm p}$ to predict the flux as a function of E and L, using the weighting matrices produced in this study.

6. PRELIMINARY RESULTS

6.1. MEA

The background response of each channel was determined for 0.25 L value increments from 3.0 to 8.0 (the previous background correction was averaged over the outer zone $L > 2.8$). The instrumental pitch-angle response of each channel was determined and an algorithm developed for efficiently determining the pitch-angle distribution. Fluxes were averaged over bins of 0.25 L and 1.0 hr in LT. An intermediate data base containing time, magnetic indices, ephemeris parameters, and raw data for each pass through each L, LT bin was generated. This data base was used to generate another data base in which all data were corrected for background and converted to flux values.

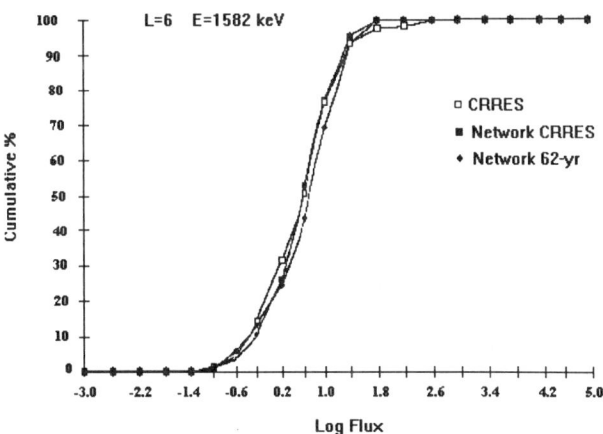

Figure 5. Comparison of CRRES MEA frequency-of-distribution with the neural network prediction for CRRES and for the period 1932 to 1993.

This data base was then used to determine average fluxes as a function of L and energy (integrating over LT). It was also used to generate a list of average pitch-angle distributions at each L. Equatorial fluxes were obtained by extrapolating the local pitch-angle distribution back to the equator. The average pitch-angle coefficients were used to transform local

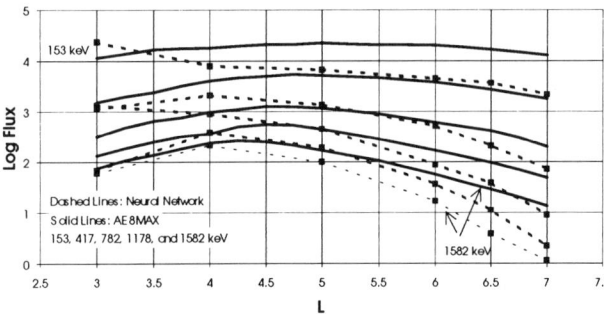

Figure 6. Comparison of the neural network 62-year average fluxes with AE-8 MAX. Equatorial unidirectional differential fluxes at five energies are plotted.

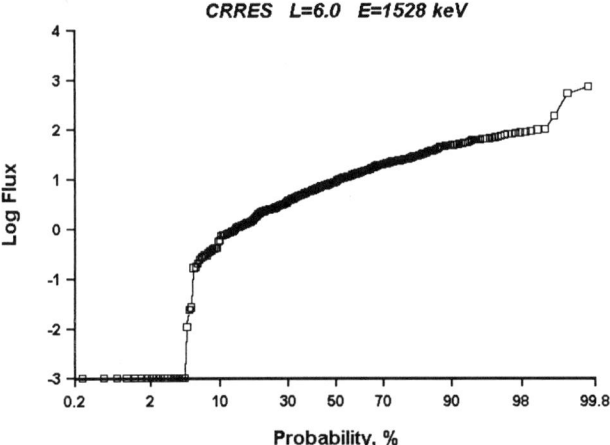

Figure 7. CRRES MEA daily average fluxes at $L = 6$, $E = 1582$ keV plotted on a probability scale. The three highest points do not fit into a smooth distribution with the rest of the points.

fluxes to equatorially-mirroring flux equivalents when the local distribution was a "butterfly" distribution. Probability distributions of flux intensities were generated for all energy channels at 0.25 L intervals from 3.0 to 8.0 (Figure 4). Typically, about 1500 flux values were used in generating each curve.

6.2. Neural network analysis

A commercial program, Professional Version of Brainmaker for Windows, was used as the neural network. Results are very encouraging. The daily average flux data were merged with $\sum K_{\rm p}$, $\sum A_{\rm p}$, and $\sum D_{\rm st}$ and divided into two approximately equal sets (one for training and the other for testing), each of which covers the entire 16-month period. $\sum A_{\rm p}$ and $\sum D_{\rm st}$ did not prove to be useful indices. The ten daily $\sum K_{\rm p}$ prior to and including the day of the data were used as the inputs to the network. A single energy channel flux value is used as the target output. An idiosyncrasy of the network is that while the residuals from the training set continue to decrease, the residuals from the test set first decrease, then increase. This "over training" appears to be the result of the network eliminating the highest and lowest values in the training set and patterning itself to the specific cases in the set. To avoid over training, we use the minimum in the test set residuals to select the optimum network.

Figure 5 compares the network results with the actual data at $L = 6.0$ for $E = 1582$ keV. The parameters presented are the cumulative probability for observing a flux exceeding the specified flux. We actually use the logarithm of the flux. The open squares are the actual distribution observed in the 428 days of CRRES MEA data. The solid squares are the prediction of the trained network for the same days using $\sum K_{\rm p}$ as input (only half of these days were used to train the network). The third curve, solid diamonds, is the output of the neural network for the $K_{\rm p}$ historical base from 1932 to 1993. Inspection of this chart shows that the network does an excellent job of predicting the fluxes observed by the CRRES MEA except at the very highest flux levels, where the cumulative probability prediction is half an order of magnitude lower than the measured distribution. We will discuss this discrepancy in the next section. Figure 5 also shows that the CRRES mission saw, on average, about a factor of two more flux than the 62-year average. The last 6 months of the CRRES mission was characterized by high geomagnetic activity.

Figure 6 compares the neural network 62-year-averaged fluxes with AE-8 MAX. AE-8 MIN is identical to AE-8 MAX at high energies and high L values. The output of the neural network is the logarithm of the unidirectional, differential flux at the magnetic equator, $B/B_0 = 1$. To obtain a similar parameter for AE-8 MAX, we obtained the integral flux at $B/B_0 = 1$ for energy thresholds 10% below and above the desired energy, differenced them, and averaged over the energy interval. To transform to unidirectional fluxes, we divided the omnidirectional flux by 3.5π. The figure shows that AE-8 MAX is about an order of magnitude too high at high energy in the outer region of the outer zone. This explains why spacecraft in geosynchronous orbit which have been designed to the AE-8 model have not had failures from dose damage.

7. UNSOLVED PROBLEM

Figure 5 showed that the network did an excellent job of predicting the MEA fluxes, except at the highest flux levels. The final plot, Figure 7, shows the distribution of CRRES flux intensities, again for $L = 6.0$, $E = 1582$, plotted on a probability scale. We see three high points (1% of the total) which are not part of the remainder of the distribution. These anomalously high values are real, but the neural network cannot handle them. It always discards them as not being valid data points unless the limit on residuals is set so high that the trained network performs very poorly with the test set. This doesn't significantly affect averages, since only 1% of the points are involved and they are less than a order of magnitude above the average. But omitting them will affect the prediction of high flux values. The highest flux values are the ones that cause anomalies on satellites, especially those due to thick dielectric charging. We are investigating methods around this problem. Most of the distributions (from the various energies and L values used in this study) do not exhibit these non-normally-distributed fluxes.

Acknowledgements. Selection of parameters for and training of the neural networks was done by H.C. Koons. This study is funded by the European Space Agency through the Mathematics and Software Division, ESTEC.

REFERENCES

Abel, R., R.M. Thorne and A.L. Vampola, Solar Cycle Behavior of Trapped Energetic Electrons in Earth's Inner Radiation Belt, *J. Geophys. Res., 99*, 19,427, 1994.

Baker, D.N., R.D. Belian, P.R. Higbie, R.W. Klebesadel and J.B. Blake, Hostile Energetic Particle Radiation Environments in Earth's Outer Magnetosphere, in The Aerospace Environment at High Altitudes and its Implications for Spacecraft Charging and Communications, AGARD CP 406, pp. 4–1, 1986.

Johnson, M.H. and J. Kierein, Combined Release and Radiation Effects Satellite (CRRES): Spacecraft and Mission, *J. Spacecraft and Rockets, 29*, 556–563, 1992.

Koga, R., S.S. Imamoto, N. Katz and S.D. Pinkerton, Data Processing Units for Eight Magnetospheric Particle and Field Sensors, *J. Spacecraft and Rockets, 29*, 574–579, 1992.

Koons, H.C. and D.J. Gorney, A Neural Network Model of the Relativistic Electron Flux at Geosynchronous Orbit, *J. Geophys. Res., 96*, 5549–5556, 1991.

Koons, H.C., D.J. Gorney and J.B. Blake, The Long Term Variability of the Electron Radiation Dose in Geosynchronous Orbit, *J. Spacecraft and Rockets, 31*, 557–561, 1994.

Meulenberg, A., Jr., Evidence for a new Discharge Mechanism for Dielectrics in Plasma, in Spacecraft Charging by Magnetospheric Plasmas, *AIAA Progress Series, Vol. 47*, ed. A. Rosen, pp. 236–247, 1976.

Vampola, A. L., Thick Dielectric Charging on High-Altitude Spacecraft, *J. Electrostatics, 20*, 21–30, 1987.

Vampola, A.L., Effects of the March-June 1991 Magnetic Storm Period on Magnetospheric Electrons, in Solar-Terrestrial Predictions-IV, Proceedings of a Workshop at Ottawa, Canada, May 18–22, 1992, Vol. 2, NOAA/SEL, Boulder, CO., pp. 703–711, 1993.

Vampola, A.L., J. Osborn and B. Johnson, The CRRES Magnetic Electron Spectrometer, *J. Spacecraft and Rockets, 29*, 592–594, 1992.

Vampola, A.L., CRRES Medium Electrons A Results, in Proceedings of the Workshop of the Earth's Trapped Particle Environment, Taos, NM, Aug. 1994, G. Reeves, ed., (in press).

Vette, J.I., The AE-8 Trapped Electron Model Environment, NSSDC WDC-A-R&S 91-24, 1991.

A.L. Vampola, Space Environmental Effects, Vista, CA.

The Trapped Radiation Software Package UNIRAD

D. Heynderickx, M. Kruglanski and J. Lemaire

Belgisch Instituut voor Ruimte-Aëronomie/Institut d'Aéronomie Spatiale de Belgique, Brussels, Belgium

E.J. Daly and H.D.R. Evans

ESA/ESTEC, Postbus 299, NL-2200 AG Noordwijk, The Netherlands

UNIRAD is a software package developed by and for ESA to evaluate the radiation fluences and doses expected in a spacecraft from a definition of the mission characteristics. The UNIRAD suite of programs provides information about the radiation environment in an arbitrary Earth orbit, predicting satellite exposures to particle fluxes, the resulting radiation dose, and the resulting damage-equivalent fluences for solar cell degradation calculations. An orbit analysis generally consists of running one or more of the UNIRAD component programs, with communication between the programs via interface files. Both graphical and tabular output is provided. The maintenance and distribution of UNIRAD is being handled by BIRA/IASB. The software has already been installed on a variety of platforms and operating systems.

1. INTRODUCTION

In the course of a series of contracts for ESA/ESTEC, the *Belgisch Instituut voor Ruimte-Aëronomie/Institut d'Aéronomie Spatiale de Belgique* (BIRA/IASB) has developed a software package, UNIRAD, for integrated analyses of the effects of the space environment on satellite missions.

The UNIRAD package consists of the following programs:

SAPRE is an orbit generator which produces a data file of orbit ephemeris information used by the next programs in the package.

BLXTRA calculates the geomagnetic coordinates (B, L) from the geographic coordinates generated by SAPRE, using a choice of all the common and current magnetic field models.

TREP determines the radiation flux for the geographic coordinates generated by SAPRE from the NASA trapped radiation models AP-8 and AE-8 [*Vette*, 1991] and determines the solar proton flux over the mission with the models of *King* [1974] and of *Feynman and Gabriel* [1990]. It produces a data file with the energy spectra of trapped protons and electrons and of solar protons.

TREPPOS calculates the trapped radiation flux for pairs of (B, L) or $(B/B_0, L)$ coordinates interactively input by the user. It produces a data file with the energy spectra of trapped protons and electrons.

TREPAVE averages the spectra generated by TREP for different orbits.

ANISO transforms the trapped proton omnidirectional integral flux produced by TREP into unidirectional integral and differential fluxes, taking into account pitch angle and azimuthal dependence [*Kruglanski*, 1996]. The user can define a set of look directions with respect to a satellite reference frame. The resulting fluxes are averaged over the orbit.

ANISOPOS provides the angular distribution (i.e. pitch angle and azimuthal dependence) of the unidirectional integral or differential flux at a given geographic location [*Kruglanski*, 1996].

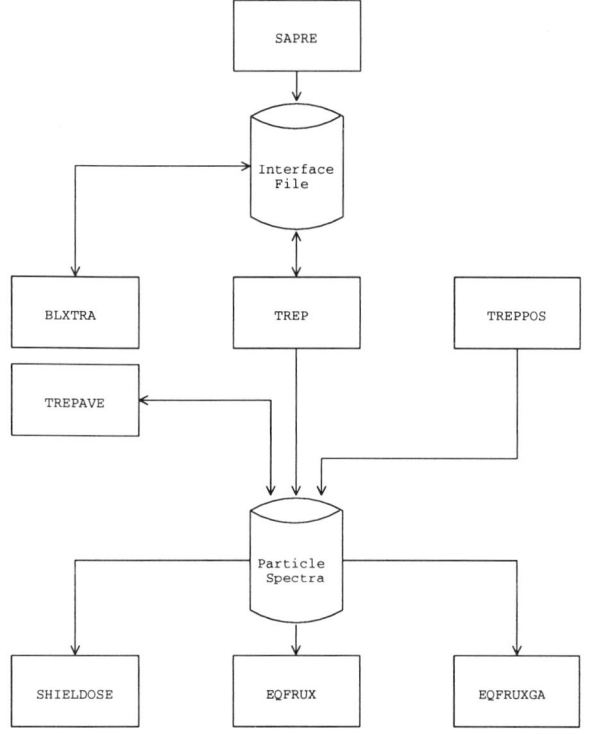

Figure 1. Flow diagram of UNIRAD

SHIELDOSE reads the energy spectra resulting from TREP, TREPAVE or TREPPOS and converts them to radiation dose-depth curves for different detector materials and simple shielding geometries [*Seltzer*, 1979, 1980].

EQFRUX determines 1 MeV electron damage equivalent fluences from the TREP spectra to evaluate degradation of Si solar cells [*Tada*, 1982].

EQFRUXGA idem as EQFRUX, but for GaAs solar cells.

A set of IDL routines to produce graphical output is provided as well.

The flow diagram of the UNIRAD package is represented in Figure 1. This diagram illustrates the interdependence of the various programs making up UNIRAD. Except for the interface files, the output files are not shown in Figure 1.

2. INPUTS AND OUTPUTS

A complete UNIRAD radiation analysis requires only one user input file: the namelist file PROJECT.NML, where PROJECT represents the project name to be chosen by the user. The namelist file contains the orbit parameters, solar activity conditions, plotting and printing options, ... All parameters in the namelist file are assigned default values when not specified, and are reset for each project. NAMELISTs for more than one project may be put in the same file. The general syntax rules for NAMELIST input follow.

Namelist file input consists of a record delimited by the dollar sign $ (except on PC, where an ampersand & should

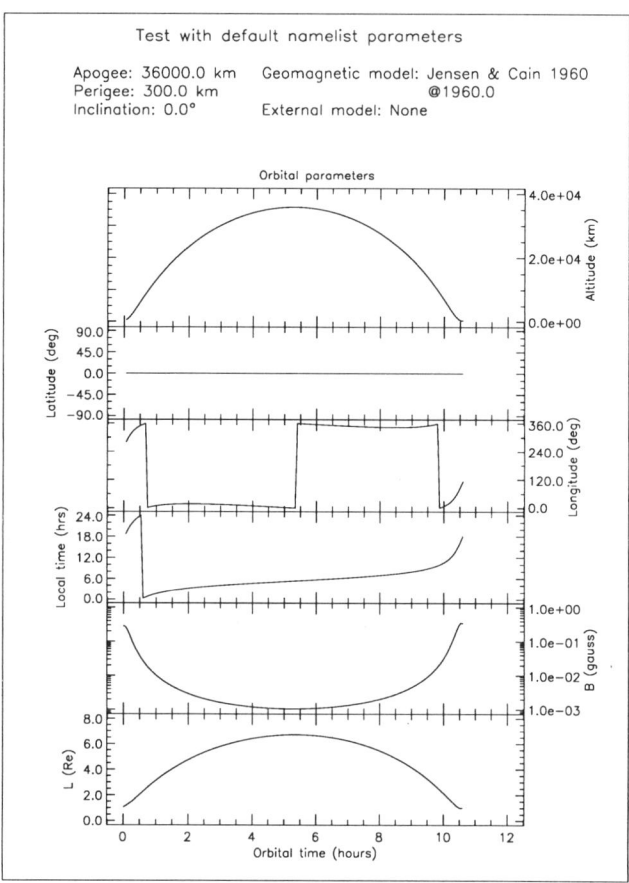

Figure 2. Geographic and magnetic coordinates for the sample orbit

be used) which starts in the second column (the first column must be blank).

Generally, namelist input has the form:

$NAME
  PARAMETER=VALUE   [, PARAMETER=VALUE, ...]
$[END]

where

- $ (or & on PC) is the special dollar sign symbol that indicates the beginning and end of input and the start of a namelist section.
- NAME is the name of the namelist file section.
- PARAMETER is the name of one of the input parameters of the program for which the namelist file provides the data. The parameter list does not have to be exhaustive, i.e. not all parameters have to be given.
- VALUE is a constant or list of constants.
- END is an optional part of the last delimiter. On PCs, a namelist should be terminated with a slash /.

A sample namelist is given in Section 4.

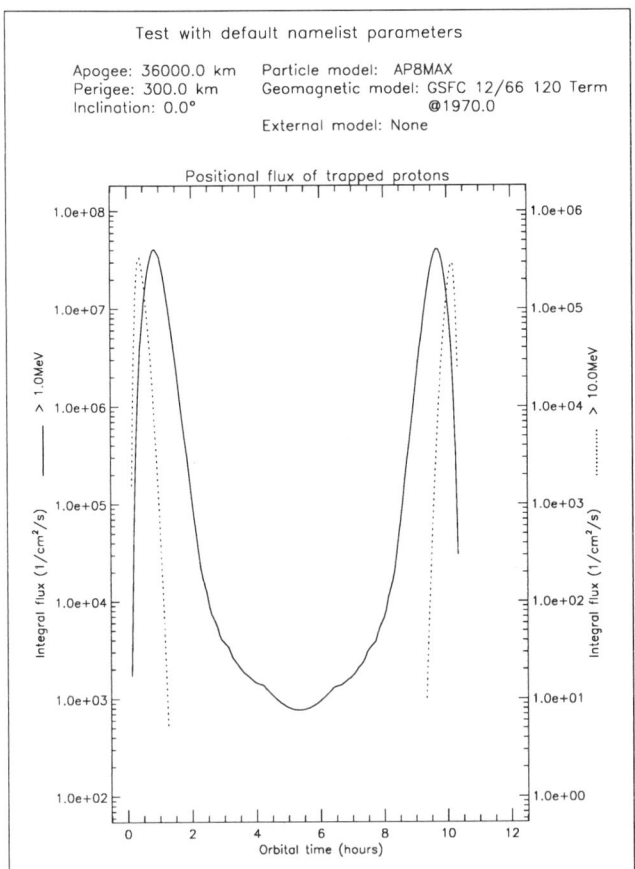

Figure 3. Integral trapped proton fluxes above 1 and 10 MeV for the sample orbit

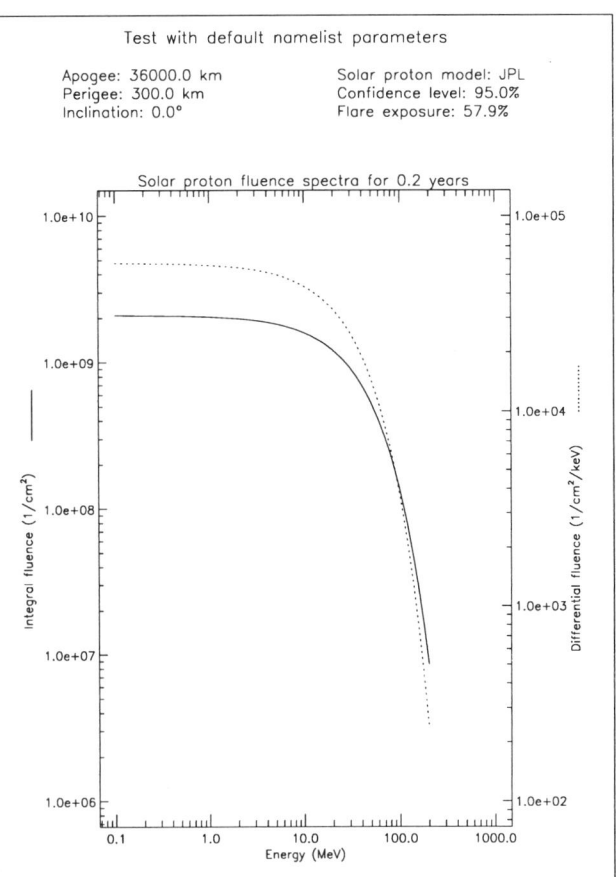

Figure 4. Integral and differential solar proton spectrum for the sample orbit

For users not familiar with the namelist construction, the interface might seem a little tedious. However, as the example in Section 4 shows, the input required to run the models is kept to a minimum, especially through the use of default values for all parameters. On the other hand, the user has full control over the models by "activating" the additional parameters for which usually default values are sufficient. In all, setting up a run with UNIRAD requires the creation and editing of only one input file, and the typing of one command per model. Another powerful feature of the namelist interface is that multiple trajectories can be specified in one parameter file with the possibility of combining the trapped particle fluxes obtained with different models (e.g. for different phases of solar activity) prior to dose calculations.

From the orbit parameters, the system will generate a detailed trajectory, magnetic coordinates, integral and differential proton and electron fluences, doses for three shield geometries in four detector materials, and solar cell degradation information, in both printed and graphical form.

If the programmes are run in the proper order, all successively needed input files are generated by UNIRAD. Alternatively, the user may supply his own input files according to the specifications given in the user manual [*Heynderickx et al.*, 1996].

The output generated by UNIRAD consists of ASCII files with file names of the form PROJECT.XXX, where XXX identifies the program generating the file and the type of information in the file. PROJECT must be chosen so that PROJECT.XXX represents a valid file name.

3. THE PLOTTING PROGRAM

The IDL routine UNIRAD.PRO is a menu driven plotting program that provides a graphical representation of the various output files produced by UNIRAD. The plots can be shown on the screen or sent to files in PostScript format.

UNIRAD.PRO will check which UNIRAD output files are available in the current directory, and present a menu with options. After making a selection from the main menu, other menus will appear depending on the UNIRAD output files available in the current directory. When a selection is made from the menu, UNIRAD.PRO produces a plot on the screen. The user then has the option to produce a PostScript version of this plot by making the appropriate menu selection. The program creates PostScript files in the current directory.

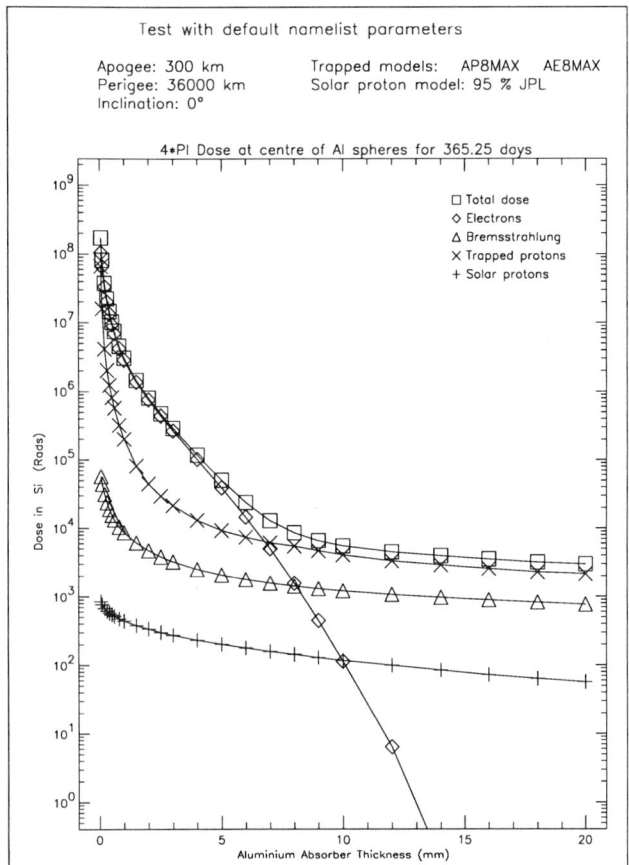

Figure 5. Dose in Si at the centre of Al spheres for the sample orbit.

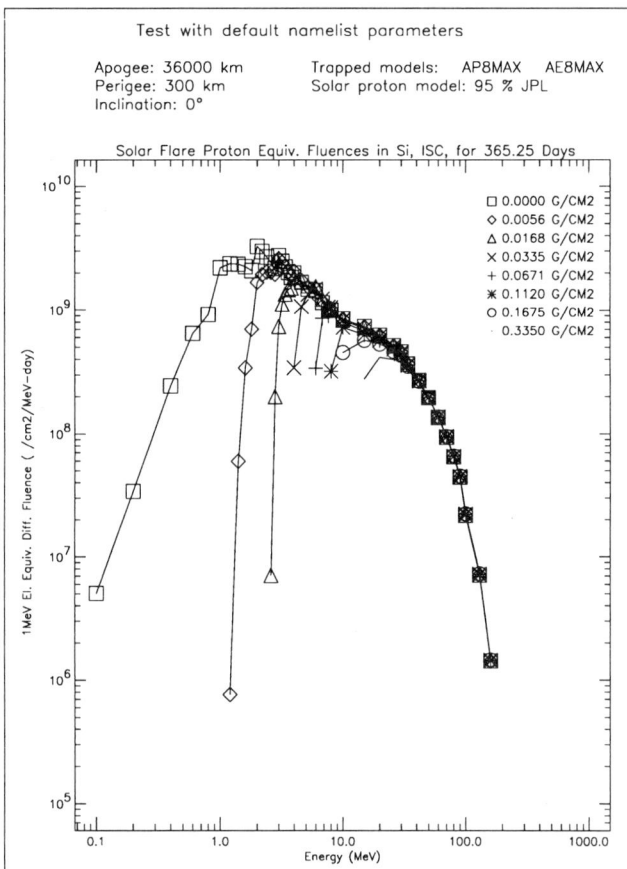

Figure 6. Solar flare proton equivalent fluences (ISC) for the sample orbit. The symbols correspond to the glass cover thicknesses listed in the graph.

4. A SAMPLE RUN

In this section, we present the output of a sample run of UNIRAD with the default values for the namelist parameters. The default orbit is a geostationary transfer orbit during conditions of solar maximum. The plots presented in this section were produced with UNIRAD.PRO.

The following namelist file was used to generate the sample run:

```
$SAPRE
  TITLE = 'Sample run with default
           parameter values'
  IAE    = 0
  HAPO   = 36000.0D0
  HPER   = 300.0D0
  RINCL  = 0.0D0
  ORBITS = 1.0D0
$END
$TREP
  SOLACT = 'MAX'
$END
$SHIELDOSE
  IPLOT = 1
$END
$EQFRUX
  IPLOT  = 2
$END
```

The parameters not in this list were assigned their default values. This illustrates that in general a UNIRAD radiation analysis requires setting only a small number of parameters. A detailed description of all namelist parameters is given in the manual [Heynderickx et al., 1996].

Figure 2 represents the orbit generated for the sample project and the B and L values calculated at each orbital point. The positional trapped proton fluxes are plotted in Figure 3, and the orbit averaged proton fluence spectra are represented in Figure 4.

Figure 5 shows the doses in Si through a spherical Al shield resulting from the trapped proton and electron fluences and the solar proton fluence.

Figure 6 shows the energy dependence of the damage equivalent solar proton fluence in a Si solar panel in ISC mode, for different glass cover thicknesses.

5. DISTRIBUTION

The distribution of UNIRAD is being handled by BIRA/IASB on a commercial basis. The software has already been installed on a variety of platforms and operating systems at different institutes.

The software package has been compiled and tested on the following operating systems: VMS, OSF, UNIX, SUN-OS, MS-DOS and MS-Windows.

6. FUTURE DEVELOPMENTS

BIRA/IASB, in collaboration with Mullard Space Science Laboratory [*Rodgers et al.*, *Johnstone*, these proceedings] and Max Planck Institut für Aeronomie [*Friedel et al.*, these proceedings], is continuing the development of new trapped radiation belt models, and the updating of existing models, for ESA/ESTEC. The products coming out of this effort will be included in future releases of UNIRAD, as well as the output of the analysis of the Radiation Environment Monitor data [*Bühler et al.*, these proceedings].

BIRA/IASB is the main contractor for the ESTEC project SPENVIS (SPace ENVironment Information System) involving the installation on the World Wide Web of the main parts of UNIRAD and additional models of the near Earth environment (including spacecraft charging models and atmospheric and ionospheric models). The full system will be operational by September 1997, but subsystems will be made available as soon as their development allows. In particular, UNIRAD is scheduled for release on the BIRA/IASB server for September 1996.

Acknowledgements. The UNIRAD package was developed under ESA/ESTEC/WMA TRP Contract Nos. 9828/92/NL/FM and 10725/94/NL/JG(SC).

REFERENCES

Feynman, J., and S.A. Gabriel, A new model for calculation and prediction of solar proton fluences, AIAA 90-0292, JPL, Pasadena, 1990.

Heynderickx, D., M. Kruglanski, and J. Lemaire, UNIRAD User Manual, BIRA/IASB, 1996.

King, J.H., Solar Proton Fluences for 1977–1983 Space Missions, *J. Spacecraft and Rockets*, *11*, 401, 1974.

Kruglanski, M., Trapped Proton Anisotropy at Low Altitudes, Technical Note 6 for ESTEC Contract No. 10725/94/JG(SC), 1996.

Seltzer, S.M., Electron, Electron-Bremsstrahlung and Proton Depth-Dose Data for Space Shielding Applications, *IEEE Trans. Nucl. Sci.*, *26*, 4896, 1979.

Seltzer, S.M., A Computer Code for Space Radiation Shielding Methods, NBS Technical Note 116, 1980.

Tada, H.Y., J.R. Carter, B.E. Anspaugh, and R.G. Downing, Solar Cell Radiation Handbook, 3rd edition, NASA JPL 82-69, 1982.

Vette, J.I., The NASA/National Space Science Data Center Trapped Radiation Environment Model Program, NSSDC/WDC-A-R&S 91-29, 1991.

D. Heynderickx, M. Kruglanski, J. Lemaire, Belgisch Instituut voor Ruimte-Aëronomie/Institut d'Aéronomie Spatiale de Belgique, Ringlaan 3, B-1180 Brussels, Belgium (E-mail: d.heynderickx@oma.be, m.kruglanski@oma.be, j.lemaire@oma.be).

E.J. Daly, H.D.R. Evans, ESA/ESTEC, Postbus 299, NL-2200 AG Noordwijk, The Netherlands.

EnviroNET Space Environment Information via the WWW: A Computer Based Demonstration

P.J. Messore

EnviroNET—University Research Foundation

M. Lauriente

NASA/Goddard Space Flight Center

NASA/GSFC's EnviroNET project has taken interactive modelling to the next level by offering its tools via the World Wide Web (WWW). Users of the EnviroNET system can now access all of the features of the EnviroNET computational tools right from their WWW browser without the need for additional software. The entire EnviroNET WWW service can be found at the address `http://envnet.gsfc.nasa.gov/`.

EnviroNET has maintained World Wide Web (WWW) services since October 1994. Over the past year, EnviroNET has been developing each of the features present in its seven-year-old telnet based service for the WWW. In order to completely convert all of EnviroNET's services to the WWW, the computational models available in the EnviroNET telnet service had to be converted to run via WWW. This development was accomplished using the WWW Hypertext Markup Language (HTML) and scriptable programming language PERL [*Wall and Schwartz*, 1992] in conjunction with the original FORTRAN code and the original Interactive Data Language (IDL) graphing routines. Over the past summer, EnviroNET staff has been able to develop seventeen of its computational models for use on the WWW. These models form a complete toolbox of programs that allow spacecraft designers and engineers to effectively model the space environment that will be encountered during the spacecraft's mission.

By making the EnviroNET computational models available over the WWW, this alleviates usability concerns for anyone with access to the Internet. The only software required to run the tools is a form capable WWW browser such as Mosaic or Netscape. The user of a WWW model will be able to:

1. run the model;
2. view on-line help that may contain helpful diagrams or plots;
3. easily navigate through the different model inputs;
4. perform varied parameter runs to generate multiple computations;
5. easily view both tabular and graphical representations of the model output, all from their WWW browser.

Table 1 shows the models available on the EnviroNET WWW Models Home Page.

Each WWW model follows the same format to aid users in becoming more familiar with using all the models. Once a user becomes comfortable with using one model, other models will also be comfortably navigated by the user. The first model screen contains the initial inputs required of the user. An example of this is shown for one of the EnviroNET WWW models in Figure 1. After all of the necessary inputs have been entered, the user will be given the opportunity to perform a single-value run or to vary one of the inputs over a range and perform a vary-parameter run. The buttons to select either option are shown at the bottom of Figure 2. The difference between the two types of runs is as follows: a single-value run will allow the user to run the model for one set of inputs and generate one set of output values and a vary-parameter run allows the user to select multiple input sets in order to generate a complete table of output values with each output set corresponding to an input set. The method of

Radiation Belts: Models and Standards
Geophysical Monograph 97
Copyright 1996 by the American Geophysical Union

Table 1. Computational Models Available via the WWW

Model	Reference
Space Environment Models	
Atomic Oxygen Fluence	*Hedin* [1987]; *NGDC* [1996]; *Schatten and Pesnell* [1996]
Geomagnetic Reference Field Model	International Geomagnetic Reference Field Revision [1987]
International Reference Ionosphere Model (IRI-90)	*Bilitza* [1990]
Marshall Engineering Thermospheric Model	*Johnson and Smith* [1985]; *Hickey* [1988]
Martian Atmosphere Model	*Nier and McElroy* [1977]
Mass Spectrometer Incoherent Scatter Model (MSIS-86)	*Hedin* [1987]
Mass Spectrometer Incoherent Scatter Model-Extended (MSISE-90)	*Hedin* [1987]
Meteoroid Model	*Gruen et al.* [1985]
Orbital Debris Model	*Kessler et al.* [1989]
Orbital Debris Probability of Impact Model	*Kessler et al.* [1989]; *Jung et al.*; *Fish and Summers* [1965]
Orbital Decay Model	*Mueller* [1980]
Solar Flux Data Retrieval Model	*NGDC* [1996]; *Schatten and Pesnell* [1993]
Thermal Analysis Model	*Little* [1976]
Trapped Radiation Model	*Teague et al.* [1979]
Launch Environment Models	
Acoustic Vent Effect Model	*On* [1991]
Payload Fill Effect Model	*Lee et al.* [1992]; *Manning* [1991]
Percentile Value Model	*Lee et al.* [1985]

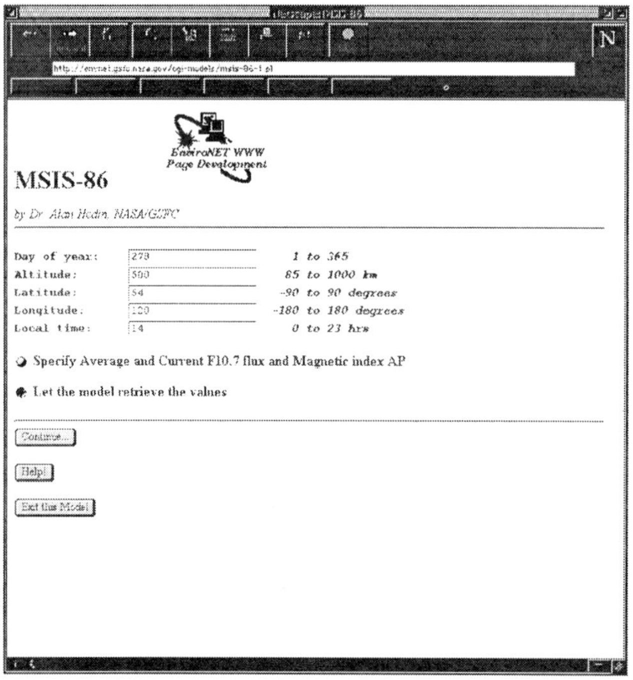

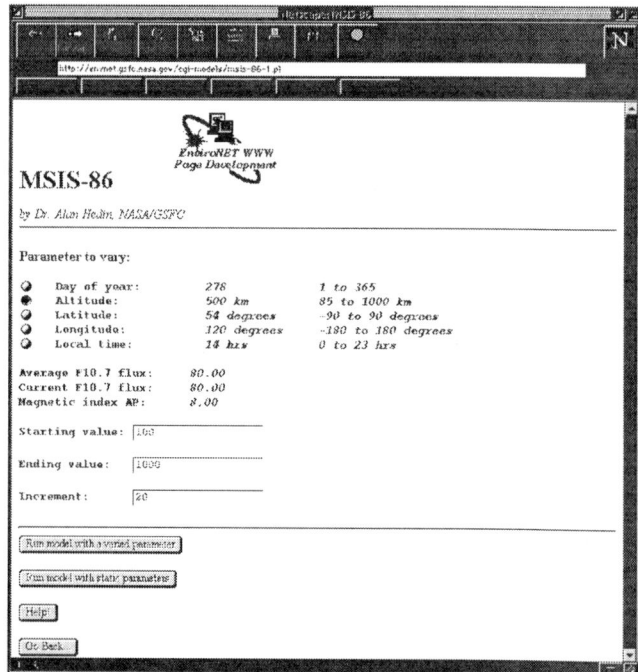

Figure 1. Sample model input entry page

Figure 2. Run option and a vary parameter range sample

Figure 3. Output graph selection screen sample

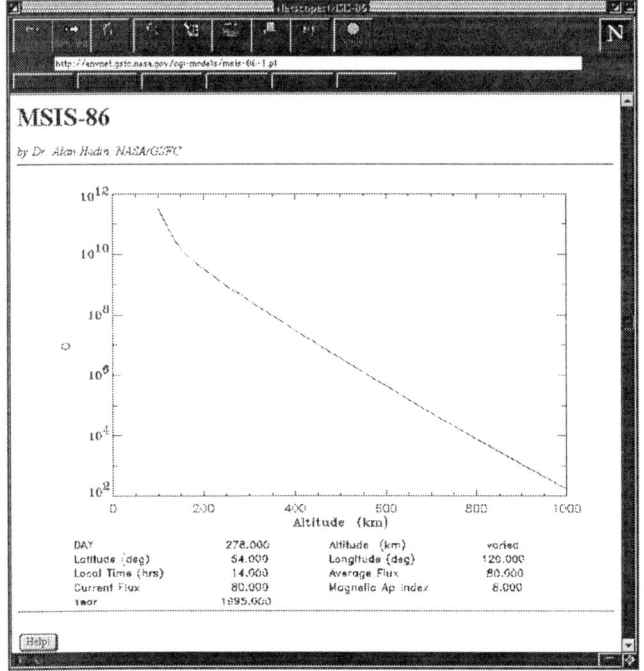

Figure 4. Graphical output sample

selecting which parameter to vary and entering the range for that input is also shown in Figure 2.

Finally, after the model has been run, the output of the model can be viewed by the user. If the user chose to vary a parameter the user will be asked to chose to view the output in a table or as a graph. If the graph option is chosen the user will be able to generate a graph for each of the different outputs computed by the model. Which output the user wants to graph can be selected from the table provided to the user. An example of this is in Figure 3. Also, a sample graph is shown in Figure 4.

The EnviroNET site is located at the following URL: http://envnet.gsfc.nasa.gov/. From there, any of the EnviroNET services are available for use. In order to access the computational models, each user must have his/her own username and password. To request a username and password combination, send an e-mail message to the address techmgr@envnet.gsfc.nasa.gov with the following information: (1) your full name; (2) affiliation; (3) postal and e-mail address; (4) requested username; (5) requested password; and (6) brief description of the purpose for which the models will be used. If users are interested in obtaining a user guide which explains in depth the use of both the EnviroNET telnet and WWW services, either request one via the WWW or send an e-mail message to journ@envnet.gsfc.nasa.gov with your name and mailing address included.

REFERENCES

Bilitza, D., IRI90, *NSSDC Report 90-20*, 1990.

Fish, R.H. and Summers, J.L., The Effects of Material Properties on Threshold Penetration, *Proceedings of the 7th Hypervelocity Impact Symposium, vol. II*, Feb. 1965.

Gruen, E., Zook, H.A., Fechtig, H. and Giese, R.H., Collisional Balance of the Meteoritic Environment, *Icarus 62*, 244-72, 1985.

Hedin, A.E., MSIS-86 Thermospheric Model, *J. Geophys. Res., 92*, 4649-4662, 1987.

Hickey, M.P., An Improvement in the Integration Procedure Used in the Marshall Engineering Thermosphere Model, NASA CR-179389, Washington, D.C., 1988.

International Geomagnetic Reference Field Revision 1987, *J. Geomagn. Geoelectr., 39*, 773-779, 1987 and *EOS Trans., 69*, 559, 1988.

Johnson, D.L. and Smith, R.E., The MSFC/J70 Orbital Atmosphere Model and the Data Bases for the MSFC Solar Activity Prediction Technique, NASA TM-86522, Washington, D.C., 1985.

Jung, L., Falco, P. and Malloy, W., Preliminary Assessment of Effects of Meteoroid and Space Debris Impacts on EOS-A, NASA TM EOS-DN-ENV-032.

Kessler, D.J., Reynolds, R.C. and Anz-Meador, P.D., Orbital Debris Environment for Spacecraft Designed to Operate in Low Earth Orbit, NASA Technical Memorandum TM 100 471, April 1989.

Lee, Y. A., Henricks, W. and Park, D., NASA Contractor Report 177905, September 1985.

Lee, Y.A., Henricks, W. and Woolley, J.P., Payload Fairing Fill Factor Prediction Methodology, NASA Contractor Report 189280, June 1992.

Little, A.D., Thermal Flux Model, Cambridge, MA, NASA/GSFC, 1976.

Manning, J.E., Analysis and Evaluation of the Fill Factor, Cambridge Collaborative, Inc., CC Report 91-6-12104-1, January,

1991.

Mueller, A.C., Decay Model for Use in Space Object Collision Studies, ACM Memorandum No. 250, July 31, 1980.

National Geophysical Data Center, Ottawa 2800 MHz Solar Flux Data, February 1996.

Nier, A.O., and M.B. McElroy, Data from Vikings 1 and 2, Journal of Geophysical Research, September 30, 1977.

On, F., Acoustic Vent Effect Model, Internal NASA Memo, February 12, 1991.

Schatten, K.H. and Pesnell, W.D., An early solar dynamo prediction cycle 23 - cycle 22, *Geophys. Res. Lett., 20*, 2275, 1993.

Teague, M.T., N.J. Schfield, and J.I. Vette, A Study of Inner Zone Electron Data and their Comparison with Trapped Radiation Models, NSSDC/WDC-A-R&S 79-06, 1979.

Wall, L. and R.L. Schwartz, *Programming PERL*, O'Reilly and Associates Inc., March 1992.

P.J. Messore, EnviroNET—University Research Foundation

M. Lauriente, NASA/Goddard Space Flight Center

Radiation Belt Models for the PC: RADMODLS

A.L. Vampola

Space Environmental Effects, P.O. Box 10225, Torrance, California

RADMODLS is a merger of three public domain software packages: ORB, ORP, and SHIELDOSE. ORB is the orbit integrator, ORP accesses the standard NASA magnetospheric particle models such as AE-8 and AP-8, and SHIELDOSE calculates the absorbed dose behind aluminum shields. RADMODLS combines these programs and their subroutines into a single program which can be executed on a PC or clone. The FORTRAN software can also be compiled for use on other platforms. RADMODLS passes data from one program to another so that a complete task, from orbit generation to dose calculation, can be finished in a single run. Both a proton model and an electron model can be used in a single run. The software accesses electron models AE-4 MIN through AE-8 MAX, and all of the versions of AP8—MIN, MAX, MIC, and MAC. A low energy inner zone proton model is also included. The models can also be accessed at discrete B, L locations without an orbital integration. This freeware code is available from the National Space Science Data Center or the author.

1. OVERVIEW

In 1964, J. Vette at The Aerospace Corporation, under contract from NASA, produced the first standardized magnetospheric trapped particle models, AE-1 and AP-1. At the time, the "A" in the designation referred to the corporation. Subsequently, Dr. Vette headed up the National Space Science Data Center at NASA's Goddard Space Flight Center and continued the updating of the radiation belt models there [*Vette*, 1991a]. To avoid confusion, the model designations were retained, but the "A" became a generic "aerospace" designation. Initial models were issued as written reports with graphical data. Later, the models were issued as decks of computer data cards with instructions for writing software to access the data bases. The models were organized in terms of the logarithm of the flux as a function of energy, L, and B/B_0, where B is the local field intensity and B_0 is the equatorial field intensity. Ultimately, NSSDC began issuing software for accessing the data bases and even included orbital integration programs which included routines for expediting the calculation of McIlwain's L parameter [*McIlwain*, 1961].

The most used version of the NSSDC software packages included two major programs, ORB and ORP. ORB was an orbit generator which returned a table of B, L parameters as a function of time along the orbit. To determine the flux as a function of time along the orbit, this table had to be used as an input data file for the second program, ORP, which actually accessed the particle models and returned the flux as a function of time. Within ORP, TRARA1 selects the energy maps and will calculate B_0 if the call was made with B. TRARA2 does linear interpolations in L and B/B_0 to extract the fluxes. Updated versions of these software packages are still in use today. If the user was interested in dose instead of flux, he/she had to use the output of ORP as input to yet another program, such as SHIELDOSE which was issued in 1982. The requirement of successively running several programs was an onerous chore. With the advent of capable personal computers on virtually every desk, there was strong motivation for translating these software packages into PC-compatible versions. While making the conversion, it was natural to merge the packages into a single user-friendly code. Several such conversions are in use, one of them being RADMODLS.

RADMODLS is a merger of three software packages, ORB, ORP, and SHIELDOSE. The versions of ORB and ORP which are used in RADMODLS were issued by NSSDC in 1983 (their latest update, although they trace their origins

Radiation Belts: Models and Standards
Geophysical Monograph 97
Copyright 1996 by the American Geophysical Union

to earlier programs such as MODELS [*Teague et al.*, 1972]). SHIELDOSE [*Seltzer*, 1980] was issued by the National Bureau of Standards. RADMODLS integrates these programs and their subroutines into a single program which can be executed on a PC or clone. RADMODLS passes data from one module to another so that a complete task, from orbit generation to dose calculation, can be finished in a single run. RADMODLS will also access the model maps as a function of L and B or B/B_0 if desired. A major portion of the code in RADMODLS interfaces with the user to obtain inputs from the keyboard or a saved parameter file, to check for errors in keyboard input, and to provide flexibility in the output formats.

2. DESCRIPTIONS OF MAJOR ROUTINES

2.1. *ORB*

ORB Ver. 4.0 issued in 1983 by NSSDC served as the basis for the orbit generator used in RADMODLS. ORB actually has two components for different eccentricities. For $e > 0.1$, the methodology is that of *Brouwer* [1959]. For $e < 0.1$, the methodology is that of *Lyddane* [1963]. B and L are determined by the subroutine INTEL [*Kluge*, 1970]. To expedite calculating L (which otherwise requires integrating along the local field line to the equator) a lookup table is "hardwired" into the subroutine package as a BLOCK DATA statement. The B, L calculation uses a modified version of IGRF 65, Epoch 1970. The modification is the use of a magnetic moment of 0.311653 which is actually a pre-1960 magnetic moment. To be consistent, this hybrid magnetic field model must be used when accessing the particle model maps issued by NSSDC.

2.2. *ORP*

ORP was described in detail by *Teague et al.* [1972]. At the time, computers were very limited in the size of data bases they could reasonably accommodate. For convenience in obtaining a large dynamic range in flux and B/B_0 in a small data module, the maps use the logarithm of the flux with a variable B/B_0 step size. Eighteen electron energies between 40 keV and 7 MeV are tabulated in AE-8 over the L range 1.2 to 11. AP-8 covers the energy range 0.1 to 400 MeV over the L range 1.17 to 7.0. The actual organization of the model maps is discussed by *Teague et al.* [1972] and by *Vette* [1991b]. It is a variable format in which new maps of different lengths can be accommodated. An initial parameter list provides the information needed to decode a map structure.

The actual subroutines in ORP which access the models are called TRARA1 and TRARA2. They do a linear interpolation in energy, B/B_0, and L. The intent in selecting the step sizes for the original map is that they all be sufficiently small that linear interpolation within the map is adequate. The value returned is the logarithm of the integral, omnidirectional flux. Because a logarithm is used, the minimum value of flux must be non-zero. In the NSSDC implementation, the minimum flux obtainable from any map is therefore 1.0 (in units of particles cm^{-2}s^{-1}).

2.3. *SHIELDOSE*

SHIELDOSE [*Seltzer*, 1980] is an energy transport code in which input particles, which may be electrons, protons, and bremsstrahlung, are transmitted through aluminum shielding. The absorbed dose is calculated as a function of shielding depth up to 30 g/cm^2 (the limit of the bremsstrahlung coefficients provided in a supplied data file; in order to speed up processing, dose-depth tables for monoenergetic electrons, protons, and bremsstrahlung are used). Shield thicknesses can be input as mils Al, mm Al, or g/cm^2. Three types of shielding geometry may be used: semi-infinite plane; transmission face of an infinite plane; half-dose at the center of a spherical medium.

In the semi-infinite plane case, the dose is absorbed in a thin detector at a specific depth (d, the shielding thickness) in a semi-infinite plane of aluminum. The radiation is assumed to be isotropic from the front. The radiation may be either flux, which results in dose-rate, or fluence, which provides total dose. The detector material may be aluminum, water, silicon, or silicon dioxide. In the second case, the dose is measured at the transmission face of an infinite aluminum plane of thickness d. The difference between this and the previous case is that in the first case, particles and photons can be reflected back from deeper in the infinite medium and be absorbed by the detector. In the third case, the dose is absorbed at the center of a sphere of radius d. Irradiation is from all directions, but only half the dose is returned as an output.

3. RADMODLS MODIFICATIONS TO THE CODES

The original codes were written in FORTRAN on IBM-360 machines. The IBM FORTRAN compilers and linkers were very flexible in that they kept track of the TYPE of a variable (e.g. REAL*8 or REAL*4, INTEGER*4 or INTEGER*2, etc.) and the coding itself could be quite sloppy. A variable could be one TYPE in one module and still be correctly transferred in a call to another module in which it was TYPEd differently. The linker took care of the problem and the modules would work together properly. Other FORTRAN compilers are not so flexible. The usual result is that erroneous results are returned if there is a mixture of TYPEs between modules for the same variable. The first step in making all of the subroutines in RADMODLS work properly was to ensure that each variable was explicitly TYPEd the same in all modules in which it was used.

3.1. *SHIELDOSE*

The second major modification to the codes for use in RADMODLS was a little more subtle. The PC is a 32-bit machine (as are most work stations and older medium-sized mainframes such as VAXes). The original SHIELDOSE code was constructed on an IBM machine which had a 36-bit single-precision floating point format which could specify values as small as 1.0E-60. The data files distributed with SHIELDOSE (which come in several versions) have values down to 1.0E-60. The INTEL chip in a PC uses a 32-bit single precision floating point format which can specify values down to about 1.0E-45. The math co-processor uses an internal 80-bit format which is capable of representing values down to less than 1.0E-4000 (or greater than 1.0E+4000). When very large or very small values are produced as a result of a calculation, the result cannot be properly represented in the external 32-bit floating-point format. An overflow or underflow results and is flagged. If the absolute value of the exponent is larger than 45 in the INTEL chip, the program then aborts. For VAXes, the problem is even more severe, since they are limited to 1.0E+/-38.

The problem with using very small values was recognized by the author of SHIELDOSE and the original SHIELDOSE code had a call to the IBM system subroutine ERRSET which disabled the underflow error notification (the result was treated as a zero and ignored, which was appropriate). Users on machines other than the IBM mainframe, e.g., VAXes, PC's, etc., must disable the call to ERRSET. If a similar function is not available on the other machine, underflows may interrupt some runs. RADMODLS has a function equivalent to ERRSET to control underflows. In addition, it uses a modified dose data base in which any cross-section value less than 1.0E-27 has been replaced with an arbitrary small value. This will not affect the accuracy of the output, since all dose rates less than 1.0E-9 rads/second are set to 0. (An artificial lower limit of 1 rad per 30 years should not affect any user of RADMODLS. If it does, use the original SHIELDOSE code and data tables issued by NBS instead of RADMODLS.)

There is a second problem with SHIELDOSE: extrapolations in the subroutine SPOL. The SPOL problem is as follows: The dose tables that come with SHIELDOSE cover the range 0.1 MeV to 10 MeV for electrons, .02 to 20 MeV for bremsstrahlung, and 2 to 5000 MeV for protons. Interpolations and extrapolations are done with a subroutine SPOL which performs a cubic spline interpolation within the range of the input function and a "parabolic runout" extrapolation at the lower and upper ends. SPOL takes an x, y input function with n elements and interpolates or extrapolates a value y' for an input value x'. No limit is put on x'.

The author warned that the extrapolation may be inaccurate. In fact, for significant extrapolations, it is unstable. Very small extrapolations are valid, modest extrapolations cause errors in the dose output, major extrapolations cause the program to abort with an overflow in the numeric calculations. Definitions of "modest" and "major" depend on the behavior of the input spectrum. Extrapolations as small as 10% beyond the high energy end of the spectrum have caused the program to abort. Erroneous results are obtained for smaller extrapolations which do not cause an abort.

Users of the code who have energy spectra that are more limited than the SHIELDOSE range tables might use significant extrapolations, thinking that the extrapolation will produce a more accurate dose calculation. For very limited extrapolations, they are correct. However, if the extrapolation extends into the region where SPOL becomes unstable, the output is invalid. The point at which the extrapolation becomes unstable depends on the input x, y function. To reduce the potential for serious errors from this problem in the original code, the default in RADMODLS is "no extrapolation." This can be changed by the user, but caution must be used in selecting the limits of the extrapolation. Extrapolations of 10% to 20% may be OK, but no guarantees are made. Normally, the dose calculation should be limited to the range of particle energies available from the models. All previous calculations made on other machines using SHIELDOSE which extrapolated the spectrum must be regarded as suspect if they used the original version of SPOL. The problem with the erroneous "parabolic runout" extrapolation in SPOL is that large deviations from a realistic extrapolation can occur but will be "transparent to the user" unless they cause the run to abort. Any outputs in which the dose starts to increase anomalously at energies below or above the input energy limits is probably in error.

3.2. Solar protons

RADMODLS permits the user to add solar protons to the dose calculation (the user must provide the averaged solar proton spectrum to the program). We do not recommend making use of this feature because it does not provide much accuracy. It does not provide for a time-dependent flux, nor does it use a proper geomagnetic cutoff. Since valid geomagnetic shielding calculations are beyond the capability of a small program like RADMODLS, the expedient of using L as a cutoff is used. The solar proton spectrum is given full value above an L-value selected by the user, zero below a second L-value selected by the user, and linearly decreases the solar proton flux between the two values.

3.3. Low energy protons

The data base that was used to generate AP-8 (and its predecessors) contained data from only one channel with an energy response below a few MeV. AZUR-4 was in a low altitude polar orbit and had a proton channel that was nominally 1.7 to 2.7 MeV, but apparently those data were not used for the NASA models because of concern about contamination by energetic protons penetrating the detector. The AP-8 spectrum below 3 MeV in the inner zone is an extrapolation from higher energies and is incorrect. An alternate model for protons from 80 keV to 3.2 MeV for all latitudes and for altitudes up to 8000 km exists [Vampola, 1996]. It is offered as an alternative to AP-8 for this region of space. Dose calculations are not available for outputs from this model.

The low-energy proton map was derived from a two-element proton telescope flown on the USAF satellite S3-3 in the mid-to-late '70s. This map is an average of experimental data using a single instrument. The data period covers July 1976 to April 1979, with about half of the data obtained during the first six months of that period. Although the period is just after solar minimum, it is not to be construed as representing any particular type of solar activity period, since it is the only data set that has been analyzed in this low energy range in the inner zone. In the range $1.3 > L > 1.7$, there is a very large difference between this map and AP-8. AP-8, both MIN and MAX, were generated as extrapolations from higher energy data. AP-8 and the S3-3 data agree at 3 MeV, but AP8 is as much as three orders of magnitude below the S3-3 data at lower energies.

4. VERIFICATION

Integral flux outputs from RADMODLS were verified by comparison with identical cases run with SOFIP [Stassino-

poulos et al., 1979] and MODEL [*Teague et al.*, 1972]. Since none of the calculational software had been changed other than the underflow/overflow problem, identical results were expected and were obtained. The RADMODLS dose calculations were checked by comparison with published doses for the CRRES orbit calculated by Seltzer using SHIELDOSE.

5. OUTPUTS

Several types of outputs are available. Printout can come directly to the screen, to an ASCII formatted text file, or to an ASCII comma-separated-variables (CSV) file. The CSV file is intended for use as input to a spread-sheet program or a plotting program which utilizes this type of file. It has the advantage that all variables from a single calculation step are contained in a single output record (a single "line" of output). If desired, the commas may be replaced with spaces at the end of the run. The ASCII text file is formatted for printout on an 80-column printer. As a result, the text outputs from a single calculation step may constitute a large number of lines in the output file. Outputs are very flexible, since a run may entail calculating both integral and differential fluxes and also doses from both an electron and a proton model at numerous energies. Ephemeris parameters (B, L, altitude, latitude, longitude) may be requested as part of the output.

6. USAGE

This software is intended for use on a PC with a DOS 2.0 or later operating system. A 386/486/Pentium class machine with a math co-processor is highly recommended, though not required. The presence of the co-processor increases execution speed by about a factor of 50. An orbital integration with dose calculation that takes a few minutes on a 486 DX machine can take several hours on a 386 SX machine without a co-processor.

For use on a PC, the software is distributed as a self-extracting file. To use this software, the executable module and its associated binary data files must reside in the same directory or subdirectory. The files require 1.1 MB. The executable module, RADMODLS.EXE, requires about 510 kB to run. The program is fully prompting. It accesses the following radiation belt models which were issued by the NSSDC: AE-4 MIN/MAX, AE-5 MIN, AE-6 MAX, AEI-7 LO/HI, AE-8 MIN/MAX, AP-8 MIN/MAX, and AP-8 MIC/MAC. AP-8 MIC and MAC are provided unmodified, but the user is warned that, although the differences between the two versions are less than a factor of two at the equator, up to an order of magnitude differences can occur between the MIC/MAC and MIN/MAX maps near the end of the B/B_0 interpolation space [*Heynderickx and Beliaev*, 1995]. The MIN and MAX maps are the preferred maps.

The program runs in two modes: 1) direct accession of model tables; 2) orbital integration with indirect accession of the model tables. In the direct mode the user is asked for the B (or B/B_0) and L values. If B is supplied, it is converted into B/B_0 using $B_0 = .311653/L^3$. In the orbital integration mode, B and L are obtained from the orbit integration using IGRF65, Epoch 1970. Both integral and differential flux values are available at user-selected energies. The differential values are obtained by obtaining the integral values at energies 10% below and above the desired energy, differencing, and averaging. SHIELDOSE requires a differential energy flux (or fluence) as input. RADMODLS will automatically extract these for dose runs.

In orbit integration runs, if an inner zone only (AE5, AE6) or outer zone only (AE-4, AEI-7) model is used, the complementary map is also accessed to ensure having a valid map for the entire orbit. These maps may not join smoothly at $L = 2.8$ (the breakpoint between NSSDC inner and outer zone trapped particle maps). The fluxes returned are OMNIDIRECTIONAL fluxes. To transform to unidirectional fluxes, the methodology of *Badhwar and Konradi* [1990] can be followed. For a simple expedient, the following divisors are recommended: approximately 4π at the equator, 2π at low altitude, and 3π in between. The rationale is that an isotropic flux is incident from a 4π solid angle and a "pancake" distribution presents a 2π solid angle. Any other distribution is intermediate between the two. For inner zone equatorial distributions, a factor of 3π or 3.5π is more appropriate. The accuracy of the various models is not significantly affected by this arbitrary selection of transformation factors. The output from this program is only as valid as the original model maps. In general, these are good to about a factor of 2 or 4 in intensity for long term averages. AP-8 is probably accurate to about 50% for energies above 20 MeV in the inner zone. AE-8 is also probably accurate to about a factor of 2 in the inner zone below 1 MeV. At geosynchronous orbit, AE-8 is about an order of magnitude high above 1 MeV for long-term averages.

The subroutine TRARA1 returns the log of the integral flux from the flux map in use. It limits the minimum value to 0. Since it is a logarithm, it returns a minimum integral flux of $1.0\,\mathrm{cm^{-2}s^{-1}}$ above the threshold. In RADMODLS, values of all fluxes, integral or differential, are set to zero if the point is outside of the L range of validity of the map: $1.17 < L > 7.0$ for AP-8 and $1.2 < L > 11.0$ for the electron maps. This artificial limitation of $1.0\,\mathrm{cm^{-2}s^{-1}}$ does not affect the dose calculation, since the dose calculation uses differential fluxes. The differential fluxes, because of the way they are calculated, go to zero when the integral fluxes are set to 1. But if the integral fluxes are summed or averaged, the artificial minimum of $1\,\mathrm{cm^{-2}s^{-1}}$ may introduce a significant bias. Therefore, RADMODLS sets the artificial minimum in integral flux to zero to produce more valid averages or sums of the integral fluxes. Thus, in the region where there is no valid flux level but which is within the range of the map (typically $L = 4$ to 7 for AP-8 for high energies) the integral outputs from RADMODLS will not agree with the values obtained from other programs such as MODEL from NSSDC or other versions of ORP.

RADMODLS produces a file which retains input parameters from the console and initializes the next run with them, expediting making several runs with minor changes between them. The interpolation tables are also saved and, if appropriate, are reused instead of being recalculated. Default tables of L and B/B_0 are provided for the user initially; thereafter, the previously used values are saved in the RADMODLS.SAV file.

A modified version of RADMODLS which can use any of the standard internal magnetic field models from DGRF45 through IGRF90, called RAD_DGRF, is also available for research purposes. RAD_DGRF incorporates FELDG and SHELLG [*Kluge*, 1972], as corrected in the BILCAL/IGRF package [*Bilitza*, 1990] to use the proper magnetic moment. RAD_DGRF has a modified version of TRARA1 which uses the actual magnetic moment derived from the field model rather than the pre-1960 value of .311653 used elsewhere. All particle maps and geomagnetic model coefficients are provided with the programs, which are available from NSSDC or the author.

REFERENCES

Badhwar, G.D. and A. Konradi, Conversion of Omnidirectional Proton Fluxes into a Pitch Angle Distribution, *J. Spacecraft and Rockets, 27*, 350, 1990.

Bilitza, D., Solar-Terrestrial Models and Application Software, NSSDC/DC-A-R&S 90-19, 1990.

Brouwer, D., Solution of the Problem of Artificial Satellite Theory Without Drag, *Astronom. J., 64*, 1959.

Heynderickx, D. and A. Beliaev, Identification of an Error in the Distribution of the NASA Model AP-8 MIN, *J. Spacecraft and Rockets, 32*, 190–192, 1995.

Kluge, G., A Generalized Method for the Calculation of the Geomagnetic Field from Multipole Expansions, European Space Operations Centre, ESOC Internal Note No. 61, Darmstadt, 1970.

Kluge, G., Direct Computation of the Magnetic Shell Parameter, *Comp. Phys. Comm., 3*, 31–35, 1972.

Lyddane, R.H., Small Eccentricities or Inclinations in the Brouwer Theory of Artificial Satellites, *Astronom. J., 68*, 1963.

McIlwain, C.E., Coordinates for Mapping the Distribution of Magnetically Trapped Particles, *J. Geophys. Res., 66*, 3681–3691, 1961.

Seltzer, S., SHIELDOSE: A Computer Code for Space-Shielding Radiation Dose Calculations, National Bureau of Standards, NBS Technical Note 1116, U. S. Government Printing Office, Washington, D. C., 1980.

Stassinopoulos, E.G., J.J. Hebert, E.L. Butler and J.L. Barth, SOFIP: A Short Orbital Flux Integration Program, NSSDC WDC-A-R&S 79-01, 1979.

Teague, M.J., J. Stein and J.I. Vette, The Use of the Inner Zone Electron Model AE-5 and Associated Computer Programs (ORP), NSSDC WDC-A-R&S 72-11, 1972.

Vampola, A.L., Low Energy Inner Zone Protons–Revisited, in Proceedings of the Workshop of the Earth's Trapped Particle Environment, Taos, NM, Aug. 1994, G. Reeves (ed.), 1996, in press.

Vette, J.I., The NASA/National Space Science Data Center Trapped Radiation Environment Model Program (1964–1991), NSSDC WDC-A-R&S 91-29, 1991a.

Vette, J.I., The AE-8 Trapped Electron Model Environment, NSSDC WDC-A-R&S 91-24, 1991b.

A.L. Vampola, Space Environmental Effects, P.O. Box 10225, Torrance, CA 90505.

Computer Animation of the TIROS/NOAA Observations of the low-altitude (850 km) Radiation Environment

H.H. Sauer

CIRES, University of Colorado, Boulder, Colorado

D.C. Wilkinson

NOAA, National Geophysical Data Center, E/GC2, 325 Broadway, Boulder, Colorado

The continuing development of a computer-based visualization system for the display of TIROS/NOAA observations of the low-altitude radiation environment is briefly outlined. This interactive data product will be made available to the user through the resources of the internet. The format visually contrasts the orbit being examined with the 3-day averaged "climatology".

1. INTRODUCTION

This note briefly describes the TIROS/NOAA observations of the 850 km radiation environment, and the continuing development of a computer visualization system for display of those observations. Using data obtained from near real-time satellite observation, such a system provides a tool in support of the radiation exposure concerns associated with manned, near-earth space activity, high altitude aircraft flight, and satellite operations. Operating off-line, on stored archival data, the system represents a means of identifying associations of the radiation environment with other geophysical parameters, on both dynamical and climatological time scales, as well as providing an important component of eventual radiation belt models.

2. THE DATA SOURCE

All satellites in the NOAA/TIROS series follow a 850 km circular trajectory inclined at 98° to the equator. The orbit plane is fixed with respect to the sun which therefore fixes each orbit in local time. As the Earth rotates inside this orbit, NOAA/TIROS satellites are able to sample the ambient energetic particle environment globally except for regions within 8° of the poles which the satellite does not reach. There have been 7 satellites in the series beginning with the launch of TIROS-N in November 1978 and continuing to the NOAA-12 and NOAA-14 platforms in operation today. The primary space weather monitoring instrument is called the Medium Energy Proton and Electron Detector or MEPED. The energy ranges for protons reach well into the range found responsible for Single Event Effects. The electron energies cover surface charging ranges and approach the lower limit required for internal charging phenomena on spacecraft.

The TIROS/NOAA MEPED observations consist of 6 data channels for electrons of energy ; $E > 30$ keV, > 100 keV, and > 300 keV; and 13 channels of protons from $E > 30$ keV to $E > 80$ MeV.

3. THE DISPLAY

Figure 1 is a gray-scale representation of the original color plot of our most recent attempt at combining current data with a background map that represents recent radiation climatology. It should be noted that the much detail is lost in the transition from color to monochrome. In this illustration, the background of that figure presents a ten-day average of the trapped $E > 300$ keV electron fluxes just prior to the large solar energetic particle event of 19 March 1989. The superposed orbit is the corresponding section from the following ten-day average of that data. In the next revision, the single orbit overlay will change in cinematographic manner to show how the radiation environment for a selected interval evolves, orbit by orbit. The background map will be made up of an average of the most recent three days data for the particle/energy

Radiation Belts: Models and Standards
Geophysical Monograph 97
Copyright 1996 by the American Geophysical Union

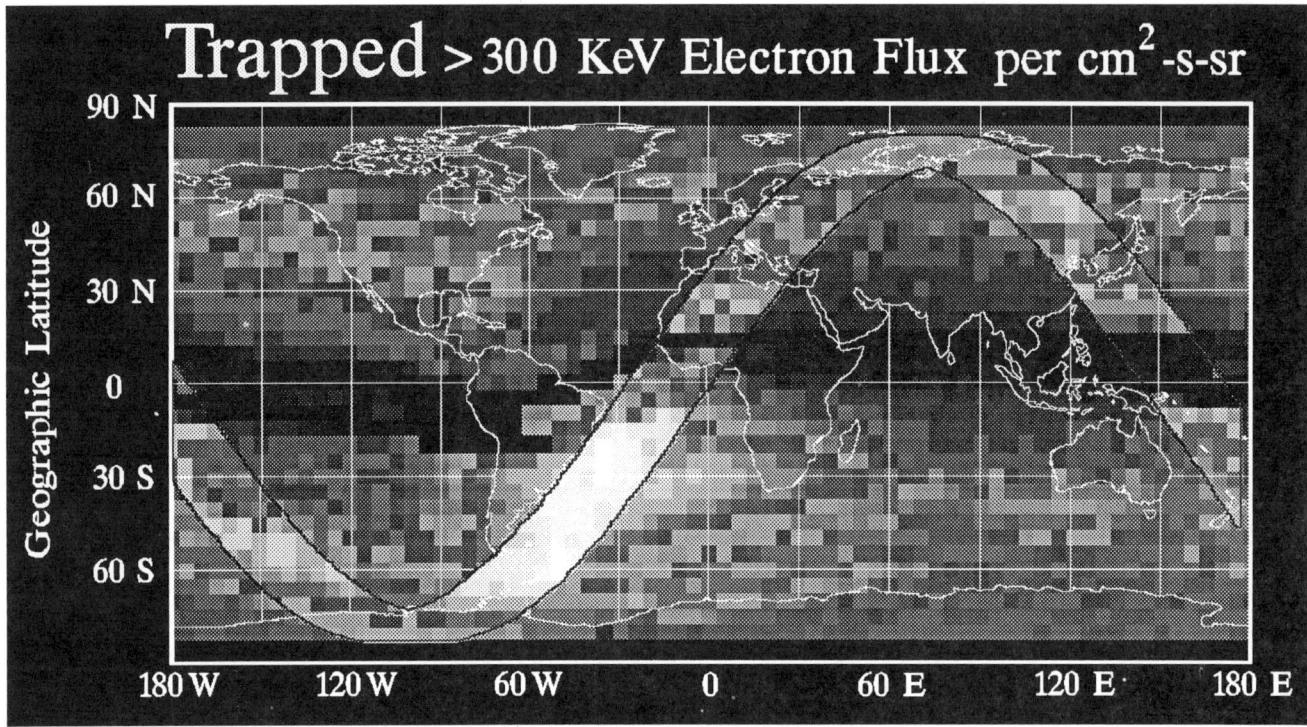

Figure 1. Gray-scale representation of the original color plot of our most recent attempt at combining current data with a background map that represents recent radiation climatology.

channel chosen for display. This display technique will give users both the timeliness inherent in the 90-minute orbit data collection interval, contrasted with the recent global radiation environment.

This data product will be made available to the user community through the Space Physics Interactive Data Resource (SPIDR) operated by World Data Center-A for Solar Terrestrial Physics. This internet resource[1] currently provides interactive access and display of a number of space physics related databases:

1. visible and infra-red imagery from DMSP satellites display Aurora, city lights, fires, and cloud cover button controls allow users to "fly" one of four DMSP satellites about the globe,
2. one-minute geomagnetic variation data from 59 observatories worldwide are displayed in daily plots,
3. maximum electron density in the ionosphere is available through four solar cycles,
4. a decade of GOES Space Environment Monitor observations are displayable at 5-minute resolution, and
5. principal solar and geomagnetic indices such as K_p, A_p, $F_{10.7}$ solar flux, and sunspot numbers are among the data sets available.

It is expected that the interactive display of the energetic particle environment observed by the TIROS/NOAA satellites will provide a useful adjunct to these available data.

H.H. Sauer, University of Colorado, Boulder, Colorado
D.C. Wilkinson, NOAA, National Geophysical Data Center, E/GC2, 325 Broadway, Boulder, Colorado

DISCUSSION

Q: M. Lauriente. What time frame are we looking at for predictions?
A: H.H. Sauer. Several characteristics (geomagnetic activity, solar activity, solar energetic particle events) are predicted for several hours, one day, and three days ahead.

---

[1] http://www.ngdc.noaa.gov/stp/stp.html